# SCIENCE FROM FISHER INFORMATION

The aim of this book is to show that information is at the root of all fields of science. These fields may be generated by use of the concept of "extreme physical information," or EPI. The physical information is defined to be the loss of Fisher information that is incurred in observing any scientific phenomenon. The act of observation randomly perturbs the phenomenon, and sets off a physical process that may be modelled as a mathematical game between the observer and a "demon" characterizing the phenomenon. The currency of the game is Fisher information. The output of the game is the distribution law characterizing the statistics of the effect and, in particular, the acquired data. Thus, in a sense, the act of measurement creates the very law that governs the measurement. It is self-realized. This second edition of *Physics from Fisher Information* has been rewritten throughout in addition to including much new material.

B. ROY FRIEDEN is a Professor Emeritus of Optical Sciences at the University of Arizona. He has held visiting professorship posts at Kitt Peak National Observatory, the Microwave Research Institute in Florence, Italy, and the University of Groningen, the Netherlands. He is known in optics for having founded the field of laser beam shaping; he also invented an optical pupil that asymptotically achieves diffraction-free imaging, invented the three-dimensional optical transfer function and founded the field of maximum entropy image processing. During the past 15 years he has been working on a unification of science through the use of Fisher information. This book is the fruits of this labor. Email address: roy.frieden@optics.arizona.edu.

# SCIENCE FROM FISHER INFORMATION

## A Unification

B. ROY FRIEDEN

*Optical Sciences Center, The University of Arizona, Tucson, AZ*

PUBLISHED BY THE PRESS SYNDICATE OF THE UNIVERSITY OF CAMBRIDGE
The Pitt Building, Trumpington Street, Cambridge, United Kingdom

CAMBRIDGE UNIVERSITY PRESS
The Edinburgh Building, Cambridge CB2 2RU, UK
40 West 20th Street, New York, NY 10011-4211, USA
477 Williamstown Road, Port Melbourne, VIC 3207, Australia
Ruiz de Alarcón 13, 28014 Madrid, Spain
Dock House, The Waterfront, Cape Town 8001, South Africa

http://www.cambridge.org



First published as *Physics from Fisher Information* 1998
Reprinted 1999 (twice),
This edition first published 2004

Printed in the United Kingdom at the University Press, Cambridge

*Typeface* Times 11/14pt. *System* 3b2 [KT]

*A catalog record for this book is available from the British Library*

*Library of Congress Cataloguing in Publication data*

Frieden, B. Roy
Science from Fisher information: a unification/B. Roy Frieden/.– [2nd ed].
p. cm.
Rev. edn of: Physics from Fisher information. 1998.
Includes bibliographical references and index.

ISBN 0 521 81079 5 – ISBN 0 521 00911 1 (pbk.)
1. Physical measurements. 2. Information theory. 3. Physics–Methodology.
I. Frieden, B. Roy Physics from Fisher information. II. Title.

QC39.F75 2004
530.8–dc22 2003064021

ISBN 0 521 81079 5 hardback
ISBN 0 521 00911 1 paperback

To my wife Sarah
and to my children Mark, Amy and Miriam

# Contents

# 0

# Introduction

## 0.1 Aims of the book

The *primary aim* of this book is to develop a *theory of measurement* that incorporates the observer into the phenomenon under measurement. By this theory, the observer becomes both a collector of data and an activator of the phenomenon that gives rise to the data. These ideas have probably been best stated by J. A. Wheeler (1990; 1994):

> All things physical are information-theoretic in origin and this is a participatory universe ... Observer participancy gives rise to information; and information gives rise to physics.

The measurement theory that will be presented is largely, in fact, a quantification of these ideas. However, the reader might be surprised to find that the "information" that is used is not the usual Shannon or Boltzmann entropy measures, but one that is relatively unknown to physicists, that of R. A. Fisher.

The measurement theory is simply a description of how Fisher information flows from a physical source effect to a data space. It therefore applies to all scenarios where quantitative data from repeatable experiments may be collected. This describes measurement scenarios of physics but, also, of science in general. The theory of measurement is found to define an analytical procedure for deriving all laws of science. The approach is called EPI, for "extreme physical information."

The *secondary aim* of the book is to show, by example, that most existing laws of science fit within the EPI framework. That is, they can be derived by its use. (Many can of course be derived by other approaches, but, apparently, no other single approach can derive *all* of them.) In this way the EPI approach unifies science under an umbrella of measurement and information. It also leads to new insights into how the laws are interrelated and, more importantly, to new laws and to heretofore unknown *analytical expressions* for physical

John A. Wheeler, from a photograph taken *c.* 1970 at Princeton University. Sketch by the author.

constants, such as for the Weinberg angle and Cabibbo angle (Chapter 11) of the weak nuclear interaction.

In this way, the usual ways and the means of science are reversed. The various branches of physics are usually derived from the *top down*, i.e., from some suitable action Lagrangian, which in turn predicts a class of measurements. By comparison, EPI derives science by viewing it, in effect, from the *bottom up*. It is based upon measurements first and foremost.

Measurements are usually regarded as merely random outputs from a particular effect. The EPI approach logically reverses this, tracking from output to input, from the data to the effect. It uses knowledge of the information flow in the measurement process to derive the mathematical form of the physical effect that gives the output measurements. The approach is subject to some caveats.

*Caveat* 1: The usual aim of theory is to form mathematical models for physical effects. This is our aim as well. Thus, the EPI approach is limited to deriving the *mathematical expression* of physical effects. It does not form, in some way, the physical effects themselves. The latter are presumed always to exist "out there" in some fixed form.

*Caveat* 2: One does not get something from nothing, and EPI is no exception to this rule. Certain things must be assumed about the unknown effect. One is *knowledge of a source.* The other is knowledge of an appropriate *invariance principle.* For example, in electromagnetic theory (Chapter 5), the source is the charge-current density, and the invariance principle is the equation of continuity of charge flow. Notice that these two pieces of information do not by themselves imply electromagnetic theory. However, they do when used in tandem with EPI.

In this way, an invariance principle plays an *active* role in deriving a physical law. Note that this is the reverse of its passive role in orthodox approaches to physics, which instead regard the invariance principle as a *derived* property from a *known* law. (Noether's theorem is often used for this purpose.) This is a key distinction between the two approaches, and should be kept in mind during the derivations.

How does one know *what* invariance principle to use in describing a given scenario?

*Caveat* 3: Each application of EPI relies upon the user's ingenuity. EPI is not a rote procedure. It takes some imagination and resourcefulness to apply. However, experience indicates that every invariance principle that is used with EPI yields a valid physical law. The approach is exhaustive in this respect.

During the same years that quantum mechanics was being developed by Schrödinger (1926) and others, the field of classical measurement theory was being developed by R. A. Fisher (1922) and co-workers (see Fisher Box, 1978, for a personal view of his professional life). According to classical measurement theory, the quality of any measurement(s) may be specified by a form of information that has come to be called Fisher information. Since these formative years, the two fields – quantum mechanics and classical measurement theory – have enjoyed huge success in their respective domains of application. Until recent times it had been presumed that the two fields are distinct and independent.

However, the two fields actually have strong overlap. The thesis of this book is that all physical law, from the Dirac equation to the Maxwell–

Boltzmann velocity dispersion law, may be unified under the umbrella of classical measurement theory. In particular, the information aspect of classical measurement theory – Fisher information – is the key to the unification.

Fisher information is part of an overall theory of physical law called the principle of EPI. The unifying aspect of this principle will be shown by example, i.e., by application to the major fields of physics: quantum mechanics, classical electromagnetic theory, statistical mechanics, gravitational theory, etc. The defining paradigm of each such discipline is a wave equation, a field equation, or a distribution function of some sort. These will be derived by use of the EPI principle. A separate chapter is devoted to each such derivation. New effects are found, as well, by the information approach.

Such a unification is, perhaps, long overdue. Physics is often considered the science of measurement. That is, physics is a quantification of *observed* phenomena, and observed phenomena contain noise, or fluctuations. The physical paradigm equations (mentioned above) define the fluctuations or errors from ideal values that occur in such observations. That is, *the physics lies in the fluctuations*. On the other hand, classical Fisher information is a scalar measure of these very physical fluctuations. In this way, Fisher information is intrinsically tied into the laws of fluctuation that define theoretical physics.

EPI theory proposes that all physical theory results from observation: in particular, *imperfect* observation. Thus, EPI is an observer-based theory of physics. We are used to the concept of an imperfect observer in addressing quantum theory, but the imperfect observer does not seem to be terribly important to classical electromagnetic theory, for example, where it is assumed (wrongly) that fields are known exactly. The same comment can be made about the gravitational field of general relativity. What we will show is that, by admitting that any observation is imperfect, one can derive both the Maxwell equations of electromagnetic theory and the Einstein field equations of gravitational theory. The EPI view of these equations is that they are expressions of fluctuation in the values of measured field positions. Hence, the four-positions $(\boldsymbol{r}, t)$ in Maxwell's equations represent, in the EPI interpretation, random excursions from an ideal, or mean, four-position over the field.

Dispensing with the artificiality of an "ideal" observer allows us to reap many benefits for purposes of *understanding* physics. EPI is, more precisely, an expression of the "inability to know" a measured quantity. For example, EPI derives quantum mechanics from the viewpoint that an ideal position cannot be known. We have found, from teaching the material in this book, that students more easily understand quantum mechanics from this viewpoint than from the conventional viewpoint of derivative operators that somehow represent energy or momentum. Furthermore, that *the same* inability to know also leads to the

Maxwell equations when applied to that scenario is even more satisfying. It is, after all, a human desire to find common cause in the phenomena we see.

Unification is also, of course, the major aim of physics, although EPI is probably not the ultimate unification that many physicists seek. Our aim is to propose a *comprehensive* approach to deriving physical laws, based upon a new theory of measurement. Currently, the approach presumes the existence of sources and particles. EPI derives major classes of particles, but not all of them, and does not derive the sources. (See Caveat 2 preceding.) We believe, however, that EPI is a large step in the right direction. Given its successes so far, the sources and remaining particles should eventually follow from these considerations as well.

At this point we want to emphasize *what this book is not about.* This is not a book whose primary emphasis is upon the *ad hoc* construction of Lagrangians and their extremization. That is a well-plowed field. Although we often derive a physical law via the extremization of a Lagrangian integral, the information viewpoint we take leads to other types of solutions as well. Some solutions arise, for example, out of *zeroing* the integral. (See the derivation of the Dirac equation in Chapter 4.) Other laws arise out of a combination of both zeroing and extremizing the integral. Similar remarks may be made about the process by which the Lagrangians are *formed.* The zeroing and extremizing operations actually allow us to *solve for* the Lagrangians of the scenarios (see Chaps. 4–9, and 11). In this way we avoid, to a large degree, the *ad hoc* approach to Lagrange construction that is conventionally taken. This subject is discussed further in Secs. 1.1 and 1.8.8. The rationale for both zeroing and extremizing the integral is developed in Chapter 3. It is one of *information transfer* from phenomenon to data.

The layout of the book is, very briefly, as follows. The current chapter is intended to derive and exemplify mathematical techniques that the reader might not be familiar with. Chapter 1 is an introduction to the concept of Fisher information. This is for single-parameter estimation problems. Chapter 2 generalizes the concept to multidimensional estimation problems, ending with the scalar information form $I$ that will be used thereafter in the applications Chapters 4–11. Chapter 3 introduces the concept of the "bound information" $J$, leading to the principle of EPI. This is derived from various points of view. Chapters 4–15 apply EPI to various measurement scenarios, in this way deriving the fundamental wave equations and distribution functions of science. Chapter 16 is a chapter-by-chapter summary of the key points made in the development. The reader in a hurry might choose to read this first, to get an idea of the scope of the approach and the phenomena covered.

## 0.2 Level of approach

The level of physics and mathematics that the reader is presumed to have is that of a senior undergraduate in physics. Calculus, through partial differential equations, and introductory matrix theory are presumed parts of his/her background. Some notions from elementary probability theory are also used. However, since these are intuitive in nature, the appropriate formula is usually just given, with reference to a suitable text as needed.

A cursory scan through the chapters will show that a minimal amount of prior knowledge of physical theory is actually used or needed. In fact, *this is the nature of the information approach taken* and is one of its strengths. The main physical input to each application of the approach is a simple law of invariance that is obeyed by the given phenomenon.

The overall mathematical notation that is used is that of conventional calculus, with additional matrix and vector notation as needed. Tensor notation is only used where it is a "must" – in Chaps. 6 and 11 on classical and quantum relativity, respectively. No extensive operator notation is used; this author believes that specialized notation often hinders comprehension more than it helps the student to understand theory. Sophistication *without* comprehension is definitely not our aim.

A major step of the information principle is the extremization and/or zeroing of a scalar integral. The integral has the form

$$K \equiv \int d\mathbf{x}\, \mathscr{L}[\mathbf{q}, \mathbf{q}', \mathbf{x}], \quad \mathbf{x} \equiv (x_1, \ldots, x_M), \quad d\mathbf{x} \equiv dx_1 \cdots dx_M, \quad \mathbf{q}, \mathbf{x} \text{ real},$$

$$\mathbf{q} \equiv (q_1, \ldots, q_N), \quad q_n \equiv q_n(\mathbf{x}),$$

$$\mathbf{q}'(\mathbf{x}) \equiv \partial q_1/\partial x_1, \partial q_1/\partial x_2, \ldots, \partial q_N/\partial x_M. \tag{0.1}$$

Mathematically, $K \equiv K[\mathbf{q}(\mathbf{x})]$ is a "functional," i.e., a single number that depends upon the values of one or more functions $\mathbf{q}(\mathbf{x})$ continuously over the domain of $\mathbf{x}$. Physically, $K$ has the form of an "action" integral, whose extremization has conventionally been used to derive fundamental laws of physics (Morse and Feshbach, 1953). Statistically, we will find that $K$ is the "physical information" of an overall system consisting of a measurer and a measured quantity. The limits of the integral are fixed and, usually, infinite. The dimension $M$ of $\mathbf{x}$-space is usually 4 (space-time). The functions $q_n$ of $\mathbf{x}$ are probability amplitudes, i.e., functions whose squares are probability densities. The $q_n$ are to be found. They specify the physics of a measurement scenario. Quantity $\mathscr{L}$ is a known function of the $q_n$, their derivatives with respect to all the $x_m$, and $\mathbf{x}$. $\mathscr{L}$ is called the "Lagrangian" density (Lagrange, 1788). It also takes on the role of an information density, by our statistical interpretation.

The solution to the problem of extremizing the information $K$ is provided by a mathematical approach called the "calculus of variations." Since the book makes extensive use of this approach, we derive it in the following.

## 0.3 Calculus of variations

### 0.3.1 Derivation of Euler–Lagrange equation

We find the answer to the lowest-dimension version $M = N = 1$ of the problem, and then generalize the answer as needed. Consider the problem of finding the single function $q(x)$ that satisfies

$$K = \int_a^b dx\, \mathscr{L}[x, q(x), q'(x)] = extrem., \qquad q'(x) \equiv dq(x)/dx. \tag{0.2}$$

A well-known example is the case $\mathscr{L} = \frac{1}{2}mq'^2 - V(q)$ of a particle of mass $m$ moving with displacement amplitude $q$ at time $x \equiv t$ in a known field of potential $V(q)$. We will return to this problem below.

Suppose that the solution to the given problem is the function $q(x)$ as shown in Fig. 0.1. Of course, at the endpoints $(a, b)$ the function has the values $q(a)$, $q(b)$, respectively. Consider any finite departure $q_\varepsilon(x, \varepsilon)$ from $q(x)$,

$$q_\varepsilon(x, \varepsilon) = q(x) + \varepsilon\eta(x), \tag{0.3}$$

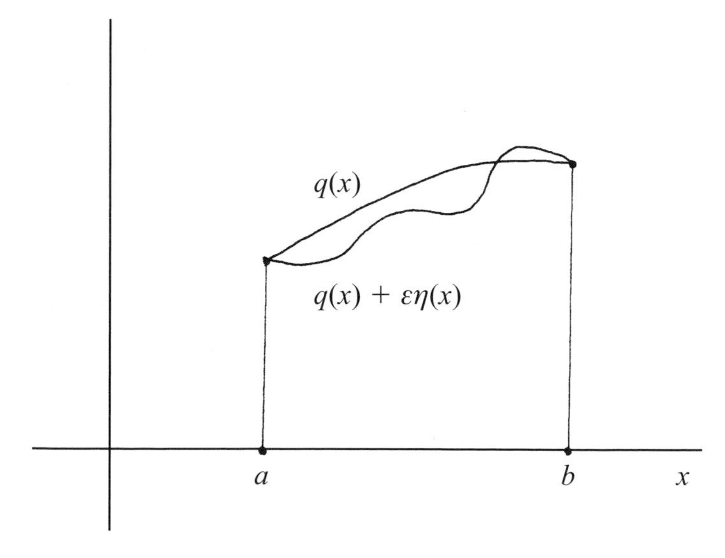


Fig. 0.1. Both the solution $q(x)$ and any perturbation $q(x) + \varepsilon\eta(x)$ from it must pass through the endpoints $x = a$ and $x = b$.

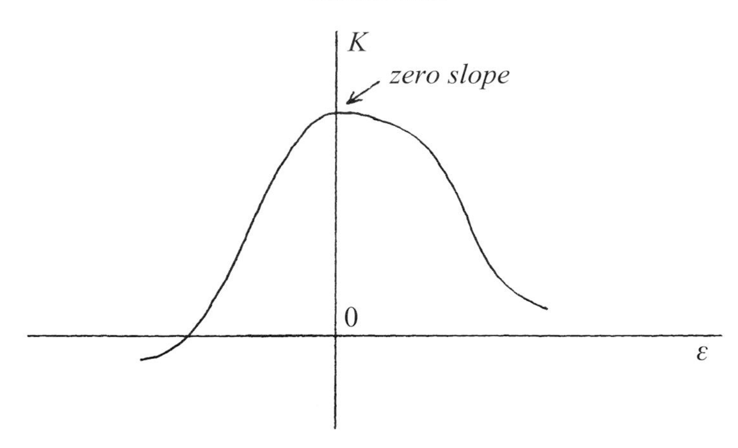


Fig. 0.2. $K$ as a function of perturbation size parameter $\varepsilon$.

with $\varepsilon$ a finite number and $\eta(x)$ any perturbing function. Any function $q_\varepsilon(x, \varepsilon)$ must pass through the endpoints so that, from Eq. (0.3),

$$\eta(a) = \eta(b) = 0. \tag{0.4}$$

Equation (0.2) is, with this representation $q_\varepsilon(x, \varepsilon)$ for $q(x)$,

$$K = \int_a^b dx\, \mathscr{L}[x,\, q_\varepsilon(x, \varepsilon),\, q'_\varepsilon(x, \varepsilon)] \equiv K(\varepsilon), \tag{0.5}$$

a function of the small parameter $\varepsilon$. (Once $x$ has been integrated out, only the $\varepsilon$-dependence remains.)

We use ordinary calculus to find the solution. By the construction (0.3), $K(\varepsilon)$ attains the extremum value when $\varepsilon = 0$. Since an extremum value is attained there, $K(\varepsilon)$ must have zero slope at $\varepsilon = 0$ as well. That is,

$$\left.\frac{\partial K}{\partial \varepsilon}\right|_{\varepsilon=0} = 0. \tag{0.6}$$

The situation is sketched in Fig. 0.2.

We may evaluate the left-hand side of Eq. (0.6). By Eq. (0.5), $\mathscr{L}$ depends upon $\varepsilon$ only through quantities $q$ and $q'$. Therefore, differentiating Eq. (0.5) gives

$$\frac{\partial K}{\partial \varepsilon} = \int_a^b dx \left[\frac{\partial \mathscr{L}}{\partial q_\varepsilon}\frac{\partial q_\varepsilon}{\partial \varepsilon} + \frac{\partial \mathscr{L}}{\partial q'_\varepsilon}\frac{\partial q'_\varepsilon}{\partial \varepsilon}\right]. \tag{0.7}$$

The second integral is

$$\int_a^b dx\, \frac{\partial \mathscr{L}}{\partial q'_\varepsilon}\frac{\partial^2 q_\varepsilon}{\partial x\, \partial \varepsilon} = \left.\frac{\partial \mathscr{L}}{\partial q'_\varepsilon}\frac{\partial q_\varepsilon}{\partial \varepsilon}\right|_a^b - \int_a^b \frac{\partial q_\varepsilon}{\partial \varepsilon}\frac{d}{dx}\left(\frac{\partial \mathscr{L}}{\partial q'_\varepsilon}\right) dx \tag{0.8}$$

after an integration by parts. (In the usual notation, setting $u = \partial\mathscr{L}/\partial q'_\varepsilon$ and $dv = \partial^2 q_\varepsilon/\partial x\,\partial\varepsilon$.)

We now show that the first right-hand term in Eq. (0.8) is zero. By Eq. (0.3),

$$\frac{\partial q_\varepsilon}{\partial\varepsilon} = \eta(x), \tag{0.9}$$

so that by Eq. (0.4)

$$\left.\frac{\partial q_\varepsilon}{\partial\varepsilon}\right|_b = \left.\frac{\partial q_\varepsilon}{\partial\varepsilon}\right|_a = 0. \tag{0.10}$$

This proves the assertion.

Combining this result with Eq. (0.7) gives

$$\frac{\partial K}{\partial\varepsilon} = \int_a^b dx \left[\frac{\partial\mathscr{L}}{\partial q_\varepsilon}\frac{\partial q_\varepsilon}{\partial\varepsilon} - \frac{\partial q_\varepsilon}{\partial\varepsilon}\frac{d}{dx}\left(\frac{\partial\mathscr{L}}{\partial q'_\varepsilon}\right)\right]. \tag{0.11}$$

Factoring out the common term $\partial q_\varepsilon/\partial\varepsilon$, evaluating (0.11) at $\varepsilon = 0$, and using Eqs. (0.3) and (0.9) give

$$\left.\frac{\partial K}{\partial\varepsilon}\right|_{\varepsilon=0} = \int_a^b dx \left[\frac{\partial\mathscr{L}}{\partial q} - \frac{d}{dx}\left(\frac{\partial\mathscr{L}}{\partial q'}\right)\right]\eta(x). \tag{0.12}$$

By our criterion (0.6) this is to be zero at the solution $q$. However, the factor $\eta(x)$ is, by hypothesis, arbitrary. The only way the integral can be zero, then, is for the factor in square brackets to be zero at each $x$, that is,

$$\frac{d}{dx}\left(\frac{\partial\mathscr{L}}{\partial q'}\right) = \frac{\partial\mathscr{L}}{\partial q}. \tag{0.13}$$

This is the celebrated *Euler–Lagrange* solution to the problem. It is a differential equation whose solution clearly depends upon the function $\mathscr{L}$, called the "Lagrangian," for the given problem. Some examples of its use follow.

*Example 1*: Return to the Lagrangian given below Eq. (0.2) where $x = t$ is the independent variable. We directly compute

$$\frac{\partial\mathscr{L}}{\partial q'} = mq' \quad \text{and} \quad \frac{\partial\mathscr{L}}{\partial q} = -\frac{\partial V}{\partial q}. \tag{0.14}$$

Using this in Eq. (0.13) gives as the solution

$$mq'' = -\frac{\partial V}{\partial q}, \tag{0.15}$$

that is, Newton's second law of motion for the particle.

It may be noted that Newton's law will not be derived in this manner in the text to follow. The EPI principle is covariant, i.e., treats time and space in the same way, whereas the above approach (0.14), (0.15) is not. Instead, the EPI approach will be used to derive the more general Einstein field equation, from which Newton's law follows as a special case (the weak-field limit). Or, see Appendix D.

The reader may well question where this particular Lagrangian came from. The answer is that it was chosen merely because it "works," i.e., leads to Newton's law of motion. It has no prior significance in its own right. This has been a well-known drawback to the use of Lagrangians. The next chapter addresses this problem in detail.

*Example 2*: What is the shortest path between two points in a plane? The integrated arc length between points $x = a$ and $x = b$ is

$$K = \int_a^b dx\,\mathscr{L}, \quad \mathscr{L} = \sqrt{1 + q'^2}. \tag{0.16}$$

Hence

$$\frac{\partial \mathscr{L}}{\partial q'} = \tfrac{1}{2}(1 + q'^2)^{-1/2} 2q', \quad \frac{\partial \mathscr{L}}{\partial q} = 0 \tag{0.17}$$

here, so that the Euler–Lagrange Eq. (0.13) is

$$\frac{d}{dx}\left(\frac{q'}{\sqrt{1 + q'^2}}\right) = 0. \tag{0.18}$$

The immediate solution is

$$\frac{q'}{\sqrt{1 + q'^2}} = const., \tag{0.19}$$

implying that $q' = const.$, so that $q(x) = Ax + B$, with $A$, $B = const.$, the equation of a straight line. Hence we have shown that the path of extreme (not necessarily shortest) distance between two fixed points in a plane is a straight line. We will show below that the extremum is a minimum, as intuition suggests.

*Example 3*: Maximum entropy problems (Jaynes, 1957a; 1957b) have the form

$$\int dx\,\mathscr{L} = max., \quad \mathscr{L} = -p(x) \ln p(x) + \lambda p(x) + \mu p(x) f(x) \tag{0.20}$$

with $\lambda$, $\mu$ constants and $f(x)$ a known "kernel" function. The first term in the integral defines the "entropy" of a probability density function (PDF) $p(x)$. (Notice we use the notation $p$ in place of $q$ here.) We will say a lot more about the concept of entropy in chapters to follow. Directly

$$\frac{\partial \mathscr{L}}{\partial p'} = 0, \quad \frac{\partial \mathscr{L}}{\partial p} = -1 - \ln p + \lambda + \mu f(x). \tag{0.21}$$

Hence the Euler–Lagrange Eq. (0.13) is

$$-1 - \ln p(x) + \lambda + \mu f(x) = 0, \quad \text{or} \quad p(x) = A \exp[\mu f(x)]. \tag{0.22}$$

The answer $p(x)$ to maximum entropy problems is always of an exponential form. We will show below that the extremum obtained is actually a maximum, as required.

*Example 4* Minimum Fisher information problems (Huber, 1981) are of the form

$$\int dx\, \mathscr{L} = min., \quad \mathscr{L} = 4q'^2 + \lambda q(x) f(x) + \mu q^2(x) h(x), \tag{0.23}$$

$\lambda$, $\mu = const.$, where $f(x)$, $h(x)$ are known kernel functions. Also, the PDF $p(x) = q^2(x)$, i.e., $q(x)$ is a "probability amplitude" function. The first term in the integral defines the Fisher information. Directly

$$\frac{\partial \mathscr{L}}{\partial q'} = 8q', \quad \frac{\partial \mathscr{L}}{\partial q} = \lambda f(x) + 2\mu q(x) h(x). \tag{0.24}$$

The Euler–Lagrange Eq. (0.13) is then

$$q''(x) - (\mu/4) h(x) q(x) - (\lambda/8) f(x) = 0. \tag{0.25}$$

That is, the answer $q(x)$ is the solution to a second-order differential equation. The particular solution will depend upon the form of the kernel functions and on any imposed boundary conditions. We will show below that the extremum obtained is a minimum, as required.

In comparing the maximum entropy solution (0.22) with the minimum Fisher information solution (0.25) it is to be noted that the former has the virtue of simplicity, always being an exponential. By contrast, obtaining the Fisher information solution always requires solving a differential equation: a bit more complicated a procedure. However, for purposes of deriving physical PDF laws the Fisher answer is actually preferred: the PDFs of physics generally obey differential equations (wave equations). We will further address this issue in later chapters.

We now proceed to generalize the variational problem (0.2) by degrees. For brevity, only the solutions (Korn and Korn, 1968) will be presented.

### 0.3.2 Multiple-curve problems

As a generalization of problem (0.2) with its single unknown function $q(x)$, consider the problem of finding $N$ functions $q_n(x)$, $n = 1, \ldots, N$ that satisfy

$$K = \int dx\, \mathscr{L}[x, q_1, \ldots, q_N, q'_1, \ldots, q'_N] = extrem. \tag{0.26}$$

The answer to this variational problem is that the $q_n(x)$, $n = 1, \ldots, N$ must obey $N$ Euler–Lagrange equations,

$$\frac{d}{dx}\left(\frac{\partial \mathscr{L}}{\partial q'_n}\right) = \frac{\partial \mathscr{L}}{\partial q_n}, \quad n = 1, \ldots, N. \tag{0.27}$$

In the case $N = 1$ this becomes the one-function result (0.13).

### *0.3.3 Condition for a minimum solution*

At this point we cannot know whether the solution (0.27) to the extremum problem (0.26) gives a maximum or a minimum value for the extremum. A simple test for this purpose is as follows.

Consider the matrix of numbers $[\partial^2 \mathscr{L}/\partial q'_i \partial q'_j]$. If this matrix is positive definite then the extreme value is a minimum; or, if it is negative definite, the extreme value is a maximum. This is called *Legendre's condition* for an extremum.

A particular case of interest is as follows.

### *0.3.4 Fisher information, multiple-component case*

As will be shown in Chapter 2, the information Lagrangian is here

$$\mathscr{L} = 4 \sum_{n=1}^{N} q'^2_n. \tag{0.28}$$

Then

$$\frac{\partial \mathscr{L}}{\partial q'_i} = 8 q'_i, \quad \text{so that} \quad \frac{\partial^2 \mathscr{L}}{\partial q'_i \partial q'_j} = 8\delta_{ij} \tag{0.29}$$

where $\delta_{ij}$ is the Kronecker delta function. Thus the matrix $[\partial^2 \mathscr{L}/\partial q'_i\, \partial q'_j]$ is diag $[8, \ldots, 8]$ so that all its $n$-row minor determinants obey

$$\det\left[\frac{\partial^2 \mathscr{L}}{\partial q'_i \partial q'_j}\right] = 8^n > 0, \; n = 1, \ldots, N. \tag{0.30}$$

Then the matrix $[\partial^2 \mathscr{L}/\partial q'_i \partial q'_j]$ is positive definite (Korn and Korn, 1968, p. 420). Consequently, by Legendre's condition the extremum is a minimum.

### *0.3.5 Exercise*

Using the Lagrangian given below Eq. (0.2), show by Legendre's condition

(Sec. 0.3.3) that the Newton's law solution (0.15) minimizes the corresponding integral $K$ in Eq. (0.2).

### 0.3.6 Nature of extremum in other examples

Return to Example 2, the problem of the *minimum* path between two points. The solution $q(x) = Ax + B$ guarantees an extremum but not necessarily a minimum. Differentiating the first Eq. (0.17) gives, after some algebra,

$$\frac{\partial^2 \mathscr{L}}{\partial q'^2} = \frac{1}{(1+q'^2)^{3/2}} = \frac{1}{(1+A^2)^{3/2}} > 0, \tag{0.31}$$

signifying a minimum by Legendre's condition.

Return to Example 3, maximum entropy solutions. The exponential solution (0.22) guarantees an extreme value to the integral (0.20) but not necessarily a maximum. We attempt to use the Legendre condition. However, the Lagrangian (0.20) does not contain any dependence upon quantity $p'(x)$. Hence Legendre's rule gives the ambiguous result $\partial^2 \mathscr{L}/\partial p'^2 = 0$. This being neither positive nor negative, the nature of the extremum remains unknown.

We need to approach the problem in a different way. Temporarily replace the continuous integral (0.20) by a sum

$$K = \sum_{n=1}^{N} \Delta x(-p_n \ln p_n + \lambda p_n + \mu p_n f_n) = max.,$$
$$p_n \equiv p(x_n), \quad f_n \equiv f(x_n), \quad x_n \equiv n\,\Delta x, \tag{0.32}$$

where $\Delta x > 0$ is small but finite. The sum approaches the integral as $\Delta x \to 0$. $K$ is now an ordinary function (not a functional) of the $N$ probabilities $p_n$ and, hence, may be extremized in each $p_n$ using the ordinary rules of differential calculus. The nature of such an extremum may be established by observing the positive- or negative-definiteness of the second derivative matrix $[\partial^2 K/\partial p_i \partial p_j]$. From Eq. (0.32) we have directly $[\partial^2 K/\partial p_i \partial p_j] = -\Delta x\,\mathrm{diag}[1/p_1, \ldots, 1/p_N]$. Hence, all its $n$-row minor determinants obey

$$\det\left[\frac{\partial^2 K}{\partial p_i \partial p_j}\right] = -\prod_{i=1}^{n} \frac{\Delta x}{p_i} < 0 \tag{0.33}$$

since all probabilities $p_i \geqslant 0$. The matrix is negative definite, signifying a maximum as required. This result obviously holds in the (continuous) limit as $\Delta x \to 0$ through positive values.

### 0.3.7 Multiple-curve, multiple-variable, problems

The most general Lagrangian (0.1) has both a multiplicity of coordinates $x_1, \ldots, x_M$ and of curves $q_1, \ldots, q_N$. The solution to the extremization problem is the $N$ Euler–Lagrange equations

$$\sum_{m=1}^{M} \frac{d}{dx_m}\left(\frac{\partial \mathscr{L}}{\partial q_{nm}}\right) = \frac{\partial \mathscr{L}}{\partial q_n}, \qquad n = 1, \ldots, N, \qquad \text{where } q_{nm} \equiv \frac{\partial q_n}{\partial x_m}. \qquad (0.34)$$

We will make much use of this result in chapters to follow. In the case $M = 1$ of one coordinate it goes over into the solution (0.27). Equation (0.34) has also a very useful covariant form, Eq. (6.18a).

### 0.3.8 Imposing constraints

In many problems of extremization there are some known constraints that must be obeyed by the unknowns $\mathbf{q}$. How may they be imposed upon the extremum solution?

Suppose that the unknown functions $\mathbf{q}$ are to obey some constraints, such as

$$F_{nk} = \int d\mathbf{x}\, q_n^{\alpha}(\mathbf{x}) f_k(\mathbf{x}), \qquad \alpha = const., \qquad n = 1, \ldots, N; \qquad k = 1, \ldots, K_o. \qquad (0.35)$$

The $f_n$ are known functions. We seek the solution $\mathbf{q}$ that obeys these constraints and, simultaneously, extremizes a Lagrangian integral (0.1). This obeys the "Lagrange method of undetermined multipliers," described as follows.

Given that $\mathscr{L}[\mathbf{q}, \mathbf{q}', \mathbf{x}]$ is the Lagrangian to be optimized, form the new problem

$$\overline{K} = \int d\mathbf{x}\, \mathscr{L}[\mathbf{q}, \mathbf{q}', \mathbf{x}] + \sum_{nk} \lambda_{nk} \left[ \int d\mathbf{x}\, q_n^{\alpha}(\mathbf{x}) f_k(\mathbf{x}) - F_{nk} \right] = extrem. \qquad (0.36)$$

Here the extremization is to be through variation of the $\mathbf{q}$ *and* the new unknowns $[\lambda_{nk}] \equiv \boldsymbol{\lambda}$. The latter are the "undetermined multipliers" spoken of above.

To show that the solution $\mathbf{q}$, $\boldsymbol{\lambda}$ obeys the constraints, first impose the condition that $\overline{K}$ is extremized with respect to the $\boldsymbol{\lambda}$, that is, $\partial \overline{K}/\partial \lambda_{nk} = 0$. From Eq. (0.36) this gives

$$\int d\mathbf{x}\, q_n^{\alpha}(\mathbf{x}) f_k(\mathbf{x}) - F_{nk} = 0, \qquad (0.37)$$

which is the same as the constraint Eqs. (0.35).

Equation (0.36) is also equivalent to the Euler–Lagrange problem of extremizing a net Lagrangian

$$\overline{\mathscr{L}} = \mathscr{L} + \sum_{nk} \lambda_{nk} q_n^\alpha(\mathbf{x}) f_k(\mathbf{x}). \tag{0.38}$$

That is, in Eq. (0.36), the terms in $F_{nk}$ do not contribute to the Euler–Lagrange solution (0.34). Hence, the problem (0.36) may be re-posed more simply as

$$K + \sum_{nk} \lambda_{nk} \int d\mathbf{x}\, q_n^\alpha(\mathbf{x}) f_k(\mathbf{x}) = extrem. \tag{0.39}$$

That is, to incorporate constraints one merely weights and adds them to the "objective" functional $K$. Some examples of interest are as follows.

With $K$ as the entropy functional, and with moment constraints, the approach (0.39) was used by Jaynes (1957a; 1957b) to estimate PDFs represented as $q^2(x)$.

Alternatively, with $K$ as the Fisher information and with a constraint of mean kinetic energy, the approach (0.39) was used to derive the Schrödinger wave equation (Frieden, 1989) and other wave equations of physics (Frieden, 1990). Historically, the former was the author's first application of a principle of minimum Fisher information to a physical problem. The questions it raised, such as why *a priori* mean kinetic energy should be a constraint (ultimate answer: most generally it shouldn't), provoked an evolution of the theory which has culminated in this book.

### 0.3.9 Variational derivative, functional derivatives

The variation of a functional, the variational derivative, and the functional derivative are useful concepts that follow easily from the preceding. We shall have occasion to use the concept of the functional derivative later on. The concept of the variational derivative is also given, mainly so as to distinguish it from the functional variety.

We first define the concept of the variation of a functional. It was noted that $K$ is a functional (Sec. 0.2). Multiply Eq. (0.12) through by a differential $d\varepsilon$. This gives

$$\left.\frac{\partial K}{\partial \varepsilon}\right|_{\varepsilon=0} d\varepsilon = \int_a^b \left[\frac{\partial \mathscr{L}}{\partial q} - \frac{d}{dx}\left(\frac{\partial \mathscr{L}}{\partial q'}\right)\right] \left.\left(\frac{\partial q}{\partial \varepsilon}\right)\right|_{\varepsilon=0} d\varepsilon\, dx. \tag{0.40}$$

We also used Eq. (0.9). Define the variation of $K$ as

$$\delta K \equiv \left.\left(\frac{\partial K}{\partial \varepsilon}\right)\right|_{\varepsilon=0} d\varepsilon. \tag{0.41}$$

This measures the change in functional $K$ due to a small perturbation away from the stationary solution $q(x)$. Similarly

$$\delta q \equiv \left(\frac{\partial q}{\partial \varepsilon}\right)\bigg|_{\varepsilon=0} d\varepsilon \tag{0.42}$$

is the first variation in $q$.

The *first variational derivative* $\delta\mathscr{L}/\delta q$ is then defined such that Eq. (0.40) goes over into the simple form

$$\delta K = \int \left(\frac{\delta \mathscr{L}}{\delta q}\right) \delta q \, dx \tag{0.43}$$

after use of definitions (0.41) and (0.42). This is the definition

$$\frac{\delta \mathscr{L}}{\delta q} \equiv \frac{\partial \mathscr{L}}{\partial q} - \frac{d}{dx}\left(\frac{\partial \mathscr{L}}{\partial q'}\right). \tag{0.44}$$

The case $\mathscr{L} = \mathscr{L}[x, q(x), q'(x)]$ (see Eq. (0.2)) is presumed. The partial derivatives $\partial/\partial q$ and $\partial/\partial q'$ in (0.44) become covariant *functional* derivatives (Lawrie, 1990) when $q$ and $q'$ are tensor coordinates such as the $g_{ij}$ of quantum gravity (Sec. 11.2.14). Then $\delta\mathscr{L}/\delta q$ in (0.44) becomes a variational covariant derivative.

*Note*: Although $K$ is a functional (see below Eq. (0.1)) $\mathscr{L}$ is an ordinary function, since it has a definite value for *each* value of $x$. Hence, we do not call $\delta\mathscr{L}/\delta q$ in (0.44) a functional derivative, preferring instead to call it a "variational" derivative. The derivative of a functional, $\delta K/\delta q$, will be defined below. (See also Feynman and Hibbs (1965), pp. 170–1.)

Interesting background reading on variational derivatives may be found in Goldstein (1950), p. 353, and Panofsky and Phillips (1955), p. 367.

The concept of a *functional* derivative will be needed in our treatment of quantum gravity (Chapter 11). One functional there is $\psi[g_{ij}(\mathbf{x})]$, the probability amplitude for the gravitational metric function $g_{ij}$ at *all* points $\mathbf{x}$ in four-space. Each possible form $g_{ij}(\mathbf{x})$ for the metric, defined over *all* $\mathbf{x}$, gives a single value for $\psi$.

We now derive the concepts of the first- and second-order *functional* derivatives. The general approach is to proceed as in derivation of Eq. (0.44). That is, we develop an expression for the variation in (now) a functional and show that it takes a simple form like (0.43) only if the integrand is taken to be the suitable *functional* derivative. In contrast with the preceding, however, we expand out to *second*-order in the changes so as to bring in the concept of a second functional derivative. (By the way, the latter concept does not seem to appear in other physical texts.)

Consider a functional

$$K = K[x, q(x, \varepsilon)], \quad \text{all } x, \tag{0.45}$$

where $q(x, \varepsilon)$ obeys Eq. (0.3). By how much does the scalar value $K$ change if function $q(x, \varepsilon)$ is perturbed by a small amount at *each* $x$?

In order to use the ordinary rules of calculus we first subdivide $x$-space as

$$x_{n+1} = x_n + \Delta x, \quad n = 1, 2, \ldots \tag{0.46}$$

in terms of which

$$K = K[x_1, x_2, \ldots, q(x_1, \varepsilon), q(x_2, \varepsilon), \ldots] \tag{0.47}$$

(cf. Eq. (0.45)). Also, Eq. (0.3) is now discretized, to

$$q(x_n, \varepsilon) = q(x_n) + \varepsilon\eta(x_n). \tag{0.48}$$

Note that the ordinary partial derivatives $\partial K/\partial q(x_n, \varepsilon)$, $n = 1, 2, \ldots$ are well defined, by direct differentiation of Eq. (0.47).

In all of the preceding, the numbers $\eta(x_n)$ are presumed to be arbitrary but *fixed*. Then the perturbations in (0.48) are purely a function of $\varepsilon$. Consequently, $K$ given by (0.47) is likewise purely a function $K(\varepsilon)$.

Next, consider the effect upon $K(\varepsilon)$ of a small change $d\varepsilon$ away from the stationary state $\varepsilon = 0$. This may simply be represented by a Taylor series in powers of $d\varepsilon$,

$$K(d\varepsilon) = dK(\varepsilon)\bigg|_{\varepsilon=0} = \frac{\partial K}{\partial \varepsilon}\bigg|_{\varepsilon=0} d\varepsilon + \frac{1}{2}\frac{\partial^2 K}{\partial \varepsilon^2}\bigg|_{\varepsilon=0} d\varepsilon^2 + \cdots. \tag{0.49}$$

The coefficients of $d\varepsilon$ and $d\varepsilon^2$ are evaluated as follows.

By the chain rule of differentiation

$$\frac{\partial K}{\partial \varepsilon} = \sum_n \frac{\partial K}{\partial q(x_n)} \frac{\partial q(x_n)}{\partial \varepsilon}. \tag{0.50}$$

(For brevity, we used $q(x_n, \varepsilon) \equiv q(x_n)$.) Also,

$$\frac{\partial^2 K}{\partial \varepsilon^2} \equiv \frac{\partial}{\partial \varepsilon}\left(\frac{\partial K}{\partial \varepsilon}\right) = \sum_m \frac{\partial}{\partial q(x_m)}\left[\sum_n \frac{\partial K}{\partial q(x_n)} \frac{\partial q(x_n)}{\partial \varepsilon}\right] \frac{\partial q(x_m)}{\partial \varepsilon}$$

after re-use of (0.50),

$$= \sum_{mn} \frac{\partial^2 K}{\partial q(x_m)\, \partial q(x_n)} \frac{\partial q(x_m)}{\partial \varepsilon} \frac{\partial q(x_n)}{\partial \varepsilon} \tag{0.51}$$

after another derivative term drops out.

If we multiply Eq. (0.50) by $d\varepsilon$ and Eq. (0.51) by $d\varepsilon^2$, evaluate them at $\varepsilon = 0$, and use definitions (0.41) and (0.42), we get

$$\frac{\partial K}{\partial \varepsilon}\bigg|_{\varepsilon=0} d\varepsilon = \sum_n \frac{\partial K}{\partial q(x_n)} \delta q(x_n)$$

and (0.52)

$$\left.\frac{\partial^2 K}{\partial \varepsilon^2}\right|_{\varepsilon=0} d\varepsilon^2 = \sum_{mn} \frac{\partial^2 K}{\partial q(x_m)\, \partial q(x_n)}\, \delta q(x_m)\, \delta q(x_n),$$

respectively.

Following the plan, we substitute these coefficient values into Eq. (0.49). Next, multiply and divide the first sum by $\Delta x$ and the second sum by $\Delta x^2$. Finally, take the continuous limit $\Delta x \rightarrow 0$. The sums approach integrals and we have

$$\left. dK(\varepsilon)\right|_{\varepsilon=0} = \int dx \left[\lim_{\Delta x \to 0} \frac{1}{\Delta x} \frac{\partial K}{\partial q(x_n)}\right] \delta q(x)$$
$$+ \frac{1}{2} \iint dx'\, dx \left[\lim_{\Delta x \to 0} \frac{1}{\Delta x^2} \frac{\partial^2 K}{\partial q(x_m)\, \partial q(x_n)}\right] \delta q(x')\, \delta q(x) \; + \cdots \tag{0.53}$$

where $x_n \rightarrow x$, $x_m \rightarrow x'$ in the limit. We demand that this take the simpler form (cf. Eq. (0.43))

$$\left. dK(\varepsilon)\right|_{\varepsilon=0} = \int dx \frac{\delta K}{\delta q(x)}\, \delta q(x) + \frac{1}{2} \iint dx'\, dx \frac{\delta^2 K}{\delta q(x')\, \delta q(x)}\, \delta q(x')\, \delta q(x) \; + \cdots. \tag{0.54}$$

By Eq. (0.53) this will be so if we define

$$\frac{\delta K}{\delta q(x)} \equiv \lim_{\Delta x \to 0} \frac{1}{\Delta x} \frac{\partial K}{\partial q(x_n)}, \qquad x_n \rightarrow x \tag{0.55}$$

and

$$\frac{\delta^2 K}{\delta q(x')\, \delta q(x)} \equiv \lim_{\Delta x \to 0} \frac{1}{\Delta x^2} \frac{\partial^2 K}{\partial q(x_m)\, \partial q(x_n)}, \qquad x_n \rightarrow x, \qquad x_m \rightarrow x'. \tag{0.56}$$

Equation (0.55) is the first functional derivative of $K$ in the case of a functional dependence (0.45). It answers the question "By how much will the number $K$ change if the function $q(x)$ changes by a small amount at all $x$?"

Equation (0.56) defines the second mixed functional derivative of $K$. As noted before, we will have occasion to use this concept in Chapter 11 on quantum gravity. The dynamical equation of this phenomenon is not the usual second-order differential equation, but, rather, a second-order *functional* differential equation. The second functional derivative is with respect to (metric) functions $\mathbf{q}(\mathbf{x})$, where $\mathbf{x}$ is, now, a *four*-position. See the following.

Although the preceding derivation was for the case of a scalar coordinate $x$, it is easily generalized to the case of a four-vector $\mathbf{x}$ as well. (One merely replaces all scalars $x$ by a four-vector $\mathbf{x}$, with all subscripts as before. Of course, $\varepsilon$ is still a scalar.) This gives a definition

$$\frac{\delta^2 K}{\delta q(\mathbf{x}')\,\delta q(\mathbf{x})} \equiv \lim_{\Delta\mathbf{x}\to 0} \frac{1}{\Delta\mathbf{x}^2} \frac{\partial^2 K}{\partial q(\mathbf{x}_m)\,\partial q(\mathbf{x}_n)}, \tag{0.57}$$

$$\mathbf{x}_n \to \mathbf{x}, \quad \mathbf{x}_m \to \mathbf{x}', \quad \Delta\mathbf{x} \equiv \Delta x\,\Delta y\,\Delta z\, c\,\Delta t,$$

where $c$ is the speed of light and $\Delta\mathbf{x}$ is the increment volume used in four-space.

Finally, we consider problems where *many* amplitude functions $q_k(\mathbf{x}, \varepsilon)$, $k = 1, 2, \ldots$ exist. (We subscripted these with a single subscript, but results will hold for any number of subscripts as well.) In Eq. (0.45), the functional $K$ now depends upon all of these functions. We now get the logical generalization of Eq. (0.57),

$$\frac{\delta^2 K}{\delta q_j(\mathbf{x}')\,\delta q_k(\mathbf{x})} = \lim_{\Delta\mathbf{x}\to 0} \frac{1}{\Delta\mathbf{x}^2} \frac{\partial^2 K}{\partial q_j(\mathbf{x}_m)\,\partial q_k(\mathbf{x}_n)}, \quad \mathbf{x}_n \to \mathbf{x}, \quad \mathbf{x}_m \to \mathbf{x}'. \tag{0.58}$$

### *0.3.10 Exercise*

Show this result, using the analogous steps to Eqs. (0.45)–(0.57). *Hint*: The right-hand functions in Eq. (0.48), and $\varepsilon$, must now be subscripted by $k$. Then Eq. (0.49) becomes a power series in changes $d\varepsilon_k$ including (now) second-order mixed-product terms $d\varepsilon_k\, d\varepsilon_j$ that are *summed* over $k$ and $j$. Proceeding, on this basis, through to Eq. (0.57) now logically gives the definition (0.58).

The procedure (0.49)–(0.56) may be easily extended to allow third, and all higher, orders of functional derivatives to be defined. The dots at the ends of Eqs. (0.53) and (0.54) indicate the higher-order terms, which may be easily added in.

### *0.3.11 Alternate form for functional derivative*

Definition (0.55) is useful when the functional $K$ has the form of a sum over discrete samples $q(x_n)$, $n = 1, 2, \ldots$. Instead, in many problems $K$ is expressed as an action integral Eq. (0.2), where $q(x)$ is *continuously* sampled. Obviously, definition (0.55) is not directly usable for such a problem. For such continuous cases one may use, instead, the equivalent definition

$$\frac{\delta K}{\delta q(y)} \equiv \lim_{\varepsilon\to 0} \frac{K[q(x) + \varepsilon\delta(x-y)] - K[q(x)]}{\varepsilon} \tag{0.59}$$

where $\delta(x)$ is the Dirac delta function (Ryder, 1987, p. 177).

### *0.3.12 An observation*

Suppose that two non-identical functionals $I[q(x)]$ and $J[q(x)]$ obey a relation

$$I - J = 0 \tag{0.60}$$

for some scalar function $q_1(x)$. Is there necessarily a solution $q_2(x)$ to the variational problem

$$I - J = extrem.? \tag{0.61}$$

The latter is a problem requiring quantity $I - J$ to be *stationary* for some $q_2(x)$. As we saw, its solution would have to obey the Euler–Lagrange Eq. (0.13). This is, of course, a very different requirement on $q_2(x)$ than the zero-condition (0.60). Moreover, even if $q_1(x)$ exists there is no guarantee that any solution $q_2(x)$ exists. Thus, the fact that a functional is zero for some solution does not guarantee that that, or any other, solution *extremizes* the functional. Analogously, the existence of an algebraic zero-point does not necessarily imply that an extremum condition is satisfied at either the zero-point or anywhere else.

## 0.4 Dirac delta function

A handy concept to use when evaluating integrals is that of the Dirac delta function $\delta(x)$. It is the continuous counterpart of the Kronecker delta function

$$\delta_{ij} = 0 \text{ for } i \neq j, \quad \sum_{j=1}^{N} \delta_{ij} = 1, \tag{0.62a}$$

$$\sum_{j=1}^{N} f_j \delta_{ij} = f_i \tag{0.62b}$$

for any $i$ on the interval $(1, N)$. Notice that $\delta_{ij}$, for a fixed value of $i$, is a function of $j$ that is a pure "spike," i.e., zero everywhere except at the single point $i = j$ where it has a "yield" of 1.

Similarly, $\delta(x)$ obeys

$$\delta(x - a) = 0 \text{ for } x \neq a, \quad \int dx\, \delta(x - a) = 1, \tag{0.63a}$$

$$\int dx\, f(x)\delta(x - a) = f(a) \tag{0.63b}$$

for any real $a$ and any function $f(x)$. (In these integrals and throughout the book, the limits are from $-\infty$ to $\infty$ unless otherwise stated.) It is useful to compare Eqs. (0.62a) with Eqs. (0.63a), and Eq. (0.62b) with Eq. (0.63b). What should function $\delta(x)$ look like?

From Eqs. (0.63a) it must be flat zero everywhere except at the point $x = a$, where it is so large that its area is still finite (at value 1). This is the continuous version of the Kronecker delta spike mentioned above.

Equation (0.63b) shows what is called the "sifting" property of the delta function. That is, the single value $f(a)$ is sifted out by the delta function during the integration over all $x$. Notice that Eq. (0.63b) *follows* from Eqs. (0.63a): because of the first property (0.63a) no value of $f(x)$ contributes to the integral (0.63b) except the value $f(a)$. Therefore it may be taken outside the integral. The remaining integral has unit area by the second property (0.63a), giving a net value of $f(a)$. The situation is sketched in Fig. 0.3.

Now a pure, discontinuous spike cannot represent a function in the ordinary sense of the word. However, it can represent the *limiting form* of a well-defined function. Functions that exist in this limit are called 'generalized functions' (Bracewell, 1965). For example, the ordinary Gaussian curve from probability theory when taken in the limit of vanishing variance obeys

$$\lim_{\sigma \to 0} p(x) = \lim_{\sigma \to 0} \frac{1}{\sqrt{2\pi\sigma^2}} \exp\left(-\frac{x^2}{2\sigma^2}\right) = \delta(x) \qquad (0.64)$$

since it obeys the requirements (0.63a), the second of which becomes simple normalization of the probability law. By the same token, *any* well-behaved probability density (outside equality in (0.64)) approaches a Dirac delta function in this limit. There are, thus, an infinite number of different representations for the delta function.

An often-used representation grows out of the Fourier transform relation

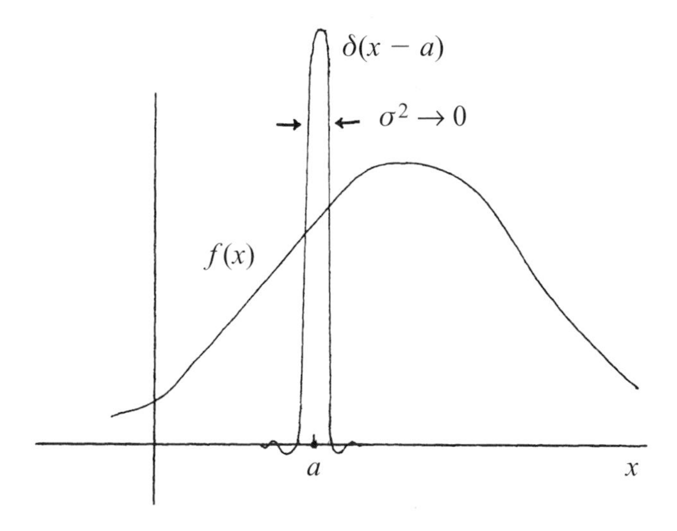


Fig. 0.3. The function $f(x)$ is sifted at value $f(a)$ by the Dirac delta function $\delta(x-a)$.

$$F(k) = \frac{1}{\sqrt{2\pi}} \int dx\, f(x) \exp(-ikx), \quad i = \sqrt{-1} \tag{0.65}$$

and its inverse relation

$$f(x) = \frac{1}{\sqrt{2\pi}} \int dk'\, F(k') \exp(ik'x). \tag{0.66}$$

Substituting Eq. (0.66) into (0.65) gives

$$F(k) = \int dx \exp(-ikx)\, \frac{1}{2\pi} \int dk' F(k') \exp(ik'x)$$

$$= \int dk'\, F(k') \frac{1}{2\pi} \int dx \exp[-ix(k - k')] \tag{0.67}$$

after switching orders of integration. Then, by the sifting property Eq. (0.63b), it must be that

$$\frac{1}{2\pi} \int dx \exp[-ix(k - k')] = \delta(k - k'). \tag{0.68}$$

Analogous properties to Eqs. (0.63b) and (0.68) exist for *multidimensional* functions $f(\mathbf{x})$, where $\mathbf{x}$ is a vector of dimension $M$. Thus there is a sifting property

$$\int d\mathbf{x}\, f(\mathbf{x}) \delta(\mathbf{x} - \mathbf{a}) = f(\mathbf{a}) \tag{0.69}$$

and a Fourier representation

$$\frac{1}{(2\pi)^{M/2}} \int d\mathbf{x} \exp[-i\mathbf{x} \cdot (\mathbf{k} - \mathbf{k}')] = \delta(\mathbf{k} - \mathbf{k}'). \tag{0.70}$$

The latter is a multidimensional delta function. These relations may be derived as easily as were the corresponding scalar relations (0.63b) and (0.68).

Another relation that we will have occasion to use is

$$\delta(ax) = \frac{\delta(x)}{|a|}, \quad a = const. \tag{0.71}$$

Other properties of the delta function may be found in Bracewell (1965).

A final relation of use is (Born and Wolf, 1959, Appendix IV)

$$\delta(x^2 - a^2) = \frac{\delta(x - a) + \delta(x + a)}{2|a|}. \tag{0.72}$$

# 1

# What is Fisher information?

Knowledge of Fisher information is not part of the educational background of most scientists. Why should they bother to learn about this concept? Surely the (related) concept of entropy is sufficient to describe the degree of disorder of a given phenomenon. These important questions may be answered as follows.

(a) The point made about entropy is true, but does not go far enough. Why not seek a measure of disorder whose variation *derives* the phenomenon? The concept of entropy cannot do this, for reasons discussed in Sec. 1.3. Fisher information will turn out to be the appropriate measure of disorder for this purpose.

(b) Why should scientists bother to learn this concept? Aside from the partial answer in (a): (i) Fisher information is a *simple* and intuitive concept. As theories go, it is quite elementary. To understand it does not require mathematics beyond differential equations. Even no prior knowledge of statistics is needed: this is easy enough to learn "on the fly". The derivation of the defining property of Fisher information, in Sec. 1.2.3, is readily understood. (ii) The subject has very little specialized jargon or notation. The beginner does not need a glossary of terms and symbols to aid in its understanding. (iii) Most importantly, once understood, the concept gives strong payoff – one might call it "phenomen-all" – in scope of application. It's simply worth learning.

Fisher information has two basic roles to play in theory. First, it is a measure of the ability to estimate a parameter; this makes it a cornerstone of the statistical field of study called parameter estimation. Second, it is a measure of the state of disorder of a system or phenomenon. As will be seen, the latter makes it a cornerstone of physical theory.

Before starting the study of Fisher information, we take a temporary detour into a subject that will provide some immediate physical motivation for it.

Ronald A. Fisher, 1929, from a photograph taken in honor of his election to Fellow of the Royal Society. Sketch by the author.

## 1.1 On Lagrangians

The Lagrangian approach (Lagrange, 1788) to physics has been utilized now for over 200 years. It is one of the most potent and convenient tools of theory ever invented. One well-known proponent of its use (Feynman and Hibbs, 1965) calls it "most elegant." However, an enigma of physics is the question of where its Lagrangians come from. It would be nice to justify and derive them from a prior principle, but none seems to exist. Indeed, when a Lagrangian is presented in the literature, it is often with a disclaimer, such as (Morse and Feshbach, 1953) "It usually happens that the differential equations for a given phenomenon are known first, and only later is the Lagrange function found, from which the differential equations can be obtained." Even in a case where the differential equations are *not* known, often candidate Lagrangians are first constructed, to see whether "reasonable" differential equations result.

Hence, the Lagrange function has been principally a contrivance for getting the correct answer. It is the means to an end – a differential equation – but with no significance in its own right. One of the aims of this book is to show, in fact, that Lagrangians do have prior significance. A second aim is to present *a systematic approach to deriving* Lagrangians. A third is to clarify the role of the observer in a measurement. These aims will be achieved through use of the concept of Fisher information.

R. A. Fisher (1890–1962) was a researcher whose work is not well known to physicists. He is renowned in the fields of genetics, statistics, and eugenics. Among his pivotal contributions to these fields (Fisher, 1959) are the maximum likelihood estimate, the analysis of variance, and a measure of indeterminacy now called "Fisher information." (He also found it likely that the famous geneticist Gregor Mendel contrived the "data" in his famous pea plant experiments. They were too regular to be true, statistically.) It will become apparent that his form of information has great utility in physics as well.

Table 1.1 shows a list of Lagrangians (most from Morse and Feshbach, 1953), emphasizing the common presence of a squared-gradient term. In quantum mechanics, this term represents mean kinetic energy, but why mean kinetic energy should be present is a longstanding mystery: Schrödinger called it "incomprehensible" (Schrödinger, 1926).

*Historical note*: As will become evident below, *Schrödinger's mysterious Lagrangian term was simply Fisher's data information.* May we presume from this that Schrödinger and Fisher, despite developing their famous theories nearly simultaneously, and with basically just the English channel between them, never communicated? If they had, it would seem that the mystery should have been quickly dispelled. This is an enigma.

In fact, Schrödinger's dilemma is a direct outgrowth of the prevailing view, both during his era and today, as to what Lagrangians physically represent. This fundamental question defines a "worldview" as well. The prevailing view was

Table 1.1. *Lagrangians for various physical phenomena. Where do these come from and, in particular, why do they all contain a squared gradient term? (Reprinted from Frieden and Soffer, 1995.)*

| Phenomenon | Lagrangian |
|---|---|
| Classical mechanics | $\frac{1}{2}m\left(\frac{\partial q}{\partial t}\right)^2 - V$ |
| Flexible string or compressible fluid | $\frac{1}{2}\rho\left[\left(\frac{\partial q}{\partial t}\right)^2 - c^2\,\nabla q\cdot\nabla q\right]$ |
| Diffusion equation | $-\nabla\psi\cdot\nabla\psi^* - \cdots$ |
| Schrödinger wave equation | $-\frac{\hbar^2}{2m}\,\nabla\psi\cdot\nabla\psi^* - \cdots$ |
| Klein–Gordon equation | $-\frac{\hbar^2}{2m}\,\nabla\psi\cdot\nabla\psi^* - \cdots$ |
| Elastic wave equation | $\frac{1}{2}\rho\dot{q}^2 - \cdots$ |
| Electromagnetic equations | $4\sum_{n=1}^{4}\Box q_n\cdot\Box q_n - \cdots$ |
| Dirac equations | $-\frac{\hbar^2}{2m}\,\nabla\psi\cdot\nabla\psi^* - \cdots = 0$ |
| General relativity (equations of motion) | $\sum_{m,n=1}^{4} g_{mn}(q(\tau))\,\frac{\partial q_m}{\partial\tau}\,\frac{\partial q_n}{\partial\tau}$ ↑ metric tensor |
| Boltzmann law | $4\left(\frac{\partial q(\mathrm{E})}{\partial \mathrm{E}}\right)^2 - \cdots,\quad p(\mathrm{E})\equiv q^2(\mathrm{E})$ |
| Maxwell–Boltzmann law | $4\left(\frac{\partial q(v)}{\partial v}\right)^2 - \cdots,\quad p(v)\equiv q^2(v)$ |
| Lorentz transformation (special relativity) | $\partial_i q_n\,\partial_i q_n$ (invariance of integral) |
| Helmholtz wave equation | $-\nabla\psi\cdot\nabla\psi^* - \cdots$ |

that they are *energies*, and their integrals are "action integrals." On this basis the Lagrangian for classical mechanics, shown at the top of Table 1.1, is the difference between a kinetic energy term $m(dq/dt)^2/2$ and a potential energy term $V$. However, consider the following counterpoint.

Lagrangians exist whose terms have *no explicit connection with energy.* Examples are those describing genetic evolution (Chapter 14), macroeconomics (Chaps. 13 and 14), and cancer growth (Chapter 15). (These Lagrangians were of course not known in Schrödinger's day.) There is no denying the law of conservation of energy, but, evidently, the concept of energy does not suffice for forming Lagrangians for *all* fields of science. Is there a concept that does?

There is no science without observation. Therefore a common denominator of all science is *measurement.* This views science from the bottom up (Sec. 0.1). On this basis the terms of the Lagrangian should describe in some way the *process* of measurement and the information flow it incurs. In fact, measurement sets in motion a flow of Fisher information (Chaps. 3, 10). On this basis the mysterious squared-gradient term of Schrödinger turns out to be the amount of Fisher information that resides in the measurement (Eq. (2.19) of Chapter 2). In particular, it is the amount of Fisher information residing in a variety of data called *intrinsic data.* The remaining terms of the Lagrangian will be seen to arise out of the information residing in the *phenomenon* that is under measurement. Thus, all Lagrangians consist entirely of two forms of Fisher information – data information and phenomenological information.

The concept of Fisher information is a natural outgrowth of classical measurement theory, as follows.

## 1.2 Classical measurement theory

### *1.2.1 The "smart" measurement*

Consider the basic problem of estimating a single parameter of a system (or phenomenon) from knowledge of some measurements. See Fig. 1.1. Let the parameter have a definite, but unknown, value $\theta$, and let there be $N$ data values $y_1, \ldots, y_N \equiv \mathbf{y}$, in vector notation, at hand. The system is specified by a conditional probability law $p(\mathbf{y}|\theta)$ called the "likelihood law."

The data obey

$$\mathbf{y} = \theta + \mathbf{x}, \tag{1.0}$$

where the $x_1, \ldots, x_N \equiv \mathbf{x}$ are added noise values. The data are used in an estimation principle to form an estimate of $\theta$ which is an *optimal* function $\hat{\theta}(\mathbf{y})$ of all the data; e.g., the function might be the sample mean $N^{-1}\sum_n y_n$. The overall measurement procedure is "smart" in that $\hat{\theta}(\mathbf{y})$ is on average a better estimate of $\theta$ than is any one of the data observables.

The noise $\mathbf{x}$ is assumed to be *intrinsic* to the parameter $\theta$ under measure-

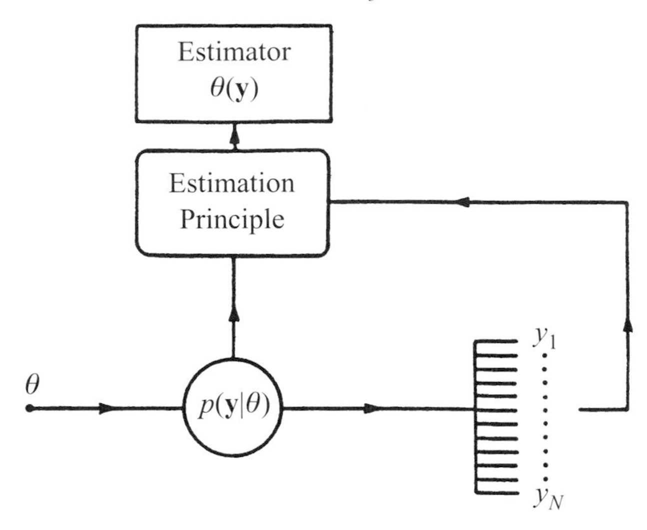


Fig. 1.1. The parameter estimation problem of classical statistics. An unknown but fixed parameter value $\theta$ causes intrinsic data $\mathbf{y}$ through random sampling of a likelihood law $p(\mathbf{y}|\theta)$. Then, the likelihood law and the data are used to form the estimator $\hat{\theta}(\mathbf{y})$ via an estimation principle. (Reprinted from Frieden, 2001, by permission of Springer-Verlag Publishing Co.)

ment. For example, $\theta$ and $\mathbf{x}$ might be, respectively, the ideal position and quantum fluctuations of a particle. Data $\mathbf{y}$ are, correspondingly, called *intrinsic data*. No additional noise effects, such as noise of detection, are assumed present here. (We later allow for outside sources of noise in Sec. 3.8 and Chaps. 10 and 14. The system consisting of quantities $\mathbf{y}$, $\theta$, $\mathbf{x}$ is a *closed*, or physically isolated, one. Those systems are then *open* or non-isolated.)

However, no particle is after all truly isolated. How, then, can assuming a state of isolation give the correct answer? Surprisingly, the answer is largely one of mathematics. Let amplitude functions $q_n(\mathbf{x})$, $n = 1, \ldots, N$ describe $N$ different phenomena. Regardless of how the Lagrangian $\mathscr{L}$ depends upon the $q_n(\mathbf{x})$, the Euler–Lagrange solution (0.34) consists of one differential equation for each phenomenon $n$. If the phenomenon described by $q_n(\mathbf{x})$ is *physically* isolated from all others, it will contribute additively to the total Lagrangian $\mathscr{L}$, and *only* through a single term $\mathscr{L}_n[q_n, q'_n]$. The resulting solution (0.34) for phenomenon $n$ will not, then, depend upon other amplitudes $q_m(\mathbf{x})$, $m \neq n$. By the additivity, the same solution would have occurred if the other phenomena $q_m(\mathbf{x})$, $m \neq n$, were ignored in forming the Lagrangian for phenomenon $n$. Ignoring these other effects is in fact what we mean by isolation.

### 1.2.2 Fisher information

This information arises as a measure of the expected error in a smart measurement. Consider the class of "unbiased" estimates, obeying $\langle\hat{\theta}(\mathbf{y})\rangle = \theta$; these are correct "on average." The mean-square error $e^2$ in such an estimate $\hat{\theta}$ obeys a relation (Van Trees, 1968; Cover and Thomas, 1991)

$$e^2 I \geqslant 1, \tag{1.1}$$

where $I$ is called the Fisher "information." In a particular case of interest $N = 1$ (see below), this becomes

$$I = \int dx\, p'^2(x)/p(x), \qquad p' \equiv dp/dx. \tag{1.2}$$

(Throughout the book, integration limits are infinite unless otherwise specified.) Quantity $p(x)$ denotes the probability density function (PDF) for the noise value $x$. If $p(x)$ is Gaussian, then $I = 1/\sigma^2$ with $\sigma^2$ the variance (see derivation in Sec. 8.3.1).

Equation (1.1) is called the Cramer–Rao inequality. It expresses *reciprocity* between the mean-square error $e^2$ and the Fisher information $I$ in the intrinsic data. Hence, it is an expression of *intrinsic* uncertainties, i.e., in the absence of outside sources of noise. It will be shown in Eq. (4.53) that the reciprocity relation goes over into the Heisenberg uncertainty principle, in the case of a single measurement of a particle position value $\theta$. Again, this ignores the possibility of noise of detection, which would add in additional uncertainties to the relation (Arthurs and Goodman, 1988; Martens and de Muynck, 1991).

The Cramer–Rao inequality (1.1) shows that estimation quality increases ($e$ decreases) as $I$ increases. Therefore, $I$ is a quality metric of the estimation procedure. This is the essential reason why $I$ is called an "information." Equations (1.1) and (1.2) derive quite easily, as is shown next.

### 1.2.3 Derivation

We follow Van Trees (1968). Consider the class of estimators $\hat{\theta}(\mathbf{y})$ that are unbiased, obeying

$$\langle\hat{\theta}(\mathbf{y}) - \theta\rangle \equiv \int d\mathbf{y}\, [\hat{\theta}(\mathbf{y}) - \theta] p(\mathbf{y}|\theta) = 0. \tag{1.3}$$

PDF $p(\mathbf{y}|\theta)$ describes the fluctuations in data values $\mathbf{y}$ in the presence of the parameter value $\theta$. PDF $p(\mathbf{y}|\theta)$ is called the "likelihood law." Differentiate Eq. (1.3) $\partial/\partial\theta$, giving

$$\int d\mathbf{y}\, (\hat{\theta} - \theta) \frac{\partial p}{\partial \theta} - \int d\mathbf{y}\, p = 0. \tag{1.4}$$

Use the identity

$$\frac{\partial p}{\partial \theta} = p \frac{\partial \ln p}{\partial \theta} \tag{1.5}$$

and the fact that $p$ obeys normalization. Then Eq. (1.4) becomes

$$\int d\mathbf{y}\, (\hat{\theta} - \theta) \frac{\partial \ln p}{\partial \theta} p = 1. \tag{1.6}$$

Factor the integrand as

$$\int d\mathbf{y} \left[ \frac{\partial \ln p}{\partial \theta} \sqrt{p} \right] [(\hat{\theta} - \theta)\sqrt{p}] = 1. \tag{1.7}$$

Square the equation. Then the Schwarz inequality gives

$$\left[ \int d\mathbf{y} \left( \frac{\partial \ln p}{\partial \theta} \right)^2 p \right] \left[ \int d\mathbf{y}\, (\hat{\theta} - \theta)^2 p \right] \geqslant 1. \tag{1.8}$$

The left-most factor is defined to be the Fisher information $I$,

$$I \equiv I(\theta) \equiv \int d\mathbf{y} \left( \frac{\partial \ln p}{\partial \theta} \right)^2 p \equiv \left\langle \left( \frac{\partial \ln p}{\partial \theta} \right)^2 \right\rangle, \qquad p \equiv p(\mathbf{y}|\theta). \tag{1.9}$$

The notation $\langle\,\rangle$ describes averaging brackets. In the case of discrete probabilities (Chapter 14) the average becomes a sum rather than the integral shown. The second factor in (1.8) defines the mean-squared error

$$e^2 \equiv \int d\mathbf{y}\, [\hat{\theta}(\mathbf{y}) - \theta]^2 p \equiv \langle [\hat{\theta}(\mathbf{y}) - \theta]^2 \rangle. \tag{1.10}$$

This proves Eq. (1.1).

The likelihood law $p(\mathbf{y}|\theta)$ in the foregoing derivation is assumed to incorporate *all* noise effects that influence the data $\mathbf{y}$. Therefore the Cramer–Rao Eqs. (1.8)–(1.10) hold for noise of any type, including noise from *inside* the system (the *intrinsic* noise assumed in Sec. 1.2.1) and noise from *outside*. Intrinsic noise is exemplified by quantum mechanical uncertainty in position in Chapter 4, exterior noise by random inputs of heat, radiation, moisture, etc., to the biological growth systems of Chapter 14.

It is noted that $I = I(\theta)$ in Eq. (1.9), i.e., in general $I$ depends upon the (fixed) value of parameter $\theta$. But note the following important exception to this rule.

### *1.2.4 Important case of shift invariance*

Suppose that there is only $N = 1$ data value taken so that $p(\mathbf{y}|\theta) = p(y|\theta)$. Also, suppose that the PDF obeys a property of shift invariance

$$p(y|\theta) = p_X(y - \theta), \qquad x = y - \theta. \tag{1.11}$$

This means that the fluctuations in $y$ from $\theta$ are invariant to the size of $\theta$, a kind of shift invariance. (This becomes an expression of *Galilean invariance* when random variables $y$ and $\theta$ are 3-vectors instead.) Using condition (1.11) and identity (1.5) in Eq. (1.9) gives

$$I = \int dy \left[\frac{\partial p(y-\theta)}{\partial(y-\theta)}\right]^2 \bigg/ p(y-\theta), \tag{1.12a}$$

since by the chain rule $\partial/\partial\theta = (\partial/\partial(y-\theta))\,\partial(y-\theta)/\partial\theta = -\partial/\partial(y-\theta)$. Parameter $\theta$ is regarded as fixed (see above), so that a change of variable $x = y - \theta$ gives $dx = dy$. Equation (1.12a) then becomes Eq. (1.2), as required, with $p_X(x) \equiv p(x)$ as simpler notation. Note that $I$ no longer depends upon $\theta$. This is convenient since $\theta$ was unknown.

Shift invariance (1.11) holds quite often. Consider a scalar case $N = 1$ of Eq. (1.0) and temporarily regard $x$ as a "noise" value. By Eq. (1.0), since $\theta$ is fixed, each time a fluctuation $x$ occurs a corresponding $y$ value occurs. Then the frequency of occurrence of a value of $y$ in the presence of a fixed $\theta$ equals that for a corresponding value of $x$, or

$$p(y|\theta) = p_X(x|\theta) = p_X(y-\theta|\theta) \quad \text{since} \quad x \equiv y - \theta. \tag{1.12b}$$

Next, consider any effect whereby the noise fluctuation $x$ is independent of the size of $\theta$. By definition of independence

$$p_X(x|\theta) = p_X(x). \tag{1.12c}$$

Using this in (1.12b) with $x \equiv y - \theta$ then gives (1.11) as required.

This derivation required that the noise fluctuation $x$ be independent of the size of $\theta$. When does this occur physically? In fact the most fundamental physical effects obey this property. A few are considered next. In these examples, all coordinates $x$, $y$, $\theta$ are measured, as usual, from a fixed origin in the laboratory.

Suppose that a particle, of mass $m$ and at general linear position $y$, is undergoing oscillatory linear motion about a *fixed* rest position $\theta$ along $X$. Denote its general displacement from $\theta$ as $x$, so that Eq. (1.0) is again obeyed. The particle is attached to one end of an elastic spring whose other end is fastened at $\theta$. The spring exerts a restoring force $-Kx$ upon the particle, $K = const$. As is well known, the motion of the particle is governed by Newton's second law, in the form

$$-Kx = m\,d^2x/dt^2. \tag{1.12d}$$

This says that the motion of the particle is completely described by the time dependence of $x$. The value of $\theta$ simply does not enter in. It results that, if the observer keeps track of the particle's trajectory values $x$ and bins them at a constant time subdivision to form a histogram of relative occurrences $p_X(x)$, then this histogram (or probability law) is *likewise* found to be independent of the size of $\theta$. That is, Eq. (1.12c) is obeyed. Then by the argument below Eq. (1.12c), the required effect (1.11) is likewise obeyed.

The condition Eq. (1.11), or equivalently (1.12c), is also called one of "shift invariance." This is for the following reasons. Suppose that the origin of laboratory coordinates were shifted in $X$ by a finite amount $\Delta x$. Then *both coordinates* $y$ and $\theta$ are so shifted. Denote by subscript s the shifted coordinates. Thus $y_s = y + \Delta x$ and $\theta_s = \theta + \Delta x$. But then $x_s \equiv y_s - \theta_s = y - \theta \equiv x$ (by Eq. (1.11)); that is, each new value $x_s$ of the displacement equals the old value $x$. Consequently, if the new *displacement* values are binned, the new probability law $p_{X_s}(x_s) = p_X(x)$, the old. The law is invariant to shift.

In many applications the invariance holds only over a finite range of shifts. This occurs, for example, for probability laws which are the "point spread functions" of optics (Born and Wolf, 1959). These are shift-invariant over only a finite area called the "isoplanatic patch." However, we shall not explicitly consider such finite-area cases in this book.

The previous argument holds as well for *any* left-hand force term in (1.12d), so long as it depends only upon $x$. That is, it holds so long as the force term depends only upon the particle's displacement from the center of potential located at $\theta$. More generally, it holds for any isolated quantum mechanical system subject to such a potential. Here, ignoring the time for simplicity, a particle obeys a probability law $p_X(x) \equiv |\psi(x)|^2$, where $x$ is again the displacement of the particle from the center of potential (or from some fixed point in the laboratory if there is no potential). The probability law is *not* of the form $|\psi(x|\theta)|^2$. The absolute position $\theta$ of the center of potential does not enter into the Schrödinger wave equation, which governs $\psi(x)$. Equivalently, irrespective of the position of the origin of coordinates in the laboratory, or indeed of the laboratory in the *universe*, the particle obeys the same Schrödinger wave equation.

Finally, we should consider cases where shift invariance *does not* hold. Possibly the most well-known example is that of the Poisson law,

$$P(y|\theta) = e^{-\theta}\frac{\theta^y}{y!}, \qquad y = 0, 1, 2, \ldots .$$

The right-hand side is visibly *not* a function of $y - \theta$, as was needed for shift invariance (Eq. (1.12b)). The Poisson law was originally used to describe the number $y$ per year of Prussian cavalry officers kicked to death by their horses. It also describes many physical situations, e.g., the random number of photons of ambient light counted during a finite detection time, where the mean count is $\theta$. Data $y$ exhibit what is called "signal-dependent noise," since their fluctuations depend upon the *absolute size* of the "signal" value $\theta$. As an example of the signal dependence, the variance of the fluctuations equals $\theta$.

Most of the probability laws that are derived in this book are presumed to obey shift invariance (1.11) or its multidimensional form Eq. (2.16). Notable exceptions are the PDFs derived in Chaps. 12 and 14.

### 1.2.5 Use of principal value in evaluating I

Certain probability laws, such as the exponential,

$$p(x) = \begin{cases} a^{-1}\exp(-x/a), & a>0, \quad x \geqslant 0 \\ 0, & x<0 \end{cases} \tag{1.12e}$$

have discontinuities in $x$. For (1.12e) it is the point $x = 0$. Then the slope $dp/dx$ becomes indefinitely large in Eq. (1.2) for the information, so that the latter blows up. This must be avoided if $I$ is to be a practical tool of the approach. Here is where the Cauchy principal value idea enters in. The point of discontinuity is simply *avoided*, by redefining the information $I$ in (1.2) such that isolated points of discontinuity are skipped over during the integration. For example, in the case (1.12e) of the exponential law, the redefined information obeys

$$I = \lim_{\delta \to 0} \int_{0+\delta}^{\infty} dx\, p'^2(x)/p(x), \quad \delta>0 \tag{1.12f}$$

(cf. Eq. (1.2)). This gives the well-defined answer $I = 1/a^2$ for the law (1.12e).

In general the redefined information also still obeys the Cramer–Rao inequality, since the resulting error

$$e^2 \equiv \lim_{\delta \to 0} \int_{0+\delta}^{\infty} dx\, [\hat{\theta}(y) - \theta]^2 p(x), \quad y \equiv \theta + x, \quad \delta>0 \tag{1.12g}$$

differs from that (Eq. (1.10)) of the non-principal value approach by an isolated point of finite value. Such a point contributes negligibly to the integral.

From this point on, by information $I$ we shall mean the *principal value* of the information.

### 1.2.6 Solutions p(x) of problem I = extrem.

In the case of a *flat* probability law of any width, the principal value of the information is, from Eq. (1.12f), identically $I = 0$. Does this make sense?

Consider the problem of finding the probability law $p(x)$ that *extremizes* the information functional (1.12f). By the Legendre condition of Secs. 0.3.3 and

0.3.4 the extremum is a *minimum*. What then should be the nature of the solution $p(x)$ to the problem?

The following tendencies will be found in Sec. 1.7. The more spread out and smooth the law $p(x)$ is the more random is $x$, and the more disordered the system is; therefore, the smaller the information $I$ should be. Thus, solving a problem $I = min.$ should give the smoothest law $p(x)$ possible. Let us see whether using the principal value of the information gives this kind of result.

*Unconstrained problem*: Suppose that $x$ is restricted to the interval $(x_1, x_2)$. The problem is to find $p(x)$ obeying

$$I \equiv \lim_{\delta \to 0} \int_{x_1+\delta}^{x_2-\delta} dx\, p'^2(x)/p(x) = min., \qquad \delta > 0. \tag{1.12h}$$

With a Lagrangian $\mathscr{L} = p'^2(x)/p(x)$, we see that $\partial\mathscr{L}/\partial p' = 2p'/p$ and $\partial\mathscr{L}/\partial p = -p'^2/p^2$. Then the Euler–Lagrange Eq. (0.13) for the solution obeys $2f' + f^2 = 0$, $f \equiv p'/p$. The solution for $f$ is $f = 2(x + a)^{-1}$, $a = const$. Then $p$ is found from this to be

$$p(x) = (bx + c)^2, \quad x_1 \leqslant x \leqslant x_2, \quad b, c = const., \tag{1.12i}$$

a truncated parabola. Back substituting this result into (1.12h) gives a minimized information of size

$$I = 4b^2(x_2 - x_1). \tag{1.12j}$$

This shows that the absolute minimum value for $I$, $I = 0$, is attained when $b = 0$. Then by (1.12i)

$$p(x) = c^2 = const., \quad x_1 \leqslant x \leqslant x_2. \tag{1.12k}$$

Hence $p(x)$ is a rectangle function. This is indeed the smoothest law *within* the fixed interval. Hence our requirement that $I$ monotonically decrease as $p(x)$ gets smoother is satisfied by use of the principal value definition of $I$. Not having had to evaluate the infinite slope values $p'(x)$ at the endpoints was essential to the calculation. To avoid such points is the reason the principal value will be implicit in all calculations of the Fisher information and Fisher channel capacity (Chapter 2) in this text.

*Constrained problem*: Next, consider a family of PDFs that are constrained to obey $p(0) = 0$, and to have the form of a power law, $p(x) = Cx^\gamma$ (cf. Eq. (1.12i)), $0 \leqslant x \leqslant x_2$, $C = const.$, $\gamma = const$. These PDFs arise in the analysis of cancer growth with time (Chapter 15), where $x$ is a value of the time. The condition $p(0) = 0$ means that the time origin is the time of inception of the cancer. The PDF (1.12i) gave an absolute minimized $I$ value of zero. *Now how small a value of I can be attained?*

Since $p(x)$ must obey normalization and $p(0) = 0$, a constant solution $\gamma = 0$ is no longer possible, so that the absolute minimum value $I = 0$ is impossible to attain. It follows that the cancer growth curve $p(x) = Cx^{\gamma}$ will monotonically increase, i.e., will exhibit growth. However, since $I$ is minimized for the given growth curve (Chapter 15), the growth is minimal. The answer, given in Eqs. (15.34)–(15.36), is that the power that attains minimal $I$ and the value of $I$ are, respectively,

$$\gamma = \frac{1}{2}(1 + \sqrt{5}) \approx 1.62, \quad I = \frac{11 + 5\sqrt{5}}{2x_2^2}. \tag{1.121}$$

The power $\gamma$ is the Fibonacci "golden mean" (often denoted as $\phi$). Physically, the information level is not zero because, to be alive, the cancer must exhibit at least *some* degree of complexity and, hence, some information content.

## 1.3 Comparisons of Fisher information with Shannon's form of entropy

A related quantity to $I$ is the Shannon entropy (Shannon, 1948) $H$. This has the form

$$H \equiv -\int dx\, p(x) \ln p(x). \tag{1.13}$$

Like $I$, $H$ is a functional of an underlying PDF $p(x)$ and is an "information." Historically, $I$ predates the Shannon form $H$ by about 25 years (1922 versus 1948). There are some known relations connecting the two information concepts (Stam, 1959a, b; Blachman, 1965; Frieden, 1991) but these are not germane to our approach. Also, $H$ can be, but is not always, the thermodynamic, Boltzmann entropy.

The analytical properties of the two information measures are quite different. Thus, whereas $H$ is a *global* measure of smoothness in $p(x)$, $I$ is a *local* measure. Hence, when extremized through variation of $p(x)$, Fisher's form gives a differential equation whereas Shannon's always gives directly the same form of solution, an exponential function. These are shown next.

### *1.3.1 Global versus local nature*

For our purposes, it is useful to work with a discrete form of Eq. (1.13),

$$H = -\Delta x \sum_n p(x_n) \ln p(x_n) \equiv \delta H, \quad \Delta x \to 0. \tag{1.14}$$

(Notation $\delta H$ emphasizes that Eq. (1.14) represents an *increment* in information.) Of course, the sum in Eq. (1.14) may be taken in any order. Graphically, this means that, if the curve $p(x_n)$ undergoes a rearrangement of its points $(x_n,\ p(x_n))$, although the shape of the curve will drastically change the value of

$H$ remains constant. $H$ is then said to be a *global* measure of the behavior of $p(x_n)$.

By comparison, the discrete form of Fisher information $I$ is, from Eq. (1.2),

$$I = \Delta x^{-1} \sum_n \frac{[p(x_{n+1}) - p(x_n)]^2}{p(x_n)}. \tag{1.15}$$

If the curve $p(x_n)$ undergoes a rearrangement of points $x_n$ as above, discontinuities in $p(x_n)$ will now occur. Hence the local slope values $[p(x_{n+1}) - p(x_n)]/\Delta x$ will change drastically, and so the sum (1.15) will also change strongly. Since $I$ is thereby sensitive to local rearrangement of points, it is said to have a property of *locality*.

Thus, $H$ is a global measure, while $I$ is a local measure, of the behavior of the curve $p(x_n)$. These properties hold in the limit $\Delta x \to 0$, and so apply to the continuous probability density $p(x)$ as well.

This global versus local property has an interesting ramification to valuating financial securities (Sec. 13.7.1). Another is as follows.

Because the integrand of $I$ contains a squared derivative $p'^2$ (see Eq. (1.2)), when the integrand is used as part of a Lagrangian the resulting Euler–Lagrange equation will contain second-order derivative terms $p''$. Hence, a second-order differential equation results (see Eq. (0.25)). This dovetails with nature, in that the major fundamental differential equations that define probability densities or amplitudes in physics are *second-order* differential equations. Indeed, the thesis of this book is that the correct differential equations result when the information $I$-based EPI principle of Chapter 3 is followed.

By contrast, the integrand of $H$ in (1.13) does not contain a derivative. Therefore, when this integrand is used as part of a Lagrangian the resulting Euler–Lagrange equation will not contain any derivatives (see Eq. (0.22)); it will be an algebraic equation, with the immediate solution that $p(x)$ has the exponential form Eq. (0.22) (Jaynes, 1957a, 1957b). This is not, then, a differential equation, and hence cannot represent a general physical scenario. The exceptions are those distributions which happen *to be* of an exponential form, as in statistical mechanics. (In these cases, $I$ gives the correct solutions anyhow; see Chapter 7.)

It follows that, if one or the other of global measure $H$ or local measure $I$ is to be used in a variational principle in order to derive the physical law $p(x)$ describing a *general* scenario, the preference is given to the local measure $I$.

As all of the preceding discussion implies, $H$ and $I$ are two distinct functionals of $p(x)$. However, quite the contrary is true in comparing $I$ with an entropy that is closely related to $H$, namely, the Kullback–Leibler entropy. This is discussed in Sec. 1.4.

### 1.3.2 Additivity properties

It is of further interest to compare $I$ and $H$ in the special case of mutually isolated systems, which give rise to independent data. As is well known, the entropy $H$ obeys additivity in this case. Indeed, many people have been led to believe that, because $H$ has this property, it is *the only* functional of a probability law that obeys additivity. In fact, information $I$ obeys additivity as well. This will be shown in Sec. 1.8.11.

## 1.4 Relation of I to Kullback–Leibler entropy

The Kullback–Leibler entropy $G$ (Kullback, 1959) is a functional of (now) two PDFs $p(x)$ and $r(x)$,

$$G \equiv -\int dx\, p(x) \ln[p(x)/r(x)] \equiv G[p, r]. \tag{1.16}$$

This is also called the "cross-entropy" or "relative entropy" between $p(x)$ and a reference PDF $r(x)$. Note that, if $r(x)$ is a constant, $G$ becomes essentially the entropy $H$. Also, $G = 0$ if $p(x) = r(x)$. Thus, $G$ is often used as a measure of the "distance" between two PDFs $p(x)$ and $r(x)$. Also, in a multidimensional case $x \to (x, y)$ the information $G$ can be used to define the mutual information of Shannon (1948).

Now we show that the Fisher information $I$ relates to $G$. Using Eq. (1.15) with $x_{n+1} = x_n + \Delta x$, $I$ may be expressed as

$$I = \Delta x^{-1} \sum_n p(x_n) \left[ \frac{p(x_n + \Delta x)}{p(x_n)} - 1 \right]^2 \tag{1.17}$$

in the limit $\Delta x \to 0$. Now quantity $p(x_n + \Delta x)/p(x_n)$ is close to *unity* since $\Delta x$ is small. Therefore, the $[\cdot]$ quantity in (1.17),

$$p(x_n + \Delta x)/p(x_n) - 1 \equiv \nu, \tag{1.18}$$

is small. Now for small $\nu$ the expansion

$$\ln(1 + \nu) = \nu - \nu^2/2 \tag{1.19}$$

holds, or equivalently,

$$\nu^2 = 2[\nu - \ln(1 + \nu)]. \tag{1.20}$$

Then by Eqs. (1.18) and (1.20), Eq. (1.17) becomes

$$I = -2\,\Delta x^{-1} \sum_n p(x_n) \ln\left( \frac{p(x_n + \Delta x)}{p(x_n)} \right) + 2\,\Delta x^{-1} \sum_n p(x_n + \Delta x) - 2\,\Delta x^{-1} \sum_n p(x_n). \tag{1.21}$$

But each of the two far-right sums is $\Delta x^{-1}$, by normalization, so that their difference cancels out, leaving

$$I = -(2/\Delta x)\sum_n p(x_n)\ln\left(\frac{p(x_n+\Delta x)}{p(x_n)}\right) \quad (1.22a)$$

$$\to -(2/\Delta x^2)\int dx\, p(x)\ln\left(\frac{p(x+\Delta x)}{p(x)}\right) \quad (1.22b)$$

$$= -(2/\Delta x^2)G[p(x),\ p(x+\Delta x)] \quad (1.22c)$$

by definition (1.16). Thus, $I$ is proportional to the cross-entropy between the PDF $p(x)$ and a reference PDF that is its shifted version $p(x+\Delta x)$.

### *1.4.1 Historical note*

Savage (1972) first proved the equality (1.22b). It was later independently re-proved by Vstovsky (1995).

### *1.4.2 Exercise*

One notes that the form (1.22b) is indeterminate $0/0$ in the limit $\Delta x \to 0$. Show that one use of l'Hôpital's rule does not resolve the limit, but two does, and the limit is precisely the form (1.2) of $I$.

### *1.4.3 Fisher information as a "mother" information*

Equation (1.22c) shows that $I$ is the cross-entropy between a PDF $p(x)$ and its infinitesimally shifted version $p(x+\Delta x)$. It has been noted (Caianiello, 1992) that $I$ more generally results as a "cross-information" between $p(x)$ and $p(x+\Delta x)$ for a host of *different* types of information measures. Some examples are as follows:

$$R_\alpha \equiv \ln\int dx\, p(x)^\alpha p(x+\Delta x)^{1-\alpha} \to -\Delta x^2\, 2^{-1}\alpha(1-\alpha)I, \quad (1.22d)$$

for $\alpha \neq 1$, where $R_\alpha$ is called the "Renyi information" measure (Amari, 1985); and

$$W \equiv \cos^{-1}\left[\int dx\, p^{1/2}(x)p^{1/2}(x+\Delta x)\right], \quad W^2 \to \Delta x^2\, 4^{-1} I, \quad (1.22e)$$

called the "Wootters information" measure (Wootters, 1981). To derive these results, one only has to expand the indicated function of $p(x+\Delta x)$ in the integrand out to *second order* in $\Delta x$, and perform the indicated integrations, using the identities $\int dx\, p'(x) = 0$ and $\int dx\, p''(x) = 0$.

Hence, Fisher information is the limiting form of many different measures of information; it is a kind of "mother" information.

## 1.5 Amplitude form of $I$

In definition (1.2), the division by $p(x)$ is bothersome. (For example, is $I$ undefined since necessarily $p(x) \to 0$ at certain $x$?) A way out is to work with a real "amplitude" function $q(x)$,

$$p(x) = q^2(x). \tag{1.23}$$

(Interestingly, probability amplitudes were used by Fisher (1943) independently of their use in quantum mechanics. The purpose was to discriminate among population classes.) Using form (1.23) in (1.2) directly gives

$$I = 4\int dx\, q'^2(x). \tag{1.24}$$

This is of a simpler form than (1.2) (no more divisions), and shows that $I$ *simply measures the gradient content in* $q(x)$ (and hence in $p(x)$). The integrand $q'^2(x)$ in (1.24) is the origin of the squared gradients in Table 1.1 of Lagrangians, as will be seen.

Representation (1.24) for $I$ may be computed independently of the preceding. One measure of the "distance" between an amplitude function $q(x)$ and its displaced version $q(x + \Delta x)$ is the quadratic measure (Braunstein and Caves, 1994)

$$L^2 \equiv \int dx\, [q(x + \Delta x) - q(x)]^2 \to \Delta x^2 \int dx\, q'^2(x) = \Delta x^2\, 4^{-1} I \tag{1.25}$$

after expanding out $q(x + \Delta x)$ in first-order Taylor series about point $x$ (cf. Eqs. (1.22c–e) preceding).

## 1.6 Efficient estimators

Classically, the main use of information $I$ has been as a measure of the ability to estimate a parameter. This is through the Cramer–Rao inequality (1.1), as follows.

If the equality can be realized in Eq. (1.1), then the mean-square error will go inversely with $I$, indicating that $I$ determines how small (or large) the error can be in any particular scenario. The question is, then, when is the equality realized?

The left-hand side of Eq. (1.7) is actually an inner product between two "vectors" $A(\mathbf{y})$ and $B(\mathbf{y})$,

$$A(\mathbf{y}) = \frac{\partial \ln p}{\partial \theta}\sqrt{p}, \quad B(\mathbf{y}) \equiv (\hat{\theta} - \theta)\sqrt{p}. \tag{1.26a}$$

Here the continuous index $\mathbf{y}$ defines the $\mathbf{y}$th component of each such vector (in contrast to the elementary case where vector components are discrete). The inner product of two vectors $A$, $B$ is always less than or equal to its value when the two vectors are *parallel*, i.e., when all their $\mathbf{y}$-components are proportional,

$$A(\mathbf{y}) = k(\theta)B(\mathbf{y}), \quad k(\theta) = const. \tag{1.26b}$$

(Note that function $k(\theta)$ remains constant since the parameter $\theta$ is, of course, constant.) Combining Eqs. (1.26a) and (1.26b) then provides a necessary condition (i) for attaining the equality in Eq. (1.1),

$$\frac{\partial \ln p(\mathbf{y}|\theta)}{\partial \theta} = k(\theta)[\hat{\theta}(\mathbf{y}) - \theta]. \tag{1.27}$$

A condition (ii) is the previously used unbiasedness assumption (1.3).

A PDF scenario where (1.27) is satisfied causes a minimized error $e^2_{\text{min}}$ that obeys

$$e^2_{\text{min}} = 1/I. \tag{1.28}$$

The estimator $\hat{\theta}(y)$ is then called "efficient." Notice that in this case the error varies inversely with information $I$, so that the latter becomes a well-defined quality metric of the measurement process.

### 1.6.1 Exercise

It is noted that only certain PDFs $p(\mathbf{y}|\theta)$ obey condition (1.27), among them (a) the independent normal law $p(\mathbf{y}|\theta) = A\prod_n \exp[-(y_n - \theta)^2/2\sigma^2]$, $A = const.$, and (b) the exponential law $p(\mathbf{y}|\theta) = \prod_n e^{-y_n/\theta}/\theta$, $y_n \geqslant 0$. On the other hand, with $N = 1$, (c) a PDF of the form

$$p(y|\theta) = A\sin^2(y - \theta), \quad A = const., \quad |y - \theta| \leqslant \pi$$

does not satisfy (1.27). Note that this PDF arises when the position $\theta$ of a one-dimensional quantum mechanical particle within a box is to be estimated. Hence, this fundamental measurement problem does not admit of an efficient estimate. Show these effects (a)–(c).

Also show that the estimators in (a) and (b) are unbiased, as required.

### 1.6.2 Exercise

If the condition (1.27) *is* obeyed, and if the estimator is unbiased, then the estimator function $\hat{\theta}(\mathbf{y})$ that *attains* efficiency is the one that maximizes the likelihood function $p(\mathbf{y}|\theta)$ through choice of $\theta$ (Van Trees, 1968). This is called the *maximum likelihood* (ML) estimator. As an example, the ML estimators for the problems (a) and (b) preceding are both the simple average of the data. Show this.

Note the simplification that occurs if one maximizes, instead of the likelihood, the *logarithm* of the likelihood. This *log-likelihood* law is also of fundamental importance to quantum measurement theory; see Chapter 10.

## 1.7 Fisher *I* as a measure of system disorder

We showed that information $I$ is a quality metric of an efficient measurement procedure. Now we will find that $I$ is also a measure of the degree of disorder of a system. *High disorder* means a lack of predictability of values $x$ over its range, i.e., a largely uniform or "unbiased" PDF $p(x)$. Such a curve is shown in Fig. 1.2(b). The curve has small gradient content, i.e., *it is broad and smooth*. Then by (1.24) the Fisher information *I is small.*

Conversely, if a curve $p(x)$ shows bias to particular $x$ values then it exhibits *low disorder*. See Fig. 1.2(a). Analytically, the curve will be *steeply sloped* about these $x$ values, and so the value of *I becomes large*. The net effect is that $I$ measures the degree of disorder of the system.

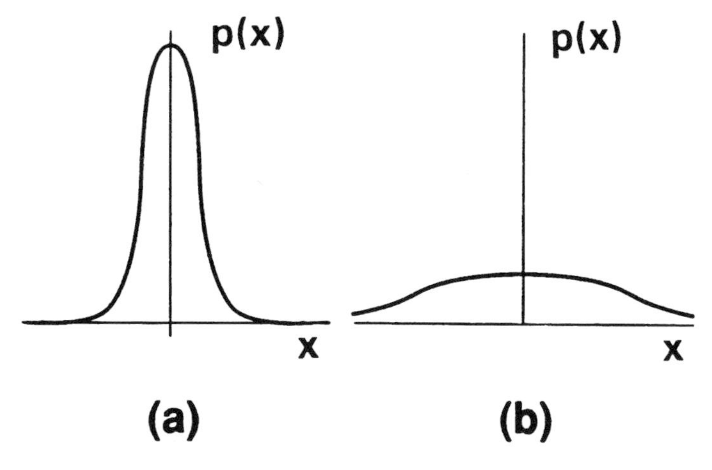


Fig. 1.2. Degree of disorder measured by $I$ values. In (a), random variable $x$ shows relatively low disorder and large $I$ (gradient content). In (b), $x$ shows high disorder and small $I$. (Reprinted from Frieden and Soffer, 1995.)

On the other hand, the ability to measure disorder is usually associated with the word "entropy." For example, the Shannon entropy $H$ is known to measure the degree of disorder of a system. (Example: By direct use of Eq. (1.13), if $p(x)$ is normal with variance $\sigma^2$ then $H = \ln \sigma + \ln \sqrt{2\pi e}$. This shows that $H$ monotonically increases with the "width" $\sigma$ of the PDF, i.e., with the degree of disorder in the system.)

Since we found that $I$ likewise measures the disorder of a system this suggests that $I$ should likewise be regarded as an "entropy." However, the entropy $H$ has another important property: When $H$ is, as well, the *Boltzmann* entropy, it obeys the second law of thermodynamics, increasing monotonically *with time*,

$$\frac{dH(t)}{dt} \geqslant 0. \tag{1.29}$$

Does $I$, then, also change monotonically with time? A particular scenario suggests that this is so.

## 1.8 Fisher $I$ as an entropy

We next show that, for a particular isolated system, $I$ monotonically decreases with time (Frieden, 1990). All measurements and PDF laws are now taken at a specified time $t$, so that the system PDF now has the form $p(x|t)$, i.e., the probability of reading $(x, x + dx)$ conditional upon a time $(t, t + dt)$.

### *1.8.1 Paradigm of the broken urn*

Consider a scenario where many particles fill a small urn. Imagine these to be ideal, point masses that collide elastically and that are not in an exterior force field. We want a smart measurement of their mean horizontal position $\theta$. Accordingly, a particle at horizontal position $y$ is observed, $y = \theta + x$, where $x$ is a random fluctuation from $\theta$. Define the mean-square error $e^2(t) = \langle[\theta - \hat{\theta}(y)]^2\rangle$ due to repeatedly forming estimates $\hat{\theta}(y)$ of $\theta$ within a small time interval $(t, t + dt)$. How should error $e$ vary with $t$?

Initially, at $t = 0$, the particles are within the small urn. Hence, any observed value $y$ should be near to $\theta$; then, any good estimate $\hat{\theta}(y)$ will likewise be close to $\theta$, and resultingly $e^2(0)$ will be small. Next, the walls of the container are broken, so that the particles are free to move away randomly. They will follow, of course, the random walk process which is called Brownian motion (Papoulis, 1965).

Consider a later time interval $(t, t + dt)$. For Brownian motion, the PDF $p(x|t)$ is Gaussian with a variance $\sigma^2 = Dt$, $D = const.$, $D \geqslant 0$. For a Gaussian PDF, $I = 1/\sigma^2$ (see derivation in Sec. 8.3.1). Then $I = I(t) = 1/(Dt)$, or $I$ decreases with $t$.

Can this result be generalized?

### 1.8.2 The 'I-theorem'

Equation (1.29) states that $H$ increases monotonically with time. This result is usually called the "Boltzmann $H$-theorem." In fact there is a corresponding "$I$-theorem"

$$\frac{dI(t)}{dt} \leqslant 0. \tag{1.30}$$

### 1.8.3 Proof

Start with the cross-entropy representation (1.22b) of $I(t)$,

$$I(t) = -2 \lim_{\Delta x \to 0} \Delta x^{-2} \int dx\, p \ln(p_{\Delta x}/p)$$
$$p \equiv p(x|t), \qquad p_{\Delta x} \equiv p(x + \Delta x|t). \tag{1.31}$$

Under certain physical conditions, e.g., "detailed balance," short-term correlation, shift-invariant statistics (Gardiner, 1985; Reif, 1965; Risken, 1996) $p$ obeys a *Fokker–Planck* differential equation

$$\frac{\partial p}{\partial t} = -\frac{d}{dx}[D_1(x,\, t)p] + \frac{d^2}{dx^2}[D_2(x,\, t)p], \tag{1.32}$$

where $D_1(x,\, t)$ is a drift function and $D_2(x,\, t)$ is a diffusion function. Suppose that $p_{\Delta x}$ also obeys the equation (Plastino and Plastino, 1996). Risken (1996) shows that two PDFs, such as $p$ and $p_{\Delta x}$, that obey the Fokker–Planck equation have a cross-entropy

$$G(t) \equiv -\int dx\, p \ln(p/p_{\Delta x}) \tag{1.33}$$

that obeys an $H$-theorem (1.29),

$$\frac{dG(t)}{dt} \geqslant 0. \tag{1.34}$$

It follows from Eq. (1.31) that $I$, likewise, obeys an $I$-theorem (1.30). Thus, the $I$-theorem and the $H$-theorem both hold under certain physical conditions.

There also is a possibility that physical conditions exist for which one theorem holds to the exclusion of the other. From the empirical viewpoint that the $I$-theorem leads to the derivation of a much wider range of physical laws (as in Chaps. 4–15) than does the $H$-theorem, such conditions must exist; however, they are yet to be found.

It should be remarked that the $I$-theorem was first proven (Plastino and Plastino, 1996) from the direct defining form (1.2) for $I$, i.e., *without* recourse to the cross-entropy form (1.22b).

### *1.8.4 Ramification to definition of time*

The $I$-theorem (1.30) is an extremely important result. It states that the Fisher information of a physical system can only decrease (or remain constant) in time. Combining this with Eq. (1.28) indicates that $e^2_{\mathrm{min}}$ must increase, so that even in the presence of efficient estimation *the quality of estimates must decrease with time*. This seems to be a reasonable alternative statement of the second law of thermodynamics. If, by the second law, the disorder of a system (as measured by the Boltzmann entropy) must increase, then the disorder of any measuring system must increase as well. This must degrade its use as a measuring instrument, causing the error $e^2_{\mathrm{min}}$ to increase. On this basis, one could estimate *the age* of an instrument by simply observing *how well* it measures.

Thus, $I$ is a measure of physical disorder that has its mathematical roots in estimation theory. By the same token, one may regard the Boltzmann entropy $H$ as a measure of physical disorder that has its mathematical roots in communication theory (Shannon, 1948). Communication theory plays a complementary role to estimation theory: the former describes how well messages can be *transmitted*, in the presence of given errors in the channel (system noise properties); the latter describes how accurately messages may be *estimated*, also in the presence of given errors in the channel.

If $I$ really is a physical measure of system disorder, it ought to relate somehow to temperature, pressure, and all other extrinsic parameters of thermodynamics. This is, in fact, the subject of Secs. 1.8.5–1.8.7.

Next, consider the concept of the flow of thermodynamic time (Zeh, 1992; Halliwell *et al.*, 1994). This concept is intimately tied to that of the Boltzmann entropy: an increase in the latter *defines* the positive passage of Boltzmann time. The $I$-theorem suggests an alternative definition to Boltzmann time: a decrease in $I$ defines an increase in "Fisher time." However, whether the two times always agree is an open question. In numerical experiments on randomly perturbed PDFs (Frieden, 1990), usually the resulting perturbations $\delta I$ went down when $\delta H$ went up, i.e., both measures agreed when disorder (and time) increased. They also usually agreed on decreases of disorder. However, there were disagreements about 1% of the time.

### *1.8.5 Ramification to temperature*

The Boltzmann temperature (Reif, 1965) $T$ is defined as $1/T \equiv \partial H_{\mathrm{B}}/\partial E$, where $H_{\mathrm{B}}$ is the Boltzmann entropy of an isolated system and $E$ is its energy. Consider two systems $A$ and $A'$ that are in thermal contact, but are otherwise isolated, and are approaching thermal equilibrium. The Boltzmann temperature has the important property that, after thermal equilibrium has been attained, a situation

$$T = T', \quad \frac{1}{T} \equiv \frac{\partial H_{\mathrm{B}}}{\partial E}, \quad \frac{1}{T'} \equiv \frac{\partial H'_{\mathrm{B}}}{\partial E'} \tag{1.35}$$

of equal temperature results. Let us now look at the phenomenon from the standpoint of information $I$, i.e., *without* recourse to the Boltzmann entropy.

Denote the total information in system $A$ by $I$, and that of system $A'$ by $I'$. The parameters $\theta$, $\theta'$ to be measured are the total energies $E$ and $E'$ of the two systems. The corresponding measurements are $Y_E$, $Y'_{E'}$. Because of the $I$-theorem (1.30), *both $I$ and $I'$ should approach minimum values as time increases*. We will show later that, since the two systems are physically separated and hence independent in their energy data $Y_E$, $Y'_{E'}$, the Fisher information state of the two is the sum of the two $I$ values. Hence, the $I$-theorem states that, after an infinite amount of time, the information of the combined system is

$$I(E) + I'(E') = min. \tag{1.36}$$

On the other hand, energy is conserved, so that

$$E + E' \equiv C, \tag{1.37}$$

$C = const.$ (Notice that this is a deterministic relation between the two ideal parameter values, not between the data; if it held for the data, then the prior assumption of independent data would have been invalid.)

The effect of (1.37) on (1.36) is

$$I(E) + I'(C - E) = min. \tag{1.38}$$

We now define a generalized "Fisher temperature" $T_\theta$ as

$$\frac{1}{T_\theta} \equiv -k_\theta \frac{\partial I}{\partial \theta}. \tag{1.39}$$

Notice that $\theta$ is any parameter under measurement. Hence, there is a Fisher "temperature" associated with any parameter to be measured. From (1.39), $T_\theta$ simply measures the sensitivity of information level $I$ to a change in system parameter $\theta$. The constant $k_\theta$ gives each $T_\theta$ value the same units. A relation between the two temperatures $T$ and $T_\theta$ is found below for a perfect gas.

Consider the case in point, $\theta = E$, $\theta' = C - \theta$. The temperature $T_\theta$ is now an energy temperature $T_E$. Differentiating Eq. (1.38) $\partial/\partial E$ gives

$$\frac{\partial I}{\partial E} + \frac{\partial I'}{\partial E'}(-1) = 0, \quad \text{or} \quad T_E = T'_{E'} \tag{1.40}$$

by (1.39). At equilibrium both systems attain a common Fisher energy temperature. This is analogous to the Boltzmann (conventional) result (1.35).

### *1.8.6 Exercise*

The right-hand side of Eq. (1.39) is impractical to evaluate (although still of theoretical importance) if $I$ is close to independent of $\theta$. This occurs in close to a shift-invariant case (1.11) where the resulting $I$ is close to the form (1.2). The key question is, then, whether the shift-invariance condition Eq. (1.11) holds when $\theta \equiv E$ and a measurement $y_E$ is made. The total number $N$ of particles comprising the system is critical here. If $N \approx 10$ or more, then (a) the PDF $p(y_E|E)$ will tend to obey the central limit theorem (Frieden, 2001) and, hence, be close to Gaussian in the shifted random variable $y_E - E$. An $I$ results that is close to the form (1.2). At the other extreme, (b) for small $N$ the PDF can typically be $\chi^2$ (assuming that the $N = 1$ law is Boltzmann, i.e., exponential). Here, shift invariance would not hold. Show (a) and (b).

### *1.8.7 Perfect gas law*

So far we have defined concepts of time and temperature on the basis of Fisher information. We now show that the perfect gas law may likewise be derived on this basis. This will also permit the (so far) unknown parameter $k_E$ to be evaluated from known parameters of the system.

Consider an ideal gas consisting of $M$ identical molecules confined to a volume $V$ and kept at Fisher temperature $T_E$. We want to know how the pressure in the gas depends upon the extrinsic parameters $V$ and $T_E$. The plan is to first compute the temporal mean pressure $\overline{p}$ within a small volume $dV = A\,dx$ of the gas and then integrate through to get the macroscopic answer.

Suppose that the pressure results from a force $F$ that is exerted normal to area $A$ and through the distance $dx$, as in the case of a moving piston. Then (Reif, 1965)

$$\overline{p} \equiv \frac{F}{A}\frac{dx}{dx} = -\frac{\partial E}{\partial V} \tag{1.41}$$

where the minus sign signifies that energy $E$ is stored in reaction to work done by the force. Using the chain rule, Eq. (1.41) becomes

$$\overline{p} = -\frac{\partial E}{\partial I}\frac{\partial I}{\partial V} = k_E T_E \frac{\partial I}{\partial V}, \tag{1.42}$$

the latter by definition (1.39) with $\theta = E$. Here $dI$ is the information in a data reading $dy_E$ of the ideal energy value $dE$. In general, quantities $\overline{p}$, $dI$, and $T_E$ can be functions of the position $\boldsymbol{r}$ of volume $dV$ within a gas. Multiplying (1.42) by $dV$ gives

$$\overline{p}(\boldsymbol{r})\, dV = k_E T_E(\boldsymbol{r})\, dI(\boldsymbol{r}) \tag{1.43}$$

with the $\boldsymbol{r}$-dependence now noted. Near equilibrium the gas should be well mixed and homogeneous, such that $\overline{p}$ and $T$ are independent of position $\boldsymbol{r}$. Then Eq. (1.43) may be directly integrated to give

$$\overline{p}V = k_E T_E I. \tag{1.44}$$

Note that $I = \int dI(\boldsymbol{r})$ is simply the total information due to many independent data readings $dy_E$. This again states that the information adds under independent data conditions.

The dependence (1.44) of $\overline{p}$ upon $V$ and $T_E$ is of the same form as the known equation of state of the gas

$$\overline{p}V = MkT, \tag{1.45}$$

where $k$ is the Boltzmann constant and $T$ is the *ordinary* (Boltzmann) temperature. Comparing Eqs. (1.44) and (1.45), exact compliance is achieved if $k_E T_E$ is related to $kT$ as

$$\frac{kT}{k_E T_E} = I/M, \tag{1.46}$$

the information per molecule. The latter should be a constant for a well-mixed gas.

These considerations seem to imply that thermodynamic theory may be developed completely from the standpoint of Fisher entropy, without recourse to the well-known properties of the Boltzmann entropy. In fact much progress is being made in this direction. It has been shown that Fisher information obeys the same Legendre transform property and concavity property as does entropy (Frieden *et al.*, 1999). Also, the use of Fisher information in place of entropy leads to a Schrödinger wave equation formulation of non-equilibrium thermodynamics (Frieden *et al.*, 2002a, b; also see Chapter 13). This formulation permits both quantum and thermodynamic effects to be analyzed simultaneously.

### 1.8.8 Ramification to derivations of physical laws

The uni-directional nature of the $I$-theorem (1.30) implies that, as $t \to \infty$,

$$I(t) = 4\int dx\, q'^2(x|t) \to min. \qquad (1.47)$$

Here we used the shift-invariant form (1.24) of $I$. The minimum would be achieved through variation of the amplitude function $q(x|t)$. It is convenient, and usual, to accomplish this through use of an Euler–Lagrange equation (see Eq. (0.13)). The result would define the form of $q(x|t)$ at temporal equilibrium.

In order for this approach to be tenable, it would need to be modified by appropriate input constraint properties of $q(x|t)$, such as normalization of $p(x|t)$. Other constraints, describing the particular physical scenario, must also be tacked on. Examples are fixed values of the means of certain physical quantities (case $\alpha = 2$ below). Such constraints may be appended to principle (1.47) by using the method of Lagrange undetermined multipliers, Eq. (0.39):

$$I + \sum_{k=1}^{K_0} \lambda_k \int dx\, q^{\alpha}(x|t) f_k(x) = extrem., \qquad (1.48a)$$

$$\int dx\, q^{\alpha}(x|t) f_k(x) = F_k, \quad k = 1, \ldots, \quad K_0, \alpha = const. \qquad (1.48b)$$

The kernel functions $f_k(x)$, constraint exponent $\alpha$, and data values $F_k$ are assumed known. The multipliers $\lambda_k$ are found such that the constraint equations (1.48b) are obeyed. This approach is taken in Chapter 13, and called the principle of minimum Fisher information (MFI). Owing to the arbitrariness of the constraints, it is Bayesian in nature, and therefore approximate. See also Huber (1981).

The most difficult step in the MFI approach is deciding what constraints to utilize (called the "input" constraints). The solution depends critically upon the choice of input constraints, and yet they cannot simply be all that are known to the user. They must be the particular subset of constraints that are *actually imposed* by nature. In general, this is difficult to know *a priori*. Our own approach – the EPI principle described in Chapter 3 and applied in subsequent chapters – is, in fact, of the Lagrange form (1.48a). However, it attempts to free the problem of the arbitrariness of the constraint terms. For this purpose, a physical rationale for the terms is utilized.

It is important to verify that a minimum (1.47) will indeed be attained in solution of the constrained variational problem. A maximum or point of inflection could conceivably result instead, defeating our present aims. For this purpose, we may use *Legendre's condition* for a minimum (Sec. 0.3.3): Let $\mathscr{L}$ denote the integrand (or Lagrangian) of the total integral to be extremized. In our scalar case, if

$$\frac{\partial^2 \mathscr{L}}{\partial q'^2} > 0 \tag{1.49}$$

the solution will be a minimum. By (1.47) and (1.48a), our Lagrangian is

$$\mathscr{L} = 4q'^2 + \sum_k \lambda_k q^\alpha f_k. \tag{1.50}$$

Using this in Eq. (1.49) gives

$$\frac{\partial^2 \mathscr{L}}{\partial q'^2} = +8, \tag{1.51}$$

showing that a minimum is indeed attained.

The foregoing assumed that coordinate $x$ is real and a scalar. However, most scenarios will involve *multiple* coordinates due to Lorentz covariance requirements. One or more of these are purely *imaginary*. (See also Sec. 1.8.14 and Appendix C.) For example, in Chapter 4 we use a space coordinate $x$ that is purely imaginary. The same analysis as the preceding shows that in this case the second derivative (1.51) is negative, so that a maximum is instead attained in this coordinate. However, others of the coordinates are real and, hence, tend to give a minimum. Obviously Legendre's condition cannot give a unique answer in this scenario. But in fact this is largely a moot point since, as will become clear, EPI is satisfied by any extremum solution.

When $I$ is minimized by the variational technique, $q(x|t)$ tends to be maximally smooth (Sect. 1.7). We saw that this describes a situation of maximum disorder. The second law of thermodynamics causes, as well, increased disorder. In this behavior, then, the $I$-theorem (1.30) acts like the second law.

### 1.8.9 Is the I-theorem equivalent to the H-theorem?

We showed using Eq. (1.34) that, if a PDF $p(x)$ and its *infinitesimally* shifted version $p(x + \Delta x)$ both obey the Fokker–Planck equation, then the $I$-theorem follows. On the other hand, Eq. (1.34) with $\Delta x$ *finite* is an expression of the $H$-theorem. Hence, the two theorems have a common pedigree, so to speak. Are, then, the two theorems equivalent? In fact they are not equivalent because the $I$-theorem is a *limiting form* (as $\Delta x \to 0$) of Risken's $H$-theorem. Taking the limit introduces the derivative $p'$ of the PDF into the integrand of what was $H$, transforming it into the form Eq. (1.2) of $I$. The presence of this derivative in $I$, and its absence in $H$, has strong mathematical implications. One example is as follows.

The equilibrium solutions for $p(x)$ that are obtained by extremizing Fisher $I$

(called the EPI principle below) are, in general, different from those obtained by the corresponding use $H = max.$ of entropy. See Sec. 1.3.1. In fact, EPI solutions and $H = max.$ solutions agree only in statistical mechanics; this is shown in Appendix A.

It is interesting that correct solutions via EPI occur even for PDFs that do not obey the Fokker–Planck equation. By its form (1.32), the time rate of change of $p$ depends only upon the present value of $p$. Hence, the process has short-term memory (see also Gardiner, 1991, p. 144). However, EPI may be used to derive the $1/f$ power spectral noise effect (Chapter 8), a law famous for exhibiting *long-term* memory. Also, the relativistic electron obeys an equation of continuity of flow $\partial p/\partial t = c\nabla \cdot (\psi^*[\alpha]\psi)$, $p \equiv \psi^*\psi$ (Schiff, 1955), where all quantities are defined in Chapter 4. This does not quite have the form of a Fokker–Planck Eq. (1.32) (compare right-hand sides). However, EPI may indeed be used to derive the Dirac equation of the electron (Chapter 4).

These considerations imply that the Fokker–Planck equation is a sufficient, but not necessary, condition for validity of the EPI procedure. An alternative condition of wider scope must exist. Such a one is the unitary condition to be discussed in Secs. 3.8.5 and 3.8.7.

### *1.8.10 Flow property*

Since information $I$ obeys an $I$-theorem Eq. (1.30), temperature effects Eqs. (1.39) and (1.40), and a gas law Eq. (1.44), indications are that $I$ is every bit as "physical" an entity as is the Boltzmann entropy. This includes, in particular, a property of temporal *flow* from an information source to a sink. This property is used in our physical information model of Sec. 3.3.2.

### *1.8.11 Additivity property*

A vital property of the information $I$ is that of additivity: the information from mutually isolated systems adds. This is shown as follows.

Suppose that we have $N$ copies of the urn mentioned in Sec. 1.8.1. See Fig. 1.3. As before, each urn contains particles that are undergoing Brownian motion. (This time the urns are not broken.) Each sits rigidly in place upon a table that moves with an unknown velocity $\theta$ in the $X$-direction, relative to the laboratory. A particle is randomly selected in each urn, and its total $X$-component laboratory velocity value $y_n$ is measured. Let $x_n$ denote the particle's *intrinsic* speed, i.e., relative to its urn, with $(x_n, n = 1, \ldots, N) \equiv \mathbf{x}$.

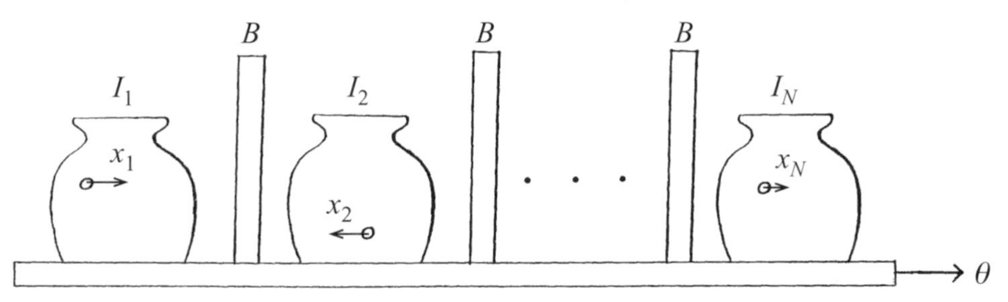


Fig. 1.3. $N$ urns are moving at a common speed $\theta$. Each contains particles in Brownian motion. Barriers B physically isolate the urns. Measurements $y_n = \theta + x_n$ of particle velocities are made, one to an urn. Each $y_n$ gives rise to an information amount $I_n$. The total information $I$ over all the data **y** is the sum of the individual informations $I_n$, $n = 1, \ldots, N$ from the urns.

The **x** are random because of the Brownian motion. Assuming non-relativistic speeds, the intrinsic data $(y_n,\ n = 1, \ldots, N) \equiv \mathbf{y}$ obey simply

$$\mathbf{y} = \theta + \mathbf{x}. \tag{1.52}$$

Assume that the urns are physically isolated from one another by the use of barriers $B$ (see Fig. 1.3), so that there is no interaction between particles from different urns. Then the data **y** are independent. This causes the likelihood law to break up into a product of factors (Frieden, 2001)

$$p(\mathbf{y}|\theta) = \prod_{n=1}^{N} p_n(y_n|\theta), \tag{1.53}$$

where $p_n$ is the likelihood law for the $n$th observation.

We may now compute the information $I$ about $\theta$ in the independent data **y**. By Eq. (1.53)

$$\ln p = \sum_n \ln p_n, \qquad p \equiv p(\mathbf{y}|\theta), \qquad p_n \equiv p_n(y_n|\theta),$$

so that

$$\frac{\partial \ln p}{\partial \theta} = \sum_n \frac{1}{p_n} \frac{\partial p_n}{\partial \theta}. \tag{1.54}$$

Squaring the latter gives

$$\left(\frac{\partial \ln p}{\partial \theta}\right)^2 = \sum_{\substack{mn \\ m \neq n}} \frac{1}{p_m} \frac{1}{p_n} \frac{\partial p_m}{\partial \theta} \frac{\partial p_n}{\partial \theta} + \sum_n \frac{1}{p_n^2} \left(\frac{\partial p_n}{\partial \theta}\right)^2 \tag{1.55}$$

where the last sum is for indices $m = n$. Then the defining Eq. (1.9) for $I$ gives, with the substitutions Eqs. (1.53), (1.55),

$$I = \int d\mathbf{y} \prod_k p_k \left[ \sum_{\substack{mn \\ m \neq n}} \frac{1}{p_m} \frac{1}{p_n} \frac{\partial p_m}{\partial \theta} \frac{\partial p_n}{\partial \theta} + \sum_n \frac{1}{p_n^2} \left( \frac{\partial p_n}{\partial \theta} \right)^2 \right]. \tag{1.56}$$

Now use the fact that, in this equation, the probabilities $p_k$ for $k \neq m$ or $n$ integrate through as simply factors 1, by normalization. The remaining factors in $\prod_k p_k$ are then $p_m p_n$ for the first sum, and just $p_n$ for the second sum. The result is, after some cancellation,

$$I = \sum_{\substack{mn \\ m \neq n}} \iint dy_m \, dy_n \frac{\partial p_m}{\partial \theta} \frac{\partial p_n}{\partial \theta} + \sum_n \int dy_n \frac{1}{p_n} \left( \frac{\partial p_n}{\partial \theta} \right)^2. \tag{1.57}$$

This simplifies, drastically, as follows. The first sum separates into a product of a sum

$$\sum_n \int dy_n \frac{\partial p_n}{\partial \theta} \tag{1.58}$$

with a corresponding one in index $m$. But

$$\int dy_n \frac{\partial p_n}{\partial \theta} = \frac{\partial}{\partial \theta} \int dy_n \, p_n = \frac{\partial}{\partial \theta} 1 = 0 \tag{1.59}$$

by normalization. Hence the first sum in Eq. (1.57) is zero.

The second sum in Eq. (1.57) is, by Eq. (1.5),

$$\sum_n \int dy_n \, p_n \left( \frac{\partial}{\partial \theta} \ln p_n \right)^2 \equiv \sum_n I_n \tag{1.60}$$

by the definition Eq. (1.9) of $I$. Hence, we have shown that

$$I = \sum_{n=1}^{N} I_n \tag{1.61}$$

in this scenario of independent data. This is what we set out to prove.

It is well known that the Shannon entropy $H$ obeys additivity, as well, under these conditions. That is, with

$$H = -\int d\mathbf{y} \, p(\mathbf{y}|\theta) \ln p(\mathbf{y}|\theta), \tag{1.62}$$

under the independence condition Eq. (1.53) it gives

$$H = \sum_{n=1}^{N} H_n, \quad H_n = -\int dy_n \, p_n(y_n|\theta) \ln p_n(y_n|\theta). \tag{1.63}$$

### 1.8.12 Exercise

Show this. *Hint*: The proof is much simpler than the preceding. One merely uses the argument below Eq. (1.56) to collapse the multidimensional integrals into the one in $y_n$ as needed.

One notes from all this that a requirement of *additivity* does not in itself uniquely identify the appropriate measure of disorder. It could be entropy or, as shown above, Fisher information. This is despite the identity $\ln(fg) = \ln(f) + \ln(g)$, which seems uniquely to imply entropy as the measure. Undoubtedly many other measures satisfy additivity as well.

### 1.8.13 $I =$ min. from statistical mechanics viewpoint

According to a basic premise of statistical mechanics (Reif, 1965), the PDF for a system that *will occur* is the one that is maximum probable to occur.

A general image-forming system is shown in Fig. 1.4. It consists of a source S of particles – any type will do, whether electrons, photons, etc. – a focussing device L of some sort and an image plane M for receiving the particles. Plane M is subdivided into coordinate positions ($x_n$, $n = 1, \ldots, N$) with a constant, small spacing $\varepsilon$. An "image event" $x_n$ is the receipt of a particle within the interval $(x_n, x_n + \varepsilon)$. The number $m_n$ of image events $x_n$ is noted, for each

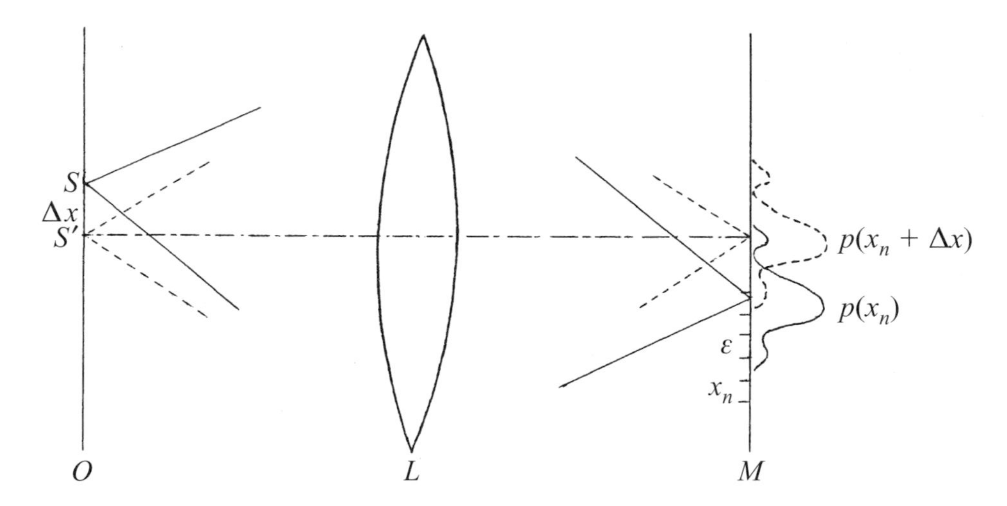


Fig. 1.4. A statistical mechanics view of Fisher information. The ideal point source position S′ gives rise to the ideal PDF $p(x_n + \Delta x)$, while the actual point source position S gives rise to the empirical PDF $p(x_n)$. Maximizing the logarithm of the probability of the latter PDF curve implies a condition of minimum Fisher information, $I[p(x_n)] = I = min$.

$n = 1, \ldots, N$. There are $M$ particles in all, with $M$ very large. What is the joint probability $P(m_1, \ldots, m_n)$?

Each image event is a possible position $x_n$, of which there are $N$. Therefore the image events comprise an $N$ary events sequence. This obeys a multinomial probability law (Frieden, 2001) of order $N$,

$$P(m_1, \ldots, m_n) = M! \prod_{n=1}^{N} \frac{r(x_n)^{m_n}}{m_n!}. \tag{1.64}$$

The quantities $r(x_n)$ are the "prior probabilities" of events $x_n$. These are considered next.

The ideal source S for the experiment would be a very small aperture that is located on-axis. This situation would give rise to ideal (prior) probabilities $r(x_n)$, $n = 1, \ldots, N$. However, in performing the experiment, we really cannot know exactly where the source is. For example, for quantum particles, there is an ultimate uncertainty in position of at least the Compton length (Sec. 4.1.17). Hence, in general, the source S will be located at a small position $\Delta x$ *off-axis.* The result is that the particles will, in reality, obey a different set of probabilities $P(x_n) \neq r(x_n)$. These can be evaluated. Assuming shift invariance (Eq. (1.11)) and 1 : 1 magnification in the system,

$$p(x_n) = r(x_n - \Delta x), \quad \text{or} \quad r(x_n) = p(x_n + \Delta x). \tag{1.65}$$

By the law of large numbers (Frieden, 2001), since $M$ is large the probabilities $p(x_n)$ agree with the occurrences $m_n$, by the simple rule

$$m_n = Mp(x_n). \tag{1.66}$$

(This takes the conventional, von Mises viewpoint that probabilities measure the frequency of occurrence of actual – not ideal – events (Von Mises, 1936).) Using Eqs. (1.65) and (1.66) in Eq. (1.64) and taking the logarithm gives

$$\ln P = C + \sum_n Mp(x_n) \ln p(x_n + \Delta x) - \sum_n \ln[Mp(x_n)]! \tag{1.67}$$

where $C$ is an irrelevant constant. Since $M$ is large we may use the Stirling approximation $\ln u! \approx u \ln u$, so that

$$\ln P \approx B + M \sum_n p(x_n) \ln\left(\frac{p(x_n + \Delta x)}{p(x_n)}\right), \tag{1.68}$$

where $B$ is an irrelevant constant. The normalization of $p(x_n)$ was also used. Multiplying and dividing Eq. (1.68) by the fine spacing $\varepsilon$ allows us to replace the sum by an integral. Also, since $P$ is to be a maximum, so will be $\ln P$. The result is that Eq. (1.68) becomes

$$\ln P \approx \int dx\, p(x) \ln\left(\frac{p(x + \Delta x)}{p(x)}\right) = max. \tag{1.69}$$

after ignoring all multiplicative and additive constants. Noticing the minus sign in Eq. (1.22b), we see that Eq. (1.69) states that

$$I[p(x)] \equiv I = min., \tag{1.70}$$

agreeing with Eq. (1.47).

This approach can be generalized. Regardless of the physical nature of coordinate $x$, there will always be uncertainty $\Delta x$ in the actual value of the origin of a PDF $p(x)$. As we saw, this uncertainty is naturally expressed as a "distance measure" $I$ between $p(x)$ and its displaced version $p(x + \Delta x)$ (Eq. (1.69)).

It is interesting to compare this approach with the derivation of the $I$-theorem in Sec. 1.8.3. That was based purely on the assumption that the Fokker–Planck equation is obeyed. By comparison, here the assumptions are that (i) maximum probable PDFs actually occur (basic premise of statistical mechanics) and (ii) the system admits of an ultimate resolution "length" $\Delta x$ of finite extent.

The two derivations may be further compared on the basis of effective "resolution lengths." In Sec. 1.8.3 the limit $\Delta x \to 0$ is rigorously taken, since $\Delta x$ is, there, just a *mathematical* artifact (which enables $I$ to be expressed as the cross-entropy via Eq. (1.22b)). Also, the approach by Plastino *et al.* to the $I$-theorem that is mentioned in that section does not even use the concept of $\Delta x$. By contrast, in the current derivation $\Delta x$ is not merely of mathematical origin. It originates physically, as an ultimate resolution length and, hence, is small but intrinsically *finite*. This means that the transition from the cross-entropy on the right-hand side of Eq. (1.69) to information $I$ via Eq. (1.22b) is, here, only an approximation.

If one takes this derivation seriously, then an important effect follows. Since $I$ is only an approximation on the scale of $\Delta x$, the use of $I[q(x)]$ in any variational principle (such as EPI) must give solutions $q(x)$ that *lose their validity at scales finer than* $\Delta x$. For example, $\Delta x$ results as the Compton length in the EPI derivation of quantum mechanics (Sec. 4.1.17). A ramification is that quantum mechanics is not represented by its famous wave equations at such scales.

This is a somewhat moot point, since then accurate observations at that scale could not be made anyhow. Nevertheless, it suggests that a different kind of mechanics ought to hold at scales finer than $\Delta x$. Such considerations of course lead one to thoughts of quantum gravity (Misner *et al.*, 1973, p. 1193); see also Chapter 11. This is a satisfying transition from a physical point of view. Also, from the statistical viewpoint, it says that EPI is a complete theory insofar as defining the limits of its range of validity.

Likewise, the electromagnetic wave equation (Chapter 5) would break down at scales $\Delta x$ finer than the vacuum fluctuation length given by Eq. (5.4); suggesting a transition to a quantum electrodynamics. And the classical gravitational field equation (Chapter 6) would break down at scales finer than the Planck length, Eq. (6.22); again suggesting a transition to quantum gravity (see preceding paragraph). The magnitudes of these ultimate resolution lengths are, in fact, predicted by the EPI approach; see Chaps. 5, 6 and Sec. 3.4.15.

However, this kind of reasoning can be repeated to endlessly finer and finer scales. Thus, the theory of quantum gravity that is derived by EPI in Chapter 11 must break down beyond a finest scale $\Delta x$ defined by the (presumed) approximate nature of $I$ at that scale. The length $\Delta x$ would be much smaller than the Planck length. This, in turn, suggests the need for a *new* "mechanics" that would hold at finer scales than the Planck length, etc., to ever-finer scales. The same reasoning applies to electromagnetic and gravitational theories. Perhaps all three theories would converge to a common theory at a finest resolution length to be determined, at which "point" the endless subdivision process would terminate.

On the other hand, if one does not take this derivation seriously then, based upon the derivation of Sec. 1.8.2, the wave equations of quantum mechanics are valid *down to all scales*, and a transition to quantum gravity is apparently not needed. Which of the two views to take is, at this time, unknown.

The statistical mechanics approach of this section is based partly on work by Shirai (1998). Other approaches utilizing Fisher information as a basis for physics are by Silver (1992), Reginatto (1998), Kahre (2000), Hall (2000), and Luo (2003).

### *1.8.14 Imaginary coordinates*

We mentioned in Sec. 1.8.8 that imaginary Fisher coordinates are sometimes used in the derivations. In Appendix C it is shown how imaginary coordinates may be incorporated into the definition of Fisher information. It is also shown that the key statistical effect (1.1), the Cramer–Rao (C–R) inequality, holds even in the case of an imaginary coordinate. This property is important since the C–R inequality is largely the basis for the EPI principle to follow.

Thus, in any EPI derivation the user is free to use either real or imaginary Fisher coordinates. The added flexibility is of benefit in broadening the scope of phenomena that can be so derived. For example, the use of mixed real and imaginary Fisher coordinates in a given application can give rise to the d'Alembertian form that is required of many wave equations, such as the

Klein–Gordon equation (Chapter 4) and the vector wave equation of electrodynamics (Chapter 5).

### 1.8.15 Multiple PDF cases

In all of the preceding, there was one, scalar parameter $\theta$ to be estimated. This implies an information Eq. (1.2) that may be used to predict a single-component PDF $p(x)$ on scalar fluctuations $x$, as sketched in Sec. 1.8.8. Many phenomena are indeed describable by such a PDF. For example, in statistical mechanics the Boltzmann law $p(E)$ defines the single-component PDF on scalar energy fluctuations $E$ (Chapter 7).

Of course, however, nature is not that simple. There are physical phenomena that require *multiple* PDFs $p_n(\mathbf{x})$, $n = 1, \ldots, N$ or amplitude functions $q_n(\mathbf{x})$, $n = 1, \ldots, N$ for their description. Also, the fluctuations $\mathbf{x}$ might be vector quantities (as indicated by the boldface). For example, in relativistic quantum mechanics there are four wave functions and, correspondingly, four PDFs to be determined (actually, we will find eight real wave functions $q_n(\mathbf{x})$, $n = 1, \ldots, 8$, corresponding to the real and imaginary parts of four complex wave functions). To derive a multiple-component, vector phenomenon, it turns out, requires use of the Fisher information defining the estimation quality of *multiple* vector parameters. This is the subject of the next chapter.

# 2

# Fisher information in a vector world

## 2.1 Classical measurement of four-vectors

In the preceding chapter, we found that the accuracy in an estimate of a single parameter $\boldsymbol{\theta}$ is determined by an information $I$ that has some useful *physical* properties. It provides new definitions of disorder, time, and temperature, and a variational approach to finding a single-component PDF law $p(x)$ of a single variable $x$. However, many physical phenomena are describable only by multiple-component PDFs, as in quantum mechanics, and for vector variables $\mathbf{x}$, since worldviews are usually four-dimensional (as required by covariance). *Our aim in this chapter, then, is to form a new, scalar information I that is appropriate to this multi-component, vector scenario.* The information should be intrinsic to the phenomenon under measurement and not depend, e.g., upon exterior effects such as the noise of the measuring device.

### *2.1.1 The "intrinsic" measurement scenario*

In Bayesian statistics, a *prior* scenario is often used to define an otherwise unknown prior probability law; see, e.g., Good (1976), Jaynes (1985), or Frieden (2001). This is a model scenario that permits the prior probability law to be computed on the basis of some ideal conditions, such as independence of data, and/or 'maximum ignorance' (see below), etc. We will use the concept of the prior scenario to define our unknown information expression.

For this purpose we proceed to analyze a particular prior scenario. This is an ideal, $4N$-dimensional measurement scenario – the vector counterpart of the scalar parameter problem of Sec. 1.2.1. Here, $N$ *four*-vectors $\boldsymbol{\theta}_n$, $n = 1, \ldots, N$ of any physical nature (four-positions or potentials, etc.) are observed as intrinsic data

$$\mathbf{y}_n = \boldsymbol{\theta}_n + \mathbf{x}_n, \quad n = 1, \ldots, N. \tag{2.1}$$

As before, the data are 'intrinsic' in that their fluctuations $\mathbf{x}_n$ are presumed to characterize solely the phenomenon under measurement. There is, e.g., no additional fluctuation due to noise in the instrument providing the measurements. An information measure $I$ of fluctuations $\mathbf{x}_n$ will likewise be intrinsic to the phenomenon, as we required at the outset.

How realistic is this model? In Chapter 10 we will find that immediately before *real* measurements are made the physics of the intrinsic fluctuations $\mathbf{x}_n$ is independent of instrument noise. Indeed, how could fluctuations prior to a measurement depend upon the measuring system? Hence, ignoring the instrumental errors in this *model, prior* measurement scenario agrees with the *real, prior* measurement scenario. We call this scenario the 'intrinsic' data scenario.

For simplicity of notation, it is convenient to define "grand" vectors $\boldsymbol{\theta}$, $\mathbf{y}$, $d\mathbf{y}$ over all $n$ as

$$\boldsymbol{\theta} = (\boldsymbol{\theta}_1, \ldots, \boldsymbol{\theta}_N), \quad \mathbf{y} = (\mathbf{y}_1, \ldots, \mathbf{y}_N), \quad d\mathbf{y} = d\mathbf{y}_1 \cdots d\mathbf{y}_N. \tag{2.2}$$

### 2.1.2 Aim of the data collection

In chapters to follow, the intrinsic data $\mathbf{y}_n$ will be analyzed presuming either of two general aims: (i) to estimate $\boldsymbol{\theta}_n$ *per se*, as when $\boldsymbol{\theta}_n$ is the $n$th four-position of a material particle; or (ii) to estimate a *function* of each $\boldsymbol{\theta}_n$, as when $\boldsymbol{\theta}_n$ is an ideal four-position and the electromagnetic four-potential $\mathbf{A}(\boldsymbol{\theta}_n)$ is required.

For either scenario (i), (ii), we want to form a scalar information measure that defines the quality of the $\mathbf{y}_n$. The answer should, intuitively, resemble the form of Eq. (1.2).

### 2.1.3 Assumption of independence

Suppose that $\boldsymbol{\theta}_n$, $\mathbf{x}_n$, and $\mathbf{y}_n$ are statistically independent. In particular, the data $\mathbf{y}_n$ are then collected efficiently. This is the usual goal in data collection. It can be accomplished by two different experimental procedures: (a) $N$ independent repetitions of the experiment under the same initial conditions, measuring $\boldsymbol{\theta}_n$ at each; or (b) in the case of measurements upon particles, one experiment upon $N$ particles, measuring the $N$ different parameters $\boldsymbol{\theta}_n$ that ensue from one set of initial conditions. In scenario (a), independence is automatically satisfied. Scenario (b) tries to induce ergodicity in the data $\mathbf{y}_n$, e.g., by measuring particles that are sufficiently separated in one or more coordinates.

### 2.1.4 *Real or imaginary coordinates*

Each coordinate $x_n$ of a four-vector $\mathbf{x}$ is, in general, either purely real or purely imaginary (Frieden and Soffer, 1995); also see Sec. 1.8.14 and Appendix C. An example of 'mixed' real and imaginary coordinates is given in Sec. 4.1.2.

## 2.2 Optimum, unbiased estimates

An observer's aim is generally to learn as much as possible about the parameters $\boldsymbol{\theta}$. For this purpose, an optimum estimate

$$\hat{\boldsymbol{\theta}}_n = \hat{\boldsymbol{\theta}}_n(\mathbf{y}) \tag{2.3}$$

of each four-parameter $\boldsymbol{\theta}_n$ may be fashioned. Each estimate is, thus, a general function of all the data. An example of such an estimator is simply $\mathbf{y}_n$, i.e. the corresponding data vector, but this will usually not be optimum. One well-known class of optimum estimators is the "maximum likelihood" estimator class discussed in Chapter 1.

### 2.2.1 *Unbiased estimators*

As with the case of "good" experimental apparatus, the estimators are assumed to be unbiased, i.e., to obey

$$\langle \hat{\boldsymbol{\theta}}_n(\mathbf{y}) \rangle \equiv \int d\mathbf{y}\, \hat{\boldsymbol{\theta}}_n(\mathbf{y}) p(\mathbf{y}|\boldsymbol{\theta}) = \boldsymbol{\theta}_n, \tag{2.4}$$

where $p(\mathbf{y}|\boldsymbol{\theta})$ is the conditional probability of all data $\mathbf{y}$ in the presence of all parameters $\boldsymbol{\theta}$. Equation (2.4) says that, although a given estimate will generally be in error, on average it will be correct. How *small* the error may be, is next established. This introduces the vital concept of information.

### 2.2.2 *Cramer–Rao inequalities*

*We temporarily suppress index n* and focus attention on the four components of any one ($n$ fixed) foursome of *scalar* values $\theta_\nu$, $y_\nu$, $x_\nu$, $\nu = 0, 1, 2, 3$. The mean-square errors from the true values $\theta_\nu$ are

$$e_\nu^2 \equiv \int d\mathbf{y}\, [\hat{\theta}_\nu(\mathbf{y}) - \theta_\nu]^2 p(\mathbf{y}|\boldsymbol{\theta}). \tag{2.5}$$

Since the data are independent, each mean-square error obeys complementarity with an 'information' quantity $I_\nu$,

$$e_\nu^2 I_\nu \geqslant 1, \tag{2.6}$$

where

$$I_\nu \equiv \int d\mathbf{y} \left[ \frac{\partial \ln p(\mathbf{y}|\boldsymbol{\theta})}{\partial \theta_\nu} \right]^2 p(\mathbf{y}|\boldsymbol{\theta}). \tag{2.7}$$

(See Appendix B for derivation, where $I_\nu \equiv F_{\nu\nu}$.) Equations (2.6) and (2.7) comprise Cramer–Rao inequalities for our vector quantities. They hold for either real or *imaginary* components $\theta_\nu$; see Appendix C. When equality is attained in (2.6), the minimum possible error $e_\nu^2$ is attained. Then the estimator is called "efficient." Quantity $I_\nu$ is the $\nu$th element along the diagonal of the Fisher information matrix [F]. The $I_\nu$ thus comprise a *vector* of informations.

## 2.3 Stam's information

We are now in a position to decide how to construct, from the vector of informations $I_\nu$, the single scalar information quantity $I$ that we seek. *Regaining subscripts* $n$ and summing on Eq. (2.6) gives

$$\sum_n \sum_\nu 1/e_{n\nu}^2 \leqslant \sum_n \sum_\nu I_{n\nu}. \tag{2.8}$$

Each term in the left-hand sum was called an "intrinsic accuracy" by Fisher. Stam (1959a) proposed using the sum as a scalar information measure $I_S$ for a vector scenario (as here),

$$I_S \equiv \sum_n \sum_\nu 1/e_{n\nu}^2 \leqslant \sum_n \sum_\nu I_{n\nu}, \tag{2.9}$$

where the inequality is due to Eq. (2.8). Stam's information is promising, since it is a scalar quantity. We adapt it to our purposes.

### *2.3.1 Exercise*

Stam's information $I_S$, in depending explicity upon the error variances, ignores all possible error cross-correlations. However, for our additive error case (2.1), where the data $y_n$ are independent and the estimators are unbiased (2.4), all error cross-correlations are zero. Show this.

### *2.3.2 Trace form, channel capacity, efficiency*

The right-hand side of Eq. (2.9) is seen to be an upper bound to $I_S$. Assume that efficient estimators are used. Then, the equality is attained in Eq. (2.6), so that each left-hand term $1/e_{n\nu}^2$ of Eq. (2.8) equals its corresponding information value $I_{n\nu}$. This means that the upper bound in Eq. (2.9) is realized. An analogous situation arises in the theory of Shannon information. There, the channel capacity, denoted as $C$, denotes the maximum possible amount of information that may be passed by a channel (Reza, 1961). Hence, we likewise define a capacity $C$ for the estimation procedure to convey Fisher information $I_s$ about the intrinsic system,

$$I \equiv C = \sum_n \int d\mathbf{y}\, p(\mathbf{y}|\boldsymbol{\theta}) \sum_\nu \left( \frac{\partial \ln p(\mathbf{y}|\boldsymbol{\theta})}{\partial \theta_{n\nu}} \right)^2, \tag{2.10}$$

the latter due to (2.7). It is interesting that this is the trace of the Fisher information matrix (see Appendix B, Eq. (B7)). This information also satisfies our goal of measuring the intrinsic *disorder* of the phenomenon under measurement (see Sec. 2.8). It also measures its *complexity* (Sec. 2.4.1).

### *2.3.3 Exercise*

Taking the trace of the Fisher information matrix ignores, of course, all off-diagonal elements. However, because we have assumed independent data, the off-diagonal elements are, in fact, zero. Show this.

The trace operation (sum over $n$) in Eq. (2.10) has many physical connotations, in particular *relativistic invariance*, as shown in Sec. 3.5.1.

## 2.4 Physical modifications

The channel capacity Eq. (2.10) simplifies, in steps, due to various physical aspects of the intrinsic measurement scenario.

### *2.4.1 Additivity of the information; a measure of complexity*

Additivity was previously shown (Sec. 1.8.11) for the case of a single parameter. The generalization to a vector of parameters is now taken up. As will be seen, because of the lack of "cross talk" between different data and parameters, the proof below is a little easier.

Since the intrinsic data are collected independently (Sec. 2.1.3), the joint probability of all the data separates into a product of marginal laws

$$p(\mathbf{y}|\boldsymbol{\theta}) = \prod_n p_n(\mathbf{y}_n|\boldsymbol{\theta}) = \prod_n p_n(\mathbf{y}_n|\boldsymbol{\theta}_n). \tag{2.11}$$

The latter equality follows since by Eq. (2.1), $\boldsymbol{\theta}_m$ has no influence on $\mathbf{y}_n$, $m \neq n$. Taking the logarithm of Eq. (2.11) and differentiating then gives

$$\frac{\partial \ln p(\mathbf{y}|\boldsymbol{\theta})}{\partial \theta_{n\mu}} = \frac{1}{p_n} \frac{\partial p_n}{\partial \theta_{n\mu}}, \quad p_n \equiv p_n(\mathbf{y}_n|\boldsymbol{\theta}_n). \tag{2.12}$$

Substitution of Eqs. (2.11) and (2.12) into Eq. (2.10) gives

$$I = \sum_n \int d\mathbf{y} \prod_m p_m(\mathbf{y}_m|\boldsymbol{\theta}_m) \sum_\nu \frac{1}{p_n^2}\left(\frac{\partial p_n}{\partial \theta_{n\nu}}\right)^2, \tag{2.13}$$

$$= \sum_n \int d\mathbf{y}_n\, p_n(\mathbf{y}_n|\boldsymbol{\theta}_n) \sum_\nu \frac{1}{p_n^2}\left(\frac{\partial p_n}{\partial \theta_{n\nu}}\right)^2 \tag{2.14}$$

after integrating out $d\mathbf{y}_m$ for terms in $m \neq n$, using normalization of each probability $p_m$. After an obvious cancellation, we get

$$I = \sum_n \int d\mathbf{y}_n \frac{1}{p_n} \sum_\nu \left(\frac{\partial p_n}{\partial \theta_{n\nu}}\right)^2. \tag{2.15}$$

This is the first simplification.

The sum in Eq. (2.15) goes from $n = 1$ to $n = N$. Each term in the sum contributes a new analytical degree of freedom to $I$. The larger $N$ is, the more degrees of freedom are needed to describe the observed effect, i.e., the more *complex* it is. See also Badii and Politi (1997). Since every term in the sum is positive or zero, the tendency is for the information $I$ to grow with the size of $N$ (although it wouldn't necessarily grow if added degrees of freedom $p_n$ affected the original ones). An example of such growth is Eq. (4.18). Thus, the size of $I$ measures the *level of complexity* of the observed effect. See also Binder (2000).

It is intuitive that independent data should also impart maximum information about a system. See Sec. 3.4.12. In summary, independent data cause both (i) additivity, and (ii) a *maximum value*, of the information capacity.

### 2.4.2 Shift invariance property

As with scalar PDFs (see Sec. 1.2.4), many vector PDFs obey "shift invariance." For example, the Schrödinger wave equation, which describes the PDF on vector position and time, is invariant to constant shifts in these coordinates. This particular application of shift invariance is called Galilean invariance (Sec. 3.5.3). Applying shift invariance to our general PDF, any fluctuation $\mathbf{y}_n - \boldsymbol{\theta}_n \equiv \mathbf{x}_n$ should occur with a probability that is independent of the absolute size of $\boldsymbol{\theta}_n$,

$$p_n(\mathbf{y}_n|\boldsymbol{\theta}_n) = p_{x_n}(\mathbf{y}_n - \boldsymbol{\theta}_n|\boldsymbol{\theta}_n) = p_{x_n}(\mathbf{y}_n - \boldsymbol{\theta}_n) = p_{x_n}(\mathbf{x}_n), \quad \mathbf{x}_n \equiv \mathbf{y}_n - \boldsymbol{\theta}_n. \tag{2.16}$$

We see from the notation that the $p_{x_n}(\mathbf{x}_n)$ are independent of absolute origins $\boldsymbol{\theta}_n$. Substituting the $p_{x_n}(\mathbf{x}_n)$ into Eq. (2.15) and changing integration variables to $\mathbf{x}_n$ gives

$$I = \sum_n \int d\mathbf{x}_n \frac{1}{p_{x_n}(\mathbf{x}_n)} \sum_\nu \left( \frac{\partial p_{x_n}(\mathbf{x}_n)}{\partial x_{n\nu}} \right)^2. \tag{2.17}$$

Observing the disappearance of absolute origins $\boldsymbol{\theta}_n$ from the expression, the information likewise obeys shift invariance.

### 2.4.3 *Use of real probability amplitudes*

Equation (2.17) further simplifies if we introduce real probability "amplitudes" $q_n(\mathbf{x}_n)$,

$$I = 4\sum_n \int d\mathbf{x}_n \sum_\nu \left( \frac{\partial q_n}{\partial x_{n\nu}} \right)^2, \quad p_{x_n}(\mathbf{x}_n) \equiv q_n^2(\mathbf{x}_n). \tag{2.18}$$

The subscript $n$ of $\mathbf{x}$ can now be suppressed, since each $\mathbf{x}_n$ ranges over the same values. Then Eq. (2.18) becomes

$$I = 4\int d\mathbf{x} \sum_n \boldsymbol{\nabla} q_n \cdot \boldsymbol{\nabla} q_n, \tag{2.19}$$

$$\mathbf{x} = (x_0, \ldots, x_3), \quad d\mathbf{x} \equiv |dx_0|\, dx_1\, dx_2\, dx_3,$$

$$\boldsymbol{\nabla} \equiv \partial/\partial x_\nu, \quad \nu = 0, 1, 2, 3.$$

(Note the boldface 'del' notation, indicating a four-dimensional gradient.) Derivation of this equation was the main aim of the chapter; see also Frieden and Cocke (1996). This is the channel capacity expression that will be used in all information-based derivations to follow of physical laws.

We observe that the integrand of the form (2.19) for the channel capacity $I$ is *a sum of squares*. This is both over discrete index values $n$ and over the continuous integration variable $\mathbf{x}$. Such a form is commonly called an $L^2$ – "length-squared" – measure or norm. In *XYZ* space, an $L^2$ norm has the obvious property that it is invariant to rigid rotation of its coordinates. Thus, in Fig. 2.1, if the axes are rotated as indicated, the squared length $L^2$ of the meter stick remains invariant, as does its square root $L$. Although this is obvious for the given three-space application, it is not so in four-space, where it holds only for specific (for example, Minkowski) coordinates (see Sec. 3.5.5).

A length-preserving rotation defines a "unitary transformation." Many physical effects obey unitarity, and these may in fact be *derived* by using EPI subject to a *condition* of unitarity. See Sec. 3.4.17 for further details on this subject. That Fisher information has fundamentally the right form (2.19) for deriving these effects is one indication of its utility.

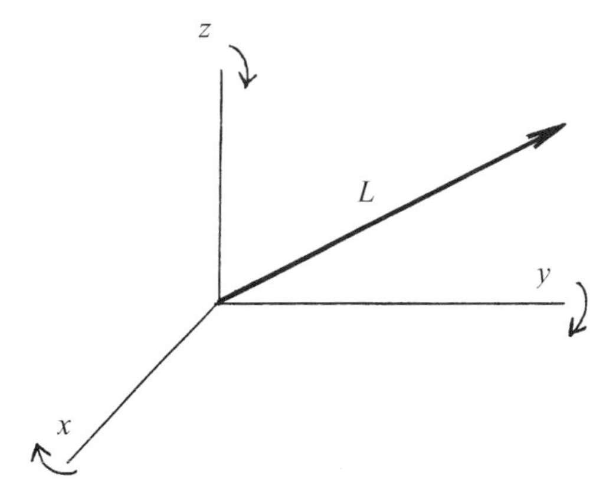


Fig. 2.1. The dark arrow represents a meter stick of length $L$. Rotation of the coordinate axes $XYZ$ leaves $L$ unaffected. A meter stick remains a meter regardless of its orientation.

### *2.4.4 Information I expressed Lorentz-covariantly*

The form (2.19) can be re-expressed as

$$I = 4\int d^4x\, q_{\mu,\nu} q_{\mu}^{,\nu} \tag{2.20}$$

by the use of covariant notation (Sec. 6.2.8). The explicit *sums* in (2.19) over indices $\mu$ and $\nu$ are implicit in (2.20), by the use of the "Einstein summation convention" (Sec. 6.2.2). This information form (2.20) will be used in Chapter 6 in derivation of the Einstein field equations of general relativity.

The property of "covariance" will play a strong role in EPI (see Chapter 3). Equation (2.20) is *manifestly Lorentz-covariant* in the index $\nu$ defining a derivative coordinate, i.e., it expresses an inner product (Sec. 6.2.8) over index values $\nu$. However, (2.20) is *not* covariant in the index $\mu$. Fortunately, despite lacking full covariance, the form (2.20) is still fine for many derivations (e.g., Chapter 6). These derivations only require covariance for (2.20) to the extent that it is a *sum* of covariant terms, that is

$$I = \int d^4x \sum_{\mu} i_{\mu}(\mathbf{x}), \qquad i_{\mu}(\mathbf{x}) \equiv 4 q_{\mu,\nu} q_{\mu}^{,\nu} \text{ (no sum on } \mu). \tag{2.21}$$

Each term $i_{\mu}(\mathbf{x})$ of the integrand is manifestly covariant in its derivative index $\nu$.

Generalizations of information form (2.20) to multiply-subscripted coordinates are given in Eqs. (11.9) and (11.17). These information expressions are suitable for problems where the amplitude functions are defined in a generally *curved space* of a multiply-subscripted metric.

### 2.5 Net probability p(x) for a particle

The intrinsic measurement scenario also makes a prediction on the overall PDF $p(\mathbf{x})$ for a single particle. The single-particle scenario was indentified in Sec. 2.1.3 as (a), where one particle is repeatedly measured. Then all particle events $x_n$ are now being "binned" in a common event space $\mathbf{x}$. Equation (2.16) now gives

$$p_n(\mathbf{y}_n|\boldsymbol{\theta}_n) = p(\mathbf{y}|\boldsymbol{\theta}_n) = p_x(\mathbf{x}|\boldsymbol{\theta}_n) = q_n^2(\mathbf{x}). \qquad (2.21c)$$

The latter is by the second Eq. (2.18). To proceed further, we have to further quantify the meaning of the $\boldsymbol{\theta}_n$. These are $N$ unknown ideal parameter values that we name $\boldsymbol{\theta}_1, \boldsymbol{\theta}_2, \ldots,$ one for each experiment (a). However, in reality we have no control over which one of the $\boldsymbol{\theta}_n$ occurs in a given experiment – the numbering choice is completely due to chance. Hence, the $\boldsymbol{\theta}_n$ obey a PDF wherein each is equally probable,

$$P(\boldsymbol{\theta}_n) = const. = \frac{1}{N}, \qquad (2.22)$$

the latter by normalization.

Knowledge of the quantities $p_x(\mathbf{x}|\boldsymbol{\theta}_n)$ and $P(\boldsymbol{\theta}_n)$ from the preceding equations suggests use of the ordinary law of total probability (Frieden, 2001),

$$p(\mathbf{x}) = \sum_{n=1}^{N} p_x(\mathbf{x}|\boldsymbol{\theta}_n)P(\boldsymbol{\theta}_n) = \frac{1}{N}\sum_{n=1}^{N} q_n^2(\mathbf{x}). \qquad (2.23)$$

We specialize, next, to the case of the relativistic electron. Define complex amplitudes as (Frieden and Soffer, 1995)

$$\psi_n \equiv \frac{1}{\sqrt{N}}(q_{2n-1} + iq_{2n}), \quad i = \sqrt{-1}, \quad n = 1, \ldots, N/2. \qquad (2.24)$$

Fluctuations $\mathbf{x}$ are now, in particular, those of the space-time coordinates of the electron. The construction of complex amplitude functions out of real amplitudes, as in Eq. (2.24), occurs as well in the theory of Higgs fields (Bhaduri, 1992, p. 176). Using Eq. (2.24) and then Eq. (2.23) gives

$$\sum_{n=1}^{N/2} \psi_n^* \psi_n = \frac{1}{N}\sum_n q_n^2 = p(\mathbf{x}). \qquad (2.25)$$

Hence, the familiar dependence (2.25) of $p(\mathbf{x})$ upon $\psi_n(\mathbf{x})$ is a straightforward expression of the law of total probability of statistics. By (2.24), the $\psi_n(\mathbf{x})$ also have the significance of being complex *probability* amplitudes. The famous Born assumption to this effect (Born, 1926; 1927) does not have to be made. A further property of the $\psi_n(\mathbf{x})$ defined in (2.24) is of course that they obey the Dirac equation, which will be established by the use of EPI in Chapter 4.

## 2.6 The two types of "data" used in the theory

We have shown that Fisher channel capacity form Eq. (2.19) follows from a scenario where $N$ "data values" are measured and utilized under ideal conditions (independence, efficiency). These were the "intrinsic" data of Sec. 2.1.1. As was noted, such data values are not actually taken in the EPI process (which is activated by *real* data). *Their sole purpose is to define the channel capacity Eq. (2.19) for the system.* The channel capacity will be seen to be the key information quantity in an approach for deriving the physical law that governs the phenomenon under measurement.

This leads one to ask how real data fit into the information approach. Real data are formed when an intrinsic data value is detected by a generally imperfect instrument. The instrument will contribute additional fluctuation to the intrinsic data value (sec. 3.8.1). *Mathematically*, real data are *appended to* the approach as data constraints, since real data are known (since Heisenberg) to constrain, or affect, the physical process that is under measurement. It is shown in Chapter 10 that the taking of real data has a dual action: it both activates the information form (2.19) *and* constrains the information by data constraints. It is this dual action that, in fact, gives rise *both* to the Schrödinger squared-gradient terms and to Mensky data terms in the Feynman–Mensky measurement model of Chapter 10.

## 2.7 Alternative scenarios for the channel capacity $I$

(This section may be skipped in a first reading.) There are alternative scenario models for arriving at the channel capacity form Eq. (2.19) of information. These will give alternative meanings to the form from a, perhaps, more familiar statistical mechanics viewpoint.

Suppose that the same system is repeatedly prepared in the same way, and measured but once at each repetition. This guarantees independence of the data, and is essentially the scenario called (a) in Sec. 2.1.3. What will be the *ensemble average* of the information in the one measurement? Equation (2.19) may be put in the form

$$I = \sum_{n=1}^{N} I_n, \qquad I_n \equiv 4\int d\mathbf{x}\, \nabla q_n \cdot \nabla q_n. \tag{2.26}$$

As in Eq. (2.23), the mean information $\overline{I}$ in a measurement is the average of the $I_n$ values weighted with respect to their probabilities $P(\boldsymbol{\theta}_n) = 1/N$ of occurring (see Eq. (2.22)). Thus

$$\overline{I} = \sum_n I_n P(\boldsymbol{\theta}_n) = I/N \tag{2.27}$$

by Eq. (2.26). We might call this the "single measurement scenario." Even though $N$ measurements were obtained, the information $\overline{I}$ represents the information that a single "average," ideal measurement would see.

An alternative scenario is where there is no measurement at all. Regard the system to be specified by a choice of real probability amplitudes $q_n \equiv q(\mathbf{x}|\boldsymbol{\theta}_n)$, $n = 1, \ldots, N$, where one is physically chosen according to an "internal degree of freedom" number $\boldsymbol{\theta}_n$. Suppose, further, that it is impossible to individually distinguish the states $\boldsymbol{\theta}_n$ by any experiment (as is the case in quantum mechanics). Then the numbers $\boldsymbol{\theta}_n$ are random events, and occur with probabilities $P(\boldsymbol{\theta}_n) = 1/N$ once again. Each such event is disjoint with each other, and defines an information $I_n \equiv I(\boldsymbol{\theta}_n)$ by the second Eq. (2.26). Then the average information obeys, once again, Eq. (2.27). Note that the disjointness of the events here replaces the independence of the data in preceding models in producing the simple sum in Eq. (2.27). Since no data were taken, the information represents an amount that is intrinsic to the system, just as we wanted. The approach resembles that of statistical mechanics, and so might be called a "statistical mechanics model."

This model also has the virtue of representing a situation of *maximum prior ignorance*. The "maximum ignorance" comes about through two effects: (1) the Fisher information Eq. (2.27) has resulted from allowing an intrinsic *indistinguishability* of the states; and (2) information (2.27) will be extremized in the derivations that follow; most often the extremum will be a minimum (Sec. 1.8.8), and minimized Fisher information defines a situation of maximum disorder (or ignorance) about the system (Sec. 1.7).

## 2.8 Multiparameter *I*-theorem

The $I$-theorem (1.30) has been found to hold for a single PDF of a single random variable $x$. We want to generalize this to the multiparameter scenario. For the present, let the joint random variable $\mathbf{x}$ define all component fluctuations of the phenomenon *except for* the time $t$. Suppose that each pair of probability densities

$$p_n(\mathbf{x}|t),\ p_n(\mathbf{x} + \Delta\mathbf{x}|t), \qquad n = 1, \ldots, N \tag{2.28}$$

obeys the multidimensional Fokker–Planck differential equation

$$\frac{\partial p_n}{\partial t} = -\sum_i \frac{\partial}{\partial x_i}[D_i(\mathbf{x},\ t)p_n] + \sum_{ij} \frac{\partial^2}{\partial x_i\,\partial x_j}[D_{ij}(\mathbf{x},\ t)p_n]. \tag{2.29}$$

Functions $D_i$ and $D_{ij}$ are arbitrary drift and diffusion functions, respectively. Risken (1996) shows that, under these conditions, the cross-entropy $G_n$ between each pair (2.28) of PDFs obeys an $H$-theorem,

$$\frac{\partial G_n(t)}{\partial t} \geqslant 0. \tag{2.30}$$

On the other hand, the derivation Eqs. (1.17)–(1.22b, c) is easily generalized to a multidimensional PDF $p_n(\mathbf{x}|t)$ and information $I_n$, $n = 1, \ldots, N$ as well, with

$$\begin{aligned} I_n &= -\frac{2}{\Delta x^2}\int d\mathbf{x}\ p_n(\mathbf{x}|t)\ln\left(\frac{p_n(\mathbf{x}+\Delta\mathbf{x}|t)}{p_n(\mathbf{x}|t)}\right) \\ &= -\frac{2}{\Delta x^2}G_n[p_n(\mathbf{x}|t),\ p_n(\mathbf{x}+\Delta\mathbf{x}|t)]. \end{aligned} \tag{2.31}$$

It follows from Eqs. (2.30) and (2.31) that *each component* $I_n$ obeys an $I$-theorem

$$\frac{\partial I_n}{\partial t} \leqslant 0. \tag{2.32}$$

Then summing Eq. (2.32) over $n$ and using the notation (2.26) gives our *multiparameter I-theorem*,

$$\frac{\partial I}{\partial t} \leqslant 0. \tag{2.33}$$

This implies that, as $t \to \infty$,

$$I(t) \to min., \tag{2.34}$$

as in Eq. (1.47) for the single-component case.

In summary, the multiparameter information $I$ *measures the physical state of disorder of the system*. This fact is important to the theory that follows.

## 2.9 Concavity, and mixing property of *I*

The Boltzmann entropy obeys a "mixing property": when two or more systems are mixed the total entropy increases, indicating increased disorder. We show below that the information capacity $I$ has a like property: Mixing the systems causes $I$ to decrease (Frieden *et al.*, 1999), likewise indicating increased disorder (Secs. 1.7 and 1.8).

The property is first established for a mixture of but two systems. We start with some preliminary work.

Let $a$, $b$ be two real constants obeying

$$a + b = 1, \quad a, b \geqslant 0. \tag{2.35}$$

Let $p_1(x)$, $p_2(x)$ be the PDFs specifying each system when unmixed. Form a new function

$$\psi \equiv \sqrt{ap_1} + i\sqrt{bp_2}, \quad \psi = \psi(x). \tag{2.36}$$

Then directly

$$|\psi|^2 = ap_1 + bp_2. \tag{2.37}$$

Also, differentiating (2.36) and modulus-squaring gives

$$|\psi'|^2 = \frac{1}{4}\left(a\frac{p_1'^2}{p_1} + b\frac{p_2'^2}{p_2}\right), \tag{2.38}$$

where a prime indicates a derivative $d/dx$. Integrating both sides of (2.38) over all $x$ gives

$$4\int dx\, |\psi'(x)|^2 = aI(p_1) + bI(p_2), \quad I(p_n) \equiv \int dx\, p_n'^2/p_n, \quad n = 1, 2. \tag{2.39}$$

We now regard $|\psi|^2$ as a probability law $P(x)$, so that by (2.37)

$$P = ap_1 + bp_2, \quad P \equiv |\psi|^2 \equiv q^2(x), \quad q(x) \equiv |\psi(x)|, \tag{2.40}$$

in terms of a real amplitude function $q(x)$. The first Eq. (2.40) has the form of the law of total probability for the system random variable $x$. Hence we identify

$$a = P_1, \quad b = P_2, \tag{2.41}$$

the *constant* prior probabilities of the two states.

By (2.40) the Fisher information (1.24) for the system is then

$$I(P) = 4\int dx \left(\frac{d|\psi|}{dx}\right)^2 = 4\int dx\, q'^2(x). \tag{2.42}$$

In order to investigate the concavity question we must find a relation between the two informations $I(P)$ and $aI(p_1) + bI(p_2)$. Aside from its representation (2.36), $\psi$ can alternatively be represented as

$$\psi(x) \equiv |\psi(x)| \exp[iS(x)] = q(x) \exp[iS(x)] \tag{2.43}$$

by the use of the third Eq. (2.40), where $S(x)$ is a new phase function. Differentiating (2.43) and taking its squared modulus gives

$$|\psi'|^2 = q'^2 + q^2 S'^2. \tag{2.44}$$

Integrating this over all $x$ and multiplying by 4 gives

$$4\int dx\,|\psi'|^2 = 4\int dx\,(q'^2 + q^2 S'^2) = aI(p_1) + bI(p_2). \tag{2.45}$$

The second identity is by Eq. (2.39). Then by the use of (2.42)

$$aI(p_1) + bI(p_2) = I(P) + 4\int dx\, q^2(x) S'^2(x). \tag{2.46}$$

Clearly the far-right integral is greater than or equal to zero. It follows that

$$aI(p_1) + bI(p_2) \geqslant I(ap_1 + bp_2) \tag{2.47}$$

by the use of (2.40). In words, the Fisher information for the total probability law of a mixed system (the right-hand side) is generally less than or equal to the weighted sum of the informations due to the unmixed system. Or, mixing a system decreases its Fisher information level, i.e., increases its level of disorder. This effect corroborates the $I$-theorem (2.33). A functional $I(ap_1 + bp_2)$ obeying an inequality (2.47) is said to be "concave." Equation (2.47) can be directly generalized to a mixture of three systems. Assume two new weights $b_1$, $b_2$ obeying

$$b_1 + b_2 = 1, \quad b_1, b_2 \geqslant 0, \tag{2.48}$$

and two new PDFs $p_{21}(x)$, $p_{22}(x)$ that provide a basis for $p_2(x)$, as

$$p_2 = b_1 p_{21} + b_2 p_{22}. \tag{2.49}$$

Then (2.47) reads, from right to left,

$$I(ap_1 + bb_1 p_{21} + bb_2 p_{22}) \leqslant aI(p_1) + bI(b_1 p_{21} + b_2 p_{22}). \tag{2.50}$$

However, because the new weights obey normalization condition (2.48), the far-right information term in (2.50) can be expanded out using our two-state result (2.47) expressed in terms of the new weights, giving

$$I(ap_1 + bb_1 p_{21} + bb_2 p_{22}) \leqslant aI(p_1) + bb_1 I(p_{21}) + bb_2 I(p_{22}). \tag{2.51}$$

Renaming weights to $A_1 \equiv a$, $A_2 \equiv bb_1$, $A_3 \equiv bb_2$, puts (2.51) in the manifestly concave form

$$I(A_1 p_1 + A_2 p_{21} + A_3 p_{22}) \leqslant A_1 I(p_1) + A_2 I(p_{21}) + A_3 I(p_{22}), \tag{2.52}$$

where, from their definitions, the new weights obey

$$A_1 + A_2 + A_3 = 1. \tag{2.53}$$

The theorem (2.52) can obviously be extended analogously to an $N$-component system, with the result

$$I\left(\sum_{n=1}^{N} A_n p_n\right) \leqslant \sum_{n=1}^{N} A_n I(p_n), \quad \sum_{n=1}^{N} A_n = 1, \tag{2.54}$$

and

$$P = \sum_{n=1}^{N} A_n p_n, \qquad P \equiv P(x). \tag{2.55}$$

Furthermore, the constants $A_n = P_n$, the prior probabilities of the states $n$ (see Eq. (2.41)). By Eq. (2.55), Eq. (2.54) takes the useful form

$$I(P) \leqslant \sum_{n=1}^{N} P_n I(p_n), \quad P = \sum_{n=1}^{N} P_n p_n, \quad P_n \equiv const., \quad p_n \equiv p_n(x). \tag{2.56}$$

The inequality states that the Fisher information for the *mixed* system (specified by the net probability law $P(x)$) is less than or equal to the average of the informations provided by the *separated* states of the system. This is for a system with discrete states weighted as $P_n$.

The result (2.56) has some important ramifications.

(1) *Non-equilibrium thermodynamics*: Because of the property Eq. (2.56) of the information $I(P)$ and a Legendre-transform property that is also obeyed by $I(P)$ (Frieden *et al.*, 1999), much of thermodynamics may be expressed in terms of $I$ in place of the usual Boltzmann entropy measure (see also Secs. 1.8.4–1.8.7, Chaps. 7 and 13, and Frieden *et al.*, 2002a, b).

(2) *Information capacity $C$ as an upper bound to the Fisher information*: In the case $P_n = 1/N$ of equal prior probabilities for the states, (2.56) gives

$$I(P) \leqslant \frac{1}{N} \sum_n I(p_n). \tag{2.57}$$

Equivalently,

$$\sum_n I(p_n) \geqslant N I(P) \geqslant I(P) \tag{2.58}$$

since $N \geqslant 1$. It was shown in Sec. 2.3.2 that the *information capacity $C$*, defined in Eq. (2.10), is an upper bound to the *Stam* information. Equation (2.58) shows a further property of the information capacity. A comparison of the *left-hand sum* in (2.58) with Eq. (2.10) shows that the sum *is* in fact the information capacity $C$ (under the hypothesis of shift invariance made in Sec. 2.4.2). Also, the right-hand side of (2.58) is the elemental Fisher information for the mixed system. Hence, the information capacity $C$ is an upper bound to *two* elementary informations: the Stam information, as previously discussed, and the *Fisher* information, by Eq. (2.58). This gives $C$ added significance.

(3) *Conditions for which $I(P)$ is a maximum*: The outside inequality (2.58) shows that the Fisher information for a system is maximized if it is separated into its constituent states. Likewise, the maximum is achieved if the constituent probabilities $p_n(x)$ do not have overlapping support regions in $x$ (Frieden, 1995). These maximization effects make sense since in either case a degree of order is being

imposed upon the system.

(4) *Dependence of the EPI derivations upon use of the channel capacity C, not I(P)*: In multiple-dimension, multiple-component problems, the channel capacity, denoted as $C$ or simply $I$ in later chapters, turns out to be more physically meaningful than the elemental Fisher information $I(P)$ for the system. That is, nature permits its amplitude laws – such as $\psi(x)$ in quantum mechanics – to be derived *as if* they supplied *maximum Fisher* information about the system parameters. This in fact agrees with Wheeler's idea of a "participatory universe" (Sec. 0.1).
Interestingly, in the oft-encountered case of a single coordinate $x$ and a single amplitude component $N = 1$ the channel capacity equals the elemental Fisher information $I(P)$. Thus, if the world were scalar and one-dimensional, elemental Fisher information would remain the key physical consideration. But the world is not scalar and one-dimensional. Therefore its intrinsically *vector nature* – the subject of this chapter – requires us to replace $I(P)$ with the physically more meaningful information quantity $C$.

(5) *Continuous systems*: For a system with a *continuum* of states, weighted by a density function $p(x)$ in place of the discrete $P_n$ in (2.56), the latter goes over into a statement that

$$I(P) \leqslant \int dx\, p(x) I(y|x), \tag{2.59}$$

where

$$P \equiv P(y) = \int dx\, p(x) P(y|x) \quad \text{and} \quad I(y|x) \equiv \int dy \frac{[dP(y|x)/dy]^2}{P(y|x)}.$$

The notation $I(y|x)$ means the information about $y$ in the presence of a state $x$.

# 3

# Extreme physical information

## 3.1 Covariance, and the "bound" information $J$

### 3.1.1 *Transition to variational principle*

Consider a system, or phenomenon, that is specified by amplitudes $q_n(\mathbf{x}|t)$ as in Chapter 2. The corresponding probabilities $p_n(x|t)$ are (temporarily) assumed to obey the Fokker–Planck Eq. (2.29) and, hence, the $I$-theorem Eq. (2.33). This theorem implies that, as time $t \to \infty$ the amplitudes $q_n(\mathbf{x}|t)$ approach equilibrium values such that $I$ is a minimum. This translates into a variational principle

$$\delta I[\mathbf{q}(\mathbf{x}|t)] = 0, \quad \mathbf{q}(\mathbf{x}|t) \equiv q_1(\mathbf{x}|t), \ldots, q_N(x|t). \tag{3.1}$$

As defined below in Eq. (0.1), $I$ is a *functional* of the amplitudes $q_n(\mathbf{x}|t)$, whose variations cause the zero in the principle. The variation could instead be with respect to the probabilities (see Eq. (2.17)), but for our purposes it is more useful to regard the amplitudes as fundamental.

It is important to ask why the information $I$ might change in the first place. In other words, what drives the $I$-theorem? As in thermodynamic theory, any interaction with the given system would cause the $q_n(\mathbf{x}|t)$ to change. Then, by Eq. (2.19), $I$ would change as well. *The interaction that is of central interest to us in this book is that due to a real measurement.* Hence, from this point on, we view the system or phenomenon under study in the context of a measurement of its parameters $\boldsymbol{\theta}$.

Looking ahead, it will be shown in Sec. 3.8 that, for certain measurement apparatus, the Fokker–Planck equation requirement mentioned above can be dropped. That is, the overall approach taken – the EPI principle – will hold independently of this requirement. This will give EPI a much wider scope of application.

### 3.1.2 Statistical covariance

A basic property of phenomena is that of Lorentz covariance (see Sec. 3.5). Covariance requires a physical law that is expressed in laboratory coordinates to keep the same mathematical form when it is evaluated in any frame that is in uniform motion relative to that of the laboratory. A law that obeys covariance must also, at the very least, depend functionally upon all of its coordinates $\mathbf{x}$ (including the time, if that is one of them) *in the same way.*

If a phenomenon is to obey covariance, then any variational principle that implies the phenomenon should likewise be covariant. Let us examine variational principle (3.1) in this context. The conditional notation $(\mathbf{x}|t)$ in the principle suggests that $t$ is *not* treated covariantly with $\mathbf{x}$. This may be shown as well. By definition of a conditional probability $p(\mathbf{x}|t) = p(\mathbf{x}, t)/p(t)$ (Frieden, 2001). This implies that the corresponding amplitudes (cf. the second Eq. (2.18)) obey $q(\mathbf{x}|t) = \pm q(\mathbf{x}, t)/q(t)$. The numerator treats $\mathbf{x}$ and $t$ covariantly, but the denominator, in depending only upon $t$, does not. Thus, principle (3.1) is not covariant.

From a *statistical* point of view, principle (3.1) is objectionable as well, because it treats time as a deterministic, or known, coordinate while treating space as random. Why should time be *a priori* known any more accurately than space?

These problems can be remedied if we simply *make* (3.1) covariant and regard $t$ as random. This may readily be done, by replacing it with the more general principle

$$\delta I[\mathbf{q}(\mathbf{x})] = 0, \quad \mathbf{q}(\mathbf{x}) \equiv q_1(\mathbf{x}), \ldots, q_N(\mathbf{x}). \tag{3.2}$$

Here $I$ is given by Eq. (2.19) and the $q_n(\mathbf{x})$ are to be varied. Coordinates $\mathbf{x}$ are, now, any *four-vector* of coordinates. In the particular case of space-time coordinates, $\mathbf{x}$ now *includes* the time.

We note that *specific* solutions of a general principle such as (3.2) *do not* have to exhibit covariance. For example, problems within a bounded space in the shape of a box often lead to separable solutions, of the form $q_1(x)q_2(y)q_3(z)q_4(t)$. This will change form when viewed from a uniformly moving frame and, hence, is not covariant.

### 3.1.3 Need for a second term J

As a principle of estimation for the $\mathbf{q}(\mathbf{x})$, however, (3.2) is lacking; $I$ is of the fixed form (2.19) so that there is no room in the principle for the injection of physical information about the measurement scenario. This dovetails with the need for Lagrange constraint terms in the one-dimensional principle (1.48a, b): constraints, after all, stand for physical prior knowledge. The question is how to do this from a physical (not *ad hoc*) standpoint. In other words, what constraints upon the principle (3.2) are actually imposed *by nature*?

This suggests that we need a revised principle of the form

$$\delta(I - J) = 0, \quad \text{or} \quad I - J = extrem., \tag{3.3}$$

where the term $J$ is to embody all constraints that are imposed by the physical phenomenon under measurement. Thus, $J$ describes, or characterizes, the physical phenomenon. Both $I$ and $J$ are functionals,

$$I[\mathbf{q}],\ J[\mathbf{q}], \tag{3.4}$$

of the unknown functions $\mathbf{q}(\mathbf{x})$, with $I[\mathbf{q}]$ given by (2.19) and $J[\mathbf{q}]$ to be found.

At this point, Eq. (3.3) is merely a hypothesis. However, it will be shown below to follow from properties for $I$ and $J$ that are analogous to those of the Boltzmann and Shannon entropies.

We note from the form of Eq. (3.3) that $J$ and $I$ have the same units. This implies that $J$ is an information too. We call it the "bound" information, in analogy to a concept of Brillouin (1956) that is discussed next.

## 3.2 An equivalence of Boltzmann and Shannon entropies

Brillouin (1956), building on the work of Szilard (1929), discovered an intriguing connection between the Boltzmann entropy and Shannon entropy. Consider a measurement made upon an isolated system. After the measurement a certain amount of Shannon entropy $\delta H$ is gained as information. But the act of measurement requires some physical interaction with the system under measurement, so that the Boltzmann entropy $H_B$ increases as well. In fact, by the second law of thermodynamics, the increase in Boltzmann entropy must exceed (or equal) the gain in Shannon entropy,

$$\delta H_B \geqslant \delta H, \quad \text{or} \quad H_B \geqslant H \tag{3.5}$$

as a sum over many such measurements. We call this "Brillouin's equivalence principle." It is exemplified in the following *gedanken* experiment (Brillouin, 1956).

### *3.2.1 Particle location experiment*

A closed cylinder of volume $V$ is maintained at a constant temperature $T$. It is known to contain a single molecule. The cylindrical volume is divided into two volumes $V_1$ and $V_2$ by means of a partition. See Fig. 3.1. An observer wants to determine whether the molecule is in $V_1$ or in $V_2$. To do this he places a

Leon Brillouin, from an undated photograph at the National Academy of Sciences. Sketch by the author.

photocell $P_1$ within $V_1$ and shines a beam of light $B_1$ into $V_1$ such that, if the particle is in $V_1$, it might deflect a photon into the photocell. In order for the photon to be distinguished from background radiation, its energy $h\nu$ must, at least, equal the background mean energy due to the ambient temperature $T$,

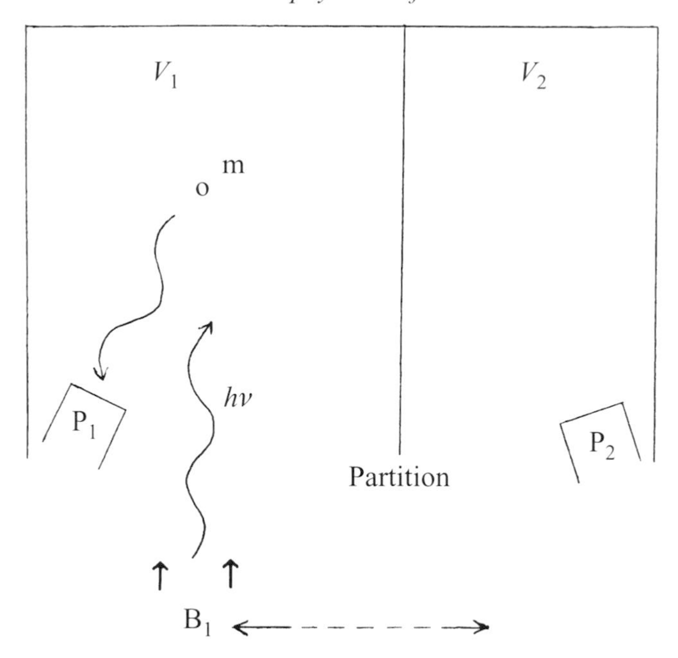


Fig. 3.1. Brillouin's *gedanken* experiment for finding the subvolume $V_1$ or $V_2$ within which a molecule m resides.

$$h\nu \geqslant kT, \tag{3.6}$$

where $h$ and $k$ are the Planck and Boltzmann constants, respectively. Alternatively, even if the particle is in $V_1$ it might scatter photons that all miss $P_1$.

Likewise a beam of light $B_2$ and a photocell $P_2$ may be used in the volume $V_2$ in order to identify the molecule as lying within *that* volume.

The molecule is located as follows. First the beam $B_1$ is used. If $P_1$ registers a photon then the particle is known to be located in $V_1$. This ends the experiment. If not, then the beam $B_2$ is used in the same way. If $P_2$ does not register a photon then beam $B_1$ is used again, etc. Eventually a photon will be detected and the particle will be located (correctly) in one or the other of $V_1$ and $V_2$.

Locating the particle causes an increase $\delta H$ in the Shannon entropy (1.14). But the location also involved the absorption of a photon into the cylinder system and hence an increase $\delta H_{\mathrm{B}}$ in its Boltzmann entropy. We calculate $\delta H$ and $\delta H_{\mathrm{B}}$ below.

The absorption of energy $h\nu$ by one of the photocells causes an entropy increase

$$\delta H_{\mathrm{B}} = h\nu/T \geqslant k, \tag{3.7}$$

the latter by (3.6).

We now compute the increase in Shannon entropy. As with Brillouin, we use a multiplier $k$ of Eq. (1.14) and $\Delta x \equiv 1$ so as to give the information the same units as the Boltzmann entropy. Then, with only two different possible location events, Eq. (1.14) becomes

$$\delta H = -k(p_1 \ln p_1 + p_2 \ln p_2), \quad p_i \equiv p(V_i), \quad i = 1, 2. \tag{3.8}$$

The probabilities $p_i$ are *a priori* probabilities, i.e., representing the state of knowledge about location before the experiment is performed. Before the experiment, the particle is equally likely to be anywhere in the cylinder, so that

$$p_i = V_i/V, \quad i = 1, 2; \quad V_1 + V_2 \equiv V. \tag{3.9}$$

This verifies that normalization $p_1 + p_2 = 1$ is obeyed. Now an important property of Eq. (3.8) is found.

### 3.2.2 Exercise

Using Eq. (3.8) and the normalization constraint, show that the maximum value of $\delta H$ over all possible $(p_1, p_2)$ (or, by Eq. (3.9), geometries $(V_1, V_2)$) obeys

$$\delta H_{\rm max} = k \ln 2 \approx 0.7k. \tag{3.10}$$

Comparing Eqs. (3.7) and (3.10) shows that Eq. (3.5) is obeyed, as was required. Brillouin called $H_B$ the "bound information," in the sense of being intrinsic, or bound, to nature. Then (3.5) states that the change in bound information must always exceed (or equal) the acquired Shannon entropy (or information).

The preceding location experiment is one example of the equivalence principle (3.5); numerous others are constructed by Brillouin (1956). Lindblad (1983) and Zeh (1992) critically discuss the principle and provide numerous other references on the subject.

Current thought seems to be that Eq. (3.5) describes, as well, the entropy changes that take place during the *erasure* of a stored message (Landauer, 1961) prior to its replacement with another message. This scenario is, thus, complementary to the Brillouin one, which deals instead with the *measurement* or acquisition of a message. Landauer's principle is that if the stored message contains an amount $H$ of Shannon information, then it requires for its erasure an entropy expenditure $H_B$ of at least the amount $H$. As Caves notes (Halliwell *et al.*, 1994), this is effectively just an application of the second law of thermodynamics to the erasure process. Overall, then, there is entropy expenditure at *both ends* of an overall information retrieval/storage process – when acquiring the message and when erasing a previous message record in order to store the acquired one.

During either the Brillouin or the Landauer phase of the information retrieval process, $H$ in Eq. (3.5) is the Shannon information contained within a message. In the Brillouin phase, the message is the one arising out of measurement. In the Landauer, the message is the one that is erased or, equivalently, its replacement (which, for a fixed storage system, must retain the same number of bits). We will be primarily interested in the Brillouin, or measurement, phase of the process.

The equivalence principle (3.5) shows a complementarity between the information in a measured message and an equivalent expenditure of Boltzmann entropy. Such complementarity is intriguing, and serves to clarify what role the quantity $J$ must play in our theory. (I thank B. H. Soffer (personal communication, 1994) for first noticing the tie-in between the Szilard–Brillouin work and the EPI formalism.)

## 3.3 System information model

At this point it is useful to review the status of our theory. We have found that $I$ is an information or channel capacity, and that $J$ is another information. The distinction between $I$ and $J$ is the distinction between the general and the specific: functional $I$ always has the same form, Eq. (2.19), regardless of the physical parameter that is being measured, whereas functional $J$ characterizes the specific physical parameter under measurement (see discussion below Eqs. (3.3) and (3.4)).

In this section, we will gradually "zero in" on the exact meaning, and form(s), to be taken by $J$, simultaneously with developing the overall EPI system model. For readers who don't mind looking ahead, Secs. 3.4.5 and 3.4.6 specify how $J$ is formed for a specific problem.

We found that $I$ relates to the cross-entropy $G$ via Eq. (1.22c). We also saw that $H_B$ is a "bound" information quantity that defines the effect of the measurement upon the physical scenario. This prompts us to make the identifications

$$I \rightarrow G \quad \text{and} \quad J \rightarrow H_B. \tag{3.11}$$

### *3.3.1 Bound Fisher information J*

The first identification (3.11) was derived in Sec. 1.4. The second, however, is new: it states that the role of $J$ is to be analogous to that of $H_B$, i.e., to define the effect of the measurement upon the physical scenario (or "parameter" or "system"; we use the terms interchangeably). This will serve to bring the particulars of the physical scenario into the variational principle, as was required above.

But, how can $J$ be actually computed? Brillouin's equivalence principle Eq. (3.5) suggests a *natural* way of doing so, i.e., without the need for *ad hoc* Lagrange constraints as in Eqs. (1.48a, b). It suggests that the intrinsic information $I$ must have a *physical equivalence* in terms of the measured phenomenon (as information $H$ and physical entropy $H_B$ were related in the preceding example). Represent by $J$ the physical effect, or manifestation, of $I$. Thus, $J$ is the information that is intrinsic (or "bound") to the phenomenon under measurement. This is why we call $J$ the "bound" Fisher information. The "boundness" also expresses a degree of *entanglement* of the space of $I$ – data space – with the space of the source $J$ of information for $I$. In cases $I = J$ there is complete entanglement. See, for example, Chapter 12.

Computationally, $J$ is to be evaluated – not directly by (2.19) – but equivalently in terms of physical parameters of the scenario (more on this later). The precise tie-in between $I$ and $J$ is quantified as follows.

### *3.3.2 Information transferral effects*

Suppose that a real measurement of a four-parameter $\boldsymbol{\theta}$ is made. The parameter is any physical characteristic of a system. The system is necessarily *perturbed* (Sec. 3.8, Chapter 10) by the measurement. More specifically, its amplitude functions $\mathbf{q}(\mathbf{x})$ are perturbed. An analysis follows.

As in Eq. (0.3), denote the perturbations in $\mathbf{q}(\mathbf{x})$ as the vector $\varepsilon\boldsymbol{\eta}(\mathbf{x})$, $\varepsilon \rightarrow 0$, with the $\boldsymbol{\eta}(\mathbf{x})$ arbitrary perturbation functions. By Eq. (2.19), the perturbations in $\mathbf{q}(\mathbf{x})$ cause a perturbation $\delta I$ in the intrinsic information (see Eq. (0.41)). Since $J$ will likewise be a functional of the $\mathbf{q}(\mathbf{x})$, a perturbation $\delta J$ also results. Is there a relation between $\delta I$ and $\delta J$?

We next find such a relation, and others, for informations $I$ and $J$. This is by the use of an *axiomatic model* for the measurement process. There are three axioms as given below.

Prior to measurement of parameter $\boldsymbol{\theta}$, the system has a bound information level $J$. Then a real measurement $\bar{\mathbf{y}}$ is initiated – say, by shining light upon the system. This causes an information transformation or transition $J \rightarrow I$ to take place, where $I$ is the intrinsic information of the system. The measurement perturbs the system (Sec. 3.8), so that informations $I$ and $J$ are perturbed by amounts $\delta I$ and $\delta J$. How are the four informations $I$, $J$, $\delta I$, $\delta J$ related?

First, the correspondence (3.11) and the Brillouin principle (3.5) suggest that $I$ and $J$ obey

$$J \geqslant I. \tag{3.12}$$

This implies some loss of information during the information transition. We also postulate the perturbed amounts of information to obey

$$axiom\ 1: \quad \delta I = \delta J. \tag{3.13}$$

This is a *conservation law*, one of conservation of information change. It is a predicted new effect. Its validity is demonstrated many times over in this book; as a vital part of the information approach that is used to derive the fundamental physical laws in Chaps. 4–15. It also will be seen to occur whenever the measurement space has a conjugate space that connects with it by unitary transformation (Sec. 3.8).

The basic premises (3.12) and (3.13) are further discussed in Secs. 3.4.5 and 3.4.7, and are exemplified in Sec. 3.8. We next build on these premises.

## 3.4 Principle of extreme physical information (EPI)

### *3.4.1 Physical information K and variational principle*

Equation (3.13) is a statement that

$$\delta I - \delta J \equiv \delta(I - J) = 0. \tag{3.14}$$

Define a new quantity,

$$K \equiv I - J. \tag{3.15}$$

Quantity $K$ is called the "physical information" of the system. As with $I$ and $J$, it is a functional of the amplitudes $\mathbf{q}(\mathbf{x})$. By its definition and Eq. (3.12), $K$ is a loss, or defect, of information. The zero in Eq. (3.14) implies that this defect of information is the same over the *range* of perturbations $\varepsilon\boldsymbol{\eta}(\mathbf{x})$. Equations (3.14) and (3.15) are the basis for the EPI principle, Eqs. (3.16) and (3.18), as shown next.

Combining Eqs. (3.14) and (3.15) gives $\delta K = 0$ or

$$K = I - J = extrem. \tag{3.16}$$

at solution $\mathbf{q}(\mathbf{x})$.

*Note*: For simplicity we use the same notation $\mathbf{q}(\mathbf{x})$ both for general amplitudes and for the solution to Eq. (3.16).

Equation (3.16) is a variational principle for finding $\mathbf{q}$. It states that, for a particular vector of amplitudes $\mathbf{q}$, the information $K$ is an extremum. As was postulated in Eq. (3.13), the variational principle is in reaction to a measurement. Variational principle (3.16) is one-half of our overall information approach.

### 3.4.2 Zero-conditions, on microlevel and macrolevel

Equation (3.12) shows that generally

$$I \neq J. \tag{3.17}$$

In fact, (3.12) shows that a "lossy" situation exists: the information $I$ in the data never exceeds the amount $J$ in the phenomenon,

$$I = \kappa J \quad \text{or} \quad I - \kappa J = 0, \quad \kappa \leqslant 1, \tag{3.18}$$

at a solution $\mathbf{q}(\mathbf{x})$. (See *Note* above.) This may be shown to follow from the $I$-theorem; see Sec. 3.4.7. Equation (3.18) comprises the second half of our overall information approach.

Interestingly, only the values $\kappa = 1/2$ or 1 have occurred in those physical-law derivations that fix $\kappa$ at specific values (see subsequent chapters). These verify that $I \leqslant J$ or, equivalently, $\kappa \leqslant 1$ in (3.18). The nature of $\kappa$ is taken up further in Sec. 3.4.5.

In all quantum applications of EPI the amplitude functions $\mathbf{q}$ will obey the zero-condition (3.18) with $\kappa = 1$. Since, by (3.12), $J$ is in fact the extreme (maximum) possible value of $I$, in these cases the zero *is* the extreme value demanded in (3.16). Then any solution $\mathbf{q}$ to the zero condition $I - J = 0$ must also obey the extremum condition (3.16). For example, the Dirac Eq. (4.42) follows from the zero-condition, and the Klein–Gordon Eq. (4.28) follows from the extremum condition. Therefore any solution $\mathbf{q}$ to the former equation must satisfy the latter. This is in fact the case (Schiff, 1955, p. 324). See also the material following Eq. (4.26a).

Returning to the question of what information $J$ represents, Eq. (3.18) provides a partial answer. It shows that $J$ is proportional to $I$, in numerical value. However, since $J$ characterizes the physical source of the data, in contrast to $I$ its *functional form* will depend upon physical parameters of the scenario (e.g., $c$ or $\hbar$), not solely upon the amplitude functions $\mathbf{q}$. This is further taken up in Sec. 3.4.5.

Equation (3.18) is a zero-condition on the *macroscopic* level, i.e., in terms of the information in physical observables (the data). On the other hand, (3.18) may be derived from viewing the information transfer procedure on the *microscopic* level. Postulate the existence of *information densities* $i_n(\mathbf{x})$ and $j_n(\mathbf{x})$ such that

$$\textit{axiom 2:} \quad I \equiv \int d\mathbf{x} \sum_n i_n(\mathbf{x}) \quad \text{and} \quad J \equiv \int d\mathbf{x} \sum_n j_n(\mathbf{x}), \tag{3.19}$$

$$\text{where} \quad i_n(\mathbf{x}) = 4\,\nabla \mathrm{q}_n \cdot \nabla \mathrm{q}_n$$

by Eq. (2.19). Also, $j_n(\mathbf{x})$ is a function that depends upon the particular scenario (see subsequent chapters). In contrast with $I$ and $J$, densities $i_n(\mathbf{x})$ and $j_n(\mathbf{x})$ exist on the microscopic level, i.e., within local intervals $(\mathbf{x}, \mathbf{x} + d\mathbf{x})$.

Let us demand a zero-condition on the microscopic level,

$$\textit{axiom } 3: \quad i_n(\mathbf{x}) - \kappa j_n(\mathbf{x}) = 0, \quad \text{all } \mathbf{x}, n. \tag{3.20}$$

Summing and integrating this $d\mathbf{x}$, and using Eqs. (3.19), verifies Eq. (3.18).

Condition (3.20) is a stronger requirement than (3.18), and is often used to implement physical-law derivations in later chapters. Sometimes quantities $[i_n(\mathbf{x}) - \kappa j_n(\mathbf{x})]$ comprise the *net* integrands in Eq. (3.19) after partial integrations are performed. See, e.g., Eq. (7.17).

Information densities $i_n(\mathbf{x})$ and $j_n(\mathbf{x})$ have the added physical significance of defining the field dynamics of the scenario; see Sec. 3.7.3.

A corollary to the preceding three axioms is given in Eq. (3.21).

### 3.4.3 EPI principle

In summary, the information model consists of the axioms (3.13), (3.19), and (3.20). As we saw, these imply the variational principle (3.16) and the zero-condition (3.18). *The variational principle and the zero-condition comprise the overall EPI principle we will use to derive major physical laws.* In addition, Eq. (3.21) will be used to determine certain constants of the laws.

Since, by Eq. (3.16), physical information $K$ is extremized the principle is called "extreme physical information" or EPI. In some derivations, the microlevel condition Eq. (3.20) (which, we saw, implied condition (3.18)) will be used as well.

### 3.4.4 Role of the Fisher coordinates

The physics of the measurement scenario lies in the fluctuations $\mathbf{x}$. These are called the "Fisher coordinates" of the problem. The choice of coordinates $\mathbf{x}$ (or equivalently, parameter $\boldsymbol{\theta}$) in EPI is crucial to its use. A restriction of special relativity is that $\mathbf{x}$ be a four-vector (Sec. 3.5).

EPI solutions $\mathbf{q}$ for scenarios in which $\mathbf{x}$ is purely real are fundamentally different from scenarios ("mixed" cases) where some components of $\mathbf{x}$ are real and some are imaginary (Secs. 1.8.8, 1.8.14, Appendix C). Solutions $\mathbf{q}$ in the purely real $\mathbf{x}$ case obey maximum disorder (minimum $I$). But in a mixed real–imaginary case a state of maximum disorder need not result. Usually, the mixed case leads to a wave equation in the coordinate space (Chaps. 4–6 and 11). Basically, the two cases differ in the way that a diffusion equation differs from a wave equation. All-real coordinates lead to diffusion-type equations, where

Self-portrait of the author, 1998.

disorder must increase, whereas mixed real–imaginary cases lead to wave equations, where the state of disorder cycles with time.

That the approach permits the use of imaginary coordinates is therefore a benefit, giving the user an extra degree of freedom for finding the appropriate Fisher coordinates for modelling a given problem. Imaginary coordinates are used in many applications in the text.

### *3.4.5 Defining J from a statement of invariance; nature of* $\kappa$

The use of EPI in a given measurement scenario requires definition of the information functional $J[\mathbf{q}]$. In fact, since $I[\mathbf{q}]$ is of a fixed form (Eq. (2.19)), the functional $J[\mathbf{q}]$ *uniquely determines* the solution $\mathbf{q}$. How may it be found?

EPI provides a *framework*, rather than a rigid prescription, for deriving physical laws. Thus, $J[\mathbf{q}]$ cannot in general be found in cookbook fashion. (See Caveat 3 of Chapter 0.) The observer must exercise some flexibility and mathematical skill in the quest. For example, there is not one, but two overall ways of finding $J[\mathbf{q}]$, denoted as (a) and (b) in Sec. 3.4.6, and only one is applicable to a given scenario. However, as an aid in selecting the approach (a) or (b), if the observed effect occurs on the quantum level then approach (b) is usually the proper choice (an exception is in Chapter 12). The approaches (a) and (b) are special cases of a general principle discussed next.

In general, the functional form $J[\mathbf{q}]$ follows from a statement of *invariance* about the system. Our overall stance is that the measuring instrument affects the physical law. Hence, ideally, the statement of invariance should be suggested by the internal workings of the measuring instrument (see Sec. 3.8 for an example). If this cannot be accomplished, then at least the invariance principle should be as "weak" as is necessary to define the system or phenomenon. For example, it might be the expression of continuity of flow (Chaps. 5, 6). Overly strong invariance principles lead, in fact, to departures from correct EPI answers; see Sec. 7.4.11.

Examples of invariance principles, as used in later chapters, are (i) a unitary transformation, such as that between direct- and momentum-space in quantum mechanics; (ii) the gauge invariance of classical electromagnetic theory or gravitational theory; and (iii) an equation of continuity (invariance) of flow, usually involving the sources.

Information $J[\mathbf{q}]$ and coefficient $\kappa$ are always *solved for* by the combination of EPI Eqs. (3.16) and (3.18), and the invariance principle. It is interesting that the resulting forms for $J[\mathbf{q}]$ turn out to be *the simplest* possible functionals of the sources (e.g., linear) that are consistent with the invariance principle. This simplicity is not forced by, e.g., recourse to Ockham's razor. Rather, it is a natural consequence of the EPI analytical approach.

The invariance principle plays a central role in the implementation of EPI. EPI is a principle for defining the physics of the system through its amplitudes $\mathbf{q}$. As mentioned before, the answer $\mathbf{q}$ that EPI gives for a problem is completely dependent upon the particular $J[\mathbf{q}]$ for that problem. This, in turn, depends completely upon the invariance principle that is postulated. Hence, the answer $\mathbf{q}$ that EPI gives can only be as valid as the presumed invariance principle. If the principle is not sufficiently "strong" in defining the system, then we can expect the EPI output $\mathbf{q}$ to be only approximately correct. Information $J$ reflects this situation, in that $J$ is the level that information $I$ acquires when the invariance principle is optimally strong. Hence, $I \leqslant J$ generally but $I = J$ with an optimally strong invariance principle.

We can discuss the meaning of the coefficient $\kappa$ in this context as well. By its definition (3.18) $\kappa = I/J$, so that $\kappa$ measures the efficiency of the EPI process in *transferring* Fisher information from the phenomenon (specified by $J$) to the output (specified by $I$). Thus, $\kappa$ is an efficiency coefficient, $0 \leqslant \kappa \leqslant 1$. From the preceding paragraph, a value of $\kappa < 1$ should indicate that the answer $\mathbf{q}$ is only approximate. We will verify, in subsequent chapters, that such $\kappa$ values occur only in EPI uses for which the invariance principle is *explicitly* incomplete: use of a non-quantum theory (Chaps. 5, 6). In all such cases, the lost information is therefore associated with ignored quantum effects. From Sec. 3.3.1, $\kappa$ also measures the extent to which data space is *entangled* with source space. A case $\kappa = 1$ indicates complete entanglement of the two spaces.

When the invariance principle is the statement of a unitary transformation between measurement $\mathbf{x}$ space and a conjugate coordinate space, then the solution to requirement Eq. (3.18) is that *functional $J$ be simply the re-expression of $I$ in the conjugate space*. We will later prove (Sec. 3.8) that, due to the unitary transformation, *identically* $I = J$. This satisfies Eq. (3.18) with the particular value of $\kappa = 1$. In this case, the invariance principle is optimally strong as well (see preceding paragraphs). Then, as was discussed, the output $\mathbf{q}$ of the invariance principle will be "correct" (i.e., not explicitly incorrect due to ignored quantum effects). Unitary transformations are used in Chapter 4 to define relativistic quantum mechanics and in Chapter 11 to define quantum gravity.

Conversely, it will turn out that, when the invariance principle *is not* that of a unitary transformation, then, depending upon the phenomenon, often the efficiency value is $\kappa < 1$. Hence, a unitary transformation seems to be the hallmark of an accurate EPI output.

Interestingly, there are also *non-quantum* (and non-unitary) theories for which $\kappa$ is unity (Chapter 7) or, even, any value in the continuum (Chaps. 8 or 9). It seems, then, that a unitary theory is sufficient, but not necessary, for producing an optimally accurate output $\mathbf{q}$. The nature of $\kappa$ is still not fully understood. It is further discussed in the context of the applications. See, e.g., Secs. 4.4, 5.1.18, and 6.3.18.

### *3.4.6 Unique or multiple solutions?*

The EPI conditions (3.16) and (3.18) are to be satisfied through variation of $\mathbf{q}$. Depending upon the nature of the invariance principle just discussed, the two conditions will either give (a) a unique solution $\mathbf{q}$ or (b) two generally different solutions.

A solution of type (a) occurs when the invariance principle is not of the *direct* equality form Eq. (3.18), e.g., *is not* the statement of a unitary transformation between coordinate spaces $I$ and $J$. Instead, it might be a statement of continuity of flow for the $\mathbf{q}$ and/or the sources. Since, in any event, Eq. (3.18) must be satisified, *here we must solve for the parameter* $\kappa$ *and the functional* $J$ *such that (3.18) (or more fundamentally, microlevel Eq. (3.20)) is true*. This is done by seeking *a self-consistent solution* $\mathbf{q}$ to conditions (3.16) and (3.20). Examples of the common solution case are found in Chaps. 5–9, 12, 14, and 15.

Finding the self-consistent solution to a given problem is not a trivial exercise. There are usually many possible ways to combine conditions (3.16) and (3.20); a resulting solution $\mathbf{q}$ of this type is called a "combined" solution. However, a combined solution does not necessarily satisfy Eqs. (3.16) and (3.20) *after back substitution*. The latter is a surprising result, and was brought to our attention by the physicist M. H. Brill (1999). Thus, a combined solution is not necessarily a self-consistent solution.

Also, some combined solutions are unacceptable because they correspond to unphysical values of $\kappa$, that is, to values lying outside the feasible region (0, 1). In summary, then, in order for a candidate solution $\mathbf{q}$ to be considered self-consistent, it must satisfy Eqs. (3.16) and (3.20) after back substitution, *and* correspond to feasible values of $\kappa$.

These considerations imply that any combined solution $\mathbf{q}$ should be checked by back substitution into (3.16) and (3.20) to verify that it is indeed a self-consistent solution. In the applications of EPI to follow, we carry through this type of check on all solutions of type (a).

A solution of type (b) occurs when the invariance principle is precisely in the form of Eq. (3.18). *An example is when the invariance principle is that of a unitary transformation (see above)*. The parameter $\kappa$ and information functional $J[\mathbf{q}]$ are then known *a priori*, and do not have to be solved for as in (a) preceding. Then Eq. (3.18) *by itself* gives a solution $\mathbf{q}$, and this solution is generally different from that of the variational principle (3.16). However, it is as valid a physical solution as the solution to (3.16). An example is in Chapter 4, where the solution to (3.16) is the Klein–Gordon equation while the solution to (3.18) is the Dirac equation. Each is, of course, equally valid within its domain of application.

### 3.4.7 Independent routes to EPI

It is important to note that the EPI Eq. (3.18) was arrived at by analogy with the Brillouin effect. It would be better yet to confirm (3.18) from other points of view.

Suppose that an object (say, a particle) is to be measured for a parameter value. The particle is in the input space to a measuring instrument. Before the measurement is initiated, the particle has "bound" information level $J$ about the parameter. Then the measurement is initiated, by the use of a probe particle that interacts with the particle (see Sec. 3.8). The interaction perturbs the particle and causes it to have the new information level $I$. The result is an information transition $J \to I$ in measurement space. Next, the $I$-theorem states that $\Delta I \equiv I - J \leqslant 0$ over the duration $\Delta t$ of the interaction. *This implies the EPI condition (3.12) or (3.18).*

The other equation of EPI, Eq. (3.16), was seen to arise out of axiom 1, Eq. (3.13). Is axiom 1 purely an assertion, or can it be shown to be physically obeyed? In fact, this axiom is obeyed by the class of measuring instruments described in Sec. 3.8. The unitary transformations that occur during internal operation of the instruments directly lead to axiom 1 and, hence, EPI Eq. (3.16). It also will be shown in Sec. 3.8 that the other half of EPI – Eq. (3.18) – is obeyed as well in this measurement scenario. This means that the EPI principle will follow independently of the validity of the $I$-theorem, greatly widening its scope of application.

Thus, the overall EPI principle may be confirmed by alternative routes.

Of course, the ultimate test of a theory is the validity of its consequences. On this basis, the EPI Eqs. (3.16) and (3.18) constitute one of the most well-verified principles of physics. It has as its outputs the Schrödinger wave equation (Chapter 4 and Appendix D), Dirac and Klein–Gordon equations (Chapter 4), Boltzmann and Maxwell–Boltzmann distribution functions (Chapter 7), and many other phenomena as found in corresponding chapters.

### 3.4.8 Physical laws as a reaction by nature to measurement

The solution to the extremum principle (3.16) is an Euler–Lagrange equation (0.34) in the amplitudes $\mathbf{q}$. This differential equation mathematically expresses the law of physics (e.g., the Klein–Gordon equation) governing the real measurement of $\boldsymbol{\theta}$. The equation holds in the input space of the measuring instrument (see Sec. 3.8 and Chapter 10). One may conclude, then, that *measurement elicits physical law.* This is a most interesting effect, and, in fact, is one of the ideas expressed by Wheeler in the Introduction, that of "observer participancy." Physics has been called the "science of measurement" (Kelvin, 1889) in the sense that a phenomenon is not understood until it is measured and quantified. Now we find that this statement can be strengthened to the extent that a mathematical description of the phenomenon is formed *in reaction* to the measurement.

### 3.4.9 The intrinsic scenario defines independent degrees of freedom

Equation (2.19) shows that information $I$ is defined by $N$ amplitude functions $\mathbf{q}$. By Eqs. (3.16) and (3.18) so are informations $J$ and $K$. The $\mathbf{q}$ comprise "degrees of freedom" of the observed phenomenon (Sec. 2.7). Alternatively, the theory of Chapter 2 models the same number $N$ of artificial measurements to have been made in an ideal, "intrinsic" scenario. The result is that *each such intrinsic measurement gives rise (via EPI) to a new degree of freedom* $q_n$ *of the physical theory.* (For example, in non-relativistic quantum mechanics where the wave amplitude is described by an amplitude pair $(q_1, q_2)$, exactly two intrinsic measurements define the physical theory.) This is an interesting significance for these model measurements.

### 3.4.10 The question of the size of N

The number $N$ of degrees of freedom for a given problem appears to be arbitrary: the intrinsic scenario model (Chapter 2) does not fix $N$ as a specific value. On this basis, the number of amplitude components $\mathbf{q}$ needed to describe a given physical phenomenon could conceivably be of unlimited value. However, in many applications the value of $N$ is fixed by prior knowledge about the measured entity. For example, in Chapter 4 the particle whose space-time coordinates are being measured is known to be a vector particle of order $N$. This ultimately fixes the particle's spin as well. In the *absence* of such prior knowledge, the *game corollary* (see below) is used to fix $N$.

### 3.4.11 A mathematical game

In certain scenarios the EPI principle may be regarded as the workings of a "mathematical game" (Morgenstern and von Neumann, 1947). In fact the simplest such game arises. This is a game with *discrete*, deterministic moves $i$ and $j$ defined below. The transition to continuous values of $i$ and $j$ may be made without changing the essential results.

Those two arch-rivals A and B are vying for a scarce commodity (information). See Table 3.1. Player A can make either of the row value moves, $i = 1$ or

Table 3.1. *A* $2 \times 2$ *payoff matrix.*

| $i$ \ $j$ | 1 | 2 |
|---|---|---|
| 1 | 3.0 | 2.0 |
| 2 | 5.0 | 4.0 |

2; and player B can make either of the column value moves, $j = 1$ or 2. Each move pair $(i, j)$ constitutes a play of the game. Each such move pair defines a location in the table, and a resulting payout of the commodity to A, *at the expense of* B. ("Thus A is happy, and B is not.") For example, the move (1, 2) denotes a gain by A of 2.0 and a loss by B of 2.0. Such a game is called "zero-sum," since at any move the total gain of both players is zero. Assume that both players know the payoff table. What are their respective optimum moves?

Suppose that A moves first. If he chooses a row $i = 1$ then he can gain either 3.0 or 2.0. Since these are at the expense of B, B will choose a column $j = 2$ so as to minimize his loss to 2.0. So, A knows that if he chooses $i = 1$ he will gain 2.0. Or, if A chooses the row $i = 2$ then he will gain, in the same way, 4.0. Since this exhausts his choices, to maximize his gain he chooses $i = 2$, certain that he will gain 4.0.

Or, suppose that B moves first. If he chooses a column $j = 1$ then he will pay out either 3.0 or 5.0. But then A will choose $i = 2$ so as to maximize his gain to 5.0. So, if B chooses $j = 1$, he will lose 5.0. Or, if B chooses column $j = 2$ then he will lose, in the same way, 4.0. Since this exhausts his choices, to minimize his loss he chooses $j = 2$, certain that he will lose 4.0.

Hence the optimum play of the game, regardless of who moves first, is the move (2, 2). This is the extreme lower-right item in the table. This item results because the point (2, 2) is locally a *saddle point* in the payouts, i.e., a local minimum in $j$ and a maximum in $i$. Also, note that there is nothing random about this game; every time it is played the result will be the same. Such a game is called "fixed-point."

### 3.4.12 EPI as a game of knowledge acquisition

We next construct a mathematical model of the EPI process that has the form of such a mathematical game (Frieden and Soffer, 1995). Note that this game model is an explanatory device, not a distinct, physical derivation of EPI. Also, delegating human-like attributes to the players is merely part of the anthropomorphic model taken. The "game that is played" is merely a descriptive device. EPI is a physical process (Chapter 10).

The game model can also be regarded as an epistemological model of EPI. It will show that EPI arises, in certain circumstances, *as if it were* the result of a quest for information and knowledge. However, despite its epistemological origin, the game has practical outcomes. These are the EPI outputs $\mathbf{q}(\mathbf{x})$ and a method of analytically determining certain physical constants (Sec. 3.4.14).

In many problems the Fisher coordinates are mixed real and imaginary quantities. Denote by $\mathbf{x}$ the subset of real coordinates of the problem. We want to deal, in this section, with problems having only real coordinates. Thus, let the amplitudes $\mathbf{q}(\mathbf{x})$ of the problem represent either the fully dimensioned amplitudes for an all-real case, or the *marginal* probability amplitudes in the real $\mathbf{x}$ for a mixed case. Equation (2.19) shows that, for real coordinates $\mathbf{x}$, $I$ monotonically *decreases* as the amplitudes $\mathbf{q}$ are monotonically broadened or blurred. This effect should hold for any fixed state of correlation in the intrinsic data.

On the other hand, we found at Eq. (2.58) that the form Eq. (2.19) for $I$ represents a model scenario of *maximized* information due to efficient estimation and independent data. The latter effect is illustrated by the following example.

*Example*: Suppose that there are two Fisher variables $(x, y)$ and these obey a Gaussian bivariant PDF, with a common variance $\sigma^2$, a *general correlation coefficient* $\rho$, and a common mean value $\theta$. Regard the latter as the unknown parameter to be estimated. Then the Fisher information Eq. (1.9) gives $I = 2\sigma^{-2}(1+\rho)^{-1}$. This shows how the information in the variables $(x, y)$ depends upon their degree of correlation. Suppose that the variables are initially correlated, with $\rho > 0$. Then as $\rho \to 0$, $I \to 2/\sigma^2 = max.$ in $\rho$. As a check, this is twice the information in a single variable (see Sec. 1.2.2), as is required by the additivity of independent data (Sec. 2.4.1). Hence, independent data have maximal $I$ values.

This effect holds in the presence of any fixed state of blur, or half-width, of the joint amplitude function $\mathbf{q}(\mathbf{x})$ of all the data fluctuations.

We have, then, determined the qualitative dependence of $I$ upon the data correlation and the state of blur. This is summarized in Fig. 3.2. As in Table 3.1, the vertical and horizontal coordinates are designated by $i$ and $j$, respectively, although they are continuous variables here. Coordinate $i$ increases with the degree of independence (by any measure) of the intrinsic data. Let each coordinate $j$ represent a possible trial solution $\hat{\mathbf{q}}(\mathbf{x})$ to EPI in the particular sequence defined below. Since, as we saw, the $I$ values increase with increasing independence, they must increase with increasing $i$.

Consider a particular solution $\mathbf{q}$ to EPI. Equation (2.19) for $I$ presumes maximal independence (vertical coordinate) of the intrinsic data (Sec. 2.1.3).

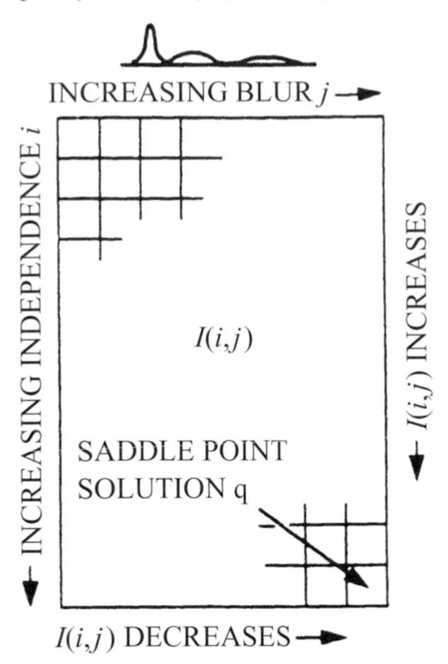


Fig. 3.2. The information game. The observer bets on a state of independence $i$ (a row value). The information demon bets on a state of blur $j$ (a column value). The payoff from the demon to the observer is the information at the saddle point, defining the EPI solution $\mathbf{q}$. (Reprinted from Frieden and Soffer, 1995.)

Then the EPI solution $\mathbf{q}$ is represented by a particular coordinate value $j$ located along the *bottom row* of Fig. 3.2.

The trial solutions $\hat{\mathbf{q}}$ are sequenced according to the coordinate $j$ as follows. Let each $\hat{q}_n(\mathbf{x})$ function of a given vector $\hat{\mathbf{q}}$ monotonically decrease in blur from the solution "point" $j \to \mathbf{q}(\mathbf{x})$ on the far right to an initial sharp state at the far left, corresponding to $j = 1$. Then, by the form Eq. (2.19) of $I$, the values $I$ decrease along that row as $j$ *increases* to the right, i.e., as the blur increases. On the other hand, we found before that, for a given $j$, values $I$ increase with increasing $i$ (independence) values. Then the solution point $\mathbf{q}$ designates a local maximum in $i$ but a minimum in $j$, i.e., a *saddle point*.

We found, by analysis of the game in Table 3.1, that a saddle point represents the outcome of a zero-sum mathematical game. Hence, the EPI solution point $\mathbf{q}$ represents the outcome of such a game. Here, the commodity that the two players are vying for is the amount of Fisher information $I$ in a trial solution $\hat{\mathbf{q}}$. Then the game is one of Fisher "information hoarding" between a player A, who controls the vertical axis choice of *correlation*, and a player B, who controls the horizontal axis choice of *the degree of blur*. Who are the opposing players?

The choice of correlation is made by the *observer* (Sec. 2.1.3) in a prior scenario, so this identifies player A. Player B is associated with the degree of blur. This is obviously out of the hands of the observer. Therefore it is adjusted by "nature." This may be further verified by the zero-sum aspect of the game. By Eq. (3.13) or (3.18) we notice that any information $I$ that is acquired by the observer is at the expense of the phenomenon under measurement. Hence, player B is identified to be the phenomenon, or nature in general. Then, according to the rules of such a game, nature has the "aim" of increasing the degree of blur so as to minimize its payout of information.

Hence, for real coordinates $\mathbf{x}$ the EPI principle represents a game of information hoarding between the observer and nature. The observer wants to maximize $I$ while nature wants to minimize it. Each different physical scenario, defined by its dependence $J[\mathbf{q}]$, defines a particular play of the game, leading to a different saddle point solution $\mathbf{q}$ along the bottom row of Fig. 3.2. This is the game-theoretic aspect of EPI that we sought.

Such real coordinates $\mathbf{x}$ occur in the Maxwell–Boltzmann law derivation in Chapter 7, in the time-independent Schrödinger wave equation derivation in Appendix D, and in the $1/f$ power-law derivation in Chapter 8. These EPI derivations are, then, equivalent to a play of the knowledge acquisition game.

It is interesting that, because the information transfer efficiency factor $\kappa \leqslant 1$ in Eq. (3.18), $I \leqslant J$ or $K \leqslant 0$. This indicates that, of the two protagonists, nature always wins, or at least breaks even, in the play of the game. This effect agrees in spirit with Brillouin's equivalence principle Eq. (3.5). That is, in acquiring data information there is an inevitable *larger* loss of phenomenological ("bound" in his terminology) information. You can't win.

*Exercise*: The scope of application of the game can be widened. Suppose that a particle has a complex amplitude function $\psi(\boldsymbol{r}, t) \equiv q_1(\boldsymbol{r}, t) + iq_2(\boldsymbol{r}, t)$, by the use of Eq. (2.24). Suppose that the particle is subjected to a conservative force field, due to a time-independent potential function $V(\boldsymbol{r})$. Then energy is conserved, and we may regard the particle as being in a definite energy state $E$. It results that the amplitude function separates, as $\psi(\boldsymbol{r}, t) = u(\boldsymbol{r})\exp(iEt/\hbar)$. Suppose that all time fluctuations $t$ are constrained to lie within a *finite*, albeit large, interval $(-\Delta T, \Delta T)$. Show that, under these circumstances, the time-coordinate contribution to information $I$ is merely an additive constant, and that the space coordinates give rise to positive information terms, so that the game may be played with the space coordinates. Hence, the knowledge acquisition game applies, more broadly, to the scenario of a quantum mechanical particle in a conservative force field.

### 3.4.13 The information "demon"

The solution $\mathbf{q}$ to the game defines a physical law (e.g., the Maxwell–Boltzmann law). The preceding shows, then, that a physical law is the result of a lossy information transfer game between an observer and the phenomenon under observation. Since we can sympathize with the observer's aims of increased information and knowledge, and since he always loses the game to the phenomenon (or at best breaks even), it seems fitting to regard the phenomenon as an all-powerful, but malevolent, force: an information "demon." (*Note*: This demon is not the Maxwell demon.)

In summary, for real Fisher coordinates $\mathbf{x}$ the EPI process amounts to carrying through a game. The game is a zero-sum contest between the observer and the information demon for a limited resource $I$ of intrinsic information. The demon generally wins.

We note with interest that the existence of such a demon implies the $I$-theorem, Eq. (1.30). (This is the converse of the proof in Sec. 3.4.7.) Information $I - J = K$ represents the change of Fisher information $\Delta I$ due to the EPI process $J \to I$ that takes place over an interaction time interval $\Delta t$. Since the game is played and the demon wins, i.e., $K \leqslant 0$, and since $\Delta I = I - J \equiv K$, necessarily $\Delta I \leqslant 0$. Then, since $\Delta t \geqslant 0$, the $I$-theorem follows.

### 3.4.14 Absolute nature of extrema; the physical constants; analytically determining unitless constants; the game corollary

One can expect the extreme value that is attained in principle (3.16) at its solution to depend upon the nature of the fields and sources for the problem. Conversely, in their absence (i.e., *free-field* situations) the extremum will be an absolute number that can only be a function of whatever free parameters the functional $K$ contains (e.g., $h$, $k$, $c$). Everything else is integrated out in Eq. (3.16). These parameters occur, in particular, within the functional $J$ part of $K$. The absolute nature of the scenario leads us to postulate that the extreme value of $J$ is a *universal physical constant* (Hughes, 1994). That is, its value is fixed, independently of boundary values. This has an unusual "bootstrap" effect.

Suppose, e.g., that in a given scenario the extremum in $J$ is a known function of parameter $c$, the speed of light in vacuum. Then, since the extremum is a universal constant, this implies that $c$ is a universal constant. Next, proceed to another scenario whereby the extremum is a known function of (say) $c$ and another parameter, e.g., $h$. Since $c$ was already fixed this, in turn, fixes $h$. In this manner, EPI and the smart measurer (Sec. 1.2.1) proceed through physical phenomena, fixing one constant after another.

A corollary is that the physical constants do not change in time. The contrary had been suggested by Dirac (1937; 1938).

Although EPI demands that these quantities be constants, it (somewhat paradoxically) regards their values as having been chosen *a priori* from a probability law. The form of this law is found in Chapter 9.

A particular class of universal constants consists of the real, *unitless* ones. An example is the constant exponent $\alpha$ found in Chapter 15 to describe cancer growth. This constant is, at first, unknown. *How can the value of a unitless constant such as $\alpha$ be found*?

Toward this end, recall that

(a) $J - I$ is an extremum at solution $q_n(\mathbf{x})$; and
(b) if the measured coordinate is purely real the action of the demon in the *information game* is to *minimize* information $I$ (Secs. 3.4.12 and 3.4.13) at solution. To review, this is by increasing the disorder in coordinates $\mathbf{x}$, that is, increasing the width or blur of the amplitude functions $q_n(\mathbf{x})$. Conversely, if the measured coordinate is imaginary, increasing the blur will minimize the *negative* of the information $I$. The two coordinate cases may be combined by the statement that the $|I|$ of the EPI output is minimized by the demon through blur of the $q_n(\mathbf{x})$.

Returning to the question of how to fix a unitless constant such as $\alpha$, a logical *extension* of the demon's activity (b) is to vary the value of $\alpha$ such that $|I|$ is further minimized, i.e., once by the EPI procedure (a), (b) and secondarily through choice of $\alpha$. An apt name for this further minimization step is the

$$\textit{game corollary}: \quad |I| = min., \tag{3.21}$$

the minimum attained through choice of the constant. The minimization should, of course, be constrained by whatever physical conditions are further imposed upon the unitless constant. Corollary (3.21) is, as we see, an effect of the information game, not directly following from axioms 1, 2, and 3 of the EPI principle.

Consider the related problem of determining the (unitless) constant $N$ defining the number of degrees of freedom $q_1, \ldots, q_N$ for a given problem (see also Sec. 3.4.10). If there is no prior physical consideration that *fixes* the value of $N$ then the game corollary should be used for its determination. Specifically, since $I$ tends to monotonically increase with $N$ (Sec. 2.4.1), the corollary requires $N$ to be the *minimum* number of independent solutions that are necessary to describe the output EPI probability law. For example, the corollary is used in Sec. 7.3.6 to fix the probability law on energy as uniquely the Boltzmann law (a case $N = 1$).

If there *is* prior knowledge that uniquely fixes the value of $N$ then the game corollary can be used to predict the *most probable* or dominant solution over the class of solutions considered. This is repeatedly applied in Chapter 4 to predict the various wave equations for elementary bosons and fermions. Also, the corollary is used in Sec. 7.4.8 to predict that the Maxwell–Boltzmann law is the preferred, i.e. equilibrium, law within a family of general non-equilibrium laws (higher values of $N$).

Does the game corollary give correct results? Aside from the preceding applications, it is used in Chapter 12 to predict important constants of the EPR–Bohm experiment. So far, all these predictions agree with either known theory or known data. See further discussion in Sec. 12.10.

### 3.4.15 Ultimate resolution lengths

The maximized nature (Eq. (2.58)) of the information $I$ suggests the possibility that ultimate resolution lengths could be determined from EPI solutions. This is the case. The solution $\mathbf{q}$ to EPI for a given measurement scenario allows $I[\mathbf{q}]$ to be calculated in terms of physical constants; see preceding section. By the inverse relation (1.28) between $I$ and the error $e^2_{\text{min}}$, this permits the latter to be calculated. If the parameter $\boldsymbol{\theta}$ under measurement is a length, the result is an expression for the ultimate resolution length $e_{\text{min}}$ intrinsic to the measured phenomenon. In this way, EPI computes the Compton resolution length of quantum mechanics (Sec. 4.1.17) and the Planck length of quantum gravity (Sec. 6.3.22).

### 3.4.16 Unified probabilistic view of phenomena

The basic elements of EPI theory are the *probability amplitudes* $\mathbf{q}$. All phenomena that are derived by EPI in this book are expressed in terms of these amplitudes. This includes intrinsically statistical phenomena such as quantum mechanics but, also, ostensibly non-statistical phenomena such as classical electromagnetic theory and classical gravitational theory. EPI regards all phenomena as being fundamentally statistical in origin.

The Euler–Lagrange solution (0.34) for each phenomenon is a differential equation in function $\mathbf{q}$. Since $\mathbf{q}$ is a probability amplitude, *EPI regards each such equation as describing a kind of "quantum mechanics" for the particular phenomenon.* Among other things, this viewpoint predicts that the classical electromagnetic four-potential $\mathbf{A}$ can be regarded as a probability amplitude for photons (Chapter 5); and that the classical gravitational metric tensor $g_{\mu\nu}$ can be regarded as a probability amplitude for gravitons (Chapter 6). This viewpoint is, of course, consistent with field operator (sometimes called "second quantized") theory. In fact, the simple correspondence Eq. (11.63) of Chapter 11 allows EPI output solutions to be regarded as field operator solutions.

### *3.4.17 Vital role played by unitarity*

The form of Eq. (2.19) expresses $I$ as generally a squared length, that is, an $L^2$ norm. This single mathematical fact provides a basis for all the EPI quantum derivations in this book, as well as special relativity (Sec. 3.5). Any transformation that preserves, in its space, the value of an $L^2$ norm from direct space is called a "unitary transformation." An example is the Fourier transformation, which gives rise to Parseval's theorem – a statement of the preservation of the $L^2$ norm from direct space to frequency space (see Sec. 3.8.7 for details). Unitarity underpins much of theoretical physics, so it is perhaps not too surprising that it likewise underpins many EPI applications. See also the discussion in Sec. 12.7.

*Each EPI derivation of type* (*b*) (Sec. 3.4.6) *follows from a particular statement of unitarity for its information I functional.* The unitary spaces can be coordinate spaces (Sec. 3.5), function spaces (Sec. 4.1.7), functional spaces (Sec. 11.2.11), etc. A basic prediction of EPI is the following:

*Every physically observable transform space pair gives rise, in this way, to a physical law.*

That is, EPI is exhaustive in this regard. A particular example is afforded by the rules of special and general relativity. These underlie all laws of physics. We show next how special relativity follows from one particular length-preserving transformation for $I$. (General relativity is taken up in Chapter 6.)

## 3.5 Derivation of Lorentz group of transformations

It may now be shown that the particular use of the trace operation in definition (2.19) of $I$ leads, via a statement of unitarity (Eq. (3.22) below) to the Lorentz transformation of special relativity; and to the requirement that the individual components of the Lagrangians (i.e., the integrands) of all information quantities $I$, $J$, and $K$ be covariant.

Consider the intrinsic measurement scenario mapped out in Sec. 2.1.1. Suppose these measurements to be carried through, and viewed, in a flat space, laboratory coordinate system O. The same measurements also are viewed in a reference system O′ that is moving with constant velocity $u$ along the $x$-direction. As usual, we denote with primes quantities that are observed from the moving system. Thus, the extremized Fisher information as viewed from O is value $I$, but as viewed from O′ is value $I'$.

### 3.5.1 Is there a preferred reference frame for estimation?

The channel capacity $I$ defines the ultimate ability to estimate from given data (see Secs. 2.1 and 2.6). Consider, next, the following basic question. Should there be a preferred frame speed for optimally estimating parameters? Should the accuracy with which parameter $\boldsymbol{\theta}$ can be estimated depend upon the speed $u$? It seems plausible that the answer is no. Let's find where this leads.

Assume then that the estimation errors $e_{n\nu}^2$ obey invariance to reference frame, as expressed by invariance of the total accuracy (2.9), $\sum_{n\nu} 1/e'^2_{n\nu} = \sum_{n\nu} 1/e^2_{n\nu}$. Next, by Eqs. (2.9) and (2.10), this expression of error invariance becomes a statement that information $I$ obeys invariance to reference frame,

$$I'[\mathbf{q}'] = I[\mathbf{q}]. \tag{3.22}$$

This is a particular statement of unitarity for the $L^2$ length $I$. It also is an example of the *EPI zero-condition* Eq. (3.18) with bound information $J \equiv I'$ and efficiency constant $\kappa = 1$. The latter describes *all exact* theories (see, e.g., Sec. 3.4.5 of Chapter 3 and Sec. 5.1.18 of Chapter 5). The exact theory that will be derived here is that of the Lorentz group of transformations. It will be a direct outgrowth of use of the EPI zero-condition (3.22).

A special case of this group is the Lorentz transformation *per se*, as discussed below. In this way the Lorentz transformation will follow from use of the EPI zero-condition.

### 3.5.2 Requirement on amplitude functions

From representation (2.19) for $I$, requirement (3.22) becomes

$$\int d\mathbf{x}' \sum_{n\nu} \left(\frac{\partial q'_n}{\partial x'_\nu}\right)^2 = \int d\mathbf{x} \sum_{n\nu} \left(\frac{\partial q_n}{\partial x_\nu}\right)^2. \tag{3.23a}$$

Since the data are independent, this will hold only if it holds for each measurement number $n$. Suppressing $n$, the new requirement is

$$\int d\mathbf{x}' \sum_{\nu} \left(\frac{\partial q'}{\partial x'_\nu}\right)^2 = \int d\mathbf{x} \sum_{\nu} \left(\frac{\partial q}{\partial x_\nu}\right)^2. \tag{3.23b}$$

At this point it becomes convenient to use the Einstein implied summation notation whereby repeated indices connote a summation. Also, derivatives follow the notation

$$\partial_\nu \equiv \partial/\partial x_\nu, \quad \nu = 0, \ldots, 3. \tag{3.24}$$

Requirement (3.23a) becomes

$$\int d\mathbf{x}'\, \partial'_\nu q'\, \partial'_\nu q' = \int d\mathbf{x}\, \partial_\nu q\, \partial_\nu q. \tag{3.25}$$

How can this be satisfied?

To be definite, we now regard coordinates **x** of the problem as the position and time of a particle. The nature of **x** is later generalized to that of any physical "four-vector" (as defined) whose components are *a priori* independent.

### 3.5.3 The transformation sought is linear

As in conventional relativity theory, assume that relation (3.25) will hold because of the special nature of some relation between coordinates **x** and coordinates $\mathbf{x}'$. In ordinary Galilean (non-relativistic) mechanics, that relation would be

$$t' = t, \quad x' = x - ut, \quad y' = y, \quad z' = z. \tag{3.26}$$

This may be placed in a convenient matrix form by the use of new, "Galilean coordinates"

$$x_0 = ct, \quad x_1 = x, \quad x_2 = y, \quad x_3 = z \tag{3.27}$$

and corresponding primed coordinates. The matrix form of Eqs. (3.26) is then

$$\mathbf{x}' = [\mathrm{B}_0]\mathbf{x}, \quad [\mathrm{B}_0] = \begin{bmatrix} 1 & 0 & 0 & 0 \\ -u/c & 1 & 0 & 0 \\ 0 & 0 & 1 & 0 \\ 0 & 0 & 0 & 1 \end{bmatrix}. \tag{3.28}$$

This emphasizes that the relation between $\mathbf{x}'$ and **x** is a linear one.

The Galilean coordinates (3.27) will turn out not to suffice for satisfying our overall requirement (3.22). Instead, a different set of coordinates – "Minkowski coordinates" (3.34c) – will. This is shown in the next two sections. It is, unfortunately, customary to use the notation **x** for both sets of coordinates. We hope that this will not unduly confuse the reader.

### 3.5.4 Reduced problem

As in Eqs. (3.28), we seek a matrix [B] of elements $B_{\mu\nu}$ that connects $\mathbf{x}'$ and **x** by a linear relation

$$\mathbf{x}' = [\mathrm{B}]\mathbf{x}, \quad B_{\mu\nu} \equiv \partial x'_\mu/\partial x_\nu. \tag{3.29}$$

By direct substitution, Eq. (3.25) will be true if [B] obeys

$$d\mathbf{x}' = d\mathbf{x} \quad \text{and} \quad \partial'_\nu q' \, \partial'_\nu q' = \partial_\nu q \, \partial_\nu q. \tag{3.30}$$

The latter, in particular, states that the (squared) length of a vector $\mathbf{v}$ with components $v_\nu \equiv \partial_\nu q$ should be invariant in the two spaces, i.e.,

$$v'^2 = v^2. \tag{3.31}$$

The two requirements (3.30), (3.31) constitute our reduced problem. To enforce these requires that we first establish how vectors $\mathbf{v}$ and $\mathbf{v}'$ are related in view of Eq. (3.29).

Probability amplitudes $q$, $q'$ correspond to probabilities $p \equiv q^2$, $p' \equiv q'^2$. From elementary probability theory the two probabilities are related as

$$p(\mathbf{x}) = p'(\mathbf{x}')|J(\mathbf{x}'/\mathbf{x})|, \quad J(\mathbf{x}'/\mathbf{x}) = \det[\mathbf{B}] = 1, \tag{3.31a}$$

where $J(\mathbf{x}'/\mathbf{x})$ is by definition the determinant of the matrix of derivatives $[\partial x'_\nu / \partial x_\mu]$ of the transformation (3.29) from unprimed to primed coordinates. Differentiating Eq. (3.29) gives the second equality (3.31a). We will later verify the third equality $\det[\mathbf{B}] = 1$. Taking the square root of the first Eq. (3.31a) gives

$$q(\mathbf{x}) = \pm q'(\mathbf{x}'). \tag{3.31b}$$

Then, by the chain rule of differentiation,

$$\frac{\partial q}{\partial x_\nu} = \pm \sum_\mu \frac{\partial q'}{\partial x'_\mu} \frac{\partial x'_\mu}{\partial x_\nu} \quad \text{or} \quad v_\nu = \pm \sum_\mu v'_\mu B_{\mu\nu} \tag{3.31c}$$

by (3.29) and the definition of $v_\nu$, or, reverting to vector notation,

$$\mathbf{v} = \pm [\mathbf{B}] \mathbf{v}'. \tag{3.32}$$

The plus- or minus-sign alternatives will not matter to the results since they will depend upon the *squared amplitudes* of vectors $\mathbf{v}$, $\mathbf{v}'$.

### 3.5.5 Solution, and Lorentz transformation

The problem asks for a linear transformation (3.29) of coordinates that maintains length. Of course, a vector that is rotated to a new position maintains its length. Or equivalently, the length remains constant when instead the vector remains fixed while the coordinate system rotates about it. Hence, a rotation matrix [B] will suffice as a solution. This defines a subgroup of solutions called "proper" transformations. A second subgroup, called "improper," defines *inversion* of coordinates as a transformation (Jackson, 1975). We will ignore this type of solution.

Recall that our $I$ is the trace (2.10) of the Fisher information matrix. It is well known that the trace of a matrix is invariant to rotation of coordinates.

This is the essential reason for the rotation matrix solution mentioned previously.

The general solution is the group of *rotations* called the homogeneous *Lorentz group* (Jackson, 1975). Thus, requiring the EPI zero-condition (3.22) leads to the Lorentz group of transformations, as we required. A further requirement Eq. (3.35a) below requires a special case of this group, the "proper" Lorentz transformation. There is a most important *special case* of this transformation, considered next.

A rotation transformation group is defined by the rotation angle $\theta$. Let this be the Fisher parameter of the problem. Recall that we may take a Fisher parameter to be imaginary (Appendix C). Accordingly, consider the case of a rotation angle $\theta = \tan^{-1}(iu/c)$, $i = \sqrt{-1}$. The proper Lorentz transformation resulting from this imaginary rotation angle is the *Lorentz transformation matrix* (Eisele, 1969; or Jackson, 1975, p. 540).

$$[\mathrm{B}] = \begin{bmatrix} \gamma & -(\gamma u/c) & 0 & 0 \\ -(\gamma u/c) & \gamma & 0 & 0 \\ 0 & 0 & 1 & 0 \\ 0 & 0 & 0 & 1 \end{bmatrix}, \tag{3.33}$$

where $\gamma \equiv (1 - u^2/c^2)^{-1/2}$ and $c$ is a constant. We next discuss the nature of $c$.

Since the argument $iu/c$ of the $\tan^{-1}$ preceding must be unitless, and since $u$ is a speed, $c$ must likewise be a speed. From the generality of this problem, $c$ is a universal constant. Hence it must be a universal speed. By Eq. (3.33) and the definition of $\gamma$ below it, for the transformation matrix [B] to remain finite and real, no speed $u$ can exceed $c$. Finally, $c$ itself must be finite in order for the matrix [B] to be used as a transformation matrix. In summary, we have deduced that $c$ must be the largest, finite speed that any particle can attain. Further than this we cannot go, without some physical input to the problem.

All applications of EPI require an input or *ansatz* that expresses an invariance (aside from the general EPI invariance, such as Eq. (3.22) *et seq.*). Here it is the statement that the invariant speed value $c$ is a particular speed: that of an electromagnetic wave in vacuum. Thus, $c$ is the same speed $c$ as occurs in Eq. (5.51), the equation for an electromagnetic wave in vacuum.

Matrix [B] obeying Eq. (3.33) defines the Lorentz transformation. This may be shown to satisfy requirement (3.31), as follows. Generate quantities $\mathbf{v}$ in terms of the $\mathbf{v}'$ according to Eqs. (3.32) and (3.33). Squaring each component of $\mathbf{v}$, and adding, then shows that they obey

$$\sum_{\nu=1}^{3} v_\nu^2 - v_0^2 = \sum_{\nu=1}^{3} v_\nu'^2 - v_0'^2. \tag{3.34a}$$

This satisfies the requirement (3.31) of invariant length (and, hence, the second requirement of Eqs. (3.30)) if we define new quantities

$$\mathbf{v} \equiv (iv_0,\ v_1,\ v_2,\ v_3) \text{ or } (v_0,\ iv_1,\ iv_2,\ iv_3), \quad i = \sqrt{-1}, \tag{3.34b}$$

and analogously for primed quantities $v'_\nu$. Finally, by the definitions $v_\nu \equiv \partial q / \partial x_\nu$ with $x_\nu$ given by Eqs. (3.27), condition (3.34b) corresponds to the use of *new coordinates*

$$\mathbf{x} \equiv (ict,\ x,\ y,\ z) \text{ or } (ct,\ ix,\ iy,\ iz). \tag{3.34c}$$

These are the required coordinates for the problem. Call these "Minkowski coordinates," in contrast with the "Galilean coordinates" previously used in Eqs. (3.27) and (3.28).

Each vector $\mathbf{v}$ and $\mathbf{x}$ given by the last two equations defines a Minkowski space and is called a "four-vector." In summary, *the second requirement (3.30) is satisfied only by recourse to Fisher coordinates that are four-vectors of Minkowski coordinates.*

It is convenient now to return to use of the subscript $n$ in $\mathbf{x}_n$ that was suppressed above Eq. (3.23b). Having a four-vector form (3.34c) is but a *necessary* requirement for joint coordinates $\mathbf{x}_n$ to be Fisher coordinates. Other requirements follow from a basic requirement of the EPI calculation:

*Desideratum*: The onus is upon the *outputs* of EPI to predict any physically valid *relations among* the Fisher coordinates $\mathbf{x}_n$ of measurement space or of any equivalent unitary space. Such a relation is called a "holonomic constraint": see Goldstein (1950). This has two ramifications.

(1) The general intrinsic fluctuations $\mathbf{x}_n$ of Secs. 2.1.3 and 2.1.4 are presumed to be *a priori* independent degrees of freedom. Hence, the Fisher coordinates of a problem must not be connected *a priori* by a holonomic constraint equation. Instead, by the desideratum, it is up to EPI to establish any holonomic relation among either the Fisher coordinates or the coordinates of any *equivalent* space (such as momentum space in Chapter 4).

   For example, the momentum and energy coordinates of a free particle are known *not* to be independent degrees of freedom (see point (2) below). Hence, *momentum and energy cannot be used as the Fisher coordinates of a measurement space* (see Chapter 7).

(2) Of course, *a fortiori* to its derivation by EPI, momentum and energy are known to be connected by Eq. (4.17). This *is* a holonomic constraint equation in that space. However, this is a *derived* result: *a priori* to the derivation the precise form of this dependence is presumed to be unknown to the observer, since, by the *desideratum* above, it is to be *derived* by EPI. The observer must merely know whether or not *any* such dependence exists for candidate Fisher coordinates.

As an example of point (1) above, the space and time coordinates of a particle *are* independent degrees of freedom *a priori*. That is, they do not obey a generally valid holonomic equation linking space and time coordinates that is analogous to Eq. (4.17). The observer in Chapter 4 is presumed to know that the space and time coordinates are independent, and also that momentum–energy coordinates are *not* (see point (1)). Therefore he chooses space-time as the Fisher coordinates of the measurement space.

In summary, general four-vector coordinates qualify as Fisher coordinates if the additional requirements (1) and (2) are met. Four-vector Fisher coordinates will be used in most of the EPI applications to follow. Likewise, the individual subscripts $i$, $j$, $k$, $l$, etc. in *multiply*-subscripted quantitities such as the gravitational metrics $g_{ij}$ and $G_{ijkl}$ used in Chaps. 6 and 11 will be four-vectors, ranging over values 0, 1, 2, 3. As discussed above, to qualify as Fisher coordinates any such four-vectors must additionally be *a priori independent* degrees of freedom of the problem, i.e., not be connected by a holonomic relation.

### 3.5.6 Volume invariance requirement

We return to satisfying the first requirement of Eqs. (3.30). All proper matrices [B] have a unit determinant,

$$\det[\mathrm{B}] = 1. \tag{3.35a}$$

Also, the four-space volumes transform as

$$d\mathbf{x}' = \det[\mathrm{B}]\, d\mathbf{x}. \tag{3.35b}$$

By Eqs. (3.35a, b) the first requirement (3.30) is satisfied as well.

With both requirements (3.30) now satisfied, we have satisfied the original information requirement (3.22). Also, we found that the latter implies the Lorentz group of transformations, as required. Finally, in the particular case where the Fisher coordinate for the problem is an imaginary rotation angle $\theta = \tan^{-1}(iu/c)$, and with the *ansatz* that $c$ is the speed of electromagnetic waves, the particular Lorentz group member called the *Lorentz transformation* results.

### 3.5.7 Invariance of the physical information K

We have found that the Lorentz transformation (3.33) satisfies the requirement (3.22) that the Fisher information $I$ in the measurements be invariant to reference frame. It follows from Eq. (3.18), then, that the bound information $J$ likewise obeys such invariance. In turn, by definition (3.16), $K$ likewise obeys the invariance. The upshot is that all information components of the EPI approach obey invariance to reference frame, respectively

$$I', J', K' = I, J, K. \tag{3.36}$$

As we found, such invariance is consistent with the EPI zero-condition and with the assumption that there is no preferred frame for accuracy.

### *3.5.8 Ramifications to EPI*

The requirement of invariance of errors to reference frame has hinged upon the use of Fisher coordinates $\mathbf{x}$ that constitute a *four*-vector. Then, by Eq. (2.1) all coordinates $\mathbf{x}$, $\mathbf{y}$, and $\boldsymbol{\theta}$ of EPI theory must be four-vectors. These, of course, place the time coordinate on an equal basis with the space coordinates. As we saw, a benefit of this is that both time and space transform by a common rule, Eq. (3.29).

EPI theory extends the common treatment of space and time to, now, regard both as *random variables*. Thus, in the EPI use of the Fisher theory of Chaps. 1 and 2, a space measurement is always accompanied jointly by a time measurement; and the time component is allowed to be as uncertain as are the space components. (Why should one regard time measurements as any more accurate *a priori* than space measurements?) A ramification is that *conditional* PDFs such as $p(\boldsymbol{r}|t)$, which regard time $t$ as fundamentally different from space coordinates $\boldsymbol{r}$ (deterministic versus random), are not directly derivable by EPI.

We showed the need for Fisher coordinates that are four-vectors of space and time, in particular. But, in the same way, it can be shown that Fisher coordinates must be *four-vectors* generally. In each application, all four coordinates are placed on an *equal footing*, both regarding transformation properties *and insofar as being fundamentally statistical quantities* (see Sec. 3.1.2). It follows that the knowledge of any prior relation among such coordinates (as in Sec. 3.5.5) rules out their use as Fisher coordinates.

Four-vectors also comprise what is called a "maximal measurement" (Roman, 1961). This is defined to be the largest number of variables that can be simultaneously measured with *arbitrary precision*. For example, the four space-time coordinates of a particle define a maximal measurement of the particle, since, if an observer tried to complement these measurements with measurements of the momentum or energy of the particle, then, by the Heisenberg uncertainty principles (4.53) and (4.54), one or more of the space-time measurements would no longer be known with arbitrary precision.

The invariance property (3.36) that is obeyed by all information quantities $I$, $J$, $K$ of the theory implies that their corresponding information *densities* (Lagrangians) should be *covariant*, i.e., keep the same functional form under

Lorentz transformation (3.29), (3.33). It is interesting that this requirement followed, ultimately, from a postulate (3.22) that errors of estimation should be equal, regardless of reference frame. By comparison, the covariance requirement conventionally follows from the "first postulate" of *relativity theory* (Einstein, 1956), that the laws of nature should keep the same form regardless of reference frame. Correspondingly, then, Eq. (3.22) is a relativistic postulate of *EPI theory*.

This relativistic postulate could be obeyed, in the first place, because the Fisher information density in Eq. (2.20) is mathematically covariant in form (at fixed index $\mu$). An information density that was not covariant could not have satisfied the Lorentz transformation requirement in Sec. 3.5.5.

The sense by which information $I$ (and consequently informations $J$ and $K$) should be covariant is of interest. The requirements (3.30) were to hold for each distinct value of index $n$ (suppressed); see below Eq. (3.23a). Hence, the Lorentz invariance/covariance requirements that were implied must, likewise, hold for each value of $n$. Therefore, the requirement of covariance that was previously discussed is to hold separately for each component number $n$ of the information densities of $I$, $J$, and $K$. That is, it is to hold for components $i_n$, $j_n$, and $k_n \equiv i_n - j_n$ (see Eq. (3.19)).

It is emphasized that the need (Eq. (3.34c)) for either imaginary time or imaginary space coordinates arises out of these purely relativistic considerations. Such mixed real–imaginary coordinates will be seen to be decisive in forming the d'Alembertian operator $\Box$ (Eq. (5.12b)) that is at the root of all wave equations derived by EPI (Chaps. 4–6, 10 and 11).

In conclusion, the postulate (3.22) of invariance to frame requires the entire EPI approach to be covariant. As a result, the output physical laws of EPI, which have the general form of wave equations, will likewise obey covariance. This satisfies the first postulate of special relativity (Einstein, 1905), that the laws of physics remain invariant under a change of reference frame. Covariance also requires mixed real and imaginary coordinates (3.34c), which will give rise to the all-important d'Alembertian operator in the derived wave equations.

## 3.6 Gauge covariance property

### *3.6.1 Is there a preferred gauge for estimation?*

We found in the preceding that a condition of Lorentz invariance follows from the demand (3.22) that information be invariant to reference frame. As with reference frame, *the choice of gauge* (Sec. 5.1.21) is arbitrary for a given physical scenario. Gauges are constructed so that their choice has no effect upon observable field quantities (Jackson, 1975). Now, mean-square errors are observable quantities. It is therefore reasonable to postulate, as well, that mean-square errors be insensitive to such choice. Then, by Eq. (2.9), the Fisher information should likewise be invariant to choice of gauge.

### *3.6.2 Ramification to I*

Now, with $\boldsymbol{\theta}$, $\mathbf{y}$, and $\mathbf{x}$ representing four-positions, in the presence of arbitrary electromagnetic potentials $\boldsymbol{A}$, $\phi$ the form (2.19) for $I$ is not gauge-invariant (Lawrie, 1990). However, the replacement of all derivatives as

$$\nabla \to \nabla - ie\boldsymbol{A}/(c\hbar), \qquad \partial/\partial t \to \partial/\partial t + ie\phi/\hbar \tag{3.37}$$

renders it gauge-covariant. Here, $e$ is the particle charge and the non-boldface gradient $\nabla$ indicates a *three*-dimensional gradient. Replacements (3.37) will often be made in output EPI wave equations in order to render them gauge-covariant. See, e.g., Chapter 4. The replacements (3.37) are often called the "gauge principle," and the right-hand sides are said to comprise a "gauge-covariant derivative."

For convenience, use of the replacements (3.37) in EPI can be delayed until after $J[\mathbf{q}]$ is formed; this is because functional $J[\mathbf{q}]$ never depends upon derivatives $\partial\mathbf{q}/\partial x_\nu$. Then the Euler–Lagrange solution to EPI is subjected to the replacements. As an example, see the derivations in Chapter 4.

Most generally, in all EPI problems where the Fisher coordinates form a generally curved space (e.g., the Wheeler–DeWitt problem of Chapter 11), all derivatives with respect to the coordinates are to be considered tensors, called "covariant derivatives" (Lawrie, 1990, p. 159). For brevity, we do not consider the latter derivatives in further detail.

### *3.6.3 Ramifications to J and K*

With $I$ now gauge covariant, by proportionality (3.18) $J$ will be gauge covariant. Then, by (3.15), so is $K$. Then any solution to the extremization of $K$ will be covariant. In this way, all aspects of the EPI approach become gauge covariant, which of course had to be true (Goldstein, 1950). The sister property of *reference frame* covariance was previously established (Sec. 3.5.7).

## 3.7 Field dynamics from information

Suppose that the EPI principle Eqs. (3.16) and (3.18) yields a solution $\mathbf{q}$ for a given scenario. The amplitudes $\mathbf{q}$ are often regarded as a "field." Knowledge of the field implies complete knowledge of *the dynamics* (momentum, energy) of the phenomenon (Morse and Feshbach, 1953). This suggests that the EPI information quantities $I$ and $J$ somehow define the dynamics of the scenario. Exactly how this happens is shown next.

### 3.7.1 Transition to complex field functions

Field dynamical quantities are usually expressed in terms of generally *complex* field functions, rather than the purely real $\mathbf{q}$ that we use. Hence, we temporarily express information $I$ in terms of complex amplitude functions, as follows.

Store the real $\mathbf{q}$ as the real and imaginary parts of generally complex field amplitude functions $\psi_1, \ldots, \psi_{N/2} \equiv \boldsymbol{\psi}$ according to Eq. (2.24). Although information $I$ is defined in Eq. (2.19) in terms of the real amplitudes $\mathbf{q}$, it may easily be shown that $I$ is equivalently expressed in terms of the complex $\psi_n$ as

$$I = 4N \int d\mathbf{x} \sum_{n=1}^{N/2} \boldsymbol{\nabla}\psi_n^* \cdot \boldsymbol{\nabla}\psi_n \tag{3.38}$$

for general coordinates $\mathbf{x}$. The $\psi_n$ are now our amplitude functions. Like the $\mathbf{q}$, they represent a solution to an EPI problem.

### 3.7.2 Exercise

Derive Eq. (3.38) from Eq. (2.19), noticing that the imaginary cross-terms in the product *had* to cancel out because the resulting $I$ must be real, by its definition Eq. (2.19).

Specific space-time coordinates $\mathbf{x} \equiv (ix_1, ix_2, ix_3, ct) \equiv (i\mathbf{r}, ct)$ will be used in derivation of quantum wave equations in Chapter 4 (see Sec. 4.1.2). For these coordinates, Eq. (3.38) becomes

$$I = \sum_{n=1}^{N/2} \iint d\mathbf{r}\, dt\, i_n(\mathbf{r}, t),$$

$$\text{where} \quad i_n(\mathbf{r}, t) \equiv 4Nc\left[-(\nabla\psi_n)^* \cdot \nabla\psi_n + \frac{1}{c^2}\left(\frac{\partial\psi_n}{\partial t}\right)^* \frac{\partial\psi_n}{\partial t}\right] \tag{3.39}$$

and $\psi_n = \psi_n(\mathbf{r}, ct) \equiv \psi_n(\mathbf{r}, t)$ in simpler notation.

Quantity $i_n(\mathbf{r}, t)$ is the $n$th component Fisher information *density*, as in Eqs. (3.19).

Equations (3.39) show that information component $i_n$ depends purely upon the *rates of change* $\nabla\psi_n, \partial\psi_n/\partial t$ of a corresponding field component $\psi_n$. By contrast, the bound information component $j_n$ (defined at Eq. (3.19)) always depends, instead, upon the component $\psi_n$ *directly* (see, e.g., Eq. (4.22)). The result is that, functionally, the information components obey

$$i_n = i_n[\nabla\psi_n, \partial\psi_n/\partial t] \quad \text{and} \quad j_n = j_n[\psi_n]. \tag{3.40}$$

### 3.7.3 Momentum and stress-energy

Next we turn to the physical quantities that specify the dynamics of the problem. These are the canonical momentum density $\mu_n$, $n = 1, \ldots, N/2$ (*note*: notation $p_n$ is reserved for probabilities) and the stress-energy tensor $W_{\alpha\beta}$, for $\alpha, \beta = 0, 1, 2, 3$ independently. For a general vector of amplitudes $\psi_n(\mathbf{x})$, the $\mu_n$ and $W_{\alpha\beta}$ are defined in terms of the Lagrangian $\mathscr{L}$ for the problem as

$$\mu_n \equiv \frac{\partial\mathscr{L}}{\partial(\partial\psi_n/\partial t)} \quad \text{and} \quad W_{\alpha\beta} \equiv \sum_{n=1}^{N/2}\left(\frac{\partial\psi_n}{\partial x_\alpha}\right)\frac{\partial\mathscr{L}}{\partial(\partial\psi_n/\partial x_\beta)} - \mathscr{L}\delta_{\alpha\beta}, \tag{3.41}$$

with $\delta_{\alpha\beta}$ the Kronecker delta (Morse and Feshbach, 1953, p. 319). But by Eq. (3.16) the Lagrangian for our problem is actually $\mathscr{L} = \sum_n(i_n - j_n)$ so that, by Eqs. (3.19) and (3.40), Eqs. (3.41) become

$$\mu_n = C_1\frac{\partial i_n}{\partial(\partial\psi_n/\partial t)},$$
$$W_{\alpha\beta} = C_2\sum_{n=1}^{N/2}\left[\left(\frac{\partial\psi_n}{\partial x_\alpha}\right)\frac{\partial i_n}{\partial(\partial\psi_n/\partial x_\beta)} - (i_n - j_n)\delta_{\alpha\beta}\right]. \tag{3.42}$$

Constants $C_1$ and $C_2$ were inserted so as to give correct units (their specific values are irrelevant to the argumentation). As examples, the units of momentum and energy are taken to be mass–length/time and mass–length$^2$/time$^2$, respectively. With these constants, Eqs. (3.42) may be regarded as defining equations for the $\mu_n$ and $W_{\alpha\beta}$ from the information viewpoint.

Equations (3.42) show how the dynamics of the problem are completely specified by the Fisher data information and the Fisher bound information, component by component. An interesting aspect of the results is that each momentum density $\mu_n$ only depends upon information $I$ through a corresponding component $i_n$. Also, the stress-energy $W_{\alpha\beta}$ depends only upon the $i_n$, except for the diagonal elements $W_{\alpha\alpha}$ which depend upon the $j_n$ as well.

In this way, Fisher information in its two basic forms $I$ and $J$ completely defines our knowledge of the dynamics of the measured phenomenon. In essence, the dynamics are defined by gradients of the information rather than by the usual gradients of potentials.

The connection (3.42) between the information densities $i_n$ and $j_n$ and energy $W_{\alpha\beta}$ provides as well, by the *equivalence of energy with heat*, a connection between $i_n$ and $j_n$ and *heat flow*. Recall also that $i_n$ and $j_n$ relate to mean-square *errors* of estimation, via the *Cramer–Rao inequality* (1.1). Thus, $i_n$ and $j_n$ relate the concepts of *energy and heat flow*, on the one hand, with *estimation error*, on the other. In this way heat flow is connected with (lack of) knowledge via the concept of Fisher information. Jaynes (1957a, b) had of course long ago made the connection via the concept of *entropy*. Both concepts make the connection because both are measures of system disorder.

Equations (3.41) and (3.42) show that one can analyze a problem of dynamics by the use of Fisher informations $i_n$ and $j_n$. (The usual "action-integral" alternative is, of course, to regard them as energies.) This is consistent with regarding $\mu_n$ and $W_{\alpha\beta}$ of Eqs. (3.41), (3.42) as *computed* momentum and stress-energy values, i.e., as *knowledge* of these values based upon given information. This also ties in with the EPI interpretation of the amplitude function $\psi_n(x)$ as the *observer's state of knowledge* of the system, where a narrow function $\psi_n(x)$ indicates strong knowledge and a wide function low knowledge (Sec. 1.7).

This stance is to be compared with the action-integral view, according to which these quantities $i_n$, $j_n$, $\psi_n$ instead represent properties *of the particles or fields* per se. As an example, by the latter interpretation the spread $e^2$ in $\psi_n(x)$ represents spread in the *actual* position of the *particle*. By comparison, the EPI interpretation is to regard $e^2$ as the *observer's* level of *uncertainty about* the position of the particle.

In fact the latter uncertainty specifically arises out of an attempt to gain knowledge (of particle position), as shown in Sec. 4.3.4 of Chapter 4, where it takes the form of a generalized Heisenberg uncertainty principle. Physics on the most fundamental level is knowledge-based.

### 3.8 An optical measurement device

There is a simple measurement device that clarifies and substantiates the axiomatic approach taken in Secs. 3.3.2 and 3.4.2. This follows the model shown in Fig. 3.3. It is the familiar optical "localization experiment" given in Schiff (1955) and other books on introductory quantum mechanics. Usually it is employed heuristically to derive the Heisenberg uncertainty principle. Instead, we will use it – likewise heuristically – to demonstrate the following.

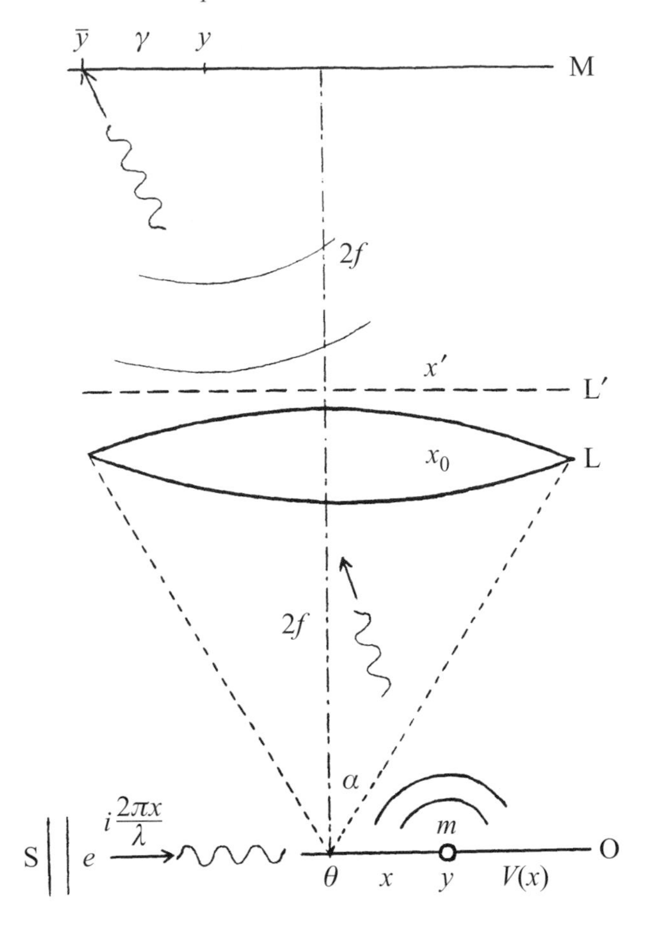


Fig. 3.3. A particle of mass $m$ in object plane O is moving under the influence of a potential $V(x)$. The particle is at position $y = \theta + x$, where random displacement $x$ is characteristic of quantum mechanics. A wave from monochromatic source S interacts with the particle. The wave traverses the lens L and Fraunhofer plane L′ before registering in measurement space M at a data position $\bar{y} = y + \gamma$. Random displacement $\gamma$ is characteristic of diffraction by the lens.

(i) *A unitary transformation of coordinates* naturally arises during the course of measurement of many (perhaps all) phenomena.
(ii) EPI, as defined by Eqs. (3.16) and (3.18), *is implied by* the unitary transformation and the perturbing effects of the measurement.
(iii) The well-known Von Neumann measurement Eq. (10.22), whereby the system wave function separates from the measurement function, naturally arises in the output image.

Hence, the localization experiment says a lot more than the Heisenberg uncertainty principle. The combination of points (i), (ii) and, then, the EPI principle, will be seen to imply (in Chapter 4 and Appendix D) that *an act of measurement gives rise to a wave equation of quantum mechanics*.

### 3.8.1 The measurement model

A particle in "object space" (using optical terminology) is being measured for its $X$-coordinate position by the use of monochromatic light source S and a diffraction-limited, thin lens L. See Fig. 3.3. The particle has mass $m$ and is moving under the influence of a potential function $V(x)$. The half-angle subtended by the lens at the particle is $\alpha$. A monochromatic light source S illuminates object space with a plane wave $\exp(2\pi i x/\lambda)$.

The classical $X$-position of the particle is at the center of the field, located at a coordinate value $\theta$. This is the "ideal" position absent of any random phenomena. However, the phenomenon of quantum mechanics causes the particle to have a position fluctuation $x$ from $\theta$ *just prior to its measurement*. Quantity $x$ is an example of an "intrinsic" fluctuation, as defined in Chapter 2. Consequently, the $X$-coordinate position of the particle at its input to the measuring system is

$$y \equiv \theta + x. \tag{3.43}$$

There is a probability amplitude on random values $y$ given a fixed value of $\theta$, called $\psi_y(y|\theta)$. By the shift-invariance property Eq. (2.16), $x$ is independent of the absolute position $\theta$ so that

$$\psi_y(y|\theta) = \psi_0(x), \quad x = y - \theta. \tag{3.44}$$

Fluctuations $x$ follow the particle probability amplitude function $\psi_0(x)$.

The optics are diffraction-limited, and operate at equal object and image distances, so that the magnification is unity. A scale is placed in "image space" (at the top of Fig. 3.3). Its origin is to the right (not shown), while that of object space O is equally off to the left (also not shown).

A wave from the source illuminates the particle. The wave and particle interact, causing the particle's intrinsic wave function $\psi_0(x)$ to be perturbed by an amount $\delta\psi(x)$, and effecting the information transition $J \rightarrow I$. These occur in object space. Also, the wave propagates through the lens, and continues on to image, or measurement, space. There, it gives rise to a light detection event (say, the sensitization of a photographic grain) at a random position

$$\overline{y} = y + \gamma = \theta + \eta, \quad \eta = x + \gamma. \tag{3.45}$$

We used Eq. (3.43) in the middle equality. Displacement $\gamma$ is random, independent of $y$ (by Eq. (2.16)), and the result of diffraction due to the finite size of the lens (see below) is effectively "outside" noise to the system defined by $\theta$ and $x$.

There is a probability amplitude on random values $\overline{y}$ in the presence of a given value of $\theta$, called $\psi_{\overline{y}}(\overline{y}|\theta)$, which obeys

$$\psi_{\overline{y}}(\overline{y}|\theta) = \psi_{\overline{y}}(\theta + \eta|\theta) \equiv \psi_i(\eta). \tag{3.46}$$

The first equality is by Eq. (3.45), and the second is again by shift invariance (2.16), now in image space. One of the aims of this analysis is to relate the image space amplitude $\psi_i(\eta)$ to the object space one $\psi_0(x)$.

To accomplish this, we have to establish the physical connection between a particle position $y$ in object space and the corresponding coordinate $\overline{y}$ of the light detection event in image space.

### 3.8.2 Optical model

For simplicity, optical effects are treated classically, i.e., by the theory of coherent wave optics (Marathay, 1982). However, the particle positions are treated statistically, i.e., obeying a random phenomenon of some sort. This is akin to the "semiclassical" treatment of radiation (Schiff, 1955) whereby the radiation is treated as obeying *classical* electromagnetic theory, while the particle with which it interacts is treated as obeying *quantum* mechanics. There is, however, an important distinction to be made: the fact that the particle obeys quantum mechanics is to be derived (by EPI), not assumed.

For further simplicity, only $X$-coordinate dependences are analyzed. Finally, only temporally stationary effects are considered. The presence of the time-independent potential $V(x)$ implies that energy is conserved, so that we may take the particle to be in a definite state of total energy. Then the time dependence of the particle wave function separates out from the space coordinate. A like property holds for the optical waves, which are monochromatic. These separated time dependences can be ignored, for the purpose of the analysis to follow. The result is that all amplitude functions (optical and quantum mechanical) of the problem are assumed to be just functions of an appropriate $X$-coordinate.

At each particle position $y$, its interaction with a plane light wave from the source causes a spherical amplitude wave to be launched toward image space. Hence, the more probable the position $y$ is for the particle, the higher will the *optical intensity* be at position $y$ in the object intensity pattern. Call the optical amplitude value at a position $y$ in *object space* $u(y|\theta)$. Then, on the basis of amplitudes, it must be that $u(y|\theta) \propto \psi_y(y|\theta)$. Define the units of optical intensity such that the area under the intensity curve $|u(y|\theta)|^2$ is unity. Then, by like normalization of the PDF $|\psi_y(y|\theta)|^2$, the proportionality constant must be unity, so that

$$u(y|\theta) = \psi_y(y|\theta). \tag{3.47}$$

By the Compton effect (Compton, 1923), the wavelength of the scattered light shifts according to the scattering angle. Assume that the particle mass is large enough and/or the range of scattering angles small enough that the shift is negligible over the range of angles considered. Then the scattered light is effectively monochromatic.

Consider, next, formation of the image. Since the light is effectively monochromatic, the object-space amplitude function $u(y|\theta)$ propagates through the lens to image space according to the rules of *coherent image formation*. Then, as with Eq. (3.47), the output *optical* amplitude, suitably normalized, can be regarded as the data-space *probability* amplitude $\psi_{\overline{y}}(\overline{y}|\theta)$ for measurement events $\overline{y}$.

By this correspondence, we can use the known rules of coherent image formation to form the measurement space law $\psi_{\overline{y}}(\overline{y}|\theta)$. These are as follows.

Designate the coherent image of a perfectly localized "point" source of radiation in object space as $w_{\overline{y}}(\overline{y}|y)$. The notation means the spread in positions $\overline{y}$ due to "source" positions at $y$. Function $w_{\overline{y}}$ is the Green function for the problem.

Coherent image formation obeys *linearity*, meaning that the output optical amplitude is a superposition of the input amplitude as weighted by the Green function,

$$\psi_{\overline{y}}(\overline{y}|\theta) = \int dy\, \psi_y(y|\theta) w_{\overline{y}}(\overline{y}|y). \tag{3.48}$$

Assume, as is usual in image formation, a condition of *isoplanatism*: the point amplitude response keeps the same shape regardless of absolute "source" position $y$,

$$w_{\overline{y}}(\overline{y}|y) = w(\overline{y} - y). \tag{3.49}$$

Optical amplitude $w$ is commonly called the "point amplitude response" of the lens. It measures the intrinsic spread due to diffraction in the coherent image of each object point.

Substituting Eqs. (3.44) and (3.49) into Eq. (3.48) gives

$$\psi_{\overline{y}}(\overline{y}|\theta) = \int dy\, \psi_0(y - \theta) w(\overline{y} - y) \tag{3.50a}$$

$$= \int dx\, \psi_0(x) w(\overline{y} - \theta - x) \tag{3.50b}$$

after an obvious change of integration variable. Using Eq. (3.46) on the left-hand side then gives

$$\psi_i(\eta) = \int dx\, \psi_0(x) w(\eta - x), \tag{3.51}$$

after again using Eq. (3.45). This relates the probability amplitude $\psi_i(\eta)$ on the *total fluctuation* $\eta$ in image, or measurement, space to the probability amplitude $\psi_0(x)$ on intrinsic fluctuation $x$ in object space and the point amplitude response $w(x)$ of the lens. The right-hand side of Eq. (3.51) takes the mathematical form of a "convolution" of $\psi_0(x)$ with $w(x)$.

### 3.8.3 Von Neumann wave function reduction

We can alternatively express the measurement wave $\psi_{\overline{y}}(\overline{y}|\theta)$ as a coherent sum

$$\psi_{\overline{y}}(\overline{y}|\theta) \equiv \int dx\, \psi(x, \overline{y}|\theta) \tag{3.52}$$

where probability amplitude $\psi(x, \overline{y}|\theta)$ expresses the *joint state* of the particle under measurement and the instrument reading. Comparing Eqs. (3.50b) and (3.52) shows that

$$\psi(x, \overline{y}|\theta) = \psi_0(x) w(\overline{y} - \theta - x). \tag{3.53}$$

The joint state of particle and instrument separates into a product of the separate states. This is the celebrated Von Neumann separation effect after a measurement.

This equation also agrees with the result Eq. (10.26a) of using a rigorous four-dimensional approach to measurement. (Note that $\overline{x}$ in Eq. (10.26a) is, from Eq. (10.6), what we called $\overline{y} - \theta$ in Eq. (3.53).) The wave function is "reduced," i.e., narrowed, by the product. This is explained further in Sec. 10.9.6.

The fact that there is such three-way agreement on measurement further validates the simple optical model.

### 3.8.4 Optical pupil space, particle momentum space

From Eqs. (3.44) and (3.47), the optical object $u(y|\theta)$ likewise obeys shift invariance and relates to the quantum mechanical amplitude $\psi_0(x)$ as

$$u(x) = \psi_0(x), \quad x = y - \theta. \tag{3.54}$$

The intrinsic quantum mechanical fluctuations of the particle define the input amplitude wave to the optics. How does the latter propagate through the optics to the measurement plane?

Since the lens L in Fig. 3.3 is a thin, diffraction-limited lens, the plane just beyond the lens (line L′ in Fig. 3.3) acts as the Fraunhofer, or Fourier, plane for amplitude functions in object space.

Hence, due to the input optical amplitude function $u(x)$, there is an optical disturbance $U(x')$ along the line L$'$ that obeys

$$\underset{|x'|\leq x_0}{U(x')} = \int dx\, u(x) \exp\left[\frac{2\pi i x x'}{\lambda R}\right], \qquad R \approx 2f. \tag{3.55}$$

Here $x'$ is the pupil spatial coordinate and $R$ is approximately the image distance. An irrelevant multiplicative constant has been ignored. Also, by the geometry of the figure, the finite pupil size $2x_0$ relates to the half-angle $\alpha$ subtended by the lens at the ideal object position $\theta$ by

$$x_0 \approx R \sin \alpha. \tag{3.56}$$

Let us suppose that we want to apply EPI to this scenario. The aim is to find the law that the particle wave functions $\mathbf{q}$ should obey. The $\mathbf{q}$ are defined to be the successive real and imaginary parts of a wave function $\psi$ (a case $N = 2$ in Eq. (2.24)). This is effectively assuming a particle without spin, as is evident from Chapter 4. Then there is one, complex wave function $\psi$ to be determined. This is, in fact, what we have been calling $\psi_0(x)$ (Eq. (3.44)).

As discussed in Sec. 3.4.5, the calculation hinges critically upon the form of the functional $J[\psi]$. This, in turn, depends upon the choice of invariance principle. It turns out that the most efficient (in the sense of $\kappa = 1$) such principle is the expression of a unitary transformation of coordinates. By Sec. 3.4.5, the transformation is to define two coordinate spaces, defining $I$ and $J$, respectively, which are connected by the information transition $J \rightarrow I$.

Ideally, the nature of the unitary transformation should not be imposed by the observer in some *ad hoc* manner. In the spirit of Wheeler's program (Sec. 0.1), *it can be defined by the internal processes of the measuring device.* The only transformations of coordinates that occur during operation of our measuring device are Fourier transformations. There are two such and, consistent with our aims, Fourier transformations are unitary (see Eq. (4.5)). One is the transition by which optical waves proceed from object space coordinates $x$ to optical pupil coordinates $x'$. This is the statement of the Fourier transform Eq. (3.55). The other is the inverse Fourier relation to Eq. (3.55). This expresses the transition from wavefront coordinates $x'$ to *output* measurement space coordinates $\eta$,

$$u_i(\eta) = \int_{-x_0}^{x_0} dx'\, U(x') \exp\left[-\frac{2\pi i x'\eta}{\lambda R}\right]. \tag{3.57}$$

Hence, the lens causes a transition of coordinates

$$x \rightarrow x' \rightarrow \eta. \tag{3.58}$$

This defines a flow of Fisher information from object space to lens space to measurement space. As is shown in Sec. 3.8.6: (a) Photons are the carriers of the information. (b) Some information is lost due to the finite $x_0$ in (3.57).

This means that the lens is here the "transducer" of the flow of information that is essential to EPI. Recall that EPI requires a unitary transformation, or some other invariance principle, as its "physical" input (Sec. 3.4.5). Then the lens itself should provide the invariance principle. In fact, the lens obeys the Fourier relation Eq. (3.55) connecting input space $x$ with another, conjugate space $x'$ (along line L$'$ in Fig. 3.3). It will become apparent later that a Fourier transformation is also a unitary one. Therefore, Eq. (3.55) defines, as well, the invariance principle that EPI needs; this has the form of a unitary transformation.

For increased generality, rather than directly using the pupil coordinate $x'$ as the conjugate coordinate, we use a coordinate proportional to $x'$, call it $\mu$, obeying

$$\mu \equiv \frac{hx'}{\lambda R}. \tag{3.59}$$

Parameter $h$ is temporarily called "Planck's parameter." It is later found to be a universal constant; see Sec. 4.1.14.

Using Eq. (3.54) and definition (3.59) in Eq. (3.55) gives

$$U\left(\frac{\lambda R}{h}\mu\right) = \int dx\, \psi_0(x) \exp\left[\frac{i\mu x}{\hbar}\right] \equiv \sqrt{h}\phi_0(\mu), \quad \hbar \equiv h/(2\pi) \tag{3.60}$$

Observing that the last equality expresses a Fourier transform relation, we can invert it to yield

$$\psi_0(x) = \frac{1}{\sqrt{2\pi\hbar}} \int d\mu\, \phi_0(\mu) \exp\left[-\frac{i\mu x}{\hbar}\right]. \tag{3.61}$$

Coordinate $\mu$ turns out to be the particle momentum (see Sec. 4.1.7). Thus, *the unitary conjugate space to x is momentum space*, and this can be pictured as being located in pupil amplitude space.

With coordinate $\mu$ that of momentum, function $\phi_0(\mu)$ becomes the *probability amplitude on momentum* (Sec. 4.1.12). Hence, the pupil amplitude function $U$ is proportional to the probability amplitude on momentum for the particle in object space, and each pupil coordinate value corresponds 1 : 1 to a momentum value.

### 3.8.5 Confirming EPI Eq. (3.18) using unitarity

To review, the instrument defines a coordinate space, along line L$'$ of Fig. 3.3, for optical amplitudes that is a Fraunhofer, or Fourier, transformation away from $x$-space. This is obviously an optical property of the instrument. But this Fourier space has, as well, special physical significance *to the measured particle* in being its momentum space. Since a Fourier transformation is unitary, the optical Fraunhofer plane of the instrument defines a particle momentum space that is unitary to the particle position space $x$. This is the unitary transformation that EPI extracts from the information flow internal to the instrument.

It follows that the Fisher information $I$, expressed equivalently in momentum space, represents the "bound" information for the particle:

$$I[\psi_0(x)] = \kappa J[\phi_0(\mu)], \quad \kappa = 1. \tag{3.62}$$

This is Eq. (3.18) of the EPI principle in the particular case $\kappa = 1$.

The functional forms of $I[\psi(x)]$ and $J[\phi(x)]$ are as found in Appendix D, Eqs. (D1) and (D3), for the case $N = 2$ (a single complex wave function):

$$I[\psi] = \int dx\, \psi'^*(x)\psi'(x), \tag{3.63}$$

and

$$J[\phi] = \frac{1}{\hbar^2}\int d\mu\, \mu^2\phi^*(\mu)\phi(\mu). \tag{3.64}$$

A superfluous factor of 8 has been dropped from both equations. Substituting in $\psi \equiv \psi_0$ and $\phi \equiv \phi_0$ identifies the $I$ and $J$ of Eq. (3.62).

We can see that the form (3.64) for $J$ follows trivially from the form (3.63) for $I$, as follows. Operating $d/dx$ on Eq. (3.61) shows that the same unitary (Fourier) transform that connects $\psi(x)$ and $\phi(\mu)$ also connects $\psi'(x)$ and $-i\mu\phi(\mu)/\hbar$. Then Eq. (3.64) follows from Eq. (3.63) as merely a statement (Eq. (4.5)) of the measure-preserving property of the unitary transformation. This is also called Parseval's theorem in the particular case at hand of a *Fourier* unitary transformation.

Notice that Eq. (3.62) is interpreted as showing equal informations $I$ and $J$ *in object space*. Hence, the transition $J \to I$ that characterizes EPI takes place, for the instrument, *right at its input port*. We will show, below, that the EPI axiom 1 of $\delta I = \delta J$ is also obeyed, again, at the input. The result is that EPI is enacted as a physical process at the input. Hence, it is the *input law* $\psi_0(x)$ that is fixed by EPI as the Schrödinger wave equation (shown in Appendix D), Klein–Gordon equation, or Dirac equation (shown in Chapter 4).

*What, then, is the need for the measurement in the first place?* (1) It is the measurement *interaction*, again at the input, that causes perturbation $\delta\psi(x)$, an activity that causes perturbations $\delta I$ and $\delta J$ and, hence, activates EPI in the first place. (2) With the input law $\psi_0(x)$ fixed by EPI, the output measurement law

$\psi_i(\eta)$ is then fixed by the convolution Eq. (3.51). In this way, *EPI generates both the input and the output amplitude laws of the instrument.*

From Fig. 3.3, some wave energy will miss the lens and, hence, not reach image space. This suggests a loss of information in transit from object space to image space. We next examine this possibility.

### 3.8.6 Output information

The output information is $I[\psi_i]$. This incorporates the possible degrading effects of instrumental noise, through function $\psi_i(\eta)$. According to Eq. (3.51), $\psi_i$ is known if the point amplitude response of the lens $w(x)$ is known. By definition of $w(x)$, it may be found by applying Eq. (3.55) and then Eq. (3.57) to an input impulse $u_\Delta(x) = \delta(x)$. This is easily done, giving

$$w(\eta) = 2x_0 \operatorname{sinc}\left(\frac{2\pi x_0}{\lambda R}\eta\right), \quad \operatorname{sinc}(u) \equiv \frac{\sin u}{u}. \tag{3.65}$$

Use of the last equation in Eq. (3.51) gives

$$\psi_i(\eta) = 2x_0 \int dx\, \psi_0(x) \operatorname{sinc}\left[\frac{2\pi x_0}{\lambda R}(\eta - x)\right]. \tag{3.66}$$

This is, again, a convolution. The output of a convolution tends to be smoother than either of its input functions. Hence, $\psi_i(\eta)$ tends to be a smoother function than $\psi_0(x)$.

The image information $I[\psi_i]$ may be found as follows. Taking the Fourier transform of Eq. (3.66) gives

$$\phi_i(\mu) = \phi_0(\mu)\, \operatorname{Rect}(\mu/\mu_0), \tag{3.67}$$

$$\text{where} \quad \operatorname{Rect}(x) \equiv 1 \text{ for } |x| \leqslant 1, \quad \text{or 0 otherwise.}$$

This shows that momentum values at the image suffer cutoff at a value $\mu_0$. By Eqs. (3.56) and (3.59), $\mu_0 = (h/\lambda)\sin\alpha$. The effect is due to the finite lens aperture size $2x_0$ or angle $\alpha$. It is interesting that $\mu_0$ is exactly the $X$-component of momentum of a "particle" of total momentum $h/\lambda$. This, of course, agrees with the discrete, or photon, aspect of light. In general, photons will radiate from object space in all directions. Thus, some will miss the lens. These constitute a loss of information. This loss effect is further examined next.

Substituting Eq. (3.67) into Eq. (3.64) gives

$$J[\phi_i] = \frac{1}{\hbar^2}\int d\mu\, \mu^2 \phi_i^*(\mu)\phi_i(\mu) = \frac{1}{\hbar^2}\int_{-\mu_0}^{\mu_0} d\mu\, \mu^2 \phi_0^*(\mu)\phi_0(\mu) \leqslant J[\phi_0] \tag{3.68}$$

because of the finite limits. Then, by correspondences (3.62),

$$I[\psi_i] \leqslant I[\psi_0]. \tag{3.69}$$

The image generally contains less Fisher information than does the object space.

This said, a special situation arises when the input particle has a maximum momentum value, call it $\mu_{\max}$. Then by correspondence (3.59) there is a corresponding pupil coordinate size

$$x'_{\max} = \frac{\lambda R \mu_{\max}}{h} \tag{3.70}$$

such that, for larger coordinates $x'$, the pupil amplitude $U(x') = 0$. By the particle picture (see above), there are no photons that radiate from object space in these directions. Hence, if the *physical* pupil size $x_0$ exceeds $x'_{\max}$, then $x_0$ may be effectively replaced by $\infty$ in Eq. (3.57). Rederiving Eq. (3.66) under these circumstances gives the equation as evaluated in the limit as $x_0 \to \infty$. By Eq. (0.68) the sinc function in (3.66) approaches a $\delta$ function, giving

$$\psi_i(\eta) = \psi_0(\eta). \tag{3.71}$$

Therefore, $I[\psi_i] = I[\psi_0]$. There is now no loss of information in the image space.

Hence, in this scenario of a finite band of momentum values and a large enough aperture size, the EPI transition $J \to I$ can be pictured as ending in the output data space. In this case, EPI directly derives $\psi_i(\eta)$ as well as $\psi_0(x)$ (since they are now the same function). Convolution Eq. (3.51) need not be used.

We return to the general scenario, where all momentum values may be present. It is shown, next, that EPI Eq. (3.16) explicitly holds for this scenario.

### 3.8.7 *Predicting EPI axiom 1 from unitarity*

Axiom 1 (Eq. (3.13)) states that if informations $I$ and $J$ are perturbed by amounts $\delta I$ and $\delta J$, then the two perturbations are equal. This assumes that $I$ and $J$ are at extremized values due to a solution $\mathbf{q}$ (the single wave function $\psi_0(x)$ here) of EPI, and that the perturbations $\delta I$ and $\delta J$ are caused by a perturbation $\delta\psi(x)$ of $\psi_0(x)$.

Hence, we must first find a mechanism for the perturbation $\delta\psi(x)$. As we found in Sec. 3.8.5, *the optical source wave provides the perturbation, by interacting with the particle.*

The resulting perturbations $\delta I$ and $\delta J$ are obviously related since $\psi_0(x)$ and $\phi_0(\mu)$ are Fourier transform mates. We next evaluate these changes, showing that they are equal.

The variations in question obey, by definition Eq. (0.41),

$$\delta I[\psi] \equiv \left.\frac{\partial I[\psi]}{\partial \varepsilon}\right|_{\varepsilon=0} d\varepsilon, \tag{3.72a}$$

$$\delta J[\phi] \equiv \left.\frac{\partial J[\phi]}{\partial \varepsilon}\right|_{\varepsilon=0} d\varepsilon. \tag{3.72b}$$

Our aim, then, is to show that

$$\left.\frac{\partial I[\psi]}{\partial \varepsilon}\right|_{\varepsilon=0} = \left.\frac{\partial J[\phi]}{\partial \varepsilon}\right|_{\varepsilon=0}. \tag{3.73}$$

The easier perturbation to compute is $\delta J$. From Eq. (3.64), this results from a perturbation of wave function $\phi_0(\mu)$. As in Eq. (0.3), let

$$\phi(\mu) = \phi_0 + \varepsilon\eta(\mu), \tag{3.74}$$

where $\eta(\mu)$ is any perturbing function of the EPI solution $\phi_0(\mu)$. Substituting Eq. (3.74) into Eq. (3.64) gives $J$ as a function of $\varepsilon$,

$$J(\varepsilon) = \frac{1}{\hbar^2}\int d\mu\, \mu^2[\phi_0^* + \varepsilon^*\eta^*][\phi_0 + \varepsilon\eta]. \tag{3.75}$$

Then directly

$$\frac{\partial J}{\partial \varepsilon} = \frac{1}{\hbar^2}\int d\mu\, \mu^2[\phi_0^* + \varepsilon^*\eta^*]\eta. \tag{3.76}$$

Hence

$$\left.\frac{\partial J}{\partial \varepsilon}\right|_{\varepsilon=0} = \frac{1}{\hbar^2}\int d\mu\, \mu^2\phi_0^*(\mu)\eta(\mu). \tag{3.77}$$

Multiplication by $\delta\varepsilon$ gives $\delta J$, as required.

We now proceed to $\delta I[\psi]$. The perturbation (3.74) in momentum space implies a perturbation in $x$ space obeying

$$\psi(x) = \psi_0(x) + \varepsilon\alpha(x), \tag{3.78}$$

where, by Eq. (D2),

$$\psi_0(x) = \frac{1}{\sqrt{2\pi\hbar}}\int d\mu\, \phi_0(\mu)\exp(-i\mu x/\hbar)$$

and (3.79)

$$\alpha(x) = \frac{1}{\sqrt{2\pi\hbar}}\int d\mu\, \eta(\mu)\exp(-i\mu x/\hbar).$$

(Note that $\varepsilon\alpha(x)$ is what we previously called $\delta\psi(x)$.) Use of Eq. (3.78) in Eq. (3.63) yields

$$I(\varepsilon) = \int dx\, [\psi_0'^* + \varepsilon^*\alpha'^*][\psi_0' + \varepsilon\alpha']. \tag{3.80}$$

Then, by direct differentiation,

$$\left.\frac{\partial I}{\partial \varepsilon}\right|_{\varepsilon=0} = \int dx\, \psi_0'^*\alpha'. \tag{3.81}$$

Differentiating Eqs. (3.79) gives

$$\psi_0'^*(x) = \frac{1}{\sqrt{2\pi\hbar}} \frac{i}{\hbar} \int d\mu\, \phi_0^*(\mu)\mu \exp(-i\mu x/\hbar)$$

and (3.82)

$$\alpha'(x) = \frac{1}{\sqrt{2\pi\hbar}} \frac{-i}{\hbar} \int d\mu'\, \eta(\mu')\mu' \exp(-i\mu' x/\hbar).$$

Substituting these equations into Eq. (3.81) and interchanging orders of integration gives

$$\left.\frac{\partial I}{\partial \varepsilon}\right|_{\varepsilon=0} = \frac{1}{2\pi\hbar} \frac{i}{\hbar} \frac{-i}{\hbar} \int d\mu\, \phi_0^*(\mu)\mu \int d\mu'\, \eta(\mu')\mu' \int dx \exp\left[\frac{ix}{\hbar}(\mu - \mu')\right]. \quad (3.83)$$

The far-right integral is $2\pi\hbar\delta(\mu - \mu')$ by Eqs. (0.68) and (0.71). Then, by use of the sifting property Eq. (0.63b), Eq. (3.83) implodes to

$$\left.\frac{\partial I}{\partial \varepsilon}\right|_{\varepsilon=0} = \frac{1}{\hbar^2} \int d\mu\, \phi_0^*(\mu)\eta(\mu)\mu^2. \quad (3.84)$$

This is identical to Eq. (3.77), as we set out to show.

This result depended upon the use of a particular unitary transformation, the Fourier, connecting coordinates $x$ and $\mu$. It can be generalized to cases where *any* unitary transformation connects coordinates $x$ and $\mu$ (Frieden, 2001, pp. 428–429).

Finally, by Eqs. (3.72a, b),

$$\delta I[\psi] = \delta J[\phi], \quad \text{or} \quad \delta\{I[\psi] - J[\phi]\} = 0. \quad (3.85)$$

The second Eq. (3.85) is an obvious factoring of the first. This is what we set out to prove.

However, we can proceed further, toward a goal of "deriving," in some sense, the extremum principle (3.16). By the second Eq. (3.85),

$$I[\psi(x)] - J[\phi(\mu)] = \textit{extrem}. \quad (3.86)$$

That is, if the first variation of something is zero, that something must have an extreme value. However, Eq. (3.86) is not quite the extremum principle (3.16) we need, since the term $I[\psi]$ is an integral over $x$-space of an unknown function $\psi \equiv \psi(x)$, while the term $J[\phi]$ is an integral over $\mu$-space of the unknown function $\phi \equiv \phi(\mu)$. The integrations are over *different spaces*. A *bona fide* variational principle such as Eq. (0.36) incorporates only integrals over *the same* space. For example, if $J[\phi] \equiv J[\phi(\mu)]$ could be re-expressed in $x$-space, as $J[\psi(x)]$, Eq. (3.86) would become the *usable* variational principle

$$I[\psi(x)] - J[\psi(x)] = \textit{extrem}. \quad (3.87)$$

for an amplitude law $\psi(x)$. To what extent can this be done?

Equations (3.85) and (3.86) are, at this point, purely mathematical identities.

They hold for all functions $\psi$, $\phi$ that are unitary (e.g., Fourier-) transform mates. However, they also constitute a prediction:

*A usable variational principle* (3.87) *results from* (3.86) *if the "coordinate"* $\mu$ *has a physical meaning that allows the functional* $J[\phi] \equiv J[\phi(\mu)]$ *to be re-expressed in x-space. This is as the equal functional* $J[\psi(x)]$.

For example, in the application to quantum mechanics in Appendix D, the "coordinate" $\mu$ is the momentum, and the functional $J[\phi(\mu)]$ is linear in the expectation of the squared momentum (see Eqs. (D3)–(D6)). The particular variational principle (D7) results.

Thus, we have not "derived" Eq. (3.87) in the sense of a purely mathematical derivation but, rather, have shown that it can arise as a *physical application* of the mathematical identity Eq. (3.86). The application is based upon physical knowledge that links the two spaces $x$ and $\mu$. Thus, Eq. (3.87) constitutes a *prediction* that EPI holds when such knowledge as available. The prediction is substantiated by its many uses in this book.

In summary, we have found that a physically meaningful variational principle (3.16) or (3.87) can be formed for an unknown physical effect, provided that the latter can be characterized by two conjugate unitary spaces whose coordinates have a *physical significance* that permits both functionals $I$ and $J$ to be expressed *in the same space*.

The preceding described an EPI solution called "type (b)" in Sec. 3.4.6. In cases where *no* such pair of conjugate spaces exists, an alternative route to the use of EPI can be taken – the "type (a)" or "self-consistent" approach discussed in Sec. 3.4.6. This route to EPI is likewise substantiated by its many uses in this book.

### 3.8.8 Summary

In general, *the EPI procedure is directly implied by the existence of a unitary transformation* between object space and a physically meaningful conjugate space (Secs. 3.8.5 and 3.8.7). In such a measurement scenario, EPI holds independently of the axiomatic approach and, hence, any assumption that the system PDF obeys the Fokker–Planck equation or (even) the $I$-theorem. This acts to widen the scope of application for the EPI principle, e.g., to deriving the wave equations for particles with spin (as in Chapter 4).

Both working equations of EPI, Eq. (3.16) and Eq. (3.18), were found to be physically obeyed during use of a measurement device. These followed from the existence of a Fourier (unitary) transformation between $X$-coordinate space and momentum space. This unitary transformation, in turn, followed from the internal dynamics of the measurement device. The measurement gave rise, as well, to the transition $J \to I$ and the perturbation $\delta\psi$ that EPI requires in order to implement the variational principle Eq. (3.16). Hence, for the measurement system at hand, all tenets of EPI were found to hold physically. An interesting result is that *the nature of the measurement instrument defined the physics of the object under measurement.*

Was this a coincidence? In fact, the optical measuring device is a prototype for a wide class of measuring instruments. Any particle location scheme requires a "probe" particle (or photon) for interaction with the measured particle. This will provide the required wave function perturbation for initiating EPI. Also, any linear measuring instrument has a Fourier transform plane, somewhere. (Even an "optical" system without a lens has a Fourier plane at infinity.) As we found, these were the requirements for EPI to be enacted.

As an example of the use of a real probe particle, consider the case of a probe electron. Under the assumption that the probe particle interacts with the measured particle via a *weak* potential energy function $V_{12}(x)$, the probability rate for scattering of the electron in a given direction is the *Fourier transform of* $V_{12}(x)$. This is by use of the Born approximation Eq. (4.30c) of Chapter 4 (see, e.g., Eisberg, 1961, p. 527). Here, once again, the measuring instrument operates through the use of a Fourier plane that is conjugate to $X$-coordinate space. Moreover, as in the lens model of preceding sections, the space of the output Fourier transform is a momentum space. Hence, it again corresponds to the coordinate space along line L$'$ of Fig. 3.3.

An EPI solution defines the wave function – here called $\psi_0(x)$ – for the object under measurement. This, in turn, forms the amplitude law $\psi_i(\eta)$ from which the observed data value $\overline{y}$ is sampled; an example is Eq. (3.66). The upshot is that *the measurement procedure elicits, or "creates" in some sense, the probability law (the physics) from which the measurement is sampled.*

It should be mentioned that the existence of a unitary transformation implies the validity of the EPI approach, but does not, in itself, guarantee that the approach can be *carried through.* As discussed above Eq. (3.87), the implementation of EPI requires its two functionals $I[\boldsymbol{\psi}(\mathbf{x})]$ and $J[\boldsymbol{\phi}(\boldsymbol{\mu})]$ to ultimately be expressed *in the same coordinate space* in order to define a usable Lagrangian. For example, in the application to quantum mechanics (Appendix D, Chapter 4), the functional $J[\boldsymbol{\phi}(\boldsymbol{\mu})]$ is essentially the mean-square momentum, and this can be re-expressed as the mean kinetic energy, a mean that can be taken *in X-coordinate space.* Since functional $I[\boldsymbol{\psi}(\mathbf{x})]$ is already expressed in $X$-coordinate space, the entire Lagrangian for the problem is now in $X$-space, allowing the EPI approach to be implemented. Conversely, if the functional $J[\boldsymbol{\phi}(\boldsymbol{\mu})]$ could not be re-expressed in $X$-space, EPI could not have been implemented.

In general, the re-expression of a $J[\boldsymbol{\phi}(\boldsymbol{\mu})]$ in $X$-space hinges on two effects: (i) that $I[\boldsymbol{\phi}(\boldsymbol{\mu})]$ is a statistical average, and statistical averages can be re-expressed in various spaces; and (ii) the quantity being averaged – $\mu^2$ – has an equivalent form in $X$-space, here, essentially the kinetic energy. Such a pair of effects also hold in the derivation of quantum gravitational effects in Chapter 11.

It is not obvious that the "equivalence" effect (ii) holds in general, i.e., for any unitary transformation of physical coordinates. Effect (ii) is not satisfied, e.g., by merely re-transforming $I[\boldsymbol{\phi}(\boldsymbol{\mu})]$ back to $I[\boldsymbol{\psi}(\mathbf{x})]$ via the known unitary transformation. In this case the two functionals $I$ and $J$ become identical and, so, information $K = 0$ identically (for *all* choices of amplitude functions $\mathbf{q}$). This is a mere tautology.

Instead, equivalence effect (ii) must be a distinct, *physical input* into the problem, as in Appendix D and Chapter 4 (Secs. 4.1.13–4.1.16).

However, in some measurement scenarios there is not an obvious unitary transformation connecting $X$-space with another space. Such cases occur in Chaps. 5–9. But even in such cases there is still an invariance principle of some kind that is obeyed by the measured particle. An example is continuity of flow (Chaps. 5 and 6). In this situation the EPI procedure may not be directly predicted, as in Sec. 3.8.7, but rather is taken to rest upon the axiomatic approach of Secs. 3.3.2 and 3.4.2.

As will be seen, the EPI process is shaped, or constrained, by the form of the particular invariance principle for the scenario. This may be shown explicitly for phenomena that obey the knowledge "game" of Fig. 3.2. All EPI solutions lie along the bottom row, and each phenomenon has a generally different solution point along that row. Each such solution point is defined by its invariance principle.

We cannot stress strongly enough the importance of the unitary transformation to applications of EPI. Experience shows that to each unitary transformation there is a corresponding EPI output solution $\psi_n$ that represents valid physics. This is an example of the exhaustive nature of EPI. It implies that new physics can be generated by the simple expedient of inventing new unitary transformations (again, agreeing with Wheeler's idea of the participatory universe).

## 3.9 EPI as a state of knowledge

According to EPI, there is a hierarchy of *physical knowledge* present. At *the top* are:

(A) the Fisher $I$-theorem (Sec. 1.8.2), which states that $I$, like entropy $H_{\mathrm{B}}$, is a physical entity that *monotonically* changes with time and, also, can be transferred, or can "flow," from one system to another (Sec. 1.8.10);

(B) the concept of a level $J$ of Fisher information that is intrinsic to, or 'bound' to, each phenomenon (Secs. 3.3.1 and 3.4.5); and

(C) the invariance, or symmetry, principle (Sec. 3.4.5) governing each phenomenon.

The laws (A)–(C), which we call the "top laws," exist prior to, or independently of, any explicit measurements. They can possibly be *verified* (or nullified) by measurement, but that's another matter.

At the second rung down the knowledge ladder are the *three axioms*:

(i) conservation of information perturbation, Eq. (3.13), during a measurement;
(ii) Eq. (3.19) defining information densities $i_n(\mathbf{x})$, $j_n(\mathbf{x})$ on the microlevel; and
(iii) Eq. (3.20) governing the efficiency of information transition, on the microlevel, from phenomenon to intrinsic data.

At the third rung down the ladder is the EPI principle. This follows (as we found) either from the axioms or from the existence of a physically meaningful unitary transformation space.

Finally, at the fourth rung down the ladder, is the carrying through of EPI as a *calculation*. This requires the EPI principle, as augmented by top law (C). The output of the calculation is the law governing formation of the amplitudes $\mathbf{q}$ for that scenario. For example, in Chapter 4 it is the Klein–Gordon "law" governing formation of the amplitude $\psi$.

The question of what should be regarded as the laws of physics is of interest. Should they, e.g., be the "top" laws (A)–(C) mentioned above, or, as is conventionally assumed, the output laws, such as the Klein–Gordon equation? We can expect, and the chapters ahead will verify, that some invariance principles (C) do double (or more) duty in implying physical laws. For example, the continuity of flow condition is used by EPI to derive both Maxwell's equations (Chapter 5) and the Einstein field equations (Chapter 6). Therefore, there are more physical laws than there are invariance conditions (C) for their derivation. Clearly it is desirable to have to make the fewest assumptions about nature. On this basis, the EPI output laws can be regarded as subsidiary to the top laws. They are also subsidiary in being subject to a contingency situation – measurement – for their existence, as is clarified next.

## 3.10 EPI as a physical process

It is alternatively possible to interpret EPI as a *physical process*. By this view, EPI is regarded as a physical sequence of real events, numbered (1)–(10) in Sec. 10.10. These comprise a definite causal progression over time.

A somewhat enigmatic effect results, one that might be called "reality on demand." Or, "measurement creates the thing it measures" (Caianiello, 1992). Let us carefully examine this Escher-like possibility.

(1) The EPI process cannot *form* or *create*, in some way, the physical effect that is the subject of its measurement. Every physical effect is presumed to exist "out there" in some fixed and unknown form (Sec. 0.1).
(2) The output of the EPI process (step (8) of Sec. 10.10) is the very probability law that governs formation of the sought-after data. In that EPI correctly finds the probability law, it appears to create its own *local reality*, thus "reality on demand" (see above). However, this "reality" is not the physical effect (see point (1)); it is a sensory-based manifestation *of* the effect. Also, what we experience as reality are the *data*, not the probability law. Although EPI fixes the latter, it does not fix the former since they are random in nature and therefore beyond our control. Actually, the probability law exerts random control over a meter that provides the required measurement. Of course, the measurement is not predictable since it is a *random reading* from the meter.
(3) What the process of measurement does induce deterministically is a flow of Fisher information that we describe as EPI. EPI amounts to a set of "rules of the game" (Secs. 3.4.11–3.4.13). Carrying through the measurement activates this game. The game, in turn, defines local physics, and the output of the physics is the probability law governing the data. Thus, the measurement uniquely creates the rules of a game whose final outputs are the random data that we call local reality. *Establishing the rules of the game is the closest we come to creating local reality.*

The latter statement is somewhat modified in cases where the law of large numbers holds. Then each observable is a sum of a colossal number of samples from the EPI-output probability law. An example is where the observed data are the total intensities due to many detected photons at each pixel in a bright spatial image. By the law of large numbers, as the scene becomes increasingly bright the signal/noise ratio at each pixel approaches infinity, and the total intensity at each pixel approaches direct proportionality to the theoretical image. The theoretical image is the EPI output probability law. Thus, in this case *the data do approach the true probability law*, so that the measurer approaches control of his local reality. However, total control is still impossible since it would require infinitely "strong" (many) data. Also, the physical effect itself would still not be *created* by the measurement process (point (1) preceding). Rather, its sensory manifestation would be.

Almost everything we sense with unaided detection is a "signal image" of this summation type. In fact, such data serve a useful psychological purpose. In most instances signals are strong, giving an observer the comforting impression that things are deterministic, i.e., "controllable."

However, as we saw, this is an illusion. Also, this illusion has the down side that it makes it hard for us to believe the main premise of quantum mechanics, that observables are intrinsically random, not controllable. Finally, we also would lose something that most people highly value – free will – since without randomness everything is predictable.

Making a measurement is a quantitative way of asking a question – "What is the *size* of this effect?" The idea of measurement begetting phenomenon seems to be the physical counterpart to the adage that a well-posed mathematical problem – again, a question – contains the seeds of its solution. It is interesting to consider whether asking a *qualitative* question, as well, leads in some sense to a physical phenomenon. This is in fact partially addressed in Eqs. (10.39), (10.40) *et vecin.* in Chapter 10.

## 3.11 On applications of EPI

In each of the following chapters, the EPI principle is applied to a different measurement scenario. Each such scenario leads to the derivation of a different physical law. The ordering of the chapters is, in the main, arbitrary so that they may be read in any order. However, the chapters are grouped as to similarity of approach or of application.

The flow of operations in each chapter's derivation follows those in Fig. 3.4. A parameter $\boldsymbol{\theta}$ is chosen to be measured. The measurement is to be carried through with an instrument that has a given "instrument function" (Sec. 3.8, Chapter 10). The measurement is initiated. The measurement process interferes, and interacts, with the phenomenon governing the parameter. This results in the perturbation of all the probability amplitudes $\mathbf{q}$ describing the phenomenon in the input (object) space to the instrument.

The phenomenon is identified by a suitable invariance principle. The principle can, by Wheeler's proposal of Sec. 0.1, be identified by the *internal processes of the measuring instrument.* An example was the unitary transformation suggested by the optical device in Sec. 3.8. An alternative to a unitary transformation is a property of continuity of flow for *the sources*. This could likewise be implied by the operation of a measuring device that *obeys* continuity of flow. The invariance principle is *the only physical input* to the procedure and, ultimately, allows the bound information $J$ to be solved for.

The continuity of flow and unitary transformation principles are, respectively, designated as types (a) and (b) in Sec. 3.4.6. Type (a) principles give rise to a unique EPI output law, whereas type (b) principles give rise to two distinct EPI laws in the form of wave equations. It is interesting that type (b) scenarios

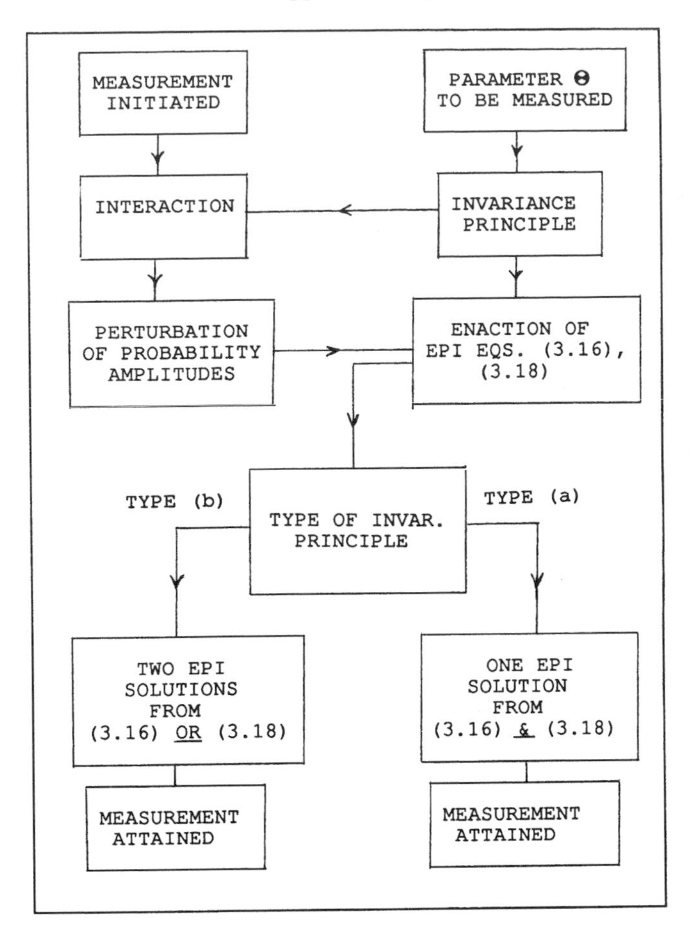


Fig. 3.4. EPI flow of operations.

occur only for quantum phenomena (Chaps. 4, 10, 11 and Appendix D). All other phenomena that are derived in this book are of type (a).

The perturbed probability amplitudes $\mathbf{q}$ perturb, in turn, the channel capacity $I$ (Eq. (2.19)) and information $J$ (Eq. (3.13)). This activates the steps (3.13)–(3.20) defining the EPI process.

The EPI solutions define the phenomenon in the *input space* to the measuring instrument. Solutions at the output, or measurement, space must be obtained by other means. As examples: the output solution is obtained by a simple convolution of the EPI solution with the instrument function (Eqs. (3.51) and (10.26b)).

We begin in Chapter 4 with the application to (relativistic) quantum mechanics, since it is probably of major interest to more readers than any other application. Followups to this topic are Appendix D, where the Schrödinger wave equation is derived, and Chapter 10, where the wave equation *at* each measurement is found (the counterpart to Feynman–Mensky measurement theory). Also, there is a very close resemblance between the development in Chapter 4 and that in Chapter 11 on quantum gravity. For example, each utilizes a unitary invariance principle (type (b)) and, hence, should give two distinct output law solutions $\mathbf{q}$. This correspondence and others are pointed out in Chapter 11.

Strong similarities also exist between the development in Chapter 5 of classical electromagnetic theory and that in Chapter 6 of the Einstein field equations of classical gravity theory. This is why the chapters follow one another. The correspondences are pointed out in Chapter 6.

Since EPI Eq. (3.16) is a variational principle, the general solution to an EPI problem is an Euler–Lagrange Eq. (0.34). This represents a second-order differential equation, whose solutions can generally be divided into two classes: *intrinsic* wave phenomena, for which the solution is a true wave equation, and *non-wave* phenomena. As will be seen, the latter is specified by a differential equation that can be immediately solved in final form for the distribution function.

For example, in Chapter 7 it is found that the EPI differential equation for classical particle velocity is soluble, in the form of the Maxwell–Boltzmann velocity distribution. Or, in Chapter 8 the differential equation for the power spectrum of a certain type of noise is immediately soluble as the ubiquitous $1/f$ power noise phenomenon. Likewise, in Chapter 9 a differential equation that is found for the probability amplitude describing the universal physical constants is immediately solved, giving a $1/x$ law for the PDF.

The fact that many of the EPI differential equations are immediately soluble in this way extends the applicability of EPI to a wider scope of phenomena than simply wave phenomena. We believe it to be applicable to *all* observable phenomena and therefore all of science.

# 4

# Derivation of relativistic quantum mechanics

## 4.1 Derivation of Klein–Gordon equation

### *4.1.1 Goals and prior knowledge*

The main wave equations of relativistic quantum mechanics are the *Klein–Gordon, Dirac, Weyl*, and *Rarita–Schwinger* equations. The principal aim of this chapter is to derive these, by means of the EPI principles (3.16) and (3.18). The derivations generally follow Frieden (1995), or Frieden and Soffer (1995), with numerous additions.

In addition, the Schrödinger wave equation (SWE), which is *non*-relativistic, is also to be derived. Since rigorous use of EPI gives only relativistic wave equations (Chapter 3), EPI can only derive the SWE in some approximate sense. This is directly carried through in Appendix D, or by taking the well-known non-relativistic limit of the Klein–Gordon equation in Appendix G.

Also, in Appendix D a simple EPI derivation of *classical mechanics* is obtained. In particular, in the limit $h \rightarrow 0$ the EPI extremum condition (3.16) is found to give *Newton's second law*, and the zero-condition (3.18) gives the *virial theorem*, of classical mechanics.

Other derived results are Eq. (4.17) – the equivalence of mass, momentum, and energy; the constancy of the Planck constant $h$ (Sec. 4.1.14); the Compton wavelength as an ultimate resolution length (Sec. 4.1.17); and the Heisenberg uncertainty principle (Sec. 4.3).

It is important to state what is assumed *a priori*. The relativistic properties of space and time coordinates were derived in Sec. 3.5. Hence, these are assumed here. Also, appropriate *units* for the concepts of mass and energy are assumed to be known, although the basic relation Eq. (4.17) between them *is not* presumed. In fact it must be derived, by the *desideratum* of Sec. 3.5.5 of Chapter 3. The concept of *mass* will be quantified in Sec. 4.1.15.

In all EPI derivations appropriate source functions are assumed to exist (Caveat 2 of Sec. 0.1). Here they are the vector and scalar electromagnetic potential functions $\boldsymbol{A}$ and $\phi$, respectively.

To start the EPI process, we need to first identify the four-parameter $\boldsymbol{\theta}$ (Sec. 3.5) that is to be measured. In this chapter this is the space-time coordinates of a known type of particle (Sec. 3.4.14). Space and time coordinates are valid Fisher coordinates, since they are *a priori* independent (Sec. 3.5.5). The relativistic wave equations will be shown to follow as reactions to the measurement.

### 4.1.2 Choice of measured parameters

Let the observer measure, once, the space-time coordinates of a particle of mass $m$. The particle has a known rank $N$, i.e., a known number of amplitude components $q_1, \ldots, q_N$. Since $N$ relates to spin (see below), the particle is known to be of a specific type (whether electron, photon, meson, etc.). Thus the particle can be a fermion or a boson (see below). The single, real measurement perturbs (Sec. 3.8, Chapter 10) the $N$ amplitude functions $q_n(\mathbf{x})$. Coordinates $\mathbf{x}$ are defined in terms of space-time coordinates $x, y, z, t$ as

$$\begin{gathered} x_1 = ix, \quad x_2 = iy, \quad x_3 = iz, \quad x_4 = ct \\ \boldsymbol{r} = (x, y, z), \quad \mathbf{x} = (x_1, x_2, x_3, x_4). \end{gathered} \tag{4.1}$$

This choice invites explanation. In the case of mixed real and imaginary coordinates, $I$ can be positive *or negative* (Appendix C). As it turns out, in this application if we make space coordinates real and the time coordinate imaginary then $I$ becomes negative at the solution. Although this would permit the required wave equations to be derived by EPI, some awkwardness would arise in Sec. 4.1.17, where a negative resolution length would result. Conversely, we found that, if we instead make the space coordinates purely imaginary and the time coordinate real, then $I$ becomes positive. This is the sole reason for the coordinate choice (4.1). This choice is, in fact, the usual one (Jackson, 1975), and is consistent with so-called Bjorken–Drell notation (Bhaduri, 1992).

### 4.1.3 Form of I

According to the plan of Sec. 3.6.2, replacements (3.37) must be imposed upon Eq. (2.19) for $I$ in order for the latter to be expressed gauge covariantly. But, as also mentioned in Sec. 3.6.2, we can defer doing this until after $J[\mathbf{q}]$ has been found. Hence, at this point in the derivation we stick with Eq. (2.19) as it stands, i.e., a field-free form.

### 4.1.4 Construction of complex amplitude functions

The use of complex amplitudes $\psi_n$ in quantum mechanics is more a convenience than a necessity. One can instead work with the purely real amplitude functions that are the real and imaginary parts of the $\psi_n$. These are, in fact, the $\mathbf{q}$ obeying Eq. (2.24). In being the components of the $\psi_n$ they are, in a sense, more fundamental than the $\psi_n$. However, for purposes of comparison with standard results, and because they are indeed more convenient to use, we now regard the $\psi_n$ as the new unknowns of the problem.

What are our basic quantities $p$, $I$, etc., in terms of these new complex amplitudes? We have already found that probability $p(\mathbf{x})$ has the usual squared-modulus form (2.25) in terms of the $\psi_n$. We proceed now to $I$.

### 4.1.5 $I$ in terms of the complex amplitudes $\psi_n$

We established above that we may defer the field replacements (3.37) in definition (2.19) of $I$ until after the Euler–Lagrange solution to EPI has found. In essense, we first use a field-free scenario. The choice of coordinates (4.1) gives an information capacity Eq. (3.39) of

$$I = 4Nc \sum_{n=1}^{N/2} \iint d\mathbf{r}\, dt \left[ -(\nabla\psi_n)^* \cdot \nabla\psi_n + \left(\frac{1}{c^2}\right)\left(\frac{\partial\psi_n}{\partial t}\right)^* \frac{\partial\psi_n}{\partial t} \right]. \tag{4.2}$$

(See the related Ex. 3.7.2.)

### 4.1.6 Finding $J = J[\psi]$ by an invariance principle

By the general approach of Sec. 3.4.5, we need to find an invariance principle involving the $\psi_n$ or $I$. We have already found this, in one-dimensional form, for the position location problem of Sec. 3.8. This is the invariance of information $I$ under a Fourier transformation from the space of coordinate $x$ to that of a conjugate coordinate $\mu$. As we found in Sec. 3.8, *the existence of a physically meaningful coordinate* $\mu$ *predicts the validity of the EPI approach.* We now suitably generalize the one-dimensional transformation to our four-dimensional problem.

### 4.1.7 Definition of Fourier coordinate space

Define new coordinates that are the Fourier conjugates to position-time,

$$\begin{array}{ccc} \text{old} & & \text{new} \\ (i\boldsymbol{r},\ ct) & \overset{\text{F.T.}}{\leftrightarrow} & (i\boldsymbol{\mu}/\hbar,\ E/c\hbar). \end{array} \tag{4.3}$$

Denote the amplitude function $\psi_n(\boldsymbol{r},\ ct)$ as $\psi_n(\boldsymbol{r},\ t)$. It may be represented as

$$\psi_n(\boldsymbol{r},\ t) = \frac{1}{(2\pi\hbar)^2} \iint d\boldsymbol{\mu}\, dE\, \phi_n(\boldsymbol{\mu},\ E) e^{i(\boldsymbol{\mu}\cdot\boldsymbol{r} - Et)/\hbar} \tag{4.4}$$

(Roman, 1961, p. 103). Thus, $\psi_n$ and the new functions $\phi_n$ are Fourier transform (F.T.) mates. (Note that the *subscripted* functions $\phi_n$ are distinct from the scalar potential $\phi$ defined in Eqs. (3.37).) The F.T. operation is unitary, obeying the $L^2$ measure-preserving requirement

$$\iint d\boldsymbol{r}\, dt\, \psi_m^* \psi_n = \iint d\boldsymbol{\mu}\, dE\, \phi_m^* \phi_n, \qquad \text{all } m,\ n. \tag{4.5}$$

*Exercise*: Show that Eq. (4.5) follows from direct substitution of Eq. (4.4) into its left-hand integral. *Hint:* Switch orders of integration so as to use the Fourier integral representation Eq. (0.70) of a Dirac delta function; then use the sifting property Eq. (0.69) of the delta function.

Note that the argument of $\psi_n$ in Eq. (4.4) is not $(i\boldsymbol{r},\ t)$ as might be expected from the left-hand side of correspondence (4.3). By convention, probabilities are of *real* events only. Hence, the probability of a complex quantity is, by definition, the *joint* probability of two real events – the real and imaginary parts of the complex quantity (also see Appendix C). Then, since probabilities are generated by probability amplitudes such as the $\psi_n$, probability amplitudes likewise follow this rule.

In Eq. (4.4) the coordinates $\boldsymbol{\mu}$, $E$ are taken to represent momentum and energy values, respectively, for the measured particle, where the *units* of momentum and energy are known: mass–length/time and mass–length$^2$/time$^2$, respectively (Sec. 3.7.3). However, the relation (4.17) *connecting* $\boldsymbol{\mu}$ and $E$ is *not* presumed to be known. In fact the *desideratum* of Sec. 3.5.5 requires it to be *derived* by the EPI approach. With these identifications, Eq. (4.4) becomes of course a continuous representation of the de Broglie wave expansion for $\psi_n(\boldsymbol{r},\ t)$. This wave expansion by itself, i.e., without the relation (4.17), does not give the Klein–Gordon equation; hence the need for its derivation by EPI or some other approach.

### *4.1.8 Nature of Planck's parameter*

Equation (4.4) introduces a new parameter $\hbar$. At this point $\hbar$ is merely a parameter that is constant for a particular problem with its particular boundary conditions. It is not the same constant over *all* boundary conditions. That is, it is not necessarily a *universal* constant. EPI theory will later fix it as a universal constant (also see Sec. 3.4.14).

### *4.1.9 Exercise*

Show that, if Eq. (4.17) is a given, then the Klein–Gordon Eq. (4.26) simply follows by multiple differentations of Eq. (4.4). This would avoid the need for using the EPI extremum condition as the route to the Klein–Gordon equation. However, this is a moot point. As we discussed in Sec. 4.1.7, Eq. (4.17) is *not* a given: in fact, it is required to be derived by the EPI approach, and *is* so in Sec. 4.1.15.

### *4.1.10 Correspondence between derivatives and product functions*

By differentiating Eq. (4.4) we find the correspondences

$$(\nabla\psi_n,\ \partial\psi_n/\partial t) \overset{\text{F.T.}}{\leftrightarrow} (i\boldsymbol{\mu}\phi_n/\hbar,\ -iE\phi_n/\hbar). \tag{4.6}$$

That is, the left-hand derivative functions in $\mathbf{x}$-space are Fourier transforms of the corresponding product functions in conjugate space.

### *4.1.11 Use of Parseval's theorem*

Since $\nabla\psi_n$ and $i\boldsymbol{\mu}\phi_n/\hbar$ are seen to be Fourier mates, by Parseval's theorem their squared areas are equal,

$$\iint d\mathbf{r}\,dt\,(\nabla\psi_n)^* \cdot \nabla\psi_n = (1/\hbar^2)\iint d\boldsymbol{\mu}\,dE\,|\phi_n(\boldsymbol{\mu},\,E)|^2\mu^2, \tag{4.7}$$

and

$$\iint d\mathbf{r}\,dt \left(\frac{\partial\psi_n}{\partial t}\right)^* \frac{\partial\psi_n}{\partial t} = \frac{1}{\hbar^2}\iint d\boldsymbol{\mu}\,dE\,|\phi_n(\boldsymbol{\mu},\,E)|^2E^2. \tag{4.8}$$

Using these two relations in Eq. (4.2) gives

$$I = \left(\frac{4Nc}{\hbar^2}\right)\sum_{n=1}^{N/2}\iint d\boldsymbol{\mu}\,dE\,|\phi_n(\boldsymbol{\mu},\,E)|^2\left(-\mu^2 + \frac{E^2}{c^2}\right) \equiv J. \tag{4.9}$$

This is the invariance principle for the given scenario. Thus, the value of $I$ is re-expressed in the new space ($\boldsymbol{\mu}$, $E$), where it is called $J$ (see Sec. 3.4.5). $J$ is the bound information for the scenario. Note also that $\kappa = 1$ implicitly (see Sec. 3.4.2), and that we are seeking a solution of type (b) (Sec. 3.4.6).

### 4.1.12 Physical properties of $\phi_n$

Taking the special case $m = n$ in the unitarity Eq. (4.5) gives a Parseval's theorem

$$c \iint d\boldsymbol{r}\, dt\, |\psi_n|^2 = c \iint d\boldsymbol{\mu}\, dE\, |\phi_n|^2, \qquad n = 1, \ldots, N/2. \tag{4.10}$$

The left-hand integral obeys a property of normalization (see Appendix I) for each value of index $n$. Summing both sides over $n$, and usng the normalization and correspondence (2.25), gives

$$1 = \iint d\boldsymbol{\mu}\, dE\, P(\boldsymbol{\mu}, E), \qquad P(\boldsymbol{\mu}, E) \equiv c \sum_{n=1}^{N/2} |\phi_n(\boldsymbol{\mu}, E)|^2. \tag{4.11}$$

Thus, the new quantity $P$ obeys $P \geqslant 0$ and normalization. This implies that $P$ is a PDF in ($\boldsymbol{\mu}$, $E$) space.

### 4.1.13 The bound information $J$

Following the lead of Sec. 4.1.6, Eqs. (4.9) and (4.11) show that

$$I \equiv J = \frac{4N}{\hbar^2} \iint d\boldsymbol{\mu}\, dE\, P(\boldsymbol{\mu}, E)\left(-\mu^2 + \frac{E^2}{c^2}\right). \tag{4.12}$$

This is now an expectation

$$J = \left(\frac{4N}{\hbar^2}\right)\left\langle -\mu^2 + \frac{E^2}{c^2}\right\rangle. \tag{4.13}$$

Hence $J$ is an information that is expressed in terms of physical attributes of the problem (energy, momentum, etc.). Also, the equality (4.12) of $I$ and $J$ means that the constant $\kappa$ in the condition Eq. (3.18) of EPI has the value unity in this scenario.

Result (4.13) has some important implications, discussed next.

### 4.1.14 Planck's parameter as a constant

By Sec. 3.4.14, in this free-field scenario $J$ is regarded as a universal constant. Since the two factors in (4.13) are independent, each must be a constant. Then parameter $\hbar$ must be a universal constant.

### 4.1.15 Equivalence of matter and energy

Next, consider the second factor in Eq. (4.13). The character of the statistical fluctuations of $E$, $\boldsymbol{\mu}$ necessarily changes from one set of boundary conditions to another. This would make $J$ a variable, contrary to our aims, unless

$$-\mu^2 + \frac{E^2}{c^2} = const. \equiv f^2(m, c) \tag{4.14}$$

where $f$ is some function of the rest mass $m$ and the speed of light $c$ (the only other parameters of the free-field scenario). Solving for $E$ gives

$$E^2 = c^2\mu^2 + f^2(m, c)c^2. \tag{4.15}$$

Since the momentum $\mu$ has units of mass–length/time (Sec. 4.1.7), the function $f(m, c)$ in Eq. (4.15) must have units of mass–length/time. Then, since $c$ has units of length/time (Sec. 3.5.5), the function $f(m, c)$ is proportional to product $mc$. A proportionality constant of unity is taken *to quantify* the unit of mass $m$,

$$f \equiv mc. \tag{4.16}$$

The use of this relation in Eq. (4.15) gives the famous equivalence relation

$$E^2 = c^2\mu^2 + m^2c^4 \tag{4.17}$$

linking mass, momentum, and energy.

### 4.1.16 Rest mass m as universal constant

Plugging Eq. (4.17) into Eq. (4.13) gives directly

$$I = 4N\left(\frac{mc}{\hbar}\right)^2 \equiv J. \tag{4.18}$$

Hence, the intrinsic information $I$ in the four-position of a particle is proportional to the square of its intrinsic energy $mc^2$.

Since $J$ is to be a universal constant, $c$ is fixed as a constant in Sec. 3.5.5, and $\hbar$ has already been so fixed, we conclude from (4.18) that the rest mass $m$ of the particle must be a universal constant. For example, if the particle is an electron, this condition fixes its rest mass as a universal constant.

### *4.1.17 Compton wavelength as an ultimate resolution length*

By Sec. 2.3.2, $I$ actually measures the *capacity* of the observed phenomenon to provide information about (in this case) four-length. Then intuitively $I$ should fix the ultimate fluctuation (resolution) length that is intrinsic to quantum mechanics; see also Sec. 3.4.15. Information (4.18) is identically

$$I = \frac{4N}{L^2}, \quad L \equiv \frac{\hbar}{mc}. \tag{4.19}$$

$L$ is the "reduced" (divided by $2\pi$) Compton wavelength of the particle. By Eq. (2.9) and the efficiency of the intrinsic scenario,

$$I = 4N/e^2_{\text{min}}. \tag{4.20}$$

We also made the reasonable assumption that each of the $N$ estimates has the same accuracy. Combining Eqs. (4.19) and (4.20) gives

$$e_{\text{min}} = L, \tag{4.21}$$

the reduced Compton length. This relates to a *resolution length* as follows.

Equation (4.4) shows that, if the particle has a unique (fixed) momentum value $\boldsymbol{\mu} = m\mathbf{v}$, where $\mathbf{v}$ is the particle velocity, then $\psi_n(\boldsymbol{r}, t)$ is periodic, obeying $\psi_n(\boldsymbol{r} + \lambda_{\text{DB}}, t) = \psi_n(\boldsymbol{r}, t)$, $\lambda_{\text{DB}} = 2\pi\hbar/\boldsymbol{\mu} = h/m\mathbf{v}$. Wavelength $\lambda_{\text{DB}}$ is called the de Broglie wavelength and is, by its definition, a measure of the uncertainty in position for the particle. Comparison with $L$ given in (4.19) shows that, since $v \leqslant c$, the Compton length $2\pi L$ is the *smallest* de Broglie wavelength that is possible for a particle, regardless of speed. This is, then, the ultimate uncertainty in position for a particle. Then, by Eq. (4.21), the error $e_{\text{min}}$ is the ultimate uncertainty divided by $2\pi$, defining an ultimate resolution length. We recall, from Chaps. 1 and 2, that this result allows for the possibility of optimal data processing. Thus, *it represents one statement of the ultimate ability "to know."* Another statement of the limiting ability to know is the Heisenberg uncertainty principle; this is derived in Sec. 4.3.

### *4.1.18 Physical information*

We can now proceed to form the physical information for the problem. Placing Eq. (4.17) in Eq. (4.12), using Eq. (4.11) and then Eq. (4.10), gives

$$\begin{aligned} J &= (4Nm^2c^3/\hbar^2)\iint d\boldsymbol{\mu}\, dE \sum_{n=1}^{N/2} \phi_n^* \phi_n \\ &= (4Nm^2c^3/\hbar^2)\iint d\boldsymbol{r}\, dt \sum_{n=1}^{N/2} \psi_n^* \psi_n. \end{aligned} \tag{4.22}$$

The latter equality is again by use of Parseval's theorem. Then, by Eqs. (4.2) and (4.22), the physical information (3.16) is

$$K \equiv I - J = 4Nc \sum_{n=1}^{N/2} \iint d\mathbf{r}\, dt$$
$$\times \left[ -(\nabla\psi_n)^* \cdot \nabla\psi_n + \left(\frac{1}{c^2}\right)\left(\frac{\partial\psi_n}{\partial t}\right)^* \left(\frac{\partial\psi_n}{dt}\right) - \frac{m^2c^2}{\hbar^2}\psi_n^*\psi_n \right]. \quad (4.23)$$

### *4.1.19 Non-unique nature of the solutions*

Recall that the EPI principle consists of two conditions: the extremum condition Eq. (3.16) and the zero-condition Eq. (3.18). Since our invariance principle (4.9) is directly of the form Eq. (3.18) the latter *by itself* gives a distinct solution $\mathbf{q}$ (see Sec. 3.4.6). Note that this $\mathbf{q}$ does not necessarily satisfy the extremum requirement (3.16) as well. Hence, in the following, we seek two generally different solutions $\mathbf{q}$ to the problem: those that satisfy the extremum condition Eq. (3.16) and those that satisfy the zero-condition Eq. (3.18). As we will see, these correspond to the classes of particles called bosons and fermions, respectively.

### *4.1.20 Vector Klein–Gordon equation without fields*

By EPI extremum condition (3.16), information $K$ is to be extremized. This is through variation of the $\boldsymbol{\psi}$, since the latter have replaced the purely real amplitudes $\mathbf{q}$ via Eq. (2.24). The variational solution is found in the usual way. The integrand of Eq. (4.23) is used as the Lagrangian $\mathscr{L}$ in the Euler–Lagrange Eq. (0.34), which is here

$$\frac{d}{dx}\left(\frac{\partial\mathscr{L}}{\partial\psi_{nx}^*}\right) + \frac{d}{dy}\left(\frac{\partial\mathscr{L}}{\partial\psi_{ny}^*}\right) + \frac{d}{dz}\left(\frac{\partial\mathscr{L}}{\partial\psi_{nz}^*}\right) + \frac{d}{dt}\left(\frac{\partial\mathscr{L}}{\partial\psi_{nt}^*}\right) = \frac{\partial\mathscr{L}}{\partial\psi_n^*}, \quad (4.24)$$

where

$$\psi_{nx}^* \equiv \partial\psi_n^*/\partial x, \quad (4.25)$$

etc., for $y$, $z$, and $t$. After multiplying through by $c^2\hbar^2$, the result is

$$-c^2\hbar^2\nabla^2\psi_n + \hbar^2\frac{\partial^2\psi_n}{\partial t^2} + m^2c^4\psi_n = 0, \quad n = 1, \ldots, N/2. \quad (4.26)$$

This is the free-field Klein–Gordon equation (Schiff, 1955) for a vector particle of rank $N$. EPI gives this as the physical law that the amplitudes $\boldsymbol{\psi}$ will obey.

### *4.1.21 Exercise*

The derivation of this answer as described above is a straightforward exercise in use of the Euler–Lagrange approach. The motivated reader should try it out.

### *4.1.22 Significance of Klein–Gordon solutions for different values of N*

Recall that the EPI measurement is of a *known* vector particle of mass $m$ (Sec. 4.1.2). Thus the amplitude function $\psi_n(\boldsymbol{r}, t)$, $n = 1, \ldots, N/2$ described by Eq. (4.26) has a fixed, *known* rank $N/2$. The rank is at this point arbitrary. For example, if the particle is a $\pi$ meson the value of $N = 2$ (see below). With $N$ *known* in this application, the *game corollary* (Sec. 3.4.14) is not used to *fix* the unitless parameter $N$. However, it is used to predict *preferred* values of $N$, in the sense of particle abundances, in Secs. 4.2.9 and 4.2.10 below.

A general EPI output, such as the result (4.26), describes the amplitude function $\psi_n(\boldsymbol{r}, t)$ of the measured particle. Thus, Eq. (4.26) constitutes a prediction by EPI that *vector particles, of arbitrary rank N (or spin), obey a Klein–Gordon equation*. The latter is known to be true (Roman, 1961, p. 157), provided that an appropriate subsidiary condition (Roman, 1961, p. 105) is enforced upon the $\psi_n(\boldsymbol{r}, t)$. The purpose of the subsidiary condition is to rule out solutions of lower rank than $N$. Consider the case $N = 8$, where the vector amplitude function is $\psi_1, \psi_2, \psi_3, \psi_4$. Since the subsidiary condition must of course be Lorentz-covariant, it turns out to be *the Lorentz condition*,

$$\frac{\partial \psi_1}{\partial x} + \frac{\partial \psi_2}{\partial y} + \frac{\partial \psi_3}{\partial z} - \frac{1}{c}\frac{\partial \psi_4}{\partial t} = 0. \tag{4.26a}$$

Particles are categorized into two families, according to their spin values: "bosons" have spins that are generally integer values, and "fermions" have spins that are odd multiples of $1/2$. Thus, EPI has correctly predicted that both bosons *and* fermions obey the Klein–Gordon equation (4.26), in this free-field case. This general property of particles was predicted on more general grounds in Sec. 3.4.2 as well. It will be found by further use of EPI, in Sec. 4.2, that fermions obey as well *another* wave equation, Eq. (4.41). Thus, fermions obey two wave equations – (4.26) and (4.41) – whereas bosons obey one, Eq. (4.26). Some interesting special cases of (4.26) follow.

In the case $N = 2$ of a single complex amplitude function $\psi_1 \equiv \psi$, and with the mass $m$ finite, the measured particle is a boson of spin 0 called the $\pi$ meson (Eisberg, 1961). Thus, the $\pi$ meson has been found to obey Eq. (4.26). The choice $N = 8$ (mentioned previously) defines bosons of spin 1. Here there are $N/2 = 4$ complex amplitude functions $\psi_n$. Thus, Eq. (4.26) describes bosons of spin 1 (Roman, 1961, p. 105) as well. If additionally the rest mass $m = 0$ the spin 1 boson is the photon (Eisberg, 1961), and the Klein–Gordon equation (4.26) becomes the wave equation for photons. The latter becomes the Helmholtz wave equation of optics in the case of a harmonic time dependence (Morse and Feshbach, 1953).

We paid particular attention to vector bosons in the preceding. Vector *fermions* are discussed in Sec. 4.2.11.

### 4.1.23 Vector Klein–Gordon equation with fields

Replacements (3.37) have been shown to make the theory gauge-covariant. Making these replacements in Eqs. (4.23) and (4.26) gives a field-dependent information

$$K = 4Nc \iint d\mathbf{r}\, dt \sum_{n=1}^{N/2} \left[ -\left(\nabla + \frac{ie\mathbf{A}}{c\hbar}\right)\psi_n^* \cdot \left(\nabla - \frac{ie\mathbf{A}}{c\hbar}\right)\psi_n \right.$$
$$\left. + \frac{1}{c^2}\left(\frac{\partial}{\partial t} - \frac{ie\phi}{\hbar}\right)\psi_n^* \left(\frac{\partial}{\partial t} + \frac{ie\phi}{\hbar}\right)\psi_n - \frac{m^2c^2}{\hbar}|\psi_n|^2 \right] \tag{4.27}$$

and a field-dependent solution

$$-c^2\hbar^2\left(\nabla - \frac{ie\mathbf{A}}{c\hbar}\right) \cdot \left(\nabla - \frac{ie\mathbf{A}}{c\hbar}\right)\psi_n + \hbar^2\left(\frac{\partial}{\partial t} + \frac{ie\phi}{\hbar}\right)^2 \psi_n + m^2c^4\psi_n = 0. \tag{4.28}$$

This wave equation describes the same vector particles as mentioned in Sec. 4.1.22 but, now, in the presence of electromagnetic fields.

### 4.1.24 Schrödinger wave equation limit

The non-relativistic limit of (4.28) can be readily taken. Following Schiff (1955), we take the limit in Appendix G. The result, Eq. (G8), is the well-known Schrödinger wave equation (SWE). Hence, the SWE is arrived at by EPI via the Klein–Gordon, equation, i.e., indirectly. Rigorous use of EPI cannot derive the SWE directly because EPI is a Lorentz-covariant approach and can only derive covariant phenomena. The SWE is, of course, not covariant.

### 4.1.25 SWE from an approximate EPI approach

Nevertheless, the SWE can follow from an approximate use of EPI. See Appendix D. This use abandons Lorentz covariance in ignoring the time coordinate $t$. Although it produces the correct stationary SWE result, because the approach is approximate it does not derive many of the other effects that were previously found: (a) the time-dependence of the SWE is not found; (b) the mass–energy relation (4.17) has to be assumed, rather than derived as here; and (c) the constancy of $\hbar$ and $m$ is not proved. Evidently, the covariance requirement is powerful enough that to ignore it means losing many benefits.

### 4.1.26 The question of normalization

Equations (4.10) and (4.11), with $N = 2$, express a condition of normalization over space and time,

$$c\iint d\boldsymbol{r}\, dt\, |\psi(\boldsymbol{r}, t)|^2 = 1. \tag{4.29}$$

This normalization property is established in Appendix I.

Equation (4.29) can be recast as

$$1 = c\int dt \int d\boldsymbol{r}\, |\psi(\boldsymbol{r}, t)|^2 = c\int_{-T/2}^{T/2} dt\, p_t(t) \tag{4.30a}$$

$$\text{where} \quad p_t(t) \equiv \int d\boldsymbol{r}\, |\psi(\boldsymbol{r}, t)|^2. \tag{4.30b}$$

For simplicity, we have assumed a symmetric time interval $-T/2 \leqslant t \leqslant T/2$ for detection. $T$ is finite in any real experiment (Appendix I).

Now, normalization in quantum theory is conventionally taken over space $\boldsymbol{r}$ alone. That is, the right-hand side of Eq. (4.30b) is taken to be a constant (unity) in time. For example, such is the case for a $\psi$ obeying the *non-relativistic* Schrödinger wave Eq. (G8) (Schiff, 1955, pp. 22, 23). However, it is known that a solution $\psi$ to the *Klein–Gordon* equation (4.28) *does not necessarily* have a constant normalization integral (4.30b) in time (Morse and Feshbach, 1953, p. 256). (Also see the overview Sec. 4.4.) Does our four-dimensional theory give constant normalization?

Yes, since our $\psi$ has a four-dimensional argument and obeys *four-dimensional* normalization Eq. (4.29) the latter can only result in a constant. Also Eq. (4.30b), which is the conventional normalization equation, becomes merely a marginal PDF $p_t(t)$ in EPI theory. If this is time-varying, there is no inconsistency. In fact, this PDF has an interesting interpretation.

The PDF $p_t(t)$ represents, by its definition, the probability that the particle is measured somewhere (anywhere) within measurement space at the time $(t, t + dt)$. A small $p_t(t)$, for example, signifies that the particle has a low probability of being measured anywhere in the detection volume at that time.

This is effectively the same as saying that the particle has a low probability of *existing*, a property we call *probabilistic annihilation*. Such a situation can occur, e.g., if the particle is absorbed or otherwise exits the volume. Conversely, a high value for $p_t(t)$ connotes *creation* of the particle. The particle has either been re-emitted or has re-entered the volume at the time $(t,\ t+dt)$.

Note that $p_t(t)$ is actually a function $p(ct)$ (see Eq. (4.3)). In cases where $p(ct)$ turns out to be a constant, Eq. (4.30a) gives the constant as $p(ct) = (cT)^{-1}$ or, equivalently, $p_t(t) = 1/T$ for $|t| \leqslant T/2$. This states that the probability of detecting the particle is the same at all times $|t| \leqslant T/2$, and varies inversely with the size of the detection time interval. If the constant $p_t(t)$ is small, then there is a constant tendency toward annihilation of the particle, etc. for the other extreme.

Do our four-dimensionally normalized wave functions obey different kinematics than do the standard, three-dimensionally normalized Klein–Gordon wave functions? Since the four-dimensional wave functions $\psi(\boldsymbol{r},\ t)$ obey the standard Klein–Gordon equation (4.28) the answer is obviously no. The energy values and wave functions must be the standard answers. The only difference between the two solutions is the nature of the normalization factor as previously discussed. (See also Appendix I.) The same will be true of solutions to the Dirac equation, derived below.

In summary, a benefit of the four-dimensional approach is that it overcomes the problem of a time varying three-dimensional normalization integral that occurs in the standard theory. In doing so, it allows for the probabilistic annihilation and creation of particles.

### 4.1.27 Derivation of Yukawa potential; probability amplitudes as potentials

Probability amplitudes such as the scalar $\psi(\boldsymbol{r})$ may often be regarded as effective potentials. This follows because of the "Born approximation" (Martin and Shaw, 1992): The probability amplitude for a particle to be scattered from an initial momentum $\boldsymbol{\mu}_1$ to a final momentum $\boldsymbol{\mu}_2$ by a weak potential $V(\boldsymbol{r})$ is proportional to

$$\phi(\boldsymbol{\mu}) = \int d\boldsymbol{r}\, V(\mathbf{r}) \exp(-i\boldsymbol{\mu}\cdot\mathbf{r}/\hbar), \quad \boldsymbol{\mu} \equiv \boldsymbol{\mu}_1 - \boldsymbol{\mu}_2. \tag{4.30c}$$

Comparing Eq. (4.30c) with Eq. (4.4), in the case of a static amplitude function $\psi(\boldsymbol{r},\ t) = \psi(\boldsymbol{r})$, shows that

$$V(\boldsymbol{r}) \propto \psi(\boldsymbol{r}), \tag{4.30d}$$

i.e., *the potential for the scattering is effectively the probability amplitude function*. Such a proportionality between a probability amplitude and a potential function will later be used in deriving Maxwell's equations in Sec. 5.1.21 and the Einstein gravitational field equations in Sec. 6.3.20. A famous potential of the type (4.30d) is derived next.

In the case of a static amplitude function, a *spherically symmetric* solution to the free-field Eq. (4.26) gives the potential (Martin and Shaw, 1992)

$$V(r) \equiv g^2\psi(\mathbf{r}) = -\frac{g^2}{4\pi}\frac{e^{-r/R}}{r}, \qquad R \equiv \hbar/(mc). \tag{4.30c}$$

Parameter $R$ is seen to be the effective range of the function $\psi(\mathbf{r})$. The proportionality constant $g^2$ between $V(r)$ and $\psi(\mathbf{r})$ is taken to define a coupling constant. The potential arises out of a single-particle exchange of virtual bosons, presumably gluons (Yndurain, 1999). The solution (4.30c) is the only spherically symmetric one that vanishes as $r$ approaches infinity. This zero limit is of course required of any probability amplitude $\psi(\mathbf{r})$ whose square is to be a normalized PDF. The potential $V(r)$ is the celebrated "Yukawa potential." In this way the Yukawa potential follows from EPI.

## 4.2 Derivation of vector Dirac equation

### *4.2.1 Alternative solutions to EPI*

We have previously satisfied the extremum condition (3.16) of EPI. This gave rise to the Klein–Gordon equation as the solution. The other "half" of EPI is condition (3.18). Because we had information $J = I$, with $\kappa = 1$, condition (3.18) becomes simply

$$I[\psi] - J[\psi] = 0. \tag{4.31}$$

We now want to solve this condition for the complex amplitudes $\psi$. This corresponds to a case (b) as discussed in Sec. 3.4.6. The solution to (4.31) will not necessarily be coincident with the one that extremized Eq. (4.27).

We have previously satisfied the extremum condition (3.16) of EPI. This gave rise to the Klein–Gordon wave Eq. (4.28) defining vector particles of arbitrary rank $N$ or spin. As was discussed, these describe either bosons (for integer spins) or fermions (for half-integer spins).

On the other hand, it is known that fermions obey, along with Eq. (4.28), a special wave equation of their own – the famous Dirac equation for spin-1/2 particles in particular, or, for *arbitrary* half-integer spin, the *Rarita–Schwinger* equations. Can these specifically fermionic wave equations perhaps be derived from the *other* EPI condition, the EPI zero-condition (3.18)? If so, this would show a valuable symmetry in the use of EPI.

Results, below, are in the affirmative: the lowest-component ($N = 4$) solution to (4.31) will define the wave equation for the massless neutrino of spin 1/2; the next lowest ($N = 8$) solution will define wave equations for spin-1/2 particles with finite mass: the electron, deconfined quark (which is only lately known to exist), neutron, proton, their anti-particles, etc.; and higher-order values of $N$ will give the wave Eqs. (4.46l) for fermions of arbitrary spin. (*Note*: Eq. (4.46h) relates a general rank $N$ to a corresponding spin $s$.)

### 4.2.2 Free-field case

By Eqs. (4.23) and (4.31), we require the solution $\boldsymbol{\psi}$ to obey

$$4Nc\iint d\mathbf{r}\,dt\sum_{n=1}^{N/2}\left[-(\nabla\psi_n)^*\cdot\nabla\psi_n+\lambda^2\left(\frac{\partial\psi_n}{\partial t}\right)^*\frac{\partial\psi_n}{\partial t}-\eta^2|\psi_n|^2\right]=0, \quad (4.32)$$

where we have introduced two parameters

$$\lambda = 1/c, \quad \eta = mc/\hbar. \quad (4.33)$$

It may be noted that the ensuing derivation will not follow Dirac's historic procedure. Thus, we do not start from an *ad hoc* Hamiltonian operator equation where derivatives represent momentum or energy. The EPI formalism is a stand-alone procedure, not needing such assumptions. Instead, we simply seek the roots of Eq. (4.32). This will be implemented by a factorization procedure that uses Dirac's matrices (Dirac, 1928; 1937).

### 4.2.3 Dirac matrices introduced

Regard a sequence of matrices $[\alpha_x]$, $[\alpha_y]$, $[\alpha_z]$, $[\beta]$, each dimensioned $(N/2)\times(N/2)$, with constant elements that are to be determined. Length $N$ is, as usual, left undetermined until it is fixed by sufficiency of solution at the end. For convenience of notation, define a vector of matrices

$$[\alpha] \equiv ([\alpha_x]\ [\alpha_y]\ [\alpha_z])^{\mathrm{T}}, \quad (4.34)$$

where T denotes the transpose. Thus the dimensions of $[\alpha]$ are $(3N/2)\times(N/2)$. Also, regard $\boldsymbol{\psi}$ as a vector (see Eq. (2.24))

$$\boldsymbol{\psi} = (\psi_1, \ldots, \psi_{N/2})^{\mathrm{T}}. \quad (4.35)$$

Hence, $\boldsymbol{\psi}$ is a dimension $(N/2)\times 1$ vector.

Define the inner product of $[\alpha]$ with $\nabla\boldsymbol{\psi}$ as

$$[\alpha]\cdot\nabla\boldsymbol{\psi} \equiv [\alpha]^{\mathrm{T}}\nabla\boldsymbol{\psi} \equiv [\alpha_x]\,\partial\boldsymbol{\psi}/\partial x + [\alpha_y]\,\partial\boldsymbol{\psi}/\partial y + [\alpha_z]\,\partial\boldsymbol{\psi}/\partial z. \quad (4.36)$$

Notice that each right-hand term is of dimension $(N/2)\times 1$.

### 4.2.4 Factorization vectors introduced

Introduce two "helper" vectors of dimension $(N/2) \times 1$,

$$\mathbf{v}_1 \equiv i[\alpha] \cdot \nabla\psi - [\beta]\eta\psi + i\lambda\, \partial\psi/\partial t, \tag{4.37a}$$

$$\mathbf{v}_2 \equiv i[\alpha^*] \cdot \nabla\psi^* + [\beta^*]\eta\psi^* - i\lambda\, \partial\psi^*/\partial t. \tag{4.37b}$$

*Exercise*: From these definitions, if $\psi_1(\boldsymbol{r}, t)$ is a solution of $\mathbf{v}_1 = 0$ then

$$\psi_2^*(\boldsymbol{r}, t) \equiv \psi_1^*(\boldsymbol{r}, -t) \tag{4.38}$$

is a solution of $\mathbf{v}_2 = 0$. Show this. *Hint*: Take minus the complex conjugate of the equation $\mathbf{v}_1[\psi_1(\boldsymbol{r}, t)] = 0$ and evaluate it at $t = -t$. The result should be $\mathbf{v}_2[\psi_2^*(\boldsymbol{r}, t)] = 0$ for a $\psi_2^*$ defined by Eq. (4.38).

### 4.2.5 Essential property

Vectors $\mathbf{v}_1, \mathbf{v}_2$ have an important property of factorization. If matrices $[\alpha_x]$, $[\alpha_y]$, $[\alpha_z]$, $[\beta]$ are Hermitian and anticommute with one another, then

$$\iint d\boldsymbol{r}\, dt\, \mathbf{v}_1 \cdot \mathbf{v}_2 = \iint d\boldsymbol{r}\, dt \sum_{n=1}^{N/2} \left[ -(\nabla\psi_n)^* \cdot \nabla\psi_n + \lambda^2 \left(\frac{\partial\psi_n}{\partial t}\right)^* \frac{\partial\psi_n}{\partial t} - \eta^2 |\psi_n|^2 \right]$$
$$+ i \iint d\boldsymbol{r}\, dt\, (S_4 + S_5). \tag{4.39}$$

Except for the presence of new functions $S_4$ and $S_5$, the helper vectors $\mathbf{v}_1, \mathbf{v}_2$ *factor the Klein–Gordon information form.* This is shown in Appendix E.

### 4.2.6 Vector Dirac equations without fields

Comparing result (4.39) with requirement (4.32) of EPI, the solution is a pair of vectors $\mathbf{v}_1, \mathbf{v}_2$ obeying

$$\iint d\boldsymbol{r}\, dt\, [\mathbf{v}_1 \cdot \mathbf{v}_2 - i(S_4 + S_5)] = 0. \tag{4.40}$$

A microlevel solution Eq. (3.20) to this problem makes the integrand zero. It is shown in Appendix E that this is satisfied by either of $\mathbf{v}_1$ or $\mathbf{v}_2$ being zero. By definition (4.37a), the result of setting $\mathbf{v}_1 = 0$ is

$$\mathbf{v}_1 \equiv i[\alpha] \cdot \nabla\psi - [\beta]\eta\psi + i\lambda \frac{\partial\psi}{\partial t} = 0. \tag{4.41}$$

This is the Dirac equation for a vector particle.

By result (4.38), setting $\mathbf{v}_2 = 0$ gives the conjugate solution $\psi^*$ to Eq. (4.41) as evaluated at negative time. Its interpretation is taken up below.

### 4.2.7 Introduction to Rarita–Schwinger wave equations, Dirac equations

Recall that the rank $N$ of the particle obeying $\mathbf{v}_1 = 0$ in Eq. (4.41) is arbitrary. For special values of $N$ this equation will be shown in Sec. 4.2.12 to be equivalent to the Rarita–Schwinger equation, describing particles of general half-integer spin – *fermions*. Electrons have the particular rank $N = 8$, and for this case the Rarita–Schwinger equation becomes the famous *Dirac equation of the electron* (see Sec. 4.2.9). Quarks and other particles listed at the end of Sec. 4.2.1 likewise have this rank value and, hence, likewise obey the Dirac equation.

Also, it was indicated in the Exercise in Sec. 4.2.4 that $\psi^*(\boldsymbol{r}, -t)$ obeys the condition $\mathbf{v}_2 = 0$, with $\mathbf{v}_2$ defined by Eq. (4.37b). The negative sign of the time indicates *backward* evolution, i.e., a particle moving backward in time. Thus, the condition $\mathbf{v}_2 = 0$ represents, as in the preceding, a "conjugate" Rarita–Schwinger equation, for a fermion moving *backward* in time. This defines an *anti-fermion*, by the well-known Feynman interpretation of anti-particles in general. For the particular particle of rank $N = 8$ the conjugate Rarita–Schwinger equation becomes the *Dirac equation for the positron* (anti-electron). It also describes anti-quarks and, more generally, all anti-particles to those listed at the end of Sec. 4.2.1.

### 4.2.8 Vector Dirac equation with fields

The replacements (3.37) make the theory covariant. Making these replacements in Eq. (4.41) gives a field-dependent solution

$$i[\alpha] \cdot \left(\nabla - \frac{ie\boldsymbol{A}}{c\hbar}\right)\psi - \eta[\beta]\psi + i\lambda\left(\frac{\partial}{\partial t} + \frac{ie\phi}{\hbar}\right)\psi = 0. \tag{4.42}$$

This is the wave equation for vector fermions, including (now) the effects of electromagnetic fields. It describes the probability amplitudes $\psi$ of a vector particle of half-integer spin $1/2, 3/2, \ldots$. The size of the spin is related to the rank $N$ of $\psi$ by Eq. (4.46h). Thus, for the particular case $N = 8$ the spin is $1/2$. Then Eq. (4.42) is called the "Dirac equation" (*sans* the word "vector") with fields. The spin is embedded in matrices $[\alpha]$, as described next.

### 4.2.9 Spin-1/2 particles with mass

The presence or absence of a rest mass $m$ will sharply delineate two types of spin-1/2 particles. First, let the particle have finite rest mass. Then parameter $\eta \neq 0$ in Eqs. (4.37a, b). The solution (4.41) is expressed in terms of the matrices $[\alpha]$, $[\beta]$ of the theory. It is required that they be Hermitian and mutually anticommute (Appendix E). The smallest $N$ that allows these properties to be obeyed is value $N = 8$, i.e., four complex wave functions $\psi_n$ (Schiff, 1955, p. 326). This describes a spin-1/2 particle (by Eq. (4.46h)), e.g., the electron. Explicit representations of the matrices are

$$[\alpha_x] = \begin{bmatrix} 0 & \sigma_x \\ \sigma_x & 0 \end{bmatrix}, \quad [\alpha_y] = \begin{bmatrix} 0 & \sigma_y \\ \sigma_y & 0 \end{bmatrix},$$
$$[\alpha_z] = \begin{bmatrix} 0 & \sigma_z \\ \sigma_z & 0 \end{bmatrix}, \quad [\beta] = \begin{bmatrix} 1 & 0 \\ 0 & -1 \end{bmatrix}. \tag{4.43}$$

The "elements" of these matrices are themselves $2 \times 2$ matrices

$$[\sigma_x] = \begin{bmatrix} 0 & 1 \\ 1 & 0 \end{bmatrix}, \quad [\sigma_y] = \begin{bmatrix} 0 & -i \\ i & 0 \end{bmatrix},$$
$$[\sigma_z] = \begin{bmatrix} 1 & 0 \\ 0 & -1 \end{bmatrix}, \quad [1] = \begin{bmatrix} 1 & 0 \\ 0 & 1 \end{bmatrix}, \tag{4.44}$$

with [0] a matrix of all zeros. Matrices $[\sigma_x]$, $[\sigma_y]$, $[\sigma_z]$ are called the Pauli spin matrices (Pauli, 1927).

As mentioned previously, this value $N = 8$ of the rank is the *minimum* one that allows matrices $[\alpha]$, $[\beta]$ to satisfy their required properties of Hermiticity and anticommutation. By Eq. (4.18) the information $I$ is linear in the rank $N$. Then $I = min.$ for this choice of the rank. Hence the *game corollary* of EPI is satisfied by this choice of the unitless parameter $N$. Of course, the corollary cannot be used *to fix N* in this application because $N$ is fixed by observation of the measured particle. Instead, the corollary indicates that, of all fermions with finite mass, those with the lowest possible spin value of 1/2 are *preferred*, which is borne out by the dominance of spin-1/2 particles among fermions.

A similar statement will next be made about the "*massless*" *neutrino* of spin 1/2.

### 4.2.10 Dimension $N = 4$ case

Consider, next, the case of a particle with close to *zero* rest mass so that, by (4.33), $\eta \approx 0$. Now the Klein–Gordon information (4.23) no longer has an end term in mass $m$. As before, this is to be factored. We may again use the approach of Appendix E to achieve the factorization. Obviously the problem is a special case $m = 0$ of our previous solution Eqs. (4.41), (4.38). Hence, the resulting requirements are again those of anticommutation, but now only for the matrices $[\alpha_x]$, $[\alpha_y]$, $[\alpha_z]$. Matrix $[\beta]$ no longer enters into the problem since the coefficient $\eta$ is zero in factor (4.41).

Setting the factorization vector $\mathbf{v}_1$ in (4.41) equal to zero now gives a requirement

$$i[\alpha] \cdot \nabla \psi + i\lambda \frac{\partial \psi}{\partial t} = 0. \tag{4.45a}$$

In fact, the three Pauli spin matrices $[\sigma_x]$, $[\sigma_y]$, $[\sigma_z]$ satisfy our requirement of anticommutation, as may easily be verified from their definitions (4.44). Hence, we identify

$$[\alpha_x] = [\sigma_x], \quad [\alpha_y] = [\sigma_y], \quad [\alpha_z] = [\sigma_z]. \tag{4.45b}$$

Since the dimension of these matrices is $2 \times 2$, we see that now we have $N = 4$ as the solution dimension (i.e., $N/2 = 2$ complex wave functions $\psi_1, \psi_2$). Equation (4.45a) is now Weyl's equation (Roman, 1961) describing the massless neutrino of spin $1/2$.

We next consider whether, as with spin-1/2 fermions of *finite* mass (Sec. 4.2.9), spin-1/2 neutrinos are preferred over neutrinos of higher spin on the basis of the *game corollary*. A complication is that Eq. (4.18) is quadratic in the mass, so that, *if* $m = 0$, any value of $N$ (or spin) would suffice to minimize $I$. On this basis the game corollary argument for preference for spin 1/2 could not be made. But, in fact, to this date *all* neutrinos have been observed to have spin 1/2. This implies that the game corollary argument must again hold. That is, the corollary *applies* to neutrinos; therefore the possibility $m = 0$ is invalid, and neutrinos have finite mass. In fact, there is growing evidence that neutrinos *do* have finite mass $m$ (Kajita, 2000). In conclusion, the game corollary of EPI, when combined with the observed dominance of spin 1/2 for neutrinos, serves to predict that neutrinos have mass, and this has recently been verified.

### 4.2.11 The $\gamma$-matrix form of vector Dirac equation

It is customary nowadays to place the vector Dirac Eq. (4.41) in a more compact form. Regard $N$ as a general dimension, and matrices $[\alpha_j]$ and $[\beta]$ to be as yet undefined square matrices of size $N/2$. Multiply through Eq. (4.41) on the left by the matrix $-[\beta]$. This gives

$$-i[\beta][\alpha] \cdot \nabla \psi - \frac{i}{c}[\beta] \frac{\partial \psi}{\partial t} + \frac{mc}{\hbar} \psi = 0, \tag{4.46a}$$

where we used the property $[\beta]^2 = [1]$ (Eq. (E13)), a unit square matrix of size $N/2$. Next, define new "gamma matrices"

$$[\gamma_j] \equiv -i[\beta][\alpha_j], \quad j = 1, 2, 3, \quad \alpha_1 \equiv \alpha_x, \quad \alpha_2 \equiv \alpha_y, \quad \alpha_3 \equiv \alpha_z. \tag{4.46b}$$

(*Note*: $[\gamma_4]$ is defined below.)

Notice that the $[\gamma_j]$ must likewise be of size $N/2 \times N/2$. Also, introduce for convenience the new coordinates and derivative notation

$$(x_1, x_2, x_3, x_4) \equiv (x, y, z, ict), \quad \partial_\mu \equiv \partial/\partial x_\mu, \quad \mu = 1, 2, 3, 4. \tag{4.46c}$$

Then Eq. (4.46a) becomes simply

$$[\gamma_j]\,\partial_j\psi + [\beta]\,\partial_4\psi + \frac{mc}{\hbar}\psi = 0. \tag{4.46d}$$

(The repeated index $j$ indicates that *a sum* over $j = 1, 2, 3$ is taken. See Sec. 6.2.2.) Further define a matrix

$$[\gamma_4] \equiv [\beta]. \tag{4.46e}$$

Then the second term in (4.46d) can likewise be incorporated into the sum, as

$$\left([\gamma_\mu]\,\partial_\mu + \frac{mc}{\hbar}\right)\psi = 0. \tag{4.46f}$$

(The repeated Greek index $\mu$ indicates a sum over $\mu = 1, 2, 3, 4$.) We emphasize that at this point $\psi$ is a singly subscripted function with dimension $N/2$ (see Eq. (4.35)), and that each matrix $[\gamma_j]$ is of size $N/2 \times N/2$.

Note that, from its derivation, (4.46f) is entirely equivalent to the EPI result (4.41).

### *4.2.12 General rank N or spin; Rarita–Schwinger wave equations of general fermions*

Equation (4.46f), shown above to be equivalent to the EPI result (4.41), holds for an amplitude function $\psi$ having *any general dimension N*. We discussed in Sec. 4.2.9 that a case $N = 8$ describes a fermion with spin $1/2$. It is intuitive that higher values of $N$ will simply correspond to *higher* spin values $3/2$, $5/2$, ... of the fermion. These are known to obey the celebrated *Rarita–Schwinger* wave Eq. (4.46l). Then the EPI-based Eq. (4.46f) ought to be equivalent to this wave equation. The equivalence is shown next.

Equation (4.46f) describes a fermion with a *general* spin value

$$s \equiv k + 1/2, \quad k = 0, 1, 2, \ldots \tag{4.46g}$$

if the amplitude function $\psi$ and matrices $[\gamma_\mu]$ are chosen as follows:

(i) The amplitude function $\psi$ is a vector of length $N/2 = 2^{2s+1}$. Alternatively, a given vector length $N$ in either (4.41) or (4.46f) represents a spin case

$$s = \frac{1}{2}\log_2 N - 1, \quad N = 2^{2(s+1)} = 8, 32, 128, \ldots, \tag{4.46h}$$

the latter values from Eq. (4.46g).

(ii) Each matrix $[\gamma_\mu]$, $\mu$ fixed, is formed as $2^{2s-1}$ *repetitions, along the diagonal*, of a specifically $4 \times 4$ matrix $[\gamma_\mu]_4$

$$[\gamma_j]_4 \equiv -i[\beta]_4[\alpha_j]_4, \quad j = 1, 2, 3, \text{ and } [\gamma_4]_4 \equiv [\beta]_4. \tag{4.46i}$$

The matrices $[\alpha_j]_4$, $[\beta]_4$ are the specific $4 \times 4$ matrices $\alpha_x$, $\alpha_y$, $\alpha_z$, and $[\beta]$ defined by Eqs. (4.43) and (4.44). An example follows.

Consider the case $k = 1$, spin $s = 3/2$.
Requirement (i): the length of the vector $\psi$ is $2^{2s+1} = 16$.
Requirement (ii): Using definitions (4.46i), we form (for instance) the particular $4 \times 4$ matrix $[\gamma_1]_4$ as

$$[\gamma_1]_4 \equiv -i[\beta]_4[\alpha_1]_4 = -i\begin{bmatrix} 1 & 0 & 0 & 0 \\ 0 & 1 & 0 & 0 \\ 0 & 0 & -1 & 0 \\ 0 & 0 & 0 & -1 \end{bmatrix}\begin{bmatrix} 0 & 0 & 0 & 1 \\ 0 & 0 & 1 & 0 \\ 0 & 1 & 0 & 0 \\ 1 & 0 & 0 & 0 \end{bmatrix}$$

$$= \begin{bmatrix} 0 & 0 & 0 & -i \\ 0 & 0 & -i & 0 \\ 0 & i & 0 & 0 \\ i & 0 & 0 & 0 \end{bmatrix}. \tag{4.46j}$$

Then in the wave Eq. (4.46f) the matrix $[\gamma_1]$ consists of four repetitions down the diagonal of $[\gamma_1]_4$,

$$[\gamma_1] = \begin{bmatrix} [\gamma_1]_4 & 0 & 0 & 0 \\ 0 & [\gamma_1]_4 & 0 & 0 \\ 0 & 0 & [\gamma_1]_4 & 0 \\ 0 & 0 & 0 & [\gamma_1]_4 \end{bmatrix}. \tag{4.46k}$$

Each "element" here is the indicated $4 \times 4$ matrix. Thus $[\gamma_1]$ is a $16 \times 16$ matrix, as is needed to multiply the grand state vector $\psi$ of length 16. The other matrices $[\gamma_2]$, $[\gamma_3]$ and $[\gamma_4]$ are likewise of size $16 \times 16$, and formed analogously.

An alternative to representation (4.46f) and conditions (i), (ii) is to unpack the one long vector $\psi$ into a multiply subscripted matrix quantity $\psi_{\mu_1\mu_2\ldots\mu_k}$, where each subscript is only of dimension 4, that is, goes independently from 1 to 4. Then Eq. (4.46f) becomes mathematically equivalent to the *Rarita–Schwinger* wave equation

$$\left([\gamma_\mu]_4\, \partial_\mu + \frac{mc}{\hbar}\right)\psi_{\mu_1\mu_2\ldots\mu_k} = 0, \quad k = 0, 1, 2, \ldots \tag{4.46l}$$

(Roman, 1961). By convention, in the special case $k = 0$ there is no subscript to $\psi$. Again, the subscript 4 of $[\gamma_\mu]_4$ means that these matrices are formed via

Eqs. (4.46i). The equivalence of Eqs. (4.46f) and (4.46l) derives from the fact that packing the matrices $[\gamma_\mu]_4$ along the diagonal of the matrices $[\gamma_\mu]$, in step (ii), and then multiplying this by one long state vector $\psi$, produces exactly the same equations as does multiplying over and over in (4.46l) the one $4 \times 4$ matrix $[\gamma_\mu]_4$ by the multiply subscripted $\psi_{\mu_1\mu_2\ldots\mu_k}$. An example follows.

We continue with the case $k = 1$. Here the long state vector $\psi$ in (4.46f) was found to be of length 16. It has the elements

$$\psi \equiv \begin{pmatrix} \psi_1 \\ \psi_2 \\ \psi_3 \\ \psi_4 \end{pmatrix}, \quad \text{where} \quad \psi_\mu = \begin{pmatrix} \psi_{1\mu} \\ \psi_{2\mu} \\ \psi_{3\mu} \\ \psi_{4\mu} \end{pmatrix}. \tag{4.46m}$$

Consider the matrix product $[\gamma_1]\psi$ in (4.46f). (The derivative operations $\partial_\mu$ are ignored since they occur in both representations (4.46f) and (4.46l).) Here $[\gamma_1]$ is of size $16 \times 16$ and given by (4.46k). Thus the product $[\gamma_1]\psi$ is a vector of length 16 elements. Consider in particular the first four of these. Because of the way in which matrix $[\gamma_1]$ is formed in (4.46k), these elements are determined by the product of *sub*matrices

$$[\gamma_1]_4\psi_1 = [\gamma_1]_4 \begin{pmatrix} \psi_{11} \\ \psi_{21} \\ \psi_{31} \\ \psi_{41} \end{pmatrix}. \tag{4.46n}$$

But these are also the first four elements of the matrix product in the Rarita–Schwinger Eq. (4.46l) (again, ignoring the common derivative operations $\partial_\mu$). The matchup between the two representations continues for the next four elements in the products, etc. Thus, the two representations generate identical wave equations.

A special situation arises for fermions of spin value $s \geqslant 5/2$. For either representation (4.46f) or (4.46l) to represent a *unique* spin value of this size, an additional set of linear constraint equations must be satisfied by the amplitude functions. This is analogous to the imposing of a Lorentz condition in classical electromagnetic or gravitational theory (Eqs. (5.8) and (6.28)). See Roman (1961) for details. *Without* these additional constraint equations, at any such spin value $s$ Eq. (4.46f) or (4.46l) represents a fermion whose *maximum* spin value is $s$.

We conclude that the EPI approach, specialized to conditions (i) and (ii) above, gives the wave Eqs. (4.46f) or (4.46l) for fermions of arbitrary spin. These were derived in the absence of charge and electromagnetic fields. In their presence, the wave equations may of course be amended by making the usual replacements (3.37).

### 4.2.13 Non-relativistic limit of Dirac equation

This limit can be directly taken in Eq. (4.42) (Schiff, 1955, pp. 329–30). As is well known, it gives the Schrödinger wave Eq. (G8) plus a term involving interaction of the particle spin with the magnetic field $\boldsymbol{H}$. This term does not disappear unless the particle has zero spin.

### 4.2.14 A digression: spin level as bit level

The right-hand side of Eq. (4.46h) is intriguing in representing, as an alternative to the spin $s$, the Shannon information $\log_2(M)$ in the presence of $M$ equally probable states (Reza, 1961). Here effectively $M = \sqrt{N/4}$. This entropy is in fact a maximum, called the "Shannon information capacity," since the assumption of equal probability *maximizes* the Shannon information. This suggests a new interpretation for the meaning of the concept of spin $s$, that $s$ represents the *number of bits* of Shannon information *capacity* due to $\sqrt{N/4}$ equally probable scalar substates $\psi_n$ of a given fermion. Strangely, by Eq. (4.46h), this number of substates is not an integer.

Another maximization of Shannon information occurs for an alternative class of "particles" with spin: black holes. The Beckenstein–Hawking law of black holes states that the Shannon entropy of a black hole is proportional to its classical horizon area $A$. In fact, the latter is precisely the information *capacity* in the presence of optimal (Huffman–Shannon–Fano) coding of area events $A$ (Frieden and Soffer, 2002). Thus, the Beckenstein–Hawking law is another example of the achievement of information capacity by nature.

## 4.3 Uncertainty principles

### 4.3.1 For position–momentum measurements

The Heisenberg uncertainty principle states that, at a given time $t$, a particle's position and momentum intrinsically fluctuate by amounts $x$ and $\mu$ from ideal (classical) values $\theta_x$ and $\theta_\mu$ with variances $\epsilon_x^2$ and $\epsilon_\mu^2$ obeying

$$\epsilon_x^2 \epsilon_\mu^2 \geqslant (\hbar/2)^2. \tag{4.47}$$

This is conventionally derived from the Fourier transform relation (4.4) connecting position and momentum spaces; see, e.g., Bracewell (1965).

The relation may be shown, as well, to arise out of the use of Fisher information.

Consider a measurement scenario where the position and momentum of a particle are measured at the same time. Like all measurements, the $X$-measurements must obey the Cramer–Rao inequality (1.1),

$$e_x^2 I_x \geqslant 1, \quad e_x^2 \equiv \langle (\hat{\theta}_x(y) - \theta_x)^2 \rangle. \tag{4.48}$$

The general estimator function is $\hat{\theta}_x(y)$. Here $I_x$ is the Fisher information (not the information capacity $I$) in the *one* measurement. What is $I_x$? (See also Stam, 1959a.)

As in Eq. (2.24), define complex amplitude functions $\psi_n(x)$ as those whose real and imaginary components are our real amplitude functions $\mathbf{q}$. By Eq. (4.4), represent each $\psi_n(x)$ as the Fourier transform of a corresponding function $\phi_n(\mu)$ of momentum fluctuations $\mu$,

$$\psi_n(x) = \frac{1}{\sqrt{2\pi\hbar}} \int d\mu\, \phi_n(\mu) \exp(i\mu x/\hbar). \tag{4.49}$$

For simplicity, choose a case $N = 2$ of one complex wave function. This represents a case of $N = 2$ *intrinsic* measurements (Sec. 2.1.1). Then the information capacity $I$ obeys Eq. (D1) with $N = 2$. Using also $\psi = |\psi| \exp(iS)$ and Eq. (1.24) for $I_x$, (D1) gives $I = 2I_x + 8\langle (dS/dx)^2 \rangle$. Then $2I_x \leqslant I$ or

$$2I_x \leqslant 8 \int dx\, \psi'^* \psi', \quad \psi \equiv \psi_1, \quad \psi' \equiv d\psi/dx. \tag{4.50}$$

Substituting Eq. (4.49) into (4.50) gives, as the information *per data value*,

$$I_x \leqslant \frac{4}{\hbar^2} \int d\mu\, \mu^2 |\phi(\mu)|^2, \quad \phi \equiv \phi_1. \tag{4.51}$$

By Eqs. (4.11), $|\phi(\mu)|^2$ is just the marginal PDF $P(\mu)$, so that the integral in (4.51) is a mean value,

$$I_x \leqslant \frac{4}{\hbar^2} \langle \mu^2 \rangle \equiv \frac{4}{\hbar^2} \epsilon_\mu^2. \tag{4.52}$$

The far-right equality follows because the $\mu$ are fluctuations from the mean momentum. Using Eq. (4.52) in Eq. (4.48) gives directly

$$e_x^2 \epsilon_\mu^2 \geqslant (\hbar/2)^2. \tag{4.53}$$

This is of the form (4.47), the Heisenberg principle for the $X$-coordinate measurement, although we have allowed for a difference of interpretation by representing the $X$-position errors differently, $\epsilon_x^2$ versus. $e_x^2$. The two principles are not quite the same, as is discussed below.

### 4.3.2 For time–energy measurements

A Heisenberg principle is also obeyed by simultaneous measurements of time and energy,

$$e_t^2 \epsilon_E^2 \geqslant (\hbar/2)^2, \tag{4.54}$$

where $e_t^2$ is the mean-square time fluctuation and $\epsilon_E^2$ is the mean-square energy fluctuation. This principle may be derived from the Cramer–Rao inequality for the intrinsic time measurements, by analogous steps to the preceding.

In this regard, a Fourier transform connecting the time and energy domains exists that is completely analogous to Eq. (4.49); see Cohen-Tannoudji *et al.* (1977, Vol. I, p. 251).

### *4.3.3 Exercise*

Carry through the derivation.

### *4.3.4 Discussion*

We observed that there are (at least) two routes to the Heisenberg principle: (a) based upon Fourier complementarity (4.49) or (b) arising out of the Cramer–Rao inequality. There are conceptual differences between approaches (a) and (b).

The taking of data means a process of randomly sampling from "prior" probability laws $p(x)$ and $P(\mu)$. (In the usual statistical sense, "prior" means prior to any measurements.) However, the Fourier approach (a) does not assume the actual taking of data. The spreads $\epsilon_x^2$, $\epsilon_\mu^2$ that it defines in Eq. (4.47) are just parameters that measure the widths of the prior laws. Hence, the approach (a) derives a Heisenberg principle (4.47) that holds independently of, and *prior to*, any measurement. This, of course, violates the spirit of EPI, according to which physical laws follow *in reaction to* measurement.

By contrast, the EPI-based approach (b) states that, if an $X$-coordinate *is measured* (with an ideal detector; see Sec. 2.1.1), then there is a resulting uncertainty $\epsilon_\mu^2$ in momentum that obeys reciprocity (Eq. 4.53)) with the coordinate uncertainty. Since this interpretation of the Heisenberg principle is measurement-based, it agrees with the spirit of EPI.

In a nutshell, our disagreement with the usual Heisenberg interpretation lies in the presumed nature of the fluctuations. The conventional interpretation is that these are intrinsic to the phenomenon and independent of measurement. Our view is that they are intrinsic to the phenomenon (Sec. 2.1.1), and *arise out of* measurement, as do all phenomena.

Of course if the detector is not ideal, and contributes noise of its own to the measurements, both interpretations need to be modified (Arthurs and Goodman, 1988; Martens and de Muynck, 1991; Caves and Milburn, 1987).

The second difference between approaches (a) and (b) lies in the nature of their predictions. Equation (4.47) states that the spreads in positions will obey the principle. By contrast, Eq. (4.53) states that the spreads in *any functions* $\hat{\theta}_x(y)$ of the data positions obey the principle. The latter includes the former, then, as a particular case (where $\hat{\theta}_x(y) = y$). In this sense, version (4.53) is the more general of the two.

The Heisenberg uncertainty principle (4.47) also provides a practical lower bound to the *resolution length* $\epsilon_x$ of an observed particle. It shows that, in order for $\epsilon_x$ to be very small, the momentum spread $\epsilon_\mu$ must be very large. Hence, the smallest possible value of $\epsilon_x$ occurs in the presence of the largest possible value of $\epsilon_\mu$. The latter is determined as follows.

In order for the object particle of mass $m$ to be detected, it must be illuminated by a probe particle (Fig. 3.3). By conservation of energy and momentum, after the illumination event the object particle acquires roughly the energy of the probe particle. But if that energy exceeds roughly $mc^2$ the object particle breaks up into two or more daughter particles. This thwarts the observation and hence defines the limiting momentum fluctuation $\epsilon_\mu$. By Eq. (4.17) this is at most $E/c$. Since we had $E = mc^2$, the limiting momentum fluctuation is

$$\epsilon_\mu = E/c = mc^2/c = mc. \tag{4.54a}$$

This knowledge is also used in Appendix I to establish the finiteness of a normalization integral.

Thus, $mc$ is the limiting value of the momentum of the particle, shy of it breaking up. Using it in Eq. (4.47) gives

$$\epsilon_{x\,\min} = \frac{\hbar}{\epsilon_\mu}, \qquad \epsilon_\mu \equiv mc, \tag{4.54b}$$

where an inconsequential factor of $1/2$ has been ignored. The limiting resolution $\epsilon_{x\,\min}$ is called the "Compton resolution length." It also agrees with the resolution length $e_{\min}$ found in Sec. 4.1.17 on the basis of periodicity of the wave function.

### *4.3.5 Uncertainty principles expressed by entropies*

The uncertainty principle (4.53) has been seen to derive from the fact that Fisher information $I$ measures the spread in momentum values. Another measure of the spread in momentum is $H$, its Shannon entropy, here denoted as

$$H(\mu) = -\int d\mu\, P(\mu) \ln P(\mu), \qquad P(\mu) = \sum_n |\phi_n(\mu)|^2. \tag{4.55}$$

(See Secs. 1.3 and 4.3.1.) The entropy associated with space $x$ is, correspondingly,

$$H(x) = -\int dx\, p(x) \ln p(x), \quad p(x) = \sum_n |\psi_n(x)|^2. \tag{4.56}$$

Then, can the complementarity of width of the two PDFs $p(x)$ and $P(\mu)$ be expressed in terms of these two entropies? The answer is yes.

Hirschman's inequality (Hirschman, 1957; Beckner, 1975) states that

$$H(x) + H(\mu) \geqslant \ln(\pi e \hbar/2). \tag{4.57}$$

Parameter $e$ is here the Naperian base. This states that it is impossible for $p(x)$ and $P(\mu)$ to both be arbitrarily narrow functions. It is interesting to evaluate (4.57) for the case where the PDFs are Gaussian. Since the entropy for a Gaussian obeys $H = \ln\sigma + \ln\sqrt{2\pi e}$, Eq. (4.57) gives

$$\ln\sigma_\mu + \ln\sqrt{2\pi e} + \ln\sigma_x + \ln\sqrt{2\pi e} \geqslant \ln(\pi e \hbar) \quad \text{or} \quad \sigma_\mu^2\sigma_x^2 \geqslant (\hbar/2)^2, \tag{4.58}$$

again the Heisenberg principle (4.47). A further similarity between the two principles is that the amplitude functions that attain a minimum Heisenberg product also attain a minimum "Hirschman sum." This is the Gaussian case cited above.

## 4.4 Overview

Of central importance to the derivations is the existence of the unitary transformation space $(i\boldsymbol{\mu}/\hbar,\ E/(c\hbar))$ (see Eq. (4.3)). As we found in Sec. 3.8.7, the physical existence of such a conjugate space to the measurement space $(i\mathbf{r},\ ct)$ guarantees the validity of the EPI approach for the given problem.

Particles are known to follow either Bose–Einstein or Fermi–Dirac statistics (Jauch and Rohrlich, 1955). The former are characterized by integral spin, the latter by 1/2-integral spin (an odd number times 1/2). What we have found is that these basic particle types derive, respectively, from the two requirements (3.16) and (3.18) of EPI. Conversely, this tends to confirm these two requirements as being physically meaningful.

General vector bosons or fermions of any spin size are found to obey the vector Klein–Gordon Eq. (4.28), whereas fermions of any half-integral spin size are found to obey the vector Dirac Eq. (4.42), which for spin 1/2 becomes the usual Dirac equation. The vector Dirac equation is found to be equivalent to the Rarita–Schwinger Eq. (4.46l).

The preceding follows from setting factor $\mathbf{v}_1 = 0$ with $\mathbf{v}_1$ given by Eq. (4.41). Instead setting factor $\mathbf{v}_2 = 0$ (Sec. 4.2.6) leads to the existence of *antiparticles*. These must move backward in time, by Eq. (4.38).

Hence there is an attractive compactness by which EPI derives these particle properties. The known particle classes of boson, fermion and anti-fermion each follow from a fundamental aspect of EPI (the extremum condition, zero-condition, and factorization of the zero giving direct particles and anti-particles, respectively).

The preceding fermion results were for those with finite rest mass. For so-called "massless" fermions the vector Dirac Eq. (4.41) becomes the *Weyl* Eq. (4.45a). As we saw (Sec. 4.2.10), the lower dimensionality and spin value for this equation coupled with the game corollary amount to EPI predictions that neutrinos (i) have small but *finite* mass, and (ii) tend to have minimal spin for a fermion, i.e., spin 1/2. Properties (i) and (ii) agree with current observation (Kajita, 2000).

That parameter $\kappa = 1$ for this scenario is of interest. Recall that $\kappa$ represents, from Eq. (3.18), the ratio of the intrinsic Fisher information $I$ to the information $J$ that is bound to the phenomenon. That $\kappa = 1$ here and in Appendix D suggests, by Sec. 3.4.5, that quantum mechanics is an accurate theory. The intrinsic information $I$ contains as much information as the phenomenon can supply. As we saw at Eqs. (4.7)–(4.9), this arose out of the unitary nature of the invariance principle that was employed. Many other phenomena, specifically, non-quantum ones, will not obey such information efficiency. See later chapters.

We want to emphasize that the derivations do not rely on the well-known association of gradient operators with momentum or energy. No Hamiltonian operators are used. All gradients in the derivations arise consistently from *within* the approach: ultimately from the definition (2.19) of Fisher information.

In the case of a conservative, scalar potential $\phi(\boldsymbol{r})$, with $\boldsymbol{A}(\boldsymbol{r}, t) = 0$, the Klein–Gordon solution Eq. (4.28) may be regarded, alternatively, as a solution to the knowledge acquisition game of Sec. 3.4.12 (see *Exercise* at end). On the other hand, since the Dirac solution Eq. (4.42) resulted from the zero-principle Eq. (3.18) of EPI, rather than from the extremization principle Eq. (3.16), it does not follow from a knowledge acquisition game. This is a strange distinction between the two solutions and, consequently, between the nature of fermions and that of bosons.

The *mathematical* route to the *Klein–Gordon equation* that is taken in Sec. (4.1) is via the conventional Lagrangian for the problem. The unique contribution of the EPI approach is in *deriving* the terms of the Lagrangian from physical and information considerations. However, even mathematically, the route taken in Sec. 4.2 to the *Dirac equation* is unique. It is not a Lagrangian variational problem, but rather finds the Dirac equations for the electron and the positron as the two roots of an equation.

A further advantage of EPI over conventional theory is worth mentioning. Assume, for simplicity, a case $N = 2$ of a single complex wave function $\psi(\boldsymbol{r}, t)$. We give an argument of Schiff (1955). By multiplying the Klein–Gordon Eq. (4.26) on the left by $\psi^*$, multiplying the complex conjugate of (4.26) on the left by $\psi$, and subtracting the results, one obtains a pseudo-conservation of flow equation

$$\frac{\partial}{\partial t}\overline{P}(\boldsymbol{r}, t) + \nabla \cdot \overline{\mathbf{S}}(\boldsymbol{r}, t) = 0, \tag{4.59}$$

where

$$\overline{P}(\boldsymbol{r}, t) \equiv \frac{i\hbar}{2mc^2}\left(\psi^* \frac{\partial \psi}{\partial t} - \psi \frac{\partial \psi^*}{\partial t}\right) \tag{4.60}$$

and $\overline{\mathbf{S}}(\boldsymbol{r}, t)$ is another form which doesn't concern us here. The quantity denoted as $\overline{P}(\boldsymbol{r}, t)$ *mathematically* obeys Eq. (4.60). What does it represent physically? The usual physical interpretation is that, from the form of Eq. (4.59), it is a probability density. Then it would be our PDF $p(\boldsymbol{r}, t)$, i.e.,

$$p(\boldsymbol{r}, t) = \overline{P}(\boldsymbol{r}, t). \tag{4.61}$$

But, obviously (4.60) and (4.61) together give a generally different form $p(\boldsymbol{r}, t)$ than our presumed form (2.25), here

$$p(\boldsymbol{r}, t) = \psi^* \psi. \tag{4.62}$$

(The two agree in the non-relativistic limit.)

Which of the two candidate forms (4.61) or (4.62) is correct in representing the joint PDF $p(\boldsymbol{r}, t)$? The form (4.60) can go negative (Schiff, 1955, p. 319). Since a PDF must always obey positivity, this rules out (4.61) as the joint PDF. The form (4.62) cannot of course go negative, and so cannot be rejected on these grounds. However, if (4.62) is integrated out over all space $\boldsymbol{r}$ (*alone*) a general time dependence can remain, violating the normalization requirement of a PDF (Morse and Feshbach, 1953, p. 256). A time-varying normalization integral is not acceptable.

Fortunately, as we found, the problem of variable normalization can be overcome by the *four-dimensionality* of the EPI approach. Even if a normalization integral varies with time after three integrations, after the fourth integration over time it must be a constant! This allows us, then, to accept the squared modulus $|\psi(\boldsymbol{r}, t)|^2$ as the form for $p(\boldsymbol{r}, t)$. This is the standard EPI choice (2.25), and is mathematically consistent, since it obeys positivity.

In summary of the last two paragraphs, the covariant EPI approach leads to a mathematically consistent definition of the PDF for the problem, whereas the usual, non-covariant approach leads to serious inconsistencies. Historically, these inconsistencies of course led to the replacement of conventional quantum mechanics by quantum field theory, where ordinary functions $\psi(\boldsymbol{r}, t)$ are replaced by operators. However, as we found, EPI does not suffer these inconsistencies. Therefore it does not in itself require the transition to operators to be made. Nevertheless the concept of operators is quite important and merits further discussion.

Generalizing the concept of *non-operator* amplitude functions $\psi(\boldsymbol{r}, t)$ to include *operator* amplitude functions does not in itself represent added physics: This use of operators is *equivalent* mathematically to a many-body formulation of *non-operator* amplitude functions (Robertson, 1973). However, operators are exceedingly effective in providing solutions to many-body solutions. Hence, it would be valuable if EPI also allowed the wave equations for operator amplitude functions to be *derived*. Is this possible? The answer is yes: The usual EPI output wave equations can alternatively be regarded as defining the *unknown field operators* $\psi(\boldsymbol{r}, t)$ for a problem. This is taken up in Sec. 11.5 of Chapter 11.

These benefits ultimately follow from the way time is treated by EPI theory: on an equal footing with space (Secs. 3.1.2 and 3.5.8). Thus, the time "coordinate" $t$ and space "coordinate" $\boldsymbol{r}$ for an amplitude function $\psi(\boldsymbol{r}, t)$ are both regarded as random fluctuations from ideal values. Also, just as space is not presumed to "flow" in standard treatments of quantum mechanics, time is likewise not interpreted to "flow" in EPI theory. As with the Einstein field equations of general relativity (Chapter 6), the Klein–Gordon and Dirac equations merely provide an "arena" for the occurrence of random space and random time fluctuations from ideal values.

Also, we mentioned in Sec. 4.1.2 that either choice of coordinates $(i\boldsymbol{r}, ct)$ or $(\boldsymbol{r}, ict)$ would suffice in leading to the end products of the approach – the Klein–Gordon and Dirac equations. Such arbitrariness implies that either "space-like" or "time-like" coordinates (Jackson, 1975) describe equally well the phenomena of quantum mechanics. Again, space and time have an equal footing.

Equations (4.3) and (4.4) regard the two coordinate spaces of space-time $(i\boldsymbol{r}, ct)$ and momentum-energy $(i\boldsymbol{\mu}/\hbar, E/(c\hbar))$ *as equivalent* in defining the dynamics of the measured particle. Either amplitude function $\psi_n(\boldsymbol{r}, t)$ or $\phi_n(\boldsymbol{\mu}, E)$ will suffice in this regard. On the other hand, the EPI wave equations that were obtained in this chapter presumed a measurement to be made in one specific coordinate space, that of *space-time*. If space-time and momentum-energy really are equivalent in defining the particle, could correct wave equations for the measured particle have been acquired out of an EPI measurement of *momentum-energy instead*? As was discussed in Sec. 3.5.5, only candidate four-coordinates that are *a priori independent* qualify as Fisher coordinates of measurement space. But in fact momentum and energy are not independent, by Eq. (4.17) (in the free-field case). This, then, rules out their joint use in an EPI approach (although they are used *independently* in Chapter 7).

Quantum mechanics is well known to obey additivity of amplitudes rather than additivity of probabilities. This leads to non-classical behavior, such as in the famous two-slit experiment (Schiff, 1955, pp. 5, 6) whereby the net probability at the receiving plane is not merely the sum of the two contributions from the individual slits, but also includes a cross-term contribution due to the interference of amplitudes from the slits. This is often taken to be the signature effect that distinguishes quantum mechanics from classical mechanics. How does EPI account for this effect?

The effect originates in Eqs. (1.23) and (1.24), whereby the PDF $p(x)$ *is defined* as the square of an amplitude function $q(x)$ and, resultingly, the information $I$ is expressed directly in terms of $q(x)$. When this $I$ is used in EPI principle Eqs. (3.16) and (3.18) the result is a differential equation in $q(x)$ ($\psi(\mathbf{r}, t)$ here), not in $p(x)$ directly. Solutions $q(x)$ to this equation are often in the form of superposition integrals. Then, squaring the superposition to get $p(x)$ brings in all cross-terms of the superposition, e.g., the two-slit cross-term mentioned above.

We noted below Eq. (1.23) that such use of probability amplitudes is classical as well, tracing to work of Fisher. It is perhaps not surprising, then, that this cross-term effect occurs in classical statistical physics as well (see Eqs. (7.56) and (7.59)).

We have shown that the wave equations of quantum mechanics follow from the EPI principle. These wave equations hold at the input space to the measuring apparatus (see Secs. 3.8.1 and 3.8.5). One might ask, then, what wave equation is obeyed at the *output* of the apparatus? This topic is taken up in Sec. 3.8.2 and Chapter 10.

Other quantum applications of EPI are found in Chaps. 10–12 and Appendix D.

In Chapter 10 the affects of numerical measurements upon EPI output amplitude functions are computed.

In Chapter 11 (a) the Wheeler–DeWitt equation of quantum gravity is derived, (b) empirical formulae are developed for the Weinberg and Cabibbo angles of the weak nuclear interaction, (c) the formation of composite nuclear particles out of elementary ones is shown to follow from the zero-condition of EPI, and (d) a transition from EPI output amplitude functions to field *operator* functions is shown.

In Chapter 12 the special case of spin in the *EPR-Bohm experiment* is taken up. The entanglement effect that this experiment is well known to describe is found to arise as a property of the bound information $J$ for the problem. This suggests that, in all quantum problems, the bound information $J$ may also be regarded as an *information of entanglement*. There are two levels of such entanglement: (a) *real* entanglement, as in Chapter 12, where the state of one particle is entangled with that of another; and (b) *self*-entanglement, as in Eq. (4.9), where one particle's state in data space is trivially entangled with *its own* state in a unitary space (momentum-energy space here).

In Appendix D the Schrödinger wave equation is derived using an approximate (since non-relativistic) EPI approach. Also, a transition to classical mechanics is made, yielding Newton's second law and the virial theorem.

# 5
# Classical electrodynamics

## 5.1 Derivation of vector wave equation

### *5.1.1 Goals*

The aim of this chapter is to derive Maxwell's equations in vacuum via the EPI principle. This is done in two steps: by (1) deriving the vector wave equation; and then (2) showing that this implies the Maxwell equations. The latter is an easy task. The main task is problem (1).

We follow the EPI procedure of Sec. 3.4. This requires identifying the parameters to be measured (Sec. 3.4.4), forming the corresponding expression (2.19) for $I$, and then finding $J$ by an appropriate invariance principle (Sec. 3.4.5). These steps are followed next.

### *5.1.2 Choice of ideal parameter* $\boldsymbol{\theta}$

The four components of the electromagnetic potential $\mathbf{A}$ are defined as

$$\mathbf{A} \equiv (A_n,\ n = 1, \ldots, 4) \equiv (A_1, A_2, A_3, \phi) \equiv (\boldsymbol{A}, \phi) \tag{5.1}$$

in terms of the vector *three*-potential $\boldsymbol{A}$ and the scalar potential $\phi$. Note that Eq. (5.1) does not define the so-called *four-potential* of electromagnetic theory. In our notation (3.34c), this would require an imaginary $i$ to multiply either $\phi$ or $\boldsymbol{A}$ in the equation. We do not use a four-potential (and four-current) in this chapter, as it would merely add unnecessary complication to the approach that is taken. Vector $\mathbf{A}$ and the main output of the approach, Eq. (5.51), can easily be placed in four-vector form anyhow; see Sec. 5.1.23.

This agrees with the fact that the potential $\mathbf{A}$ and vector $\mathbf{q}$ are proportional for this problem, and $\mathbf{q}$ was found (Sec. 2.4.4) to not generally be a four-vector, i.e., not be covariant. Instead, *each component* $A_n$ and $q_n$ will be covariant, i.e., will obey a covariant differential equation; again, as implied in Sec. 2.4.4.

On the other hand, the Fisher coordinates $\mathbf{x}$ will remain a four-vector, as is generally required by EPI (Sec. 3.5.8).

The magnetic and electric field quantities $\boldsymbol{B}$ and $\boldsymbol{E}$ are defined in terms of $\mathbf{A}$ by Eqs. (5.85) and (5.86) (Jackson, 1975). Vector $\mathbf{A}$ obeys the Lorentz condition, Eq. (5.84). Fields $\boldsymbol{E}$ and $\boldsymbol{B}$ are direct observables.

Consider an experiment whose aim is to determine one of the fields $\boldsymbol{E}$ or $\boldsymbol{B}$ at an ideal four-position $\boldsymbol{\theta}$. The measuring instrument that is used for this purpose acts, as in Sec. 3.8, to perturb the amplitude functions $\mathbf{q}$ of the problem at its input space. (Note that the $\mathbf{q}$ are not identified with specifically electromagnetic quantities until Sec. 5.1.21.) Also, in the input space, random errors in position $\mathbf{x}$ occur that define the "intrinsic data" of the problem

$$\mathbf{y} = \boldsymbol{\theta} + \mathbf{x}. \tag{5.2}$$

Thus, $\boldsymbol{E}$ or $\boldsymbol{B}$ is measured, but *precisely where* is unknown. The $\mathbf{x}$ are the Fisher variables of the problem.

The Fisher variables $\mathbf{x}$ must be related to the space $\boldsymbol{r}$ and time $t$ fluctuations of the measurement location. We choose to use

$$\mathbf{x} \equiv (i\boldsymbol{r}, ct), \qquad \boldsymbol{r} \equiv (x, y, z) \equiv (x_1, x_2, x_3). \tag{5.3}$$

These are consistent with Eq. (4.1) defining the measurement problem in quantum mechanics. Actually, the coordinates $(\boldsymbol{r}, ict)$ would work as well, as in quantum mechanics (Sec. 4.1.2).

Equation (5.2) states that input positional accuracy is limited only by fluctuations that characterize the electromagnetic field. The most fundamental of these are the "vacuum fluctuations," which give rise to an uncertainty in position of amount

$$\Delta x \sim [\hbar/(m_0\omega_0)]^{1/2}. \tag{5.4}$$

Quantities $m_0$ and $\omega_0$ are the mass and frequency of an oscillator defining the electromagnetic field (see Misner *et al.*, 1973, p. 1191). (Note: At this point Eq. (5.4) is purely motivational; it is derived in Eq. (5.47) below.)

### *5.1.3 Information I*

By Eqs. (2.11) and (5.3), the intrinsic data $\mathbf{y}_n$ are assumed to be collected independently, and there is a PDF

$$p_n(\mathbf{y}_n|\boldsymbol{\theta}_n) = p_n(\boldsymbol{r}, t) \equiv q_n^2(\boldsymbol{r}, ct) \tag{5.5}$$

describing each ($n$th) four-measurement. Note that $(\boldsymbol{r}, t)$ are real coordinates: The PDF for a general complex coordinate is that of its real and imaginary parts (Sec. 4.1.7). Also, the argument $ct$ goes to $t$, as following Eq. (4.3). The Fisher information $I$ in the *position* measurements (see preceding) is to be found. The result is Eq. (2.19) which, with our choice (5.3) of coordinates, becomes

$$I = 4c \iint d\boldsymbol{r}\, dt \sum_{n=1}^{N} \left[ -\nabla q_n \cdot \nabla q_n + \frac{1}{c^2} \left( \frac{\partial q_n}{\partial t} \right)^2 \right]. \tag{5.6}$$

Note that this is very similar to the quantum mechanical answer (4.2) for $I$. Here we simply don't pack the amplitude functions $\mathbf{q}$ to form new, *complex* amplitudes.

### 5.1.4 Invariance principles

The purpose of the invariance principle for a scenario is to define the bound information $J$. In Chapter 4, this was accomplished by a principle of unitary transformation between coordinate–time space and momentum–energy space. Here, there is no unitary transformation that connects $(\boldsymbol{r}, t)$ space with another physically meaningful coordinate space. Since information $J$ is to specifically relate to (be "bound" to) the phenomenon, we instead seek as the invariance principle a defining property of the source (see Sec. 3.11). *Assuming that the system is a closed one*, the most basic such principle is a property of *continuity of charge and current flow.* This is

$$\frac{\partial \rho}{\partial t} + \nabla \cdot \boldsymbol{j} = 0, \quad \nabla \equiv (\partial/\partial x_1, \ldots, \partial/\partial x_3) \equiv (\partial/\partial x, \partial/\partial y, \partial/\partial z)$$
$$\rho = \rho(\boldsymbol{r}, t), \quad \boldsymbol{j} = \boldsymbol{j}(\boldsymbol{r}, t), \tag{5.7}$$

with the current/area $\boldsymbol{j}$ and the charge/volume $\rho$ as the sources.

As was mentioned in Secs. 3.5.7 and 3.6.3, the information $K$ should obey gauge covariance and relativistic covariance. Use of the (covariant) Lorentz condition

$$\frac{1}{c}\frac{\partial q_4}{\partial t} + \sum_{n=1}^{3} \frac{\partial q_n}{\partial x_n} = 0 \tag{5.8}$$

helps to achieve these aims. Equation (5.8) may be regarded as an auxiliary condition that is supplemental to the invariance principle (5.7). Alternatively, since (5.8) is *itself* the mathematical expression of an equation of continuity of flow, we can regard the problem to be physically defined by a single *joint* condition of continuity of flow, (5.7) and (5.8).

### 5.1.5 Fixing $N = 4$

The Lorentz condition (5.8) is the only physical input into EPI that functionally involves the unknown amplitudes $\mathbf{q}$. Since the highest index $n$ that enters into

Eq. (5.8) is value 4, we take this to mean that the sufficient number of amplitudes needed to express the theory is $N = 4$. This step could have been taken at the end of the derivation, but is more convenient to fix here once and for all.

### 5.1.6 On finding the bound information J

In the quantum mechanical scenario of Chapter 4, the invariance principle in use permitted $J$ to be directly expressed as $I$ (see Eq. (4.9)). Then we immediately knew that $\kappa = 1$ and were able to solve (3.16) and (3.18) for their distinct solutions.

By comparison, principles (5.7) and (5.8) do not directly express $I$ in terms of a physically dependent information $J$. Therefore, at this point we do not know either $J$ or the efficiency constant $\kappa$. They have to be solved for, based upon use of the conditions (5.7) and (5.8). This corresponds to a case (a) described in Sec. 3.4.6, where both EPI conditions (3.16) and (3.18) must be solved *simultaneously* for a common solution $\mathbf{q}$. We do this below.

### 5.1.7 General form for J

The bound information $J$ is a scalar *functional* of all physical aspects of the problem, i.e. quantities $\mathbf{q}(\mathbf{x})$, $\boldsymbol{j}(\mathbf{x})$, and $\rho(\mathbf{x})$ at all $\mathbf{x}$. It can be represented generally as an inner product

$$J = 4c \int\!\!\int d\mathbf{r}\, dt \sum_{n=1}^{4} E_n J_n, \qquad E_n = const_n, \qquad J_n = J_n(\mathbf{q}, \boldsymbol{j}, \rho). \tag{5.9}$$

The constants $E_n$ and functions $J_n(\mathbf{q}, \boldsymbol{j}, \rho)$ are to be found.

### 5.1.8 EPI variational solution

As mentioned in the plan above, we first find the solution for $\mathbf{q}$ to problem (3.16). Subtracting Eq. (5.9) from (5.6) gives the information Lagrangian

$$\mathscr{L} = 4c \sum_n \left[ -\nabla q_n \cdot \nabla q_n + \frac{1}{c^2}\left(\frac{\partial q_n}{\partial t}\right)^2 - E_n J_n(\mathbf{q}, \boldsymbol{j}, \rho) \right]. \tag{5.10}$$

This is used in the Euler–Lagrange Eq. (0.34) for the problem,

$$\sum_{k=1}^{3} \frac{d}{dx_k}\left(\frac{\partial \mathscr{L}}{\partial q_{nk}}\right) + \frac{d}{dt}\left(\frac{\partial \mathscr{L}}{\partial q_{n4}}\right) = \frac{\partial \mathscr{L}}{\partial q_n},$$
$$n = 1, \ldots, 4; \qquad q_{nk} \equiv \frac{\partial q_n}{\partial x_k}, \qquad q_{n4} \equiv \frac{\partial q_n}{\partial t}. \tag{5.11}$$

The result is directly

$$\Box q_n = -\frac{1}{2}\sum_m E_m \frac{\partial J_m}{\partial q_n}, \tag{5.12a}$$

$$\text{where} \quad \Box \equiv \frac{1}{c^2}\frac{\partial^2}{\partial t^2} - \nabla^2, \quad \nabla^2 \equiv \sum_{k=1}^{3} \frac{\partial^2}{\partial x_k^2}. \tag{5.12b}$$

$\Box$ is called a d'Alembertian operator and $\nabla^2$ is called a Laplacian operator.

### 5.1.9 Alternative form for $I$

The expression (5.6) for $I$ may be integrated by parts. First note the elementary result

$$\int dx \left(\frac{dq}{dx}\right)^2 = \frac{dq}{dx}\, q\Big|_{-\infty}^{\infty} - \int dx\, q \frac{d^2 q}{dx^2} = 0 - \int dx\, q \frac{d^2 q}{dx^2} \tag{5.13}$$

for a probability amplitude $q$. The zero occurs because we will assume Dirichlet or Neumann conditions to be obeyed by the potential $\mathbf{A}$ at the boundaries to the observation space (Eq. (5.55)); and each component of $\mathbf{A}$ will be made proportional to a corresponding component of $\mathbf{q}$ (Eq. (5.48)). Integrating by parts in this manner for every coordinate $(\boldsymbol{r}, t)$, information (5.6) becomes

$$I = -4c \int\int d\boldsymbol{r}\, dt \sum_n q_n \Box q_n. \tag{5.14}$$

This is used in the following.

### 5.1.10 EPI zero-root solution

The second EPI problem is to find the roots $\mathbf{q}$ of (3.18),

$$I[\mathbf{q}] - \kappa J[\mathbf{q}] = 0, \tag{5.15}$$

$I$ given by (5.14) and $J$ given by (5.9). The problem is thus

$$-4c \int\int d\boldsymbol{r}\, dt \sum_n (q_n \Box q_n + \kappa E_n J_n(\mathbf{q}, \boldsymbol{j}, \rho)) \equiv 0. \tag{5.16}$$

Its solution is the microscale Eq. (3.20), which is here

$$q_n \Box q_n = -\kappa E_n J_n. \tag{5.17}$$

### 5.1.11 Common solution q

From intermediary solutions (5.12a) and (5.17), it must be that

$$\tfrac{1}{2}q_n \sum_m E_m \frac{\partial J_m}{\partial q_n} = \kappa E_n J_n. \tag{5.18}$$

This allows us to specify more precisely the form for unknown functions $J_n(\mathbf{q}, \boldsymbol{j}, \rho)$, as follows.

Notice that the right-hand side of (5.18) has been "sifted" out of the sum on the left-hand side. Then it must be that

$$\tfrac{1}{2}q_n \frac{\partial J_m}{\partial q_n} = \kappa J_m \delta_{mn}, \tag{5.19}$$

where $\delta_{\mathrm{mn}}$ is the Kronecker delta function. Now consider two cases.

Case 1: $m \neq n$. Eq. (5.19) gives $\partial J_m / \partial q_n = 0$. This implies that

$$J_m(\mathbf{q}, \boldsymbol{j}, \rho) = J_m(q_m, \boldsymbol{j}, \rho). \tag{5.20}$$

Case 2: $m = n$. Eq. (5.19) gives $q_n \partial J_n / \partial q_n = 2\kappa J_n$. But since, by (5.20), the $\mathbf{q}$ dependence of each $J_n$ is only through the single $q_n$, it follows that $\partial J_n / \partial q_n = dJ_n / dq_n$, the full derivative. Using this in Eq. (5.19) allows the latter to be integrated,

$$\ln J_n = 2\kappa \ln q_n + D_n(\boldsymbol{j}, \rho), \quad \text{or} \quad J_n = q_n^{2\kappa} G_n(\boldsymbol{j}, \rho). \tag{5.21}$$

Quantities $D_n$ and $G_n$ are integration "constants" after the integration in $q_n$, and hence can still functionally depend upon $(\boldsymbol{j}, \rho)$ as indicated.

### 5.1.12 Resulting wave equation

Using form (5.21) for $J_n$ in wave equation (5.12a) gives

$$\Box q_n = -\tfrac{1}{2}E_n \cdot 2\kappa q_n^{2\kappa-1} \cdot G_n(\boldsymbol{j}, \rho), \quad \text{or} \quad \Box q_n = q_n^b F_n(\boldsymbol{j}, \rho), \tag{5.22a}$$

$$\text{where} \quad F_n \equiv -\kappa E_n G_n, \quad n = 1\text{–}4, \quad b \equiv 2\kappa - 1. \tag{5.22b}$$

The parameter $b$ and the new functions $F_n(\boldsymbol{j}, \rho)$ need to be found.

### 5.1.13 Where we stand so far

It is interesting to note that the information approach has allowed us to proceed as far as a wave equation, Eq. (5.22a), without the need for any more specifically electromagnetic a requirement than an unspecified dependence $F_n$ upon the electromagnetic sources $\boldsymbol{j}$ and $\rho$. To evaluate this dependence will, finally, require use of the invariance principles (5.7) and (5.8). This is done in the following sections. Note that principles (5.7) and (5.8) are actually quite weak as statements of specifically electromagnetic phenomena. That their use can lead to Maxwell's equations seems remarkable. The key is to combine them with EPI, as will become apparent.

### 5.1.14 Use of Lorentz condition and conservation of flow condition

We proceed in the following sections to find the parameter $b$ and the functions $F_n(\boldsymbol{j}, \rho)$ in Eq. (5.22a). In this section we find a set of conditions (5.25a, b) that the unknowns $F_n(\boldsymbol{j}, \rho)$ must obey. This is accomplished through use of the two invariance conditions (5.7) and (5.8).

Start by operating upon Eq. (5.22a) for $n = 1$ with $\partial/\partial x_1$. Save the resulting equation. Then operate upon Eq. (5.22a) for $n = 2$ with $\partial/\partial x_2$; etc., through $n = 4$ and operation $(1/c)\,\partial/\partial t$. Add the resulting equations. (Of course this could all be indicated more briefly, albeit with some sacrifice of simplicity, if we used tensor notation. However, for the sake of clarity we defer use of tensor notation until the chapter on general relativity, where its use is mandatory.) Because the operations $\partial/\partial x_1, \partial/\partial x_2$, etc., commute with $\Box$, the left-hand side becomes

$$\Box\left(\sum_{k=1}^{3}\frac{\partial q_k}{\partial x_k}+\frac{1}{c}\frac{\partial q_4}{\partial t}\right)=0 \tag{5.23a}$$

since the quantity within parentheses is the Lorentz form (5.8).

Then operating in the same way on the *right-hand* side of Eq. (5.22a) should likewise give zero. Operating $\partial/\partial x_1$ in this way for $n = 1$ gives

$$q_1^b\left(\sum_{m=1}^{3}\frac{\partial F_1}{\partial j_m}\frac{\partial j_m}{\partial x_1}+\frac{\partial F_1}{\partial \rho}\frac{\partial \rho}{\partial x_1}\right) \tag{5.23b}$$

plus another term, which is zero if $b = 0$ (as will be the case). Or, operating in the same way for $n = 2, 3, 4$ gives analogous equations. Adding them gives

$$\sum_{n=1}^{3} q_n^b\left(\sum_{m=1}^{3}\frac{\partial F_n}{\partial j_m}\frac{\partial j_m}{\partial x_n}+\frac{\partial F_n}{\partial \rho}\frac{\partial \rho}{\partial x_n}\right)+q_4^b\left(\sum_{m=1}^{3}\frac{\partial F_4}{\partial j_m}\frac{1}{c}\frac{\partial j_m}{\partial t}+\frac{\partial F_4}{\partial \rho}\frac{1}{c}\frac{\partial \rho}{\partial t}\right)\equiv S\equiv 0 \tag{5.24}$$

by our requirement.

We show next that this is satisfied if

$$q_n^b\frac{\partial F_n}{\partial j_m}=B\delta_{mn},\quad q_n^b\frac{\partial F_n}{\partial \rho}=0;\quad m,\ n=1,\ 2,\ 3; \tag{5.25a}$$

$$q_4^b\frac{\partial F_4}{\partial j_m}=0,\quad m=1,\ 2,\ 3;\quad q_4^b\frac{\partial F_4}{\partial \rho}=cB; \tag{5.25b}$$

$$\text{where}\quad B\equiv B(\boldsymbol{j},\ \rho), \tag{5.25c}$$

a new *scalar* function of the sources. Substituting the identities (5.25a) into (5.24) and using the sifting property of the Kronecker delta makes the first sum over $n$ collapse to

$$B\sum_{n=1}^{3}\frac{\partial j_n}{\partial x_n} \equiv B\nabla\cdot\boldsymbol{j}. \tag{5.26}$$

Operator $\nabla$ is the usual three-dimensional gradient operator (5.7). Similarly, substituting the identities (5.25b) makes the second sum collapse to

$$B\frac{\partial\rho}{\partial t}. \tag{5.27}$$

Hence, the total sum $S$ in Eq. (5.24) becomes

$$S = B\left(\nabla\cdot\boldsymbol{j} + \frac{\partial\rho}{\partial t}\right) \equiv 0 \tag{5.28}$$

by the equation of flow (5.7). This is what we set out to show. Therefore, Eq. (5.24) is satisfied by the solution Eqs. (5.25a–c).

### 5.1.15 Finding exponent b

Condition (5.25a) permits us to find the exponent $b$. Since $F_n = F_n(\boldsymbol{j}, \rho)$, the left-hand side of the first relation (5.25a) only varies in $\mathbf{q}$ as $q_n^b$. This means that the right-hand side's dependence upon $\mathbf{q}$ is most generally of this form as well. But by Eq. (5.25c) function $B$ does not depend upon $\mathbf{q}$. Hence, the only way the $\mathbf{q}$-dependence in the first relation (5.25a) could balance is by having

$$b = 0. \tag{5.29}$$

### 5.1.16 Finding function B(j, ρ)

This unknown function may be found from requirements (5.25a) *under the condition* (5.29). The second requirement (5.25a) implies that

$$F_n(\boldsymbol{j}, \rho) = F_n(\boldsymbol{j}), \quad n = 1, 2, 3. \tag{5.30a}$$

The *first* requirement (5.25a) for $m \neq n$ shows, using Eq. (5.29), that

$$F_n(\boldsymbol{j}) = F_n(j_n). \tag{5.30b}$$

Then, the first requirement (5.25a) gives, for $m = n$,

$$\frac{\partial F_n(j_n)}{\partial j_n} = B(\boldsymbol{j}, \rho), \quad n = 1, 2, 3. \tag{5.31}$$

Since the left-hand side doesn't depend upon $\rho$, necessarily

$$B(\boldsymbol{j}, \rho) = B(\boldsymbol{j}). \tag{5.32}$$

On the other hand, the first Eq. (5.25b) implies that $F_4(\boldsymbol{j}, \rho) = F_4(\rho)$ alone. Then the left-hand side of the *second* Eq. (5.25b) has no $\boldsymbol{j}$-dependence, implying that on the right-hand side $B(\boldsymbol{j}, \rho) = B(\rho)$ alone.

The only way that the latter and Eq. (5.32) can simultaneously be true is if

$$B(\boldsymbol{j}, \rho) = B \equiv const. \tag{5.33}$$

### 5.1.17 Finding functions $F_n(\boldsymbol{j}, \rho)$

Since $b = 0$, the first condition (5.25b) shows that

$$F_4(\boldsymbol{j}, \rho) = F_4(\rho). \tag{5.34}$$

Then the second condition (5.25b), combined with results (5.29) and (5.33), becomes a simple differential equation

$$\frac{dF_4(\rho)}{d\rho} = cB = const. \tag{5.35}$$

This has the elementary solution

$$F_4(\rho) = cB\rho + C_4, \quad C_4 = const. \tag{5.36}$$

Likewise, the first Eq. (5.25a) becomes, for index $m = n$ and using Eqs. (5.29) and (5.30a, b), a simple differential equation

$$\frac{dF_n(j_n)}{dj_n} = B = const., \quad n = 1, 2, 3. \tag{5.37}$$

The solution is

$$F_n(j_n) = Bj_n + C_n, \quad C_n = const., \quad n = 1, 2, 3. \tag{5.38}$$

In summary, all functions $F_n$ are linear in their corresponding currents or charge density. Constants $B$, $C_n$ remain to be found.

### 5.1.18 Efficiency parameter $\kappa$

By Eqs. (5.22b) and (5.29), for this application of EPI the efficiency parameter

$$\kappa = 1/2. \tag{5.39}$$

The complete meaning of this effect is not yet known. However, by the reasoning of Sec. 3.4.5, the implication is that classical electromagnetics is an approximation: only 50% of the bound or phenomenological information $J$ is utilized in the intrinsic information $I$. This, of course, agrees with the fact that the classical approach *is* an approximation, in ignoring quantum effects.

### 5.1.19 Constants C from photon case

Substitution of results (5.29), (5.36), and (5.38) into the wave equation (5.22a) gives

$$\Box \mathbf{q} = B\mathbf{J}_s + \mathbf{C}, \quad \mathbf{q} \equiv (q_1, \ldots, q_4),$$
$$\mathbf{C} \equiv (C_1, \ldots, C_4), \quad \mathbf{J}_s \equiv (\boldsymbol{j}, c\rho). \tag{5.40}$$

All quantities are four-component vectors as indicated (but not relativistic four-vectors; see below Eq. (5.1)). This equation would take the form of the classical vector wave equation for the electromagnetic potentials $(A_1, A_2, A_3, \phi) \equiv \mathbf{A}$ if they were substituted for the amplitudes $\mathbf{q}$ (much more on this in Sec. 5.1.21 *et seq.*). However, the vector wave equation does not have an additive constant $\mathbf{C}$ on its right-hand side. In fact, $\mathbf{C}$ must be $\mathbf{0}$, as is shown in the following.

We have not yet defined the physical origin of fluctuations $(\boldsymbol{r}, t)$. By comparison with Eq. (5.40), consider the Klein–Gordon equation (4.28) for a particle with mass $m = 0$ and charge $e = 0$,

$$\Box \psi = 0, \quad \psi = Q_1 + iQ_2, \tag{5.41}$$

where $Q_1$, $Q_2$ are real amplitude functions. This is a d'Alembert equation. It can be taken to describe the probability amplitude $\psi$ for the electromagnetic quantum (Eisberg, 1961, p. 697), i.e., the photon. This is in the following sense.

It had been conjectured (see, e.g. Akhiezer and Berestetskii, 1965, pp. 10, 11) that a wave function $\psi(\boldsymbol{r}, t)$ for localization of a photon does not exist. However, it was recently shown (Białynicki-Birula, 1996) that a wave function obeying Eq. (5.41) represents the local occurrence of photon energy over finite, but small (the order of a wavelength) regions of space. Thus, $\psi$ represents the probability amplitude for a "coarse-grained" (Cook, 1982) space of possible photon locations. See also the related work (Sipe, 1995) and (Deutsch and Garrison, 1991). It is interesting to note, as well, the corresponding *particle* case-limitation mentioned in Sec. 4.1.17, whereby particle localization can be no finer than the Compton length. Hence, in practice, *both* photons and particles can be defined only over coarse-grained spaces.

Compare Eq. (5.41) with Eq. (5.40) *under the same charge-free* conditions $\mathbf{J}_s = 0$,

$$\Box \mathbf{q} = \mathbf{C}. \tag{5.42}$$

From their common origin as electromagnetic phenomena, Eqs. (5.41) and (5.42) must be describing the same physical effect: by (5.41), a propagating photon. We see that the two equations are equivalent if $\mathbf{q}$ is linear in the components $Q_1$, $Q_2$ of $\psi$ and if $\mathbf{C} = \mathbf{0}$,

$$q_n = K_{n1}Q_1 + K_{n2}Q_2, \quad n = 1\text{–}4, \quad \text{and} \quad \mathbf{C} = \mathbf{0}. \tag{5.43}$$

This confirms our assertion that $\mathbf{C} = \mathbf{0}$. Note that, if $\mathbf{C}$ were not zero, then it would represent an artificial electromagnetic source in (5.42) – one that exists in the absence of real sources $\mathbf{J}_S$. Such a source has not been observed on the macroscopic level.

### 5.1.20 Plane wave solutions

We can conclude, then, that in the absence of real sources the fluctuations $(\boldsymbol{r}, t)$ in the measured field positions are those in the (coarse-grained) positions of photons. These trace from the d'Alembert Eq. (5.41) governing photon positions. A solution $\psi$ to (5.41) in the case of definite energy $E$ and momentum $\boldsymbol{\mu}$ is in the form of a plane wave

$$\psi = \exp\left[\frac{i}{\hbar}(\boldsymbol{\mu} \cdot \boldsymbol{r} - Et)\right], \quad E = c\mu. \tag{5.44}$$

Assume the wave to be travelling along the $X$ direction. Then (5.44) shows oscillatory behavior in $x$ with a wavelength

$$\Delta x = \frac{2\pi\hbar}{\mu}. \tag{5.45}$$

The equivalent classical oscillator of mass $m_0$ has an energy

$$E_0 = \frac{\mu^2}{2m_0} \equiv \hbar\omega_0, \quad \text{so that} \quad \mu = (2\hbar m_0\omega_0)^{1/2}. \tag{5.46}$$

Using the latter in Eq. (5.45) gives an uncertainty

$$\Delta x = \pi\sqrt{2}[\hbar/(m_0\omega_0)]^{1/2}. \tag{5.47}$$

This confirms Eq. (5.4) for the expected vacuum fluctuation.

### 5.1.21 A correspondence between probability amplitudes and potentials

We found in Sec. 5.1.19 that $\mathbf{C} = \mathbf{0}$. Then the EPI output Eq. (5.40) takes the form of a wave equation in the amplitudes $\mathbf{q}$ due to an *electromagnetic source* $\mathbf{J}_S$. Also, as discussed in Sec. 4.1.27, probability amplitudes $q_n$ are effectively *potentials* for single-particle exchange forces. This and the *photon* plane wave correspondences of Secs. 5.1.19 and 5.1.20 imply that the amplitudes $\mathbf{q}$ in (5.40) are *linear* in the *electromagnetic potentials* $(A_1, A_2, A_3, \phi)$,

$$q_n \equiv aA_n, \quad n = 1, 2, 3, \quad q_4 \equiv a\phi, \quad a = const. \tag{5.48}$$

Using $\mathbf{C} = \mathbf{0}$ and the latter in Eq. (5.40) gives

$$\Box \mathbf{A} = B\mathbf{J}_s. \tag{5.49}$$

We absorbed the constant $a$ into the constant $B$. This is the vector wave equation with an, as yet, unspecified constant multiplier $B$.

Regarding uniqueness, note that we could have added an arbitrary function $\mathbf{f}$ to the right-hand sides of definitions (5.48). Provided that $\Box \mathbf{f} = 0$, Eq. (5.49) would still result. However, for purposes of defining $\mathbf{q}$ uniquely, we do not do this. Function $\mathbf{f}$ amounts to an arbitrary choice of gauge function for potential $\mathbf{A}$. We later show that the solution $\mathbf{A}$ to Eq. (5.49) obeys Eq. (5.57), where it is uniquely defined by given sources $\mathbf{J}_s$. (By Eq. (5.48), $\mathbf{q}$ is *likewise* unique.)

### 5.1.22 The constant B

Here we find the value of the constant in Eq. (5.49). Since $B$ does not depend upon either the sources $\mathbf{J}_s$ or the potentials $\mathbf{A}$, it must be a constant. By Sec. 3.4.14 it must also be a *universal* constant. Next use dimensional analysis. From the known units for fields $\boldsymbol{B}$ and $\boldsymbol{E}$, by Eqs. (5.85) and (5.86) those for $\mathbf{A}$ are charge/length. Then the units of $\Box \boldsymbol{A}$ are charge/length$^3$. Also, those of $\mathbf{J}_s$ are charge/length$^2$–time. Hence, to balance units in (5.49) $B$ must have units of time/length. Then the universal constant is the reciprocal of a velocity. From the correspondence Eq. (5.48), this must be the velocity of propagation of electromagnetic disturbances, i.e., $c$. With the particular use of m.k.s. units as well, we then get

$$B = \frac{4\pi}{c}. \tag{5.50}$$

### 5.1.23 Vector wave equation

Then Eq. (5.49) becomes

$$\Box \mathbf{A} = \frac{4\pi}{c}\mathbf{J}_s. \tag{5.51}$$

This is the vector wave equation in the Lorentz gauge. Its derivation was the main goal of this chapter.

It is more usual to express Eq. (5.51) in terms of four-vectors rather than the four-component vectors $\mathbf{A}$ and $\mathbf{J}_s$ that we use. If the fourth component of Eq. (5.51) is merely multiplied by the imaginary number $i$ then (5.51) takes the usual form.

### 5.1.24 Probability law on fluctuations

By the use of Eqs. (5.5) and (5.48) in Eq. (2.23), we get

$$p(\boldsymbol{r},\, t) = a^2 \sum_{n=1}^{4} A_n^2, \quad p(\boldsymbol{r},\, t) \geqslant 0 \tag{5.52}$$

as a PDF on the position errors $\boldsymbol{r}$, $t$. A factor of 1/4 is absorbed into the constant $a$. The important property of positivity for the PDF is noted. To be useful, a PDF should be uniquely defined by given physical conditions. Also, it should be normalizable or, if not, at least usable for computing averages. We examine these questions, in turn.

### 5.1.25 Uniqueness property

Is the PDF $p(\boldsymbol{r}, t)$ defined by Eq. (5.52) unique? By this equation, the uniqueness of $p$ depends upon the uniqueness of $\mathbf{A}$ (up to a $\pm$ sign). The latter is defined by solution $\mathbf{A}$ to Eq. (5.51). It is known that solutions to this problem under stated boundary value conditions and initial conditions are unique (Morse and Feshbach, 1953, p. 834). Typical boundary conditions are as shown in Fig. 5.1.

Suppose that the solution $\mathbf{A}(\boldsymbol{r}, t)$ is to be determined at field positions $\boldsymbol{r}$ that are within a volume $V$ that contains the source $\mathbf{J}_s$. Field- and source-coordinates are denoted as $(\boldsymbol{r}, t)$ and $(\boldsymbol{r}', t')$, respectively. Volume $V$ is enclosed by a given surface $\boldsymbol{\sigma}$. Denote field positions $\boldsymbol{r}$ lying within $V$ as $\boldsymbol{r} \subset V$. We consider cases of both finite and infinite $V$. The solution is given by (Morse and Feshbach, 1953, p. 837)

$$\mathbf{A}(\boldsymbol{r},\, t)_{\boldsymbol{r}\subset V} = \mathbf{A}_0(\boldsymbol{r},\, t) + \frac{1}{c}\int_0^t dt' \int_V d\boldsymbol{r}'\, G\mathbf{J}_s(\boldsymbol{r}',\, t')$$

$$+ \frac{1}{4\pi}\int_0^{t^+} dt' \oint d\boldsymbol{\sigma}\cdot(G\nabla\mathbf{A} - \mathbf{A}\nabla G), \quad G \equiv G(\boldsymbol{r},\, t;\, \boldsymbol{r}',\, t'), \tag{5.53a}$$

where

$$\mathbf{A}_0(\boldsymbol{r},\, t) = -\frac{1}{4\pi c^2}\int_V d\boldsymbol{r}' \left[\left(\frac{\partial G}{\partial t'}\right)\mathbf{A}(\boldsymbol{r}',\, t') - G\,\frac{\partial \mathbf{A}(\boldsymbol{r}',\, t')}{\partial t'}\right]_{t'=0}. \tag{5.53b}$$

In Eq. (5.53a), the indicated gradients $\nabla\mathbf{A}$ and $\nabla G$ are with respect to the primed coordinates $(\boldsymbol{r}', t')$. Also, the notation $\oint d\boldsymbol{\sigma}$ denotes an integral over the surface $\boldsymbol{\sigma}$ enclosing the source, and $d\boldsymbol{\sigma}$ is a vector whose direction is the outward-pointing normal to the local element $d\boldsymbol{\sigma}$ of the surface. Finally, the

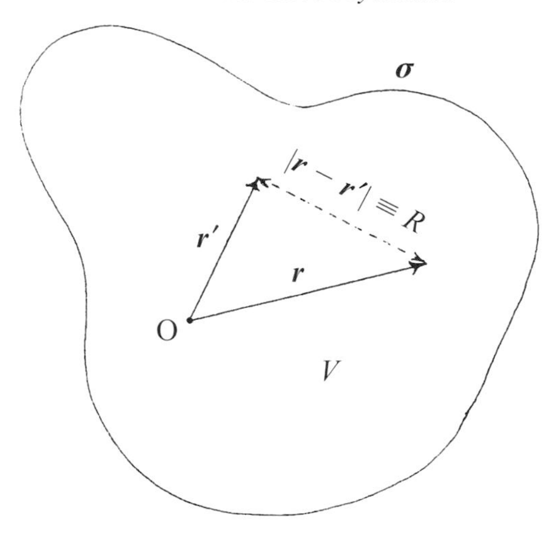


Fig. 5.1. Geometry for the electromagnetic problem. Boundary surface $\boldsymbol{\sigma}$ encloses volume $V$. With an origin at point O, position $\boldsymbol{r}$ locates a field point, and $\boldsymbol{r}'$ locates a source point, within $V$. Quantities $\boldsymbol{r}$ and $\boldsymbol{r}'$ are also random errors in position measurements, by the EPI model.

new function $G$ is called a "Green function." It is the scalar solution to Eq. (5.51) with the vector source $\mathbf{J}_s$ replaced by a scalar impulse source $\delta(\boldsymbol{r} - \boldsymbol{r}')\delta(t - t')$.

We next show that, due to the isolation requirement (Secs. 1.2.1, 1.3.2, 1.8.5, 3.2 and 5.1.4) of the problem, under typical electromagnetic boundary conditions only the second right-hand term in Eq. (5.53a) contributes to $\mathbf{A}$.

The integral (5.53b) represents the contribution of initial conditions in $\mathbf{A}$ and $G$ to $\mathbf{A}$ as evaluated at later times $t$. The source is assumed to be turned "on" at a time $t' = t_1 \gg 0$ and "off" at a time $t' = t_2 > t_1$. Thus, the integral (5.53b) is evaluated at a time $t' = 0$ in the *infinitely distant past* to $t$. Since, by correspondence (5.48) $\mathbf{A}$ represents a probability amplitude, by the *isolation requirement* of EPI it must be that $\mathbf{A}(\boldsymbol{r}', 0) = 0$: the probability for the infinitely past event vanishes. This is isolation in the temporal sense. Also, from the viewpoint of normalization, a PDF must go to zero as its argument approaches $\pm\infty$. So must its first derivative. The upshot is that both $\mathbf{A}(\boldsymbol{r}', t')$ and $\partial\mathbf{A}(\boldsymbol{r}', t')/\partial t'$ are zero at $t' = 0$, and Eq. (5.53b) obeys

$$\mathbf{A}_0(\boldsymbol{r}, t) = 0. \tag{5.54}$$

The initial potential is zero for this problem.

We next turn to evaluation of the third right-hand term in Eq. (5.53a), a surface integral. Typically, the potential $\mathbf{A}$ or its first derivatives $\nabla\mathbf{A}$ are zero on the surface,

$$\mathbf{A}\Big|_{\sigma} = 0 \quad \text{or} \quad \nabla\mathbf{A}\Big|_{\sigma} = 0. \tag{5.55}$$

These are called zero "Dirichlet" or "Neumann" boundary conditions, respectively. We can choose the Green function to obey the same boundary condition as does the potential function, so that either

$$\mathbf{A}\Big|_{\sigma} = 0 \text{ and } G\Big|_{\sigma} = 0, \quad \text{or} \quad \nabla\mathbf{A}\Big|_{\sigma} = 0 \text{ and } \nabla G\Big|_{\sigma} = 0. \tag{5.56}$$

Substituting Eqs. (5.54)–(5.56) into Eq. (5.53a) gives

$$A(\boldsymbol{r}, t)_{\boldsymbol{r}\subset V} = \frac{1}{c}\int_0^t dt' \int_V d\boldsymbol{r}'\, G(\boldsymbol{r}, t; \boldsymbol{r}', t')\mathbf{J}_{\mathrm{s}}(\boldsymbol{r}', t'). \tag{5.57}$$

This is what we set out to show.

The integral (5.57) is well defined both for continuous and for impulsive (point, line or surface) sources. The sifting property (0.69) of the delta function would merely turn the right-hand integral in (5.57) into a sum over the sources.

We next turn to evaluation of the right-hand integral. As it turns out, this depends upon whether the bounding surface $\boldsymbol{\sigma}$ is at a finite or infinite distance from the source. The former is a "closed space problem"; the latter is an "open space problem."

Consider the closed space problem. In view of the limited time interval $t_1 \leqslant t' \leqslant t_2$ for which the source is "on" (i.e., is non-zero), and the fact that the output potential $\mathbf{A}$ is proportional to the source $\mathbf{J}_{\mathrm{s}}$, potential $\mathbf{A}$ is effectively zero outside some time interval. Hence, we limit our attention to time fluctuation values $t$ obeying $\bar{t}_1 \leqslant t \leqslant \bar{t}_2$. Denote these as $t \subset T$. Since the space of positions $\boldsymbol{r} \subset V$ is closed as well (Eq. (5.53a)), we see that we have a four-dimensional *closed space* problem

$$\boldsymbol{r} \subset V, \quad t \subset T. \tag{5.58}$$

From Eq. (5.57) we see that the solution $\mathbf{A}$ is defined by the Green function $G$ for the problem. For the boundary conditions mentioned above, the solution is known to be (Morse and Feshbach, 1953, p. 850)

$$G(\boldsymbol{r}, t; \boldsymbol{r}', t') = 4\pi c^2 \sum_n \frac{\sin[\omega_n(t - t')]}{\omega_n} u(t - t')\psi_n^*(\boldsymbol{r}')\psi_n(\boldsymbol{r}), \tag{5.59}$$

where $u(t - t') = 1$ for $t > t'$ or 0 for $t < t'$ (the Heaviside function). Also, $\omega_n \equiv ck_n$, where the $k_n$ are eigenvalues, and the functions $\psi_n(\boldsymbol{r})$ are eigenfunctions, of an associated problem: The $\psi_n$ and $k_n$ are to (a) satisfy the Helmholtz equation

$$\nabla^2\psi_n + k_n^2\psi_n = 0, \quad \psi_n \equiv \psi_n(\boldsymbol{r}) \tag{5.60}$$

within the volume $V$, subject to the same Dirichlet or Neumann zero-conditions on the surface $\sigma$ as $\mathbf{A}$, and (b) form an orthonormal set within the volume $V$,

$$\int_V d\boldsymbol{r}\, \psi_n^* \psi_m = \delta_{nm}. \tag{5.61}$$

Obviously the shape of the surface $\boldsymbol{\sigma}$ is decisive in defining the orthonormal functions $\psi_n(\boldsymbol{r})$ of the problem.

Substituting Eq. (5.59) into Eq. (5.57) gives the unique solution to the closed space problem. We next briefly turn to the open space problem.

In the open space problem, the boundary $\boldsymbol{\sigma}$ recedes toward infinity. As before, assume the source $\mathbf{J}_s$ to extend over a limited space – call it $V_0$ – and to be "on" for a limited amount of time. Assume the same initial conditions as before, so that Eq. (5.54) holds. Also, the surface integral in Eq. (5.53a) vanishes once again, this time because of the isolation requirement of EPI (Secs. 1.2.1, 1.8.5, 3.2, etc.). No spatial "events" $\boldsymbol{r} \to \infty$ are allowed so that, in its role as a *probability* amplitude, $\mathbf{A}$ must obey

$$\lim_{r\to\infty} \mathbf{A}(\boldsymbol{r}, t) = 0, \qquad \lim_{r\to\infty} \nabla\mathbf{A}(\boldsymbol{r}, t) = 0. \tag{5.62}$$

That is, there is zero probability for such events.

Hence, the solution for $\mathbf{A}$ obeys Eq. (5.57) once more. This requires knowledge of the open space Green function $G$. Assume a situation of "causality," that is, that the effect of a source located a finite distance $|\boldsymbol{r} - \boldsymbol{r}'|$ away that is turned "on" at a time $t'$ is to be felt a *later* time $t > t'$. Then the Green function is (Morse and Feshbach, 1953, p. 838)

$$G(\boldsymbol{r}, t; \boldsymbol{r}', t') = \frac{\delta[(R/c) - (t - t')]}{R}, \qquad R \equiv |\boldsymbol{r} - \boldsymbol{r}'|, \tag{5.63}$$

where $t > t'$. Substituting this into Eq. (5.57) gives

$$\mathbf{A}(\boldsymbol{r}, t) = \frac{1}{4\pi c} \int_{V_0} d\boldsymbol{r}' \left[ \frac{\mathbf{J}_s(\boldsymbol{r}', t - R/c)}{R} \right]. \tag{5.64}$$

This is the well-known "retarded potential" solution, whereby the potential at a time $t$ arises from source values that occurred at an earlier time $t - R/c$. This, in turn, follows from the fact that the velocity of propagation of the disturbance is the finite value $c$. The solution for $\mathbf{A}$ is once more unique.

It may be noted that the solution (5.63) for the Green function of the open space problem depends only upon the difference of all coordinates, i.e., $(\boldsymbol{r} - \boldsymbol{r}')$, $(t - t')$. The resulting Eq. (5.57) is then in a covariant form (Jackson, 1975, p. 611). This agrees with Einstein's precept that the laws of physics should be covariant (Sec. 3.5.8). However, the Green function for the closed space problem is Eq. (5.59), and this does not seem to be expressible as a function of differences of coordinates for *arbitrary* surface shapes $\boldsymbol{\sigma}$. Hence, only certain closed space scenarios, and the open space scenario, admit of covariant solutions. This seems, at first, curious since the progenitor of these solutions – the wave Eq. (5.51) – *is* a covariant law. It also leads us to ask whether the potential **A** and its wave equation are lacking in some way as valid physical quantities.

In fact, other well-known covariant laws suffer the same fate. For example, the Klein–Gordon Eq. (4.28) of relativistic quantum mechanics is covariant. This agrees with Einstein's precept. But, the *particular solution* for which the particle is in a definite state of rest energy has a separate time factor $\exp(-imc^2 t/\hbar)$, as in Eq. (G3). Such a solution is not covariant: it ultimately gives rise to the non-relativistic, non-covariant Schrödinger wave equation (Appendix G).

One concludes that the laws of physics (electromagnetic wave equation, Klein–Gordon equation, etc.) must be covariant, as Einstein required, but their solutions under particular conditions need not be. If one knows a non-covariant solution **A** in one frame of reference and wants it in another, this can be implemented by use of the Lorentz transformation rules (3.29) and (3.33).

### 5.1.26 On covariant boundary conditions

We assumed above Eq. (5.54) that both $\mathbf{A}(\boldsymbol{r}, t)$ and $\partial\mathbf{A}(\boldsymbol{r}, t)/\partial t$ are zero at the initial time $t = 0$. These are called "Cauchy conditions" on the time coordinate. By contrast, we assumed either Dirichlet or Neumann conditions on the space coordinates (see below Eq. (5.54)). Hence, the combined space-time boundary conditions are not covariant, since they treat space and time differently. As we discussed above, the result is a non-covariant form (5.59) for the resulting Green function for the problem. This led to a non-covariant solution **A**. It is interesting to see what happens if we depart from Cauchy conditions on the time, utilizing instead the *same* Dirichlet or Neumann conditions as those that were satisfied by the space coordinates. The resulting boundary conditions would now be covariant. However, the resulting solution **A** under such conditions is not unique! This is shown in Appendix H.

Now, as we noted above Eq. (5.54), the Cauchy conditions on the time are actually *consistent with* the EPI assumption of an isolated system. Hence, in the EPI formulation, particular solutions to the wave equation cannot be covariant (although the wave equation itself, as derived by EPI in preceding sections, *is covariant*). The non-covariance problem can be avoided, as we saw, by use of the Lorentz transformation.

Does, then, the EPI approach violate the postulates of relativity? As we noted above, the Einstein requirement of covariance applies to laws of physics

– here, the electromagnetic wave equation – rather than to *particular solutions* of them. Hence, EPI does not violate the postulates of relativity.

### 5.1.27 Normalization for a closed space

Normalization requires

$$\int_T \int_V dt\, d\boldsymbol{r}\, p(\boldsymbol{r}, t) \propto \int_T \int_V dt\, d\boldsymbol{r}\, |\mathbf{A}|^2 = const., \tag{5.65}$$

based upon representation (5.52) for $p$. We consider cases of finite and, then, infinite volume $V$ for the position coordinate, i.e., closed and open spaces. The time coordinate $t$ is always within a closed temporal space $T$ (see Eq. (5.58)), as in quantum mechanics (Sec. 4.1.26).

For the closed space case, we first establish an identity. The Green function $G$ in the expansion Eq. (5.59) for $G$ is real, so the eigenfunctions $\psi_n(\boldsymbol{r})$ in that expansion can be chosen to be purely real. This is equivalent to representing a real function by a series of sin and cos functions, rather than by complex exponentials. The choice is arbitrary. Using Eq. (5.59), the orthonormality property (5.61) and the sifting property (0.62b) of the Kronecker delta function, $\delta_{\rm mn}$, gives

$$\int_V d\boldsymbol{r}\, G(\boldsymbol{r}, t; \boldsymbol{r}', t')G(\boldsymbol{r}, t; \boldsymbol{r}'', t'') = \sum_n \int_{t_1}^{t_2} dt\, \frac{\sin[\omega_n(t - t')]\sin[\omega_n(t - t'')]}{\omega_n^2} \times u(t - t')u(t - t'')\psi_n(\boldsymbol{r}')\psi_n(\boldsymbol{r}''). \tag{5.66}$$

An unimportant multiplier has been ignored. This is the identity we sought.

For simplicity of presentation we find the normalization integral (5.65) for *any one component* $A_m^2$ of $|\mathbf{A}|^2$, with subscript $m$ suppressed for convenience. This component arises from the corresponding component $J_s$ of $\mathbf{J}_s$ in Eq. (5.57). Using the latter and identity (5.66) in Eq. (5.65) gives, after rearranging orders of integration,

$$\int_T \int_V dt\, d\boldsymbol{r}\, p(\boldsymbol{r}, t) = \sum_n \omega_n^{-2} \int_{t_1}^{t_2} dt \left[\int_0^t dt' \int_V d\boldsymbol{r}' \sin[\omega_n(t - t')]\, J_s(\boldsymbol{r}', t')\psi_n(\boldsymbol{r}')\right]^2. \tag{5.67}$$

The orthonormality property (5.61) was also used, and the Heaviside functions $u$ were replaced by unity because their arguments are only positive here.

At this point, the analysis branches, depending upon whether the source component $J_s$ contains singularities; this includes poles, impulsive lines (of any shape), or surfaces. Suppose, first that there are no singularities. Then use of the Schwarz inequality on the right-hand side of Eq. (5.67) gives

$$\int_T dt \int_V d\boldsymbol{r}\, p(\boldsymbol{r}, t) \leqslant \sum_n \omega_n^{-2} \int_{t_1}^{t_2} dt\, U_n(\omega_n, t) W_n(t), \tag{5.68a}$$

$$\text{where} \quad U_n(\omega_n, t) \equiv \int_0^t dt' \int_V d\boldsymbol{r}' \sin^2[\omega_n(t - t')]\, J_\mathrm{s}^2(\boldsymbol{r}', t') \tag{5.68b}$$

$$\text{and} \quad W_n(t) \equiv \int_0^t dt' \int_V d\boldsymbol{r}'\, \psi_n^2(\boldsymbol{r}'). \tag{5.68c}$$

Factor $W_n(t)$ may be directly evaluated using orthonormality property (5.61), as

$$W_n(t) = t. \tag{5.69}$$

Also, we may again apply the Schwarz inequality, now to Eq. (5.68b), giving

$$U_n(\omega_n, t) \leqslant \int_0^t dt' \int_V d\boldsymbol{r}' \sin^4[\omega_n(t - t')] \int_0^t dt' \int_V d\boldsymbol{r}'\, J_\mathrm{s}^4(\boldsymbol{r}', t'). \tag{5.70}$$

The first double integral on the right-hand side may be evaluated analytically. The second double integral – call it $J(t)$ – is finite since volume $V$ is finite and function $J_\mathrm{s}$ is assumed to be non-singular. Substituting these results and Eq. (5.69) into Eq. (5.68a) gives

$$\int_T dt \int_V d\boldsymbol{r}\, p(\boldsymbol{r}, t) \leqslant V \sum_n \frac{1}{\omega_n^2} \int_{t_1}^{t_2} dt\, t^2 [\tfrac{3}{8} - \tfrac{1}{2}\operatorname{sinc}(2\omega_n t) + \tfrac{1}{8}\operatorname{sinc}(4\omega_n t)] J(t). \tag{5.71}$$

It is useful to examine convergence properties in the limit of $\omega_n$ large. In this limit the sinc functions contribute effectively zero to Eq. (5.71), and we get

$$\int_T dt \int_V d\boldsymbol{r}\, p(\boldsymbol{r}, t) \leqslant \tfrac{3}{8} V \left[ \int_{t_1}^{t_2} dt\, t^2 J(t) \right] \sum_n \frac{1}{\omega_n^2}. \tag{5.72}$$

The right-hand integral is finite for a well-behaved function $J(t)$ over the given interval. Also, we had $\omega_n = ck_n$, where $k_n$ is the $n$th eigenvalue. For geometries that are one-dimensional (Morse and Feshbach, 1953, p. 712) or rectangular these are proportional to $n$, with $n = 1, 2, \ldots$. Observing the summation

$$\sum_{n=1}^{\infty} \frac{1}{n^2} = \zeta(2) \approx 1.645, \tag{5.73}$$

the Riemann zeta function of argument 2, we see that the normalization integral (5.72) converges for large $n$ as the Riemann zeta function. At smaller $n$, the contributions to the integral now include terms $\operatorname{sinc}(2\omega_n t)$ and

$\mathrm{sinc}(4\omega_n t)$, and, so, depart from terms of the Riemann function, but are still finite.

Hence, the normalization integral for the single component $A_m^2$ in the sum (5.65) is finite. Since the assumed component $m$ was arbitrary (and suppressed) in the preceding analysis, each of them gives a finite contribution, so that their total contribution is finite. Hence, the total normalization integral (5.65) is finite.

The preceding was for the case of no impulses in the source component $J_s$. The same kind of analysis can be gone through for impulsive sources, including a finite number of point sources, line sources (of whatever shape), and surface sources. We indicate the approach for the case of a surface source. The other cases can be treated analogously.

Suppose that the equation of the surface is

$$f(x', y') = z' \quad \text{for} \quad z_1 \leqslant z' \leqslant z_2, \tag{5.74}$$

where $z_1$, $z_2$ are finite because $V$ is finite. Then the surface source function can be represented as

$$J_s(\boldsymbol{r}', t') = J_0(x', y', t')\delta[z' - f(x', y')] \tag{5.75}$$

where $J_0$ is a well-behaved function of its arguments, containing no singularities. Using this in Eq. (5.67) gives

$$\int_T \int_V dt\, d\boldsymbol{r}\, p(\boldsymbol{r}, t)$$
$$= \sum_n \omega_n^{-2} \int_{t_1}^{t_2} dt \left[ \int_0^t dt' \int_{\boldsymbol{A}} d\boldsymbol{\rho}' \sin[\omega_n(t - t')]\, J_0(\boldsymbol{\rho}', t')\psi_n(\boldsymbol{\rho}', f(\boldsymbol{\rho}')) \right]^2,$$
$$\boldsymbol{\rho}' \equiv (x', y'), \quad d\boldsymbol{\rho}' \equiv dx'\, dy', \quad \boldsymbol{A} \equiv (x_1 \leqslant x \leqslant x_2), (y_1 \leqslant y \leqslant y_2). \tag{5.76}$$

The sifting property (0.63b) of the Dirac delta function was used.

Equation (5.76) is again of the form (5.67). Then new functions $U_n$ and $W_n$ may be formed as in Eqs. (5.68a–c). $U_n$ is bounded above, by essentially the same arguments as were made below Eq. (5.70), where $J_0$ now replaces $J_s$. The new $W_n$ is

$$W_n(t) = \int_0^t dt' \int_{\boldsymbol{A}} d\boldsymbol{\rho}'\, \psi_n^2(\boldsymbol{\rho}', f(\boldsymbol{\rho}')) = t \int_V d\boldsymbol{r}'\, \psi_n^2(\boldsymbol{r}')\delta[z' - f(x', y')]. \tag{5.77}$$

In the last step we did the integration $dt'$ and "anti-sifted" with the delta function. Since it obeys orthonormality property (5.61), $\psi_n^2(\boldsymbol{r}')$ may be regarded as a PDF on a fluctuation $\boldsymbol{r}'$. This is consistent, as well, with the EPI interpretation that a source "position" $\boldsymbol{r}'$ represents, in fact, a random fluctuation in position. The last integral (5.77) then represents the probability density for the event that a point $\boldsymbol{r}'$ that is randomly sampled from the given PDF will lie somewhere on the surface $f(x', y')$. For any well-behaved PDF this is a finite number. As examples, if the $\psi_n$ are products of trigonometric functions or Bessel functions, then they obey $[\psi_n(\boldsymbol{r}')]^2 \leqslant 1$ and, so, can only lead to a finite integral in (5.77). Once again, the normalization integral (5.65) is finite.

### 5.1.28 Normalization for an open space

The potential $\mathbf{A}$ for this case was shown to obey Eq. (5.64), the retarded potential solution. Then the normalization integral becomes

$$\int_T \int_V dt\, d\boldsymbol{r}\, p(\boldsymbol{r}, t) = \frac{1}{(4\pi c)^2} \int_{t_1}^{t_2} dt \int_{-\infty}^{\infty} d\boldsymbol{r} \left[ \int_{V_0} d\boldsymbol{r}' \frac{J_s(\boldsymbol{r}', t - R/c)}{R} \right]^2, \tag{5.78}$$

$$R \equiv |\boldsymbol{r} - \boldsymbol{r}'|.$$

The integration region for coordinate $\boldsymbol{r}$ is of interest. The finite integration limits $(t_1, t_2)$ on $t$ mean that we are detecting the potential $\mathbf{A}$ over a finite time interval. We also have assumed that the source $J_s$ is turned "on" for a finite time interval. Because of the latter property, and because the source extends over a finite space $V_0$ as well, there is a largest and a smallest value for $\boldsymbol{r}$ that will contribute non-zero values of $J_s$ to the normalization integral. For $\boldsymbol{r}$ values outside this range, $R$ is so large that the time factor $(t - R/c)$ for $J_s$ lies outside the "on" interval. Denoting the resulting integration region for $\boldsymbol{r}$ as $V_1$, Eq. (5.78) now reads

$$\int_T \int_V dt\, d\boldsymbol{r}\, p(\boldsymbol{r}, t) = \frac{1}{(4\pi c)^2} \int_{t_1}^{t_2} dt \int_{V_1} d\boldsymbol{r} \left[ \int_{V_0} d\boldsymbol{r}' \frac{J_s(\boldsymbol{r}', t - R/c)}{R} \right]^2. \tag{5.79}$$

Because of the finite integration regions all around, it is evident that finiteness for the overall integral hinges on finiteness for the squared innermost integral,

$$Q \equiv \left[ \int_{V_0} d\boldsymbol{r}' J_s(\boldsymbol{r}', t - R/c) \frac{1}{R} \right]^2. \tag{5.80}$$

We wrote the integral this way to suggest use of the Schwarz inequality,

$$Q \leqslant \int_{V_0} d\boldsymbol{r}' J_s^2\left( \boldsymbol{r}', t - \frac{|\boldsymbol{r} - \boldsymbol{r}'|}{c} \right) \int_{V_0} d\boldsymbol{r}' \frac{1}{|\boldsymbol{r} - \boldsymbol{r}'|^2}. \tag{5.81}$$

The indicated integrals are for fixed values of $t$ and $\boldsymbol{r}$.

Assume that the source function $J_s$ is "square-integrable," i.e.,

$$\int_{V_0} d\boldsymbol{r}' J_s^2\left(\boldsymbol{r}', t - \frac{|\boldsymbol{r} - \boldsymbol{r}'|}{c}\right) \leqslant C, \tag{5.82}$$

a finite constant, for any values of $t$ and $\boldsymbol{r}$. This means that $J_s$ should be a well-behaved function, without any poles whose squared areas are unbounded. A wide class of source functions obeys Eq. (5.82).

An exception to (5.82) would be impulsive sources (points, lines, or surfaces), since the area under a squared delta function is infinity. The latter situation can be avoided by substitution of the impulsive sources into Eq. (5.79), performing the integration $d\boldsymbol{r}$ using the sifting property, *and then squaring* as indicated. The Schwarz inequality can then be used on the resulting *summation* (rather than integral) $Q$ for this case. Assuming a finite number of such sources, the sums must likewise be finite.

Also, the second integral in (5.81) can be evaluated, by change of variable to $\boldsymbol{R} \equiv \boldsymbol{r} - \boldsymbol{r}'$ ($\boldsymbol{r}$ fixed), as

$$\int_{V_0} d\boldsymbol{r}' \frac{1}{|\boldsymbol{r} - \boldsymbol{r}'|^2} \leqslant \int_{\boldsymbol{R}_0} d\boldsymbol{R} \frac{1}{R^2} = 4\pi \int_{R_1}^{R_2} dR \frac{R^2}{R^2} = 4\pi(R_2 - R_1), \tag{5.83}$$

a finite number. Due to the change of variable, the integration region for the new coordinate $\boldsymbol{R}$ – call it region $\boldsymbol{R}_0$ – is not spherical, but has some generally irregular shape. By comparison, the region $R_0$ for coordinates $R$ is defined to be the smallest radially symmetric region, defined by inner and outer radii $R_1$ and $R_2$, that *includes within it* the region $\boldsymbol{R}_0$. Thus, part of this radial region has points of zero yield in place of the values $1/R^2$ at the remaining ones. This gives rise to the inequality sign in Eq. (5.83).

Since both right-hand integrals in Eq. (5.81) are bounded from above, $Q$ is bounded from above. As we saw (below Eq. (5.79)), this means that the normalization integral (5.78) is likewise bounded. For an open space problem, the integral of the square of the four-potential $\mathbf{A}$ over four-space is finite, and the corresponding PDF $p(\boldsymbol{r}, t)$ given by Eq. (5.52) is normalizable.

This result could also have been proved by a different approach. Consider the case where the temporal and spatial parts of the source function $J_s(\boldsymbol{r}', t')$ separate into $J_1(\boldsymbol{r}')J_2(t')$ (as in the case of a harmonic source). Then the integral in Eq. (5.80) takes on the form of a convolution of $J_1(\boldsymbol{r}')$ with the function $J_2(t - |\boldsymbol{r}'|/c)/|\boldsymbol{r}'|$. Use of Young's inequality (Stein and Weiss, 1971, p. 31) then establishes the upper bound to the integral.

We found in Sec. 5.1.27 that the PDF is normalizable over the closed space as well. Hence, the PDF is normalizable for either class of observation space.

Finally, by Sec. 5.1.25, the PDF is uniquely defined under suitable conditions of physical isolation for the measurement scenario. Thus, the proposed PDF has the basic properties of uniqueness and normalizability that are required of a probability density.

I would like to acknowledge the help of Professor William Faris in establishing normalizability for the open space case.

## 5.2 Maxwell's equations

As mentioned at the outset, Maxwell's equations follow in straightforward manner from the vector wave Eq. (5.51). This is shown as follows.

### 5.2.1 Lorentz condition on potential A

By Eqs. (5.8) and (5.48), the four-potential **A** obeys the Lorentz condition

$$\sum_{n=1}^{3} \frac{\partial A_n}{\partial x_n} + \frac{1}{c}\frac{\partial \phi}{\partial t} = 0. \tag{5.84}$$

### 5.2.2 Definitions of fields

The magnetic field $\boldsymbol{B}$ and electric field $\boldsymbol{E}$ are defined in terms of the three-potential $\boldsymbol{A} \equiv (A_1, A_2, A_3)$ and scalar potential $\phi$ in the usual way:

$$\boldsymbol{B} \equiv \nabla \times \boldsymbol{A}, \qquad \boldsymbol{B} \equiv (B_x, B_y, B_z) \tag{5.85}$$

and

$$\boldsymbol{E} \equiv -\nabla\phi - \frac{1}{c}\frac{\partial \boldsymbol{A}}{\partial t}, \qquad \boldsymbol{E} \equiv (E_x, E_y, E_z). \tag{5.86}$$

### 5.2.3 The equations

The four Maxwell equations derive as follows: First,

$$\nabla \cdot \boldsymbol{B} = 0 \tag{5.87}$$

by definition (5.85) and since $\nabla \cdot (\nabla \times \boldsymbol{A}) = 0$ is a vector identity. Second,

$$\frac{1}{c}\frac{\partial \boldsymbol{B}}{\partial t} \equiv \frac{1}{c}\nabla \times \frac{\partial \boldsymbol{A}}{\partial t} = \nabla \times (-\nabla\phi - \boldsymbol{E}) = -\nabla \times \boldsymbol{E}, \tag{5.88}$$

using definitions (5.85) and (5.86) and the fact that $\nabla \times \nabla\phi \equiv 0$, respectively. Third,

$$
\begin{aligned}
\nabla \times \boldsymbol{B} &\equiv \nabla \times (\nabla \times \boldsymbol{A}) \equiv \nabla(\nabla \cdot \boldsymbol{A}) - \nabla^2 \boldsymbol{A} \\
&= -\frac{1}{c}\frac{\partial}{\partial t}\nabla\phi - \frac{1}{c^2}\frac{\partial^2 \boldsymbol{A}}{\partial t^2} + \frac{4\pi}{c}\boldsymbol{j} \\
&= \frac{1}{c}\frac{\partial}{\partial t}\left(-\nabla\phi - \frac{1}{c}\frac{\partial \boldsymbol{A}}{\partial t}\right) + \frac{4\pi}{c}\boldsymbol{j} \\
&= \frac{1}{c}\frac{\partial \boldsymbol{E}}{\partial t} + \frac{4\pi}{c}\boldsymbol{j}.
\end{aligned}
\tag{5.89}
$$

Here the first line is by definition (5.85) and a vector identity; the second line is by wave Eq. (5.51) and the Lorentz condition (5.84); the third line is a rearrangement of the second; and the fourth line is by definition (5.86). Fourth,

$$
\begin{aligned}
\nabla \cdot \boldsymbol{E} &= \nabla \cdot \left(-\nabla\phi - \frac{1}{c}\frac{\partial \boldsymbol{A}}{\partial t}\right) \\
&= -\nabla^2\phi - \frac{1}{c}\frac{\partial}{\partial t}(\nabla \cdot \boldsymbol{A}) \\
&= -\nabla^2\phi + \frac{1}{c^2}\frac{\partial^2 \phi}{\partial t^2} \\
&= \Box\phi = 4\pi\rho.
\end{aligned}
\tag{5.90}
$$

Here the first line is by definition (5.86); the second line is a rearrangement of the first; the third line is by the Lorentz condition (5.84); and the fourth line is by Eq. (5.12b) and the wave Eq. (5.51).

### 5.2.4 Checks on solution for bound information $J$

Combining Eqs. (5.17), (5.19), and (5.40), and using $\boldsymbol{C} = \boldsymbol{0}$ and $\kappa = 1/2$, shows that the $n$th component $J_n$ of the bound information $J$ obeys

$$
J_n \equiv J_n(q_n) = -2BE_n^{-1}J_{sn}q_n. \tag{5.91}
$$

Quantity $J_{sn}$ is the $n$th component of the source current density ($\boldsymbol{j}$, $c\rho$). Using (5.91) on the right-hand side of (5.12a) gives the wave Eq. (5.40) with $\boldsymbol{C} = \boldsymbol{0}$. This verifies (5.12a), the result of the EPI *variational* principle.

Solving (5.91) for product $E_nJ_n$ and using this on the right-hand side of Eq. (5.17), with $\kappa = 1/2$, again gives the wave Eq. (5.40). This verifies the EPI *zero*-condition (5.17). Hence, both EPI conditions are consistent with (5.91) as the $n$th bound information component.

## 5.3 Overview

Maxwell's equations may be derived in two steps: (1) by deriving the vector wave Eq. (5.51) under the Lorentz condition, and then (2) showing that (5.51) together with the definitions (5.85) and (5.86) of the fields imply Maxwell's Eqs. (5.87)–(5.90).

In order to accomplish step (1), we use EPI. The measurement that initiates EPI is of the electric field $\boldsymbol{E}$ or the magnetic field $\boldsymbol{B}$ at a four-position $\mathbf{y}$. The latter is in error by intrinsic noise of amount $\mathbf{x} = (ix, iy, iz, ct)$. As usual in EPI derivations, the "noise" is the physical effect to be quantified. The four-measurement is assumed to be made in the presence of an electromagnetic four-source $\mathbf{J}_S$, so that the noise is electromagnetic in origin. A typical spread $\Delta x$ in $x$-coordinate is given by Eq. (5.47). The aim of the EPI approach is to find the probability amplitudes $(q_n(\boldsymbol{r}, t), n = 1, \ldots, N) = \mathbf{q}$. These are the degrees of freedom of the measured effect.

The invariance principles that are used to find information $J$ are the Lorentz condition (5.8) and conservation of flow (5.7) for the sources. Condition (5.8) implies that the minimum value of $N$ consistent with the givens of the problem is value 4. With information $J$ determined, the four amplitudes $\mathbf{q}$ are found to obey a general vector wave Eq. (5.40) with, however, a specifically *electromagnetic* source term $\mathbf{J}_S$. The latter implies that the wave equation is the electromagnetic one, which requires the *probability* amplitudes $\mathbf{q}$ to be linear in corresponding electromagnetic *potentials* $\mathbf{A}$; see Eq. (5.48). The resulting vector wave equation in the potentials, Eq. (5.51), is the most important step of the overall derivation.

The vector wave Eq. (5.40) that is initially derived contains an additive constant source vector $\mathbf{C}$. If it were non-zero, then in perfect vacuum (no real sources) there would still be an artificial, non-zero source present. Since there is no experimental evidence for such a classical vacuum source, this implies that $\mathbf{C} = \mathbf{0}$. The constant $\mathbf{C}$ arises in electromagnetic theory precisely as Einstein's additive field source $\Lambda$ does (see next chapter).

Another constant of the theory that must be defined is $B$ (Eq. (5.33)). It must be a universal constant, have units of 1/velocity, and define electromagnetic wave propagation. Hence it is identified as being inversely proportional to the speed of light $c$, Eq. (5.50).

With the vector wave Eq. (5.51) derived, it is a simple matter to derive from it the Maxwell equations in the Lorentz gauge. This is shown in Sec. 5.2.

An interesting outcome of the approach is a new PDF $p(\boldsymbol{r}, t)$ defining four-space localization within an electromagnetic field. This is given by Eq. (5.52), and follows from the connection Eq. (5.48) between probability amplitudes $\mathbf{q}$, on the one hand, and *field amplitudes* **A**, on the other. A PDF should be unique and normalizable. The PDF $p(\boldsymbol{r}, t)$ is shown to have these properties in Secs. (5.1.25)–(5.1.28). These are obeyed for open or closed space geometries, and in the presence of continuous or impulsive sources.

An enigma is that completely covariant boundary conditions led to a non-unique solution to the wave equation (Sec. 5.1.26). To attain uniqueness, non-covariant conditions were needed: Cauchy conditions for the time coordinate and either Dirichlet or Neumann conditions for the space coordinates (Sec. 5.1.25). However, as discussed in Secs. 5.1.25 and 5.1.26, the Einstein requirement of covariance is for the *laws* (here, differential equations) of physics, and not necessarily their solutions due to particular boundary conditions.

The use of statistics, as in this chapter, to derive the Maxwell equations, is not unique. DeGroot and Suttorp (1972) derive the macroscopic Maxwell equations from an assumption of Maxwell–Lorentz field equations operating at the microscopic level. The link between the two equations is provided by methods of statistical mechanics, i.e., another statistical approach. Their assumption of the Maxwell–Lorentz field equations is, however, a rather strong one. These already have the mathematical form of the Maxwell equations. By comparison, the only physical inputs that EPI required were the two invariance conditions Eqs. (5.7) and (5.8), which are weak by comparison with an assumption of the Maxwell–Lorentz field equations. EPI seems to have the ability to derive a physical theory from a weaker set of assumptions than by the use of any other approach (when the latter exists).

EPI regards Maxwell's equations as *fundamentally* statistical entities. Where have the statistics crept in? It was in admitting that any measurement within an electromagnetic field must suffer from random errors in position. Then the link Eq. (5.48) between the probability amplitude **q** *for the errors* and the electromagnetic four-potential **A** makes the connection to electromagnetic theory. This connection is unconventional.

Classically, the potential **A** is merely a convenient mathematical tool to be used for computing the *observables* of the problem – the field strengths $\boldsymbol{B}$ and $\boldsymbol{E}$. **A** has no physical meaning *per se*, except in the rarely used Proca theory of electromagnetism (Jackson, 1975, p. 598). By contrast, EPI regards **A** as a physical entity: it is a probability amplitude. This interpretation gives some resolution to the longstanding problem of what could possibly propagate through the emptiness of a vacuum: a field of probabilities, according to EPI.

In viewing **A** as a probability amplitude, EPI regards classical electromagnetism as defining a "quantum mechanics" of its own (Sec. 3.4.16). This is in fact the usual second-quantized field operator theory of electromagnetics.

But, if so, where can the discrete effects that are the hallmarks of quantum theory enter in? In fact, solutions for the vector potential **A** are often in the form of discrete eigenfunctions and eigenvalues: we need look no further than the combination of Eqs. (5.57) and (5.59). Such solutions also, of course, occur in solution of the microwave or laser cavity resonator. In other words, discrete electromagnetic solutions arise out of boundary-value conditions, just as they do in ordinary quantum mechanics.

More generally, EPI regards all of its output differential equations as defining the physics of probability amplitudes (Sec. 3.4.16). Thus, even the classical field equation of general relativity defines a metric function that is, alternatively, a probability amplitude (on the four-position of gravitons). This is taken up next.

# 6

# The Einstein field equation of general relativity

## 6.1 Motivation

As is well known, classical gravitational effects are caused by a local distortion of space and time. The Einstein field equation relates this distortion to the local density of momentum and energy. This phenomenon seems, outwardly, quite different from the electromagnetic one just covered. However, luckily, this is not the case. As viewed by EPI they are very nearly the same.

This chapter on gravity is placed right after the one on electromagnetic theory because the two phenomena are derived by practically identical steps. Each mathematical operation of the EPI derivation of the electromagnetic wave equation has a 1 : 1 counterpart in derivation of the weak-field Einstein field equation. For clarity, we will point out the correspondences as they occur. The main mathematical difference is one of *dimensionality*: quantities in the electromagnetic derivation are vectors, i.e., singly subscripted, whereas corresponding quantities in the gravitational derivation are doubly subscripted tensors. This does not amount to much added difficulty, however.

A further similarity between the two problems is in the simplicity of the final step: in Chapter 5 this was to show that Maxwell's equations follow from the wave equation. Likewise, here, we will proceed from the weak-field approximation to the general-field result. In fact, the argumentation in this final step is even simpler here than it was in getting Maxwell's equations.

Gravitational theory uses tensor quantities and manipulations. We introduce these in the section to follow, restricting attention to those that are necessary to the EPI derivation that follows. A fuller education in tensor algebra may be found, e.g., in Misner *et al.* (1973) or Landau and Lifshitz (1951).

## 6.2 Tensor manipulations: an introduction

It may be noted that the inner product of the coordinates $\mathbf{x}$ chosen in Eq. (5.3),

$$\mathbf{x} \equiv (x_0, x_1, x_2, x_3) \equiv (ct, ix, iy, iz) \tag{6.1}$$

obeys

$$\mathbf{x} \cdot \mathbf{x} = (ct)^2 - x^2 - y^2 - z^2. \tag{6.2}$$

The imaginary unit $i$ in coordinates (6.1) is bothersome to many people, since it places the time $t$ in a different number space from the three space coordinates. The $i$ may be avoided by supplementing subscripted variables (6.1) with superscripted ones, as follows.

### *6.2.1 Contravariant coordinates, covariant coordinates*

Define two sets of coordinates,

$$x^\nu \equiv (x^0, x^1, x^2, x^3) \equiv (x^0, x^j), \quad j = 1, 2, 3, \tag{6.3}$$

and

$$x_\nu \equiv (x_0, x_1, x_2, x_3) \equiv (x_0, x_j), \quad x_0 = ct, \quad j = 1, 2, 3. \tag{6.4}$$

The former are called contravariant coordinates and the latter are called covariant coordinates. We now note an index convention that is used in Eqs. (6.3) and (6.4). A *Greek* index such as $\nu$ is implied to take on all four values 0–3. A *Latin* index such as $j$ takes on only the three values 1–3 corresponding to the three space coordinates. The inner product is defined in terms of the two sets of coordinates as

$$\mathbf{x} \cdot \mathbf{x} \equiv \sum_{\nu=0}^{3} x_\nu x^\nu. \tag{6.5}$$

This will allow us to express one set in terms of the other.

### *6.2.2 Einstein summation convention*

Before doing this, we simplify the notation by using the *Einstein summation convention*. By this convention, any repetition of an index in an expression means that that index is to be summed over. Since index $\nu$ is repeated in Eq. (6.5), the latter becomes

$$\mathbf{x} \cdot \mathbf{x} = x_\nu x^\nu. \tag{6.6}$$

This summation convention will be assumed in all equations to follow, except for those that state it to be "turned off."

### 6.2.3 Raising and lowering operators

Since Eqs. (6.2) and (6.5) must have identical right-hand sides, it must be that

$$x^\mu = \eta^{\mu\nu} x_\nu, \quad \text{where} \quad \eta^{\mu\nu} \equiv \text{diag}(1, -1, -1, -1) = \eta^{\nu\mu}. \tag{6.7}$$

Since matrix $\eta^{\mu\nu}$ is diagonal, it is symmetric, as indicated. The matrix is called the "flat field metric," for reasons that are found below. Equation (6.7) shows that $\eta^{\mu\nu}$ acts as a "raising" operator, transforming a subscripted variable $x_\nu$ into a superscripted variable $x^\mu$, that is, transforming a covariant coordinate into a contravariant one.

Likewise, there is a "lowering" operator $\eta_{\mu\nu}$ that causes

$$x_\mu = \eta_{\mu\nu} x^\nu, \quad \eta_{\mu\nu} = \eta^{\mu\nu}. \tag{6.8}$$

Combining the raising and lowering operators gives a delta function. By (6.7) and (6.8) $x_\alpha = \eta_{\alpha\nu}(\eta^{\nu\beta} x_\beta)$, so that

$$\eta_{\alpha\nu}\eta^{\nu\beta} = \delta_\alpha^\beta, \quad \text{where} \quad \delta_\alpha^\beta = 0 \text{ for } \alpha \neq \beta \text{ and } \delta_\alpha^\alpha = 1 \text{ (no summation).} \tag{6.9}$$

### 6.2.4 Metric tensor g

The inner product form (6.6) holds for differentials as well. In this way we can define a four-length $ds$ obeying

$$(ds)^2 = g_{\mu\nu}\, dx^\mu\, dx^\nu, \quad \text{where functionally } g_{\mu\nu} = g_{\mu\nu}(x^\alpha),\ x^\alpha \equiv x^0, x^1, x^2, x^3. \tag{6.10}$$

In contrast to matrix $\eta_{\mu\nu}$, matrix $g_{\mu\nu}$ is not generally diagonal. From the form of Eq. (6.10), $g_{\mu\nu}$ defines the coefficients of the components of a squared differential length for a given coordinate system in use. As an example, let spherical coordinates $t$, $r$, $\theta$, $\phi$ define the exterior four-space of a spherically symmetric star. The line element in this coordinate system, in vacuum, is (ignoring factors $c$) the Schwarzschild measure

$$(ds)^2 = \left(1 - \frac{2M}{r}\right) dt^2 - \frac{dr^2}{1 - 2M/r} - r^2\, d\theta^2 - r^2 \sin^2\theta\, d\phi^2. \tag{6.11}$$

From this, the metric tensor coefficients $g_{\mu\nu}$ may readily be "picked off."

### 6.2.5 Exercise

Find the metric tensor $g_{\mu\nu}$ in this way.

### 6.2.6 Flat field metric

As a more simple case, the geometry of space-time of special relativity incurs a differential length

$$(ds)^2 = (dx^0)^2 - (dx^1)^2 - (dx^2)^2 - (dx^3)^2. \tag{6.12}$$

Comparison with Eq. (6.10) shows that here $g_{\mu\nu} = \eta_{\mu\nu}$. This space is "flat," in that $g_{\mu\nu}(x^\alpha) = g_{\mu\nu} = const$. Hence, $\eta_{\mu\nu}$ is called the "flat field" metric.

### 6.2.7 Derivative notation

Since we have two sets of coordinates $x^\nu$ and $x_\nu$, there are two different kinds of partial derivatives,

$$\frac{\partial f_{\alpha\beta}}{\partial x^\nu} \equiv f_{\alpha\beta,\nu} \quad \text{and} \quad \frac{\partial f_{\alpha\beta}}{\partial x_\nu} \equiv f_{\alpha\beta}{}^{,\nu}. \tag{6.13}$$

Hence, a comma denotes differentiation with respect to the coordinate whose index follows it. The $\eta_{\mu\nu}$ and $\eta^{\mu\nu}$ operators lower and raise derivative indices as well. For example, the chain rule of differentiation gives

$$f_{\alpha\beta,\nu} \equiv \frac{\partial f_{\alpha\beta}}{\partial x^\nu} = \frac{\partial f_{\alpha\beta}}{\partial x_\mu}\frac{\partial x_\mu}{\partial x^\nu} = f_{\alpha\beta}{}^{,\mu}\eta_{\mu\nu} \tag{6.14}$$

by Eq. (6.8). Hence, $\eta_{\mu\nu}$ serves to lower a raised derivative index. Likewise,

$$f_{\alpha\beta}{}^{,\nu} = f_{\alpha\beta,\mu}\eta^{\mu\nu}. \tag{6.15}$$

Operator $\eta^{\mu\nu}$ serves to raise a lower derivative index.

### 6.2.8 Fisher information

As an example of the simplifications that the comma notation and preceding operations allow, consider the squared gradient with respect to coordinates (6.1). This becomes

$$\nabla q_n \cdot \nabla q_n \equiv \sum_{\lambda=0}^{3} \left(\frac{\partial q_n}{\partial x_\lambda}\right)^2 = q_{n,\lambda} q_n{}^{,\lambda} \tag{6.16}$$

after conversion to covariant coordinates (6.4), and use of the summation convention, lowering operation (6.8), and comma notation (6.13). This result allows the Fisher information Eq. (2.19) to be expressed as (2.20),

$$I = 4\int d^4x\, q_{n,\lambda} q_n{}^{,\lambda}, \quad \text{where} \quad d^4x \equiv dx^0 \cdots dx^3. \tag{6.17}$$

Regarding covariance properties, the salient feature of this equation is its covariance in the derivative indices $\lambda$. When (6.17) is used in the EPI principle, the output wave equation is generally covariant. Of course the EPI output equation of this chapter, the Einstein field equation, is manifestly covariant.

### *6.2.9 Exercise: tensor Euler–Lagrange equations*

Suppose that the Lagrangian $\mathscr{L}[q_\nu, q_{\nu,\lambda}]$, $q_\nu = q_\nu(x^\lambda)$ is known. Show that the Euler–Lagrange Eq. (0.34) for determination of $q_\nu$ is, in tensor notation,

$$\frac{\partial}{\partial x^\lambda}\left[\frac{\partial \mathscr{L}}{\partial(q_{\nu,\lambda})}\right] = \frac{\partial \mathscr{L}}{\partial q_\nu}. \tag{6.18a}$$

### *6.2.10 Exercise: tensor form of d'Alembertian*

Using Eqs. (6.7) and (5.12b), show that

$$q_\nu{}^{,\lambda}{}_{,\lambda} = \eta^{\lambda\alpha} q_{\nu,\alpha,\lambda} = \Box q_\nu. \tag{6.18b}$$

## 6.3 Derivation of the weak-field wave equation

### *6.3.1 Goals*

The aim of this chapter is to derive the Einstein field equation obeyed by $g_{\mu\nu}$. This is done in two steps: by (1) deriving the weak-field equation; and then (2) showing that this implies the field equation for a *general* field. The latter is an easy matter. The main task is problem (1).

We follow the EPI procedure of Sec. 3.4. This requires identifying the parameters to be measured, forming the corresponding expression for $I$, and then finding $J$ by an appropriate invariance principle.

This derivation follows in almost 1 : 1 fashion that of the electromagnetic wave equation in Chapter 5. Most sections in this chapter have a direct counterpart in Chapter 5: Sec. 6.3.1 corresponds to Sec. 5.1.1, Sec. 6.3.2 corresponds to Sec. 5.1.2, etc.

The main departure operationally is to replace singly subscripted quantities of the electromagnetic theory, such as vector potential $A_n$, by doubly subscripted quantities such as the weak-field metric $\overline{h}_{\mu\nu}$. This double subscripting is naturally accommodated by the EPI derivation, as will be seen.

### *6.3.2 Choice of measured parameters θ*

The gravitational metric $g_{\mu\nu}$ can be measured (Misner *et al.*, 1973, p. 324). Consider such a measurement procedure, where the gravitational field is so weak that there are only small local departures from flat space. Then the metric $g_{\mu\nu}$ may be expanded as

$$g_{\mu\nu} = \eta_{\mu\nu} + h_{\mu\nu}, \quad |h_{\mu\nu}| \ll 1. \tag{6.19}$$

The $h_{\mu\nu}$ are small metric perturbations. It is customary to work with an associated metric of perturbations

$$\overline{h}_{\mu\nu} \equiv h_{\mu\nu} - \tfrac{1}{2}\eta_{\mu\nu}h, \quad h \equiv \eta^{\mu\nu}h_{\mu\nu} \tag{6.20}$$

that is linear in $h_{\mu\nu}$. We call $\overline{h}_{\mu\nu}$ the "weak field" metric, and seek its field equation.

It has been suggested (Lawrie, 1990) that $\overline{h}_{\mu\nu}$ can be interpreted as the wave function $q_{\mu\nu}$ of a *graviton*, provided that $q_{\mu\nu}$ is regarded as a field *operator* function. (This interpretation is permitted by EPI, as shown in Sec. 11.5.) This idea is confirmed later in Secs. 6.3.20–6.3.26. Also see the corresponding Secs. 5.1.19–5.1.24 that allow the source-free electromagnetic potential $\mathbf{A}$ to be interpreted as the probability amplitude $\mathbf{q}$ for a *photon*.

Consider, then, an experiment consisting of measurements of the weak-field metric. The latter is to be found at ideal (target) four-positions $\theta_{\mu\nu}$. The double subscripting is, of course, an arbitrary naming (or ordering) of the positions.

The measuring instrument perturbs the amplitude functions $\mathbf{q}$ of the problem, which initiates the EPI process. Also, at the input space to the instrument, fluctuations $x_{\mu\nu}$ occur that define the intrinsic data (Sec. 2.1.1) of the problem

$$y_{\mu\nu} = \theta_{\mu\nu} + x_{\mu\nu}. \tag{6.21}$$

By our notation convention (Sec. 6.2.1) this amounts to 16 measurements, although, as we will see, $N = 10$ measurements suffice. This means that the position coordinates $x_{\mu\nu}$ (not the metric values $\overline{h}_{\mu\nu}$ themselves) are the Fisher variables.

Equation (6.21) states that input positional accuracy is limited only by fluctuations $x_{\mu\nu}$ that are characteristic of the gravitational metric. As is well known, these fluctuations occur on the scale of the Planck length

$$\Delta x \sim (\hbar G/c^3)^{1/2} \tag{6.22}$$

(cf. Eq. (5.4)). $G$ is the gravitational constant. At this point Eq. (6.22) is purely motivational. It is derived in Eq. (6.64) below.

### 6.3.3 Information I

By Eq. (2.11) the four-data data values $\mathbf{y}_{\mu\nu}$ are assumed to be independent, and there is a PDF

$$p_{\mu\nu}(y_{\mu\nu}|\theta_{\mu\nu}) = p_{\mu\nu}(x^\alpha) \equiv q^2_{\mu\nu}(x^\alpha) \tag{6.23}$$

for the $(\mu\nu)$th measurement. The information is then Eq. (6.17), where the probability amplitudes $\mathbf{q}$ are now doubly subscripted,

$$I = 4 \int d^4x \, q_{\mu\nu,\lambda} q_{\mu\nu}{}^{,\lambda}. \tag{6.24}$$

### *6.3.4 Invariance principles*

As in the electromagnetic problem, we will use two equations of continuity of flow for the invariance principle required by EPI.

The first equation of flow governs the sources that are driving the probability amplitudes. In Chapter 5, the general source was the vector current $\mathbf{J} = J_\mu$. Here the source is the "stress-energy tensor" $T_{\mu\nu}$, defined as follows.

Consider a two-dimensional area $A_k$ at rest in the observer's frame, with the normal to $A_k$ pointing in the direction $\hat{e}_k$. Due to moving matter in the vicinity of the frame, there will be an amount of momentum $P^\mu$ crossing the surface area $A_k$ in the direction $\hat{e}_k$ during a time $\Delta t$. The $(\mu k)$ component of the stress-energy tensor is defined as

$$T^{\mu k} = \frac{P^\mu}{A_k \, \Delta t} = T^{k\mu}. \tag{6.25}$$

The four-momentum $P^\mu$ mentioned here requires definition. A particle of mass $m$ has a momentum

$$P^\mu \equiv m u^\mu, \tag{6.26}$$

where $u^\mu$ is the particle's four-velocity $u^\mu \equiv (dt/d\tau, u^j)$ with $\tau$ its "proper" time (time as seen by the particle). The final component, $T^{00}$, is the total mass-energy measured in the observer's frame.

In general, four-momentum is conserved. This is expressed as a conservation of flow theorem (cf. Eq. (5.7))

$$T_{\mu\nu}{}^{,\nu} = 0. \tag{6.27}$$

The second equation of flow is a Lorentz condition (usually obeyed by $\overline{h}_{\mu\nu}$) for the probability amplitudes $q_{\mu\nu}$,

$$q^{\mu\nu}{}_{,\nu} = 0 \tag{6.28}$$

(cf. Eq. (5.8)). We will later show that $q_{\mu\nu}$ and $\overline{h}_{\mu\nu}$ are proportional, justifying this condition.

A final invariance principle is a demand that the weak-field equation be "form invariant," i.e., remain invariant in form to Lorentz and gauge transformations. Regarding *constant* tensors of order two, the only ones that obey such invariance are the flat-field metric and the Kronecker delta,

$$\eta_{\mu\nu}, \; \delta^\mu_\nu. \tag{6.29}$$

### 6.3.5 Fixing N

The Lorentz condition (6.28) is the only physical input into EPI that functionally involves the unknown amplitudes $q_{\mu\nu}$. Since each of indices $\mu$, $\nu$ goes from 0 to 3, it might appear that there are $N = 16$ independent amplitudes in this theory. However, because of the symmetry (6.25) in the stress-energy tensor, there are only $N = 10$ independent stress-energy values. We anticipate, at this point, that the $q_{\mu\nu}$ will obey symmetry as well so that $N = 10$ also defines the number of independent amplitudes that are sufficient for defining the theory. This step could have been taken at the end of the derivation, but it is more convenient to do here once and for all.

### 6.3.6 On finding the bound information J

The invariance principles (6.27) and (6.28) do not explicitly express information $I$ in terms of a physically parameterized information $J$. This means that we cannot at this point know $J$ or the efficiency parameter $\kappa$. They have to be solved for, based upon *use* of the two invariance principles in the two EPI conditions (3.16) and (3.18). The latter must be solved *simultaneously* for a common solution $q_{\mu\nu}$ (Sec. 3.4.6). We do this in the following sections.

### 6.3.7 General form for J

At this point the bound information $J$ is an unknown scalar functional of the $q_{\mu\nu}$ and sources $T_{\mu\nu}$. It can therefore be represented as an inner product

$$J \equiv 4 \int d^4x \, A^{\alpha\beta} J_{\alpha\beta}, \qquad A^{\alpha\beta} = const., \qquad J_{\alpha\beta} = J_{\alpha\beta}(q_{\mu\nu}, T_{\mu\nu}). \tag{6.30}$$

The notation $J_{\alpha\beta}(q_{\mu\nu}, T_{\mu\nu})$ means that each $J_{\alpha\beta}$ depends in an unknown way upon *all* $q_{\mu\nu}$ and $T_{\mu\nu}$.

### 6.3.8 EPI variational solution

As planned above, we must first find the solution for $q_{\mu\nu}$ to EPI problem (3.16). Using the $I$ form (6.17) for, now, a doubly subscripted $q_{\mu\nu}$ and using (6.30) for $J$ gives an information $K$ Lagrangian

$$\mathscr{L} = q_{\mu\nu,\lambda} q_{\mu\nu}{}^{,\lambda} - A^{\mu\nu} J_{\mu\nu}. \tag{6.31}$$

We ignored a common factor 4. This is used in the Euler–Lagrange equation for the problem, which by Eq. (6.18a) is

$$\frac{\partial}{\partial x^\lambda}\left[\frac{\partial \mathcal{L}}{\partial(q_{\mu\nu,\lambda})}\right] = \frac{\partial \mathcal{L}}{\partial q_{\mu\nu}}. \tag{6.32}$$

By Eq. (6.31) and lowering operation (6.8),

$$\frac{\partial \mathcal{L}}{\partial q_{\mu\nu,\lambda}} = 2\eta^{\lambda\alpha} q_{\mu\nu,\alpha} \quad \text{and} \quad \frac{\partial \mathcal{L}}{\partial q_{\mu\nu}} = -A^{\alpha\beta}\frac{\partial J_{\alpha\beta}}{\partial q_{\mu\nu}}. \tag{6.33}$$

Using these in Eq. (6.32) gives a formal solution

$$2\eta^{\lambda\alpha} q_{\mu\nu,\alpha,\lambda} = -A^{\alpha\beta}\frac{\partial J_{\alpha\beta}}{\partial q_{\mu\nu}} \quad \text{or} \quad \Box q_{\mu\nu} = -\tfrac{1}{2}A^{\alpha\beta}\frac{\partial J_{\alpha\beta}}{\partial q_{\mu\nu}} \tag{6.34}$$

by identity (6.18b) (cf. Eq. (5.12a)). This expresses $q_{\mu\nu}$ in terms of the unknown information components $J_{\mu\nu}$.

### 6.3.9 *Alternative form for I*

Once again use form (6.17) for $I$ with singly subscripted amplitudes replaced by doubly subscripted ones. This form may be integrated by parts. First note the elementary result (5.13). There, the zero held because we assumed that either $q = 0$ or $dq/dx = 0$ (Dirichlet or Neumann conditions) hold on the boundaries of measurement space. Assume such conditions to hold here as well. Then integrating by parts Eq. (6.17) gives an information

$$\begin{aligned} I = 4\int d^4x\, q_{\mu\nu,\lambda} q_{\mu\nu}{}^{,\lambda} &= 4\sum_\lambda q_{\mu\nu}{}^{,\lambda} q_{\mu\nu}\Big|_{-\infty}^{+\infty} - 4\int d^4x\, q_{\mu\nu} q_{\mu\nu}{}^{,\lambda}{}_{,\lambda} \\ &= 0 - 4\int d^4x\, q_{\mu\nu}\Box q_{\mu\nu} \end{aligned} \tag{6.35}$$

by identity (6.18b).

### 6.3.10 *EPI zero-root solution*

The second EPI problem is to find the roots $q_{\mu\nu}$ of Eq. (3.18),

$$I[q_{\mu\nu}] - \kappa J[q_{\mu\nu}] = 0, \tag{6.36}$$

$I$ given by (6.35) and $J$ given by (6.30). The problem is then

$$I - \kappa J = -4\int d^4x\,(q_{\mu\nu}\Box q_{\mu\nu} + \kappa A^{\mu\nu} J_{\mu\nu}) = 0. \tag{6.37}$$

The solution is the microscale Eq. (3.20), which is here (cf. Eq. (5.17))

$$q_{\mu\nu}\Box q_{\mu\nu} = -\kappa A^{\mu\nu} J_{\mu\nu} \text{ (no summation).} \tag{6.38}$$

### 6.3.11 Common solution $q_{\mu\nu}$

By Sec. 6.3.6, Eqs. (6.34) and (6.38) are to be simultaneously solved. Eliminating $\Box q_{\mu\nu}$ between them gives

$$\tfrac{1}{2}A^{\alpha\beta} q_{\mu\nu} \frac{\partial J_{\alpha\beta}}{\partial q_{\mu\nu}} = \kappa A^{\mu\nu} J_{\mu\nu} \text{ (no summation on } \mu\nu). \tag{6.39}$$

Noting that the right-hand side of (6.39) has been sifted out of the left-hand sum over $\alpha\beta$, it must be that

$$\tfrac{1}{2}q_{\mu\nu} \frac{\partial J_{\alpha\beta}}{\partial q_{\mu\nu}} = \kappa \delta^{\mu}_{\alpha} \delta^{\nu}_{\beta} J_{\mu\nu} \text{ (no summation)}, \tag{6.40}$$

where $\delta^{\mu}_{\nu}$ is the Kronecker delta (6.9). Then the case $\alpha\beta \neq \mu\nu$ gives $\partial J_{\alpha\beta}/\partial q_{\mu\nu} = 0$, assuming non-trivial solutions $q_{\mu\nu} \neq 0$. This implies that

$$J_{\alpha\beta}(q_{\mu\nu}, T_{\mu\nu}) = J_{\alpha\beta}(q_{\alpha\beta}, T_{\mu\nu}) \text{ (no summation)}. \tag{6.41}$$

That is, for the particular $\alpha\beta$, function $J_{\alpha\beta}$ depends on $q_{\alpha\beta}$ alone among all the $q$, but upon all the $T$.

Next, the opposite case $\alpha\beta = \mu\nu$ in Eq. (6.40) gives a condition

$$q_{\mu\nu} \frac{\partial J_{\mu\nu}}{\partial q_{\mu\nu}} = 2\kappa J_{\mu\nu}(q_{\mu\nu}, T_{\alpha\beta}) \text{ (no summation)}. \tag{6.42}$$

Because of the particular dependence (6.41), the left-hand side partial derivative in (6.42) is actually an ordinary derivative. This allows the equation to be integrated with respect to $q_{\mu\nu}$. The result is

$$J_{\mu\nu} = q_{\mu\nu}^{2\kappa} B_{\mu\nu}(T_{\alpha\beta}) \text{ (no summation)} \tag{6.43}$$

(cf. Eq. (5.21)). Functions $B_{\mu\nu}(T_{\alpha\beta})$ arise as the constants (in $q_{\mu\nu}$) of integration and are to be found, along with parameter $\kappa$.

### 6.3.12 Resulting wave equation

Using the result (6.43) in either wave equation (6.34) or (6.38) gives an answer for $q_{\mu\nu}$ of the same form,

$$\Box q_{\mu\nu} = q_{\mu\nu}^{b} F_{\mu\nu}(T_{\alpha\beta}) \text{ (no summation)}, \quad \text{where } b \equiv 2\kappa - 1 \tag{6.44}$$

(cf. Eq. (5.22a)). The parameter $b$ and the new functions $F_{\mu\nu}(T_{\alpha\beta})$ need to be found. In general, each $F_{\mu\nu}$ depends upon all the $T_{\alpha\beta}$.

### 6.3.13 Where we stand so far

At this point we know that the amplitude functions $q_{\mu\nu}$ obey a wave equation (6.44), but with an unknown dependence upon the right-hand source functions

$F_{\mu\nu}(T_{\alpha\beta})$. Hence, the EPI approach has taken us quite a way toward solution of the problem, even without the use of any specifically gravitational inputs of information. But we can proceed no further without one. Hence, the invariance principles (6.27) and (6.28) will now be used. These are actually quite weak as statements of specifically gravitational phenomena. (Notice that they are practically identical to the electromagnetic invariance principles (5.7) and (5.8).) Hence, the fact that their use in EPI will lead to the weak-field equation is a definite plus for the theory.

### *6.3.14 Use of Lorentz condition and conservation of matter–energy flow condition*

We proceed in the following sections to find the exponent $b$ and functions $F_{\mu\nu}$. These will be defined by the conservation law (6.27), Lorentz condition (6.28), and invariance properties (6.29).

We start by operating with $\eta^{\nu\alpha}\,\partial/\partial x^{\alpha}$ on Eq. (6.44), with implied summation over $\nu$ and $\alpha$. The left-hand side becomes

$$\eta^{\nu\alpha}\,\Box q_{\mu\nu,\alpha} = \eta^{\nu\alpha}\eta_{\mu\beta}\eta_{\nu\gamma}\,\Box q^{\beta\gamma}{}_{,\alpha} = \eta_{\mu\beta}\,\Box q^{\beta\alpha}{}_{,\alpha}. \tag{6.45}$$

(The middle equality used lowering operations (6.8). The second equality used the delta function definition (6.9).) This vanishes, by the Lorentz condition (6.28).

Operating in the same way on the *right-hand* side of Eq. (6.44) should likewise give zero. By the chain rule of differentiation, it gives

$$\eta^{\nu\alpha}\left[q^{b}_{\mu\nu}\,\frac{\partial F_{\mu\nu}(T_{\alpha\beta})}{\partial T_{\beta\gamma}}\right]T_{\beta\gamma,\alpha} \tag{6.46}$$

plus another term, which is zero if $b = 0$ (as will be the case).

Observing the resemblance of the derivative $T_{\beta\gamma,\alpha}$ in (6.46) to the term $T_{\mu\nu}{}^{,\nu}$ in conservation equation (6.27), we see that (6.46) can be made to vanish if

$$q^{b}_{\mu\nu}\,\frac{\partial F_{\mu\nu}(T_{\alpha\beta})}{\partial T_{\beta\gamma}} = \tfrac{1}{2}(B^{\beta}_{\mu}\delta^{\gamma}_{\nu} + B^{\gamma}_{\mu}\delta^{\beta}_{\nu})\ (\text{no sum on } \mu\nu) \tag{6.47}$$
$$\text{where}\quad B^{\gamma}_{\lambda} \equiv B^{\gamma}_{\lambda}(T_{\alpha\beta},\, q_{\alpha\beta}).$$

The $B^{\gamma}_{\lambda}$ are new, unknown, functions of all the $q_{\alpha\beta}$ and $T_{\alpha\beta}$. As a check, one may substitute the first right-hand term of (6.47) into (6.46) to obtain

$$B^{\beta}_{\mu}\eta^{\nu\alpha}\delta^{\gamma}_{\nu}T_{\beta\gamma,\alpha} = B^{\beta}_{\mu}\eta^{\nu\alpha}T_{\beta\nu,\alpha} = B^{\beta}_{\mu}T_{\beta\nu}{}^{,\nu} = 0, \tag{6.48}$$

as was required, after sifting with the delta function, using an index-raising operation (6.7), and using conservation property (6.27). The second right-hand side term of (6.47) may be treated similarly.

### 6.3.15 Finding exponent b

The left- and right-hand sides of Eq. (6.47) must balance in their $q$- and $T$-dependences. Because the left-hand side is a separated function $q_{\mu\nu}^b$ of $q_{\mu\nu}$ times a function of the $T_{\alpha\beta}$, the right-hand functions $B_\mu^\beta$ must likewise be so separated. For the case $\beta \neq \nu$, $\gamma = \nu$, Eq. (6.47) gives

$$B_\mu^\beta \equiv B_{1\mu}^\beta(q_{\alpha\beta}) B_{2\mu}^\beta(T_{\alpha\beta}), \tag{6.49a}$$

$$B_{1\mu}^\beta = C q_{\mu\nu}^b, \quad B_{2\mu}^\beta = \frac{\partial F_{\mu\nu}(T_{\alpha\beta})}{\partial T_{\beta\gamma}}. \tag{6.49b}$$

Quantity $C$ is a constant. The first Eq. (6.49b) says that

$$B_{1\mu}^\beta = C q_{\mu 0}^b = C q_{\mu 1}^b \cdots = C q_{\mu 3}^b. \tag{6.50}$$

Since the $q_{\mu\nu}$ are independent degrees of freedom of the theory, they cannot in general be equal. The only alternative possibility is that $B_{1\mu}^\beta$ be independent of all $q$, $B_{1\mu}^\beta = const.$, so that

$$b = 0. \tag{6.51}$$

Also, by Eq. (6.49a), $B_\mu^\beta = B_\mu^\beta(T_{\alpha\beta})$ alone.

### 6.3.16 Finding functions $F_{\mu\nu}(T_{\alpha\beta})$

Evaluating Eq. (6.47) at $\beta = \nu$, $\gamma = \mu \neq \nu$ and using (6.51) gives

$$\frac{\partial F_{\mu\nu}}{\partial T_{\mu\nu}} = \tfrac{1}{2} B_\mu^\mu(T_{\alpha\beta}). \tag{6.52}$$

We used the symmetry (6.25) of $T$. Doing the same operations with, now, $\nu \rightarrow \nu'$, gives the same right-hand side as (6.52). Therefore,

$$\frac{\partial F_{\mu\nu}}{\partial T_{\mu\nu}} = \frac{\partial F_{\mu\nu'}}{\partial T_{\mu\nu'}} \quad \text{for all } (\nu, \nu'). \tag{6.53}$$

The general solution must be that each side is independent of $\nu$ (and $\nu'$), hence $(\mu\nu)$, and hence $T_{\mu\nu}$. Therefore Eq. (6.53) may be integrated to give a general linearity between $F_{\mu\nu}$ and $T_{\mu\nu}$. The solution is

$$F_{\mu\nu} = \tfrac{1}{2}(B_\mu^\beta T_{\beta\nu} + B_\mu^\gamma T_{\nu\gamma}) + C_{\mu\nu}, \quad B_\mu^\beta = const., \quad C_{\mu\nu} = const. \tag{6.54}$$

This may be verified by back substitution into Eq. (6.47) with $b = 0$.

### 6.3.17 Exercise

Carry through the substitution, verifying that (6.54) is indeed a solution of Eq. (6.47).

### 6.3.18 Efficiency parameter $\kappa$

By Eqs. (6.44) and (6.51), the efficiency parameter is here

$$\kappa = 1/2. \tag{6.55}$$

This is the same value as for the electromagnetic case (Eq. (5.39)). Hence, the weak-field gravitational theory is inefficient, allowing only $I/J = 1/2$ of the bound information to be utilized in the intrinsic information $I$. This indicates that the theory is inefficient in the sense that it is an approximation. The use of EPI in a generally strong-field scenario, and taking into account quantum gravitational effects, would yield a value of $\kappa = 1$. This is verified in Chapter 11.

### 6.3.19 Constants $B_\mu^\beta$, $C_{\mu\nu}$ from form invariance

Using Eqs. (6.44) and (6.54) gives a wave equation in the amplitudes,

$$\Box q_{\mu\nu} = \tfrac{1}{2}(B_\mu^\beta T_{\beta\nu} + B_\mu^\gamma T_{\nu\gamma}) + C_{\mu\nu}. \tag{6.56}$$

We demand that this be form-invariant under both Lorentz transformation and gauge transformation. Noticing that $B_\mu^\beta$, $B_\mu^\gamma$ and $C_{\mu\nu}$ are constant tensors, by Eqs. (6.29) it must be that

$$B_\mu^\beta = B\delta_\mu^\beta, \quad B_\mu^\gamma = B\delta_\mu^\gamma, \quad C_{\mu\nu} = C\eta_{\mu\nu}, \quad B,\ C \text{ constants.} \tag{6.57}$$

### 6.3.20 Identification of amplitudes with weak-field metric

Using this in Eq. (6.56), along with the symmetry $T_{\mu\nu} = T_{\nu\mu}$, gives

$$\Box q_{\mu\nu} = BT_{\mu\nu} + C\eta_{\mu\nu}. \tag{6.58}$$

This has the form of a wave equation in the amplitudes $q_{\mu\nu}$ due to a *gravitational* source $T_{\mu\nu}$. Since the weak-field gravitational equations have precisely such a form, this implies that the probability amplitude $q_{\mu\nu}$ and weak-field metric $\overline{h}_{\mu\nu}$ are proportional,

$$\overline{h}_{\mu\nu}(x^\alpha) = \sqrt{\Delta V}\, q_{\mu\nu}(x^\alpha), \tag{6.59}$$

where $\Delta V$ is a constant *four-volume*. (*Note*: As with the analogous electromagnetic Eq. (5.48), for uniqueness of definition we ignore the possible presence of an added right-hand side function whose $\Box$ is zero.) The correspondence (6.59) states that the weak-field metric at a position $x_{\mu\nu}^\alpha$ may be interpreted, alternatively, as the square-root of the probability density (or, alternatively, by Sec. 6.3.2, a field operator function) for a fluctuation $x_{\mu\nu}^\alpha$ from an ideal position $\theta_{\mu\nu}^\alpha$. This correspondence is also suggested by remarks of Lawrie (1990) and by the fact that, in the source-free case, the resulting field equation will go over into the wave equation for a particle of the gravitational field, i.e., the *graviton*. See Sec. 6.3.24.

The proportionality factor $\sqrt{\Delta V}$ in Eq. (6.59) is obtained from the fact that $\overline{h}_{\mu\nu}$ is unitless whereas, by Eq. (6.23), $q_{\mu\nu}$ is the square-root of a PDF on four-position and, hence, must have units of reciprocal area.

The metric $\overline{h}_{\mu\nu}$ is a classical (macroscopic), continuous quantity. It is usual to assume that it suffers from quantum fluctuations that dominate within $x$-intervals that are on the scale of an elemental length $L$ (Misner *et al.*, 1973, pp. 1190, 1191). We show in the next section that $L$ is the Planck length. The approach will also suggest that either the cosmological constant $\Lambda = 0$ or $\Lambda \propto G$.

The *four-volume* interval corresponding to $L$ is $L^4$. Assume that within such volume intervals quantum fluctuations dominate. Hence, *fluctuations $x^\alpha$ can now be identified as quantum in origin* (cf. last paragraphs of Sec. 5.3). By correspondence (6.59), $q_{\mu\nu}$ must also randomly fluctuate within such volumes. However, a macroscopic observer cannot see these details, instead seeing an average of $q_{\mu\nu}$ over volume $L^4$. Effectively, then, $L^4$ is an uncertainty volume, and all of four-space is subdivided into contiguous four-cells of size $L^4$ within which the details of $q_{\mu\nu}$ and $\overline{h}_{\mu\nu}$ cannot be seen.

### 6.3.21 Weak-field equation

Let us simply identify $\Delta V$ with the volume $L^4$. Squaring Eq. (6.59) and using Eq. (6.23) gives

$$p_{\mu\nu}(x^\alpha)\,\Delta V = \overline{h}^2_{\mu\nu} \tag{6.60}$$

for the probability of a fluctuation $x^\alpha$ lying within the elemental four-interval. (Note that, because the left-hand side of (6.60) is essentially a differential, (6.60) tends to agree with the weak-field assumption $\overline{h}^2_{\mu\nu} \ll 1$.)

Now use Eq. (6.59) on the left-hand side of Eq. (6.58) and set $B = 16\pi G/c^4\sqrt{\Delta V}$, $C = -2\Lambda/\sqrt{\Delta V}$, where $\Lambda$ is the cosmological constant. Then Eq. (6.58) becomes the weak-field equation in the Lorentz gauge,

$$\Box\overline{h}_{\mu\nu} = \left(\frac{16\pi G}{c^4}\right)T_{\mu\nu} - 2\Lambda\eta_{\mu\nu}. \tag{6.61}$$

The symmetry (6.7) and (6.25) in $\eta_{\mu\nu}$ and $T_{\mu\nu}$ implies symmetry in an $\overline{h}_{\mu\nu}$ obeying (6.61) as well. Therefore there are 10, rather than 16, independent degrees of freedom $\overline{h}_{\mu\nu}$, as was anticipated.

### 6.3.22 Field equations in $q_{\mu\nu}$; the Planck length

The wave equation (6.58) in the probability amplitudes $q_{\mu\nu}(x)$ may be further specified. (*Note*: these are to be distinguished from probability amplitudes on the metric $g_{\mu\nu}$ itself.) Since $q_{\mu\nu}$ is purely a determinant of *quantum* fluctuations (Sec. 6.3.20), its field equation should be independent of specifically gravitational parameters such as $G$ and $\Lambda$. This is an *ansatz*. Use of Eq. (6.59) in (6.61) gives, after multiplying both sides by quantity $c\hbar/\sqrt{\Delta V}$,

$$c\hbar \Box q_{\mu\nu} = a_1 T_{\mu\nu} - a_2 \eta_{\mu\nu} \tag{6.62a}$$

$$a_1 \equiv \frac{16\pi\hbar G}{\sqrt{\Delta V} c^3}, \qquad a_2 \equiv \frac{2c\hbar\Lambda}{\sqrt{\Delta V}}. \tag{6.62b}$$

Constants $a_1$ and $a_2$ will be determined by the choice of $\Delta V$.

Each side of Eq. (6.62a) has units of energy/volume. Hence, since $T_{\mu\nu}$ also has these units, $a_1$ is unitless. Also, since Eq. (6.62a) defines the quantum quantity $q_{\mu\nu}$, by our previous *ansatz* the right-hand side must be independent of gravitational parameters, notably $G$. Then, in (6.62b), $a_1$ must be both unitless and independent of gravitational parameter $G$. The choice

$$\sqrt{\Delta V} = \hbar G/c^3 \tag{6.63}$$

accomplishes these aims, to an arbitrary numerical multiplier of the right-hand side. Next, since $\Delta V = L^4$, elemental length $L$ is

$$L = \sqrt{\frac{\hbar G}{c^3}}. \tag{6.64}$$

This is the Planck length. Thus, the Planck length follows naturally from the EPI approach and an *ansatz*. This tends to verify the *ansatz*, so we use it again next.

### 6.3.23 Cosmological constant

Next, consider the constant $a_2$. Using the solution (6.63) in the second Eq. (6.62b) gives

$$a_2 = 2c^4\left(\frac{\Lambda}{G}\right). \tag{6.65}$$

Again, by the *ansatz* $a_2$ should not depend upon gravitational parameters. Possible solutions are $\Lambda = 0$ or $\Lambda \propto G$. Recent astronomical observations (Perlmutter *et al.*, 1998) seem to require $\Lambda$ to be finite, implying the $\Lambda \propto G$ option. However, numerous theoretical arguments (Thomas, 2002) using the "holographic principle" of quantum gravity lead to the prediction $\Lambda = 0$. The question is not yet resolved.

### 6.3.24 A d'Alembert wave equation: gravitons

With use of $\Lambda = 0$ and Eqs. (6.62b) and (6.63), the field equation (6.62a) becomes

$$\Box q_{\mu\nu} = \frac{16\pi}{c\hbar} T_{\mu\nu}. \tag{6.66}$$

For a zero stress-energy source $T_{\mu\nu}$ this is a d'Alembert *wave equation*, which, as we saw in Chaps. 4 and 5, describes particles with integer values of spin, i.e., Bose–Einstein particles. Because of the double subscripting the spin is here 2, defining in the usual way the graviton. Analogously with the exchange of virtual particles that are imagined to give rise to the nuclear potential $V(r)$ of Sec. 4.1.27, an exchange of *gravitons* is imagined to give rise to the gravitational potential $q_{\mu\nu}$ here.

Hence, in the source-free scenario, $q_{\mu\nu}$ is both a field operator (see Sec. 6.3.2) amplitude, for the position of a graviton, and a potential function resulting from graviton interchange. This dual property tends to verify the identification between $q_{\mu\nu}$ and $\overline{h}_{\mu\nu}$ that was made in Sec. 6.3.20.

### 6.3.25 Graviton creation and absorption

Since $q^2_{\mu\nu} = q^2_{\mu\nu}(x, y, z, ct)$ is a four-dimensional PDF, its integral over all $(x, y, z)$ is $q^2_{\mu\nu}(ct) \equiv p_{\mu\nu}(ct)$, a PDF on time $t$ of detection. This can be zero at certain times, indicating no possible detection or, equivalently, the annihilation of a graviton. A subsequent detection event would then characterize the creation of a graviton. This implies that gravitons can be created and absorbed, in time, by sources and sinks of the gravitational field. See the analogous effects for bosons in general, as discussed in Sec. 4.1.26.

### 6.3.26 Probability on fluctuations

By the use of Eqs. (6.23) and (6.59) in Eq. (2.23), we get as a PDF on the fluctuations $x^\alpha$

$$p(x^\alpha) = \frac{1}{10L^4} \sum_{\mu\nu} \overline{h}^2_{\mu\nu}(x^\alpha), \tag{6.67}$$

where $(\mu\nu)$ range over their ten non-symmetric pairs. In the source-free case (Sec. 6.3.24) we found that this represents the PDF on the position of a graviton. However, in cases where there are sources present the fluctuations $x^\alpha$ no longer represent simply the positions of gravitons since Eq. (6.66) defining amplitudes $q_{\mu\nu}$ would no longer have the Helmholtz form. The $x^{\alpha}$ then generally represent uncertainties in field positions (Sec. 6.3.2).

### *6.3.27 Uniqueness and normalization properties*

The weak-field Eq. (6.61) with zero cosmological constant (Sec. 6.3.23) is of the same form as the electromagnetic wave Eq. (5.51). The only difference between the two is the higher dimensionality of Eq. (6.61). Therefore, by the reasoning in Secs. 5.1.25–5.1.28, the weak-field equation likewise has unique, normalizable solutions.

### *6.3.28 Newtonian mechanics*

Newtonian mechanics arises out of EPI in two different ways: (1) as the limiting form of the field Eqs. (6.61) or (6.68) for weak gravity and low velocities (Misner *et al.*, 1973, p. 412); or (2) as the classical limit $\hbar \to 0$ of an EPI derivation of the Schrödinger wave equation (Appendix D). The latter includes derivations both of Newton's second law and of the virial theorem.

### *6.3.29 Reference*

Most of the material in this chapter is from the paper by Cocke and Frieden (1997).

## 6.4 Einstein field equation and equations of motion

The transition from the weak-field Eq. (6.61) to the general field case may be made by the usual bootstrapping argument (Misner *et al.*, 1973, p. 417). That is, the only doubly subscripted tensor whose linear approximation is $\Box\overline{h}_{\mu\nu}$ is $2R_{\mu\nu} - g_{\mu\nu}R$, where $R_{\mu\nu}$ is the Ricci curvature tensor. Also, the source-free constant should now be proportional to the metric $g_{\mu\nu}$ instead of $\eta_{\mu\nu}$. Accordingly Eq. (6.61) becomes

$$R_{\mu\nu} - \tfrac{1}{2}g_{\mu\nu}R = \left(\frac{8\pi G}{c^4}\right)T_{\mu\nu} - \Lambda g_{\mu\nu}, \tag{6.68}$$

the Einstein field equation. From before, $\Lambda \propto G$ or $\Lambda = 0$.

It is known (Misner *et al.*, 1973, p. 480) that the field equation can be used as well to derive the *equations of motion* for a test particle in the field. Hence, the relativistic equations of motion also follow from the EPI approach.

## 6.5 Overview

The Einstein field equation may be derived via EPI in almost 1 : 1 fashion with the derivation of Maxwell's equations in Chapter 5. As in the latter derivation, there are two overall steps to accomplish: (1) derive the weak-field wave equation under the Lorentz condition; and (2) raise the rank of the weak-field equation to the tensor level of the Einstein equation by a simple argument.

The smart measurements are of the metric $g_{\mu\nu}$ at ideal four-positions $\theta^{\alpha}_{\mu\nu}$. The input space measurements are at positions $y^{\alpha}_{\mu\nu}$ with "noise" values $x^{\alpha}_{\mu\nu} \equiv (ct, x, y, z)_{\mu\nu}$. The noise is quantum in origin, so that the probability amplitudes $q_{\mu\nu}(x^{\alpha})$ are on fluctuations in measured four-position due to quantum effects on the scale of the Planck length. These arguments allow us to construct the Planck length in Sec. 6.3.22.

The information $I$ in the data positions obeys Eq. (6.24). The invariance principles that are used to find information $J$ are the Lorentz condition (6.28) and conservation of flow (6.27) for the stress-energy. The total number $N$ of degrees of freedom $q_{\mu\nu}(x^{\alpha})$ is fixed at ten by the two-index nature of all quantities and symmetry under interchange of the indices.

With information $J$ determined, the ten amplitudes $q_{\mu\nu}(x^{\alpha})$ are found to obey a wave equation (6.58) due, in particular, to a *gravitational source* $T_{\mu\nu}$. This has the form of the weak-field gravitational equation. Hence, it implies that the amplitudes $q_{\mu\nu}$ are linear (6.59) in the corresponding metric elements $\overline{h}_{\mu\nu}$. This results in the weak-field equation (6.61), where the cosmological constant $\Lambda$ arises as a constant of integration $C_{\mu\nu}$ in Eq. (6.54). By a dimensional *ansatz* in Sec. 6.3.23 the cosmological constant is found to be either zero or proportional to $G$, the gravitational constant. The same kind of argument fixes the Planck length $L$ as obeying proportionality to Eq. (6.64).

The combined solution (6.58) satisfies the EPI extremum requirement (6.34) and zero requirement (6.38). This may be verified as in Sec. 5.2.4.

Because of the correspondence between quantities $q_{\mu\nu}$ and $\overline{h}_{\mu\nu}$, there is a *wave equation* (6.66) for the probability amplitudes corresponding to the weak-field metric Eq. (6.61). Interpreting this probability amplitude as a field operator implies the existence of gravitons, whose PDF on four-position $x^{\alpha}$ obeys Eq. (6.67). Its *marginal* PDF on the time has properties that are consistent with the hypothesis that gravitons are created and absorbed.

The concept of a "prior measurement" was defined in Sec. 2.1.1. The derived Eqs. (6.61) and (6.66) show that each "prior measurement" $y^{\alpha}_{\mu\nu}$ defines a new degree of freedom $q_{\mu\nu}$ or $\overline{h}_{\mu\nu}$ of the theory. This is an intuitively correct result. It implies that, in the sense of acquired information Eq. (6.17), *ten* prior measurements of the metric tensor are sufficient to define classical gravitational theory.

EPI regards the Einstein field equation as having a *fundamentally statistical* nature. In analogy with the electromagnetic case in Chapter 5, the statistics are an outgrowth of admitting that any measurement of the gravitational metric must suffer from random errors in four-position of the field point. That is, the measurer does not know precisely where and when the metric is measured.

It is interesting to compare this viewpoint with that of conventional gravitation theory. The latter, of course, derives from a framework of *deterministic* Riemannian geometry. In this chapter, by contrast, we showed that gravitation grows out of the concept of Fisher information, a geometrical measure of distance in a statistical parameter space (Secs. 1.4.3 and 1.5). By this approach, gravitation may be viewed as arising from a framework of *statistical*, rather than deterministic, geometry. See also Amari (1985) on the subject of the mathematics of distance (information) measures in statistical parameter space; and Caianiello (1992) on some interesting physical interpretations of the space.

# 7
# Classical statistical physics

## 7.1 Goals

Classical statistical physics is usually stated in the non-relativistic limit, and so we restrict ourselves to this limit in the analyses to follow. However, as usual, we initiate the analysis on a covariant basis.

The overall aim of this chapter is to show that many classical distributions of statistical physics, defining both equilibrium and non-equilibrium scenarios, follow from a covariant EPI approach. Such equilibrium PDFs as the Boltzmann law on energy and the Maxwell–Boltzmann law on velocity will be derived. *Non-equilibrium* PDFs on velocity will also be found. Finally, some recently discovered inequalities linking entropy and Fisher information will be derived.

## 7.2. Covariant EPI problem

### *7.2.1 Physical scenario*

Let a gas be composed of a large number $M$ of identical molecules of mass $m$ within a container. The temperature of the gas is kept at a constant value $T$. The molecules are randomly moving and mutually interacting through forces due to potentials. The particles randomly collide with themselves and the container walls. All such collisions are assumed to be perfectly elastic. The mean velocity of any particle over many samples is zero.

The overall goal is to find the probability laws governing the energy $E$ and momentum $\boldsymbol{\mu}$ fluctuations of an arbitrary particle of the gas, at an arbitrary time $t$. The value of $t$ is not necessarily large, so the gas is not necessarily at equilibrium. Consequently we seek general *non-equilibrium* probability laws.

An alternative EPI approach to the one to be taken here is given in Chapter

13. It is called the "macroscopic approach" (of MFI in Sec. 1.8.8) and is briefly discussed in this paragraph. The macroscopic approach entails equating information $J$ *to zero*, effectively replacing it with macroscopic constraint data. However, the data are necessarily finite and of an *ad hoc* nature, and therefore insufficient to allow an exact estimate of the PDF $p(x)$. Thus the macroscopic approach is *approximate*. However, it has some notable advantages: (a) it is simple to implement, since the need to evaluate $J$ is avoided; (b) the estimate is smooth (in the Fisher sense), and therefore least biased or maximally equivocal regarding preferred values of $x$; (c) it generally describes non-equilibrium statistical mechanics; and (d) it also solves problems of economic valuation.

In contrast with the foregoing macroscopic approach, in this chapter the usual *microscopic* viewpoint of EPI is taken. This differs from the macroscopic approach in the following ways. (1) The information functional $J$ will be *computed*, replacing the arbitrary data of the macroscopic approach. (2) Since the information $J$ represents, in effect, *nature's* choice of the correct "constraint" term (Sec. 1.8.8; Sec. 7.4.11 below), the EPI approach will be *exact*. The EPI outputs will describe known stochastic effects of statistical mechanics – the Boltzmann and Maxwell–Boltzmann distributions and the wave equation for the anharmonic oscillator. We now continue that approach.

Since energy-momentum coordinates $(E, \boldsymbol{\mu})$ comprise a four-vector by Eq. (4.3), the covariant nature of EPI would *seem* to require the joint probability law $p(E, \boldsymbol{\mu})$ on the full four-vector to be sought. However, as we previously found (Sec. 3.5.5), joint momentum-energy coordinates are not appropriate Fisher four-coordinates, since they are *not a priori independent* degrees of freedom: they are dependent, by Eq. (4.17). One step removed from knowledge of the joint probability law is knowledge of the *marginal* law on each of $E$ and $\boldsymbol{\mu}$. Hence we will attempt to find these. The derivations, using EPI, are in Secs. 7.3 and 7.4, respectively. The introductory concepts for these derivations are, for brevity, developed in parallel in the next three sections.

Denote the true energy and momentum of the particle as $(\theta_E, \boldsymbol{\theta}_\mu)$, respectively, where the subscript $\mu$ signifies momentum (and is not a numerical index!), and $\boldsymbol{\theta}_\mu \equiv (\theta_{\mu 1}, \theta_{\mu 2}, \theta_{\mu 3})$ are the usual Cartesian components. By Eq. (2.1) the intrinsic data $\mathbf{y} \equiv (y_E, \mathbf{y}_\mu)$ obey

$$y_E \equiv E = \theta_E + x_E, \qquad E_0 \leqslant y_E \leqslant \infty, \tag{7.1a}$$

$$\mathbf{y}_\mu = \boldsymbol{\theta}_\mu + \mathbf{x}_\mu, \qquad \mathbf{y}_\mu = (y_{\mu 1}, y_{\mu 2}, y_{\mu 3}), \tag{7.1b}$$

$$\mathbf{x}_\mu \equiv c\boldsymbol{\mu} \tag{7.1c}$$

with fluctuations $x_E$ and $\mathbf{x}_\mu$. Subscripts 1, 2, 3 denote Cartesian components. All momentum coordinates $(\mathbf{y}_\mu, \boldsymbol{\theta}_\mu, \mathbf{x}_\mu)$ are expressed in units of the speed of light $c$, as in Eq. (7.1c), so as to share a common unit with the energy coordinates (see next section).

The particle obeys a joint PDF $p(x_E, \mathbf{x}_\mu)$ on the four-fluctuation $(x_E, \mathbf{x}_\mu)$. As discussed above, we want to know its marginal PDFs $p(x_E)$ and $p(\mathbf{x}_\mu)$ or, equivalently, their corresponding probability amplitudes $q(x_E)$ and $q(\mathbf{x}_\mu)$. After finding $p(x_E)$ we will use Eq. (7.1a) to find the required law $p(E)$ on the energy. The EPI procedure (Sec. 3.4) is used.

### *7.2.2 Fisher coordinates for the two problems*

Although $(E, \boldsymbol{\mu})$ are not appropriate joint Fisher coordinates for the problem (preceding section), they still are a four-vector in the Minkowski sense. Then, by Sec. 3.5.5, one of these must be imaginary. The proper choice for this application is an imaginary energy coordinate,

$$(ix_E, c\boldsymbol{\mu}), \quad \text{where} \quad \boldsymbol{\mu} \equiv \mu_1, \mu_2, \mu_3 \tag{7.2}$$

are purely real Cartesian coordinates.

### *7.2.3 Fisher information quantities*

The basic unknowns of the problem are the probability amplitudes

$$q_n(x_E), \quad n = 1, \ldots, N_E \quad \text{and} \quad q_n(\mathbf{x}_\mu), \quad n = 1, \ldots, N. \tag{7.3}$$

By Eq. (2.18) the Fisher information quantities $I(E)$ and $I(\boldsymbol{\mu})$ for the energy and momentum obey, respectively,

$$I(E) = -4\int dx_E \sum_{n=1}^{N_E} \left(\frac{dq_n(x_E)}{dx_E}\right)^2 \tag{7.4}$$

and

$$I(\boldsymbol{\mu}) = 4\int d\mathbf{x}_\mu \sum_{n=1}^{N} \sum_{m=1}^{3} \left(\frac{\partial q_n(\mathbf{x}_\mu)}{\partial x_{\mu m}}\right)^2. \tag{7.5}$$

The minus sign in Eq. (7.4) is a consequence of the imaginary nature of coordinate $x_E$ in Eq. (7.2); see Appendix C.

### *7.2.4 Bound information quantities*

Corresponding to each information functional $I(E)$ and $I(\boldsymbol{\mu})$ is a bound information functional $J(E)$ and $J(\boldsymbol{\mu})$ to be determined. By the EPI principle (3.16) and (3.18) we have two EPI problems to solve:

$$I(E) - J(E) = \mathit{extrem.}, \quad I(E) - \kappa J(E) = 0 \tag{7.6}$$

and

$$I(\boldsymbol{\mu}) - J(\boldsymbol{\mu}) = extrem., \qquad I(\boldsymbol{\mu}) - \kappa J(\boldsymbol{\mu}) = 0. \tag{7.7}$$

We first solve problem (7.6) for $p(E)$.

## 7.3 Boltzmann probability law

Here our goal is to determine $p(E)$. This will be found by first determining $p(x_E)$ and then using Eq. (7.1a) to go via Jacobian transformation (Frieden, 2001) from the coordinate $x_E$ to the coordinate $E$. For simplicity of notation we drop superfluous subscripts $E$, and permit the values of energy to vary over some general range $(a, b)$,

$$x \equiv x_E, \qquad E \equiv y_E, \qquad a \leqslant x \leqslant b. \tag{7.8}$$

The solution we seek is therefore denoted as $p(x)$, with amplitude function $q(x)$. We anticipate that the upper bound $b$ to the energy fluctuation $x$ must be infinite. However, to observe the dependence upon $b$, we first regard it as a large but finite value. The infinite limit is taken at Eq. (7.24c).

### *7.3.1 Fisher information*

The energy coordinate has been taken to be imaginary, by Eq. (7.2). As we saw in Eq. (7.4), in this case Eq. (2.19) becomes negative,

$$I(E) \equiv I = -4\int_a^b dx \sum_n q_n'^2(x), \qquad q_n'(x) \equiv dq_n(x)/dx. \tag{7.9}$$

All integrals in this calculation are understood to have these finite limits.

As usual, by Eq. (3.23) the PDFs for the energy fluctuations relate to $q_n$ as simply

$$p_n(x) = q_n^2(x). \tag{7.10}$$

### *7.3.2 Finding J*

As in the preceding two chapters, we form the functional $J$ and efficiency $\kappa$ by the requirement that the solution $q_n(x)$ be self-consistent, i.e., satisfy the two EPI requirements – Eqs. (3.16) and (3.18). It was noted (Sec. 3.4.6) that not every combination of the requirements (3.16) and (3.18) actually gives a feasible solution that satisfies each separately. In this problem, for example, one combined solution gives an energy efficiency $\kappa$ that lies outside the required range (0, 1) (see Exercise following Sec. 7.3.7). This is not then a feasible solution and is rejected.

The upshot is that any candidate solution has to be *verified*, by back substitution into (3.16) and (3.18), and has to be consistent with a $\kappa$ that lies within the interval (0, 1). The following solution was found to be the only one that passes these tests. (Alternative solutions that do not pass the test are pointed out as well.)

The information functional $J$, since it defines the information source, should depend upon the $q_n \equiv q_n(x)$. Thus, we let

$$J \equiv 4\int dx \sum_n j_n[q_n(x)] \equiv 4\int dx \sum_n j_n[q_n] \tag{7.11}$$

for simplicity. A more general form $J_n[q, x]$ allowing an additional dependence upon $x$ could likewise have been tried, but is not necessary for our purposes; see Sec. 7.4.10. The factor 4 is taken for convenience, and does not lessen the generality of the expression.

We first carry through on satisfying the EPI extremum requirement (3.16). By Eqs. (7.9) and (7.11), this is

$$K \equiv I - J = -4\int dx \sum_n (q_n'^2 + j_n[q_n]) = extrem. \tag{7.12}$$

The *Euler–Lagrange* Eq. (0.34) for this single-coordinate, single-component problem is

$$\frac{d}{dx}\left(\frac{\partial k}{\partial q_n'}\right) = \frac{\partial k}{\partial q_n}, \tag{7.13}$$

where $k$ is the (Lagrangian) integrand of Eq. (7.12),

$$k = -4\sum_n (q_n'^2 + j_n[q_n]). \tag{7.14}$$

Using this in (7.13) gives a solution

$$q_n'' = \frac{1}{2}\frac{\partial j_n}{\partial q_n} = \frac{1}{2}\frac{dj_n}{dq_n}. \tag{7.15}$$

The partial derivative becomes a full derivative because $j_n = j_n[q_n]$ alone.

We now turn to satisfying the *zero-condition* (3.18). It is useful to first evaluate the integral in Eq. (7.9) by a partial integration, giving

$$I = I_0 + 4\int_a^b dx \sum_n q_n q_n'', \quad I_0 \equiv 4\int_a^b dx \sum_n C_n, \tag{7.16}$$

$$C_n \equiv \frac{q_n(a)q_n'(a) - q_n(b)q_n'(b)}{b-a}.$$

The zero-condition (3.18) is, by Eqs. (7.11) and (7.16),

$$I - \kappa J = 4\int dx \sum_n [q_n q_n'' + C_n - \kappa j_n[q_n]] \equiv 0. \tag{7.17}$$

The *microscale* problem (3.20) is then

$$q_n q_n'' + C_n - \kappa j_n[q_n] = 0. \tag{7.18}$$

We now combine solutions. Amplitude functions $q_n$ are required to satisfy both Eqs. (7.15) and (7.18). Multiplying Eq. (7.15) by $q_n$ and using the resulting product $q_n q_n''$ in (7.18) gives

$$\frac{1}{2} q_n \frac{dj_n}{dq_n} = \kappa j_n[q_n] - C_n. \tag{7.19}$$

This may be placed in the convenient form

$$2\frac{dq_n}{q_n} = \frac{dj_n}{\kappa j_n[q_n] - C_n}$$

for integration, giving

$$j_n[q_n] = \frac{1}{\kappa}(\alpha_n^2 q_n^{2\kappa} + C_n), \tag{7.20}$$

where $\alpha_n^2$ is a constant of integration. It is generally complex.

With $j_n$ now the known function (7.20) of $q_n$, this may be substituted either into Eq. (7.15) or into Eq. (7.18) to get a solution. The result of either substitution is a differential equation

$$q_n'' = \alpha_n^2 q_n^{2\kappa - 1}. \tag{7.21}$$

Since $q_n$ is a real function, $\alpha_n^2$ is now seen to be necessarily fixed as real. Its sign, though, is still arbitrary.

*Exercise*: Verify that combining Eq. (7.20) with either of the two EPI conditions (7.15) and (7.18) gives rise to the *same* Eq. (7.21) in the amplitudes $q_n$.

### 7.3.3 Fixing $\kappa$

The answer for $q_n(x)$ must hold for energy values $x$ of either *quantum* or classical nature. Therefore, as in Chapter 4, the information efficiency must be unity,

$$\kappa = 1. \tag{7.22}$$

An interesting ramification of this is found after its substitution into Eq. (7.20), giving

$$j_n[q_n] = \alpha_n^2 q_n^2 + C_n, \quad \text{or} \quad J = 4\int dx \sum_n \alpha_n^2\, q_n^2(x) + 4(b-a)\sum_n C_n \tag{7.23}$$

by Eq. (7.11). The EPI extremum problem (3.16) is then

$$I - 4\sum_n \alpha_n^2 \int dx\, q_n^2(x) = extrem.,$$

where we switched orders of integration and summation, and ignored an irrelevent constant. This is equivalent to a variational problem where the objective functional $I$ is extremized subject to $N$ imposed normalization constraints on the PDFs $q_n^2(x)$, and is called the "macroscopic" MFI principle in Chapter 13. But, of course, EPI has not imposed the constraint; it has *derived* it, as Eq. (7.23). Deriving constraints is both a major problem and a major strength of EPI.

A PDF must of course obey normalization regardless of its chosen coordinate space. Normalization is then, in effect, the *invariance principle* for this application. Since any PDF $p_n(x)$ obeys this principle, it represents a weakest possible constraint to impose upon a PDF, or, equivalently, a state of maximum ignorance on the part of the observer as to the shape of the PDF.

### *7.3.4 General solution*

For the particular case (7.22), the solution Eq. (7.21) obeys

$$q_n'' = \alpha_n^2 q_n, \tag{7.24a}$$

where $\alpha_n^2$ is real and of either sign (see below Eq. (7.21). This is our "wave equation" for the problem. Since $q_n(x)$ must be real, the general solution to (7.24a) is

$$q_n(x) = \mathrm{Re}[B_n \exp(-\alpha_n x) + D_n \exp(+\alpha_n x)], \quad a \leqslant x \leqslant b, \tag{7.24b}$$

with $B_n$ and $D_n$ any real constants. This shows that $q_n(x)$ is exponential if $\alpha_n$ is real, or $q_n(x)$ is trigonometric if $\alpha_n$ is imaginary.

### *7.3.5 Trigonometric solutions ruled out*

Equation (7.24b) states that the $q_n$ are trigonometric functions if the $\alpha_n$ are imaginary. Temporarily suppose this to be the case. Are trigonometric functions acceptable probability amplitudes? One requirement of a probability density is that it obey normalization. Thus, the *square* of any such function $q_n(x)$ must obey the normalization. However, these squared values would now be rectified (all-positive) trigonometric curves. Such curves have finite area *only if* their support region $a \leqslant x \leqslant b$ is finite. However, here the upper bound to the energy is unbounded,

$$b \to \infty, \tag{7.24c}$$

as was mentioned at the outset. Therefore, trigonometric functions are not permissible solutions to this problem. This leaves exponential solutions as the only remaining possibility. That is, (7.24b) becomes

$$q_n(x) = B_n \exp(-\alpha_n x) + D_n \exp(+\alpha_n x), \quad a \leqslant x \leqslant \infty, \quad \text{with } \alpha_n \text{ real} \tag{7.25a}$$

Once more enforcing normalizability shows that the positive exponent is ruled out, so that $D_n = 0$. The result is a solution

$$q_n(x) = B_n \exp(-\alpha_n x), \quad \text{real } \alpha_n > 0, \quad a \leqslant x \leqslant \infty. \tag{7.25b}$$

### *7.3.6 Fixing $N_E$ by means of the game corollary*

The number $N_E$ of degrees of freedom to the problem is so far unknown. As discussed in Sec. 3.4.14, the value of a unitless constant, such as $N_E$, can be fixed by the requirement Eq. (3.21) that it *minimize* the absolute value of the information level $I$ for solutions $q_n(x)$. Therefore we need to establish how $I$ depends upon $N$ in this application.

From Eq. (7.25b) and the requirement of normalization over interval $a \leqslant x \leqslant \infty$ for its square, the PDF $p_n(x)$, the constant $B_n$ has value

$$B_n = \sqrt{2\alpha_n} \exp(\alpha_n a). \tag{7.25c}$$

As a result, by Eq. (7.25b), $q_n$ has the form

$$q_n(x) = \sqrt{2\alpha_n} \exp[\alpha_n(a - x)]. \tag{7.25d}$$

Using this in Eq. (7.9) gives an absolute value of the information of the very simple form

$$|I| = 4 \sum_{n=1}^{N_E} \alpha_n^2. \tag{7.25e}$$

Because the $\alpha_n$ are all real by the second Eq. (7.25b), this $|I|$ monotonically increases with the value of $N_E$.

Also, there is but one physical consideration that constrains the $\alpha_n$, this being the temperature $T$ of the container (Sec. 7.2.1). By the *game corollary* Eq. (3.21), $|I|$ should be minimized through choice of $N_E$. By the additive nature of Eq. (7.25e), the solution is

$$N_E = 1 \tag{7.25f}$$

degree of freedom to the problem.

Therefore, there is but one term to the sum (7.9) for $I$ and the sum (7.11) for

$J$, and their subsequent forms, and we may *drop the now superfluous subscript* $n$ from all quantities $j_n$, $q_n$, $\alpha_n$, etc. Thus,

$$j_n,\ q_n,\ \alpha_n,\ B_n,\ C_n \rightarrow j,\ q,\ \alpha,\ B,\ C. \tag{7.25g}$$

In particular, there is now but one $q(x)$ and one corresponding $p(x)$, from (7.25d)

$$q(x) = \sqrt{2\alpha}\exp[\alpha(a - x)], \quad a \leqslant x \leqslant \infty. \tag{7.25h}$$

Consequently

$$p(x) = 2\alpha\exp[2\alpha(a - x)], \quad a \leqslant x \leqslant \infty. \tag{7.25i}$$

Finally, for later use, by Eqs. (7.24c) and (7.25h),

$$q(b) = q(\infty) = 0 \quad \text{and} \quad q(a)q'(a) = -2\alpha^2. \tag{7.25j}$$

### *7.3.7 Checks on the solution*

It is important to check that the two EPI conditions are satisfied by the predicted solution (7.25h) for $q$, and the predicted solution

$$j = \alpha^2 q^2 + C \tag{7.26a}$$

for $j$ (the latter by Eqs. (7.23) and (7.25g)).

First consider the extremum condition (7.15). Twice differentiating (7.25h) gives

$$q'' = \alpha^2 q. \tag{7.26b}$$

Also, differentiating (7.26a) gives

$$\frac{dj}{dq} = 2\alpha^2 q = 2q'', \quad \text{or} \quad q'' = \frac{1}{2}\frac{dj}{dq}, \tag{7.26c}$$

by (7.26b). Equation (7.26c) is the same as the EPI extremum condition (7.15) under $N_E = n = 1$. Hence, the EPI extremum condition is satisfied by the solution, as required.

The next check is on satisfying the zero-condition (3.18). To do this, we compute the individual values of $I$ and $J$ at solution, checking that they are equal. As a preliminary, differentiating (7.25h) and squaring gives

$$q'^2 = \alpha^2 q^2. \tag{7.27a}$$

Next, by Eqs. (7.9), (7.25f), and (7.27a) and normalization for $q^2(x)$,

$$I = -4\int dx\, q'^2 = -4\int dx\, \alpha^2 q^2 = -4\alpha^2. \tag{7.27b}$$

Finally, by Eqs. (7.11), (7.25f), and (7.26a), the definition of $C$ in Eq. (7.16), (7.25j), and normalization for $q^2(x)$,

$$J = 4\int dx\,(\alpha^2 q^2(x) + C) = 4[\alpha^2 + (b-a)C] = 4\alpha^2 + 4q(a)q'(a)$$

$$= 4\alpha^2 - 8\alpha^2 = -4\alpha^2. \qquad (7.27c)$$

Comparing Eqs. (7.27b) and (7.27c) shows that $I$ and $J$ are equal, as required by the EPI zero-condition. Notice that both are also negative, as required by the form of Eq. (7.9).

*Exercise*: *A combined solution that is not self-consistent.* As discussed in Sec. 3.4.6, combining both EPI conditions (3.16) and (3.18) does not always result in a feasible solution $\kappa$, $q(x)$ that obeys *each* condition separately. For example, forcing conditions (3.16) and (3.18) to be obeyed can lead to a requirement that $\kappa$ be *unfeasible*. The following *alternative* approach to the given problem is of this type. Repeat steps (7.9)–(7.15). Equation (7.15) results out of the EPI extremum condition. Next, in enforcing the zero-condition, do not do step (7.16), but rather, leave $I$ in its defining form Eq. (7.9). The zero microlevel requirement (3.20) is then $q_n'^2 + \kappa J_n(q_n) = 0$. Differentiate this $d/dx$, using the chain rule $dJ_n/dx = (dJ_n/dq_n)q_n'$. Then, assuming that $q_n' \neq 0$, the result is a requirement $q_n'' = -(\kappa/2)(\partial J_n/\partial q_n)$. Comparing this with Eq. (7.15) requires $\kappa = -1$, an unfeasible value of the efficiency (Sec. 3.4.5). Not all roads lead to Rome.

### 7.3.8 PDF p(E)

By Eq. (7.1a), the random variable $E$ is simply a shifted version of $x_E = x$. Since $dE/dx = 1$, the PDF on $E$ is, from Eq. (7.25i),

$$p(E) = Ce^{-2\alpha E}, \quad C \equiv 2\alpha e^{2\alpha\theta}. \qquad (7.28)$$

The evaluation of the parameters $\alpha$, $C$ of this PDF requires more specific knowledge of the physical scenario, in particular a number for the lower bound to $E$.

### 7.3.9 Lower bound to energy

By inequality (7.1a) the domain of energy $E$ is $E_0 \leqslant E < \infty$. Suppose that we also know the mean value $\langle E\rangle$. It is easily shown that the constants $C$ and $\alpha$ in Eq. (7.28) that satisfy both normalization and $\langle E\rangle$ give a

$$p(E) = D^{-1}\exp[-(E - E_0)/D], \quad D = (\langle E\rangle - E_0), \quad \text{for } E \geqslant E_0 \qquad (7.29)$$

and $p(E) = 0$ for $E < E_0$. This is a simple exponential law. It is interesting that the exponent must always be negative, indicating monotonically decreasing probability for increasing measured energy.

The non-relativistic limit of Eq. (4.17) with fields inserted is, of course, that energy $E = E_{\text{kin}} + V$, where $E_{\text{kin}}$ is the kinetic energy of the particle and $V$ is its potential energy at the given point of detection. We can always add a constant to $V$ without affecting any physical observable. Hence, subtract the constant $E_0$ from it. Then Eq. (7.29) becomes a simple exponential form

$$p(E) = \langle E \rangle^{-1} e^{-E/\langle E \rangle}, \quad E \geqslant 0. \tag{7.30}$$

### 7.3.10 Exercise

Verify that the PDF (7.28) becomes (7.29) by imposing normalization and a known mean value $\langle E \rangle$ upon (7.28).

### 7.3.11 Boltzmann law

We now express $\langle E \rangle$ in terms of physical quantities. Designate by $E_{\text{t}}$ the total energy in the gas. As there are $M$ identical particles, it is elementary that

$$\langle E_{\text{t}} \rangle = M \langle E \rangle. \tag{7.31}$$

Energy $E$ of Sec. 1.8.7 is called $E_{\text{t}}$ here. In this notation, by Eq. (1.41) and either (1.44) or (1.45), we have

$$\overline{p}V = -\frac{dE_{\text{t}}}{dV} V = -\left\langle \frac{E_{\text{t}}}{V} \right\rangle V = -\langle E_{\text{t}} \rangle \sim T \text{ or } T_E, \tag{7.32}$$

the latter the Fisher temperature. This assumes a well-mixed, perfect gas. More specifically, if the gas particles are constrained to move without rotation within the container (three degrees of freedom per molecule) it is well known that

$$\langle E_{\text{t}} \rangle = 3MkT/2, \tag{7.33}$$

where $k$ is the Boltzmann constant. This is usually called the "equipartition of energy" law. Then, by (7.31), $\langle E \rangle = 3kT/2$, so Eq. (7.30) becomes

$$p(E) = (3kT/2)^{-1} e^{-2E/(3kT)}, \quad E \geqslant 0, \tag{7.34}$$

the Boltzmann law for a three-dimensional gas.

### 7.3.12 Transition to discrete states and discrete probabilities

But of course the energies of the gas particles are not indefinitely continuous. Depending upon the type of gas particle that is present, one of the quantum equations, (4.28) or (4.42) or (D9), governs the energy levels. The result is that any particle energy value $E$ is quantized as $E \to E_j$, $j = 0, 1, \ldots$, where integer $j$ denotes a state of the particle. Thus, by the proportionality between a probability density and its corresponding absolute probability, Eq. (7.30) now describes an absolute probability

$$P(E_j) \equiv P_j = Ce^{-E_j/\langle E \rangle}, \quad C = const., \quad j = 0, 1, \ldots. \tag{7.35}$$

If we now use result (4.17) with the usual replacement $E \to E - V$ (ignoring any vector potential $\mathbf{A}$), we get

$$P_j = Ce^{-[E_{kin}(\boldsymbol{\mu}_j) + V(\boldsymbol{r}_j)]/\langle E \rangle}, \tag{7.36a}$$

$$E_{\text{kin}} = (c^2\mu^2 + m^2c^4)^{1/2} - mc^2 \tag{7.36b}$$

$$\approx \mu^2/(2m) \tag{7.36c}$$

in the non-relativistic limit. The constant $C$ has absorbed a constant factor $\exp(-mc^2/\langle E \rangle)$. Quantity $\boldsymbol{r}_j$ is a random position of the particle, and $E_{\text{kin}}$ and $V$ are its kinetic and potential energies, respectively. Both $E_{\text{kin}}$ and $V$ are random variables, since they are functions of the random variables $\boldsymbol{\mu}_j$ and $\boldsymbol{r}_j$. Hence, Eq. (7.36a) becomes a joint probability law

$$P(E_{\text{kin}\,j}, V_j) = P_E(E_{\text{kin}\,j})P_V(V_j), \tag{7.37a}$$

where

$$P_E(E_{\text{kin}\,j}) \sim e^{-E_{\text{kin}}(\boldsymbol{\mu}_j)/\langle E \rangle} \quad \text{and} \quad P_V(V_j) \sim e^{-V(\boldsymbol{r}_j)/\langle E \rangle}. \tag{7.37b}$$

This allows us to form discrete probabilities $P_r(\boldsymbol{r}_j)$ and $P_\mu(\boldsymbol{\mu}_j)$ on momentum $\boldsymbol{\mu}_\text{j}$ and position $\boldsymbol{r}_j$ as follows.

A given potential value $V$ is associated with a fixed number of position values $\boldsymbol{r}_j$. As examples, if $V \sim 1/r$ then the association is unique; or, if $V \sim r^2$, then there are two values $\boldsymbol{r}_j$ for each $V$. The upshot is that $P_r(\boldsymbol{r}_j) \sim P_V(V_j)$. In the same way, $P(\boldsymbol{\mu}_j) \sim P_E(E_{\text{kin}\,j})$. Then, by Eqs. (7.37b),

$$P(\boldsymbol{r}_j) \sim e^{-V(\boldsymbol{r}_j)/\langle E \rangle} \tag{7.38a}$$

and

$$P(\boldsymbol{\mu}_j) \sim e^{-E_{\text{kin}}(\boldsymbol{\mu}_j)/\langle E \rangle}. \tag{7.38b}$$

The latter will act as a check on the answer we get for $p(\boldsymbol{\mu})$ below. The former is often used in a continuous limit, in Sec. 7.3.14.

### *7.3.13 Transition to phase space*

The "phase space" of a classical particle consists of its joint momentum and position values $(\mu, \mathbf{r})$ in the $j$th elemental momentum-space "box" of the space (see Sec. 14.8.1). This is denoted as $(\mu, \mathbf{r})_j$. By the product form of Eq. (7.37a), use of Eq. (7.37b) and (7.36c) gives

$$P(\boldsymbol{\mu}, \boldsymbol{r})_j = K \exp\left(-\frac{\mu_j^2 + V(\boldsymbol{r}_j)}{\langle E\rangle}\right). \tag{7.38c}$$

Here $K$ is a normalization constant, and $\langle E\rangle = 3kT/2$ if the particle is constrained to move without rotation (see below Eq. (7.33)). Equation (7.38c) is the usual phase space representation of the distribution law for such a particle.

### 7.3.14 Barometric formula

Suppose that the gas in question is in a gravitational field. We want to predict the number density of particles as a function of the altitude $z$. Assuming that the altitude is small compared with the earth's radius, the potential is $V(\boldsymbol{r}) \equiv V(z) = mgz$, where $g$ is the acceleration due to gravity. Placing this in the continuous version of Eq. (7.38a) gives

$$p(z) = p(0)e^{-mgz/\langle E\rangle}. \tag{7.39}$$

By the law of large numbers (Frieden, 2001) this defines as well the required number density of particles. Unfortunately, this formula does not hold experimentally. The problem traces from our assumption that the temperature $T$ is constant; of course it varies with altitude $z$.

## 7.4 Maxwell–Boltzmann velocity law

This distribution law $p(v)$ on the magnitude of the velocity fluctuation is found as follows. First we find $p(\mathbf{x}_\mu)$, the equilibrium PDF on $c$ times the momentum fluctuations $\boldsymbol{\mu}$. Once known, this readily gives $p(\boldsymbol{\mu})$, the PDF on momentum or, by $\boldsymbol{\mu} = m\mathbf{v}$, the desired PDF $p(v)$.

To find $p(\mathbf{x}_\mu)$, we proceed as in Sec. 7.3, solving for the unknown bound information $J$ by simultaneously solving the two EPI problems (7.7). The essential differences are that, here, (a) the information efficiency parameter $\kappa$ has already been fixed, by Eq. (7.22), at value $\kappa = 1$; (b) by Eq. (7.2) the Fisher coordinates are *real* values $c\boldsymbol{\mu}$ as compared with the imaginary coordinate $ix_E$ used previously; and (c) the bound information is represented more generally, with an unknown, explicit dependence upon $\boldsymbol{\mu}$. As will be seen, the resulting PDF $p(\boldsymbol{\mu})$ at equilibrium will confirm result (7.38b) in the non-relativistic limit (7.36c). Since most of the steps repeat those of Sec. 7.2, we can proceed at a slightly faster pace.

### 7.4.1 Bound information J

Here represent

$$J[\mathbf{q}] = 4\int d\mathbf{x} \sum_{n=0}^{N} J_n(q_n, \mathbf{x}), \qquad \text{with } \mathbf{x}_\mu \equiv (c\mu_1, c\mu_2, c\mu_3) \equiv \mathbf{x} \tag{7.40}$$

as simpler notation. (Lower index limit $n = 0$ is used for convenience.) We allow for an explicit dependence upon the momenta $\mathbf{x}$ as well as upon the amplitudes $\mathbf{q}$. This more general representation than the corresponding one (7.11) for energy will permit a wider scope of solutions to the problem, including, in fact, non-equilibrium solutions!

We now proceed to solve the two EPI problems (7.7).

### 7.4.2 EPI extremum problem

By Eqs. (7.5) and (7.40) the first problem (7.7) is

$$I(\boldsymbol{\mu}) - J(\boldsymbol{\mu}) = 4\int d\mathbf{x} \sum_{n=0}^{N}\left[\sum_{m=1}^{3}\left(\frac{\partial q_n}{\partial x_m}\right)^2 - J_n(q_n, \mathbf{x})\right] = extrem. \tag{7.41}$$

The Euler–Lagrange Eq. (0.34) for the solution is

$$\sum_m \frac{\partial}{\partial x_m}\left(\frac{\partial \mathscr{L}}{\partial q_{nm}}\right) = \frac{\partial \mathscr{L}}{\partial q_n}, \qquad n = 1, \ldots, N,$$

$$q_{nm} \equiv \frac{\partial q_n}{\partial x_m}. \tag{7.42}$$

With the Lagrangian $\mathscr{L}$ the integrand in Eq. (7.41), the solution is the differential equation

$$\sum_m \frac{\partial^2 q_n}{\partial x_m^2} = -\frac{1}{2}\frac{\partial J_n(q_n, \mathbf{x})}{\partial q_n}. \tag{7.43}$$

### 7.4.3 EPI zero-root problem

By Eqs. (7.5) and (7.40) the second problem (7.7) is

$$I - \kappa J = -4\int d\mathbf{x} \sum_n \left[q_n(\mathbf{x})\sum_m \frac{\partial^2 q_n}{\partial x_m^2} + J_n(q_n, \mathbf{x})\right] = 0, \tag{7.44}$$

where we used $\kappa = 1$ and the partial integration (5.13) of $I$. (Note that (5.13) holds here because the "boundary" is at $\mathbf{x} \to \infty$, where the $q_n(\mathbf{x}) = 0$.) The solution is the microscale Eq. (3.20), which is here

$$q_n \sum_m \frac{\partial^2 q_n}{\partial x_m^2} = -J_n(q_n, \mathbf{x}). \tag{7.45}$$

### 7.4.4 Simultaneous solution

The simultaneous solution $\mathbf{q}$ and $J_n$ to conditions (7.43) and (7.45) obviously obeys

$$\frac{1}{2} \frac{\partial J_n(q_n, \mathbf{x})}{\partial q_n} = \frac{J_n(q_n, \mathbf{x})}{q_n}. \tag{7.46}$$

This may be integrated in $J_n$ and $q_n$ to give

$$J_n(q_n, \mathbf{x}) = q_n^2 f_n(\mathbf{x}) \tag{7.47}$$

for some functions $f_n(\mathbf{x})$. The latter arose as additive "constants" during the integration above. They are discussed next.

### 7.4.5 Nature of functions $f_n(x)$

Substituting the result (7.47) into either Eq. (7.43) or (7.45) produces the same solution (as we required),

$$\nabla^2 q_n = -q_n(\mathbf{x}) f_n(\mathbf{x}), \tag{7.48}$$

where the Laplacian $\nabla^2$ is with respect to coordinates $x_i$, $i = 1, 2, 3$. This shows that the form of functions $f_n(\mathbf{x})$ directly affects that of the output amplitudes $\mathbf{q}$ and, hence, the PDF $p(\mathbf{x})$.

We have not yet input into the development anything "momentum-like" about the $x_i$. Some plausible assumptions of this kind can be made. First, the probability of a given momentum should not depend upon its direction. Hence, $p(\mathbf{x})$ should be even in each component $x_i$. Second, $p(\mathbf{x})$ should depend upon each $x_i$ in the same way. Third, we take a non-relativistic approach by which all velocities are small compared with the speed of light $c$, so that the $x_i$ are likewise small compared with $mc^2$.

These considerations imply that $f_n(\mathbf{x})$ should be expandable as a power series in even powers of the modulus $x$ of $\mathbf{x}$,

$$f_n(\mathbf{x}) = A_n + B_n x^2, \quad A_n, B_n = const. \tag{7.49}$$

It is not necessary to include terms beyond the quadratic because $x$ is small, although doing so has important ramifications (Sec. 7.4.13). The constants in (7.49) need to be defined.

### 7.4.6 Hermite–Gauss solutions

Substituting series (7.49) into Eq. (7.48) gives as a solution for $\mathbf{q}$ the differential equation

$$\nabla^2 q_n(\mathbf{x}) + (A_n + B_n x^2) q_n(\mathbf{x}) = 0. \tag{7.50}$$

A separation of variables

$$q_n(\mathbf{x}) = q_{n1}(x) q_{n2}(y) q_{n3}(z), \quad \mathbf{x} \equiv (x, y, z) \tag{7.51}$$

gives three distinct differential equations:

$$q''_{ni}(x_i) + (A_{ni} + Bx_i^2) q_{ni}(x_i) = 0, \quad i = 1, 2, 3, \quad B_n \equiv B, \quad \sum_{i=1}^{3} A_{ni} \equiv A_n. \tag{7.52}$$

The coordinate $x_i \equiv x$, $y$, or $z$, in turn. Each such equation becomes a parabolic cylinder differential equation (Abramowitz and Stegun, 1965) if the constants obey

$$A_{ni} = \frac{(n_i + 1/2)}{a_0^2}, \quad \sum_{i=1}^{3} n_i \equiv n, \tag{7.53}$$

$$A_n = \frac{n + 3/2}{a_0^2}, \quad B = -\frac{1}{4a_0^4}, \quad a_0 = const. \tag{7.54}$$

Then Eq. (7.52) has a Hermite–Gaussian solution

$$q_{ni}(x_i) = e^{-x_i^2/(4a_0^2)} 2^{-n_i/2} H_{n_i}(x_i/a_0\sqrt{2}), \quad i = 1, 2, 3. \tag{7.55}$$

### 7.4.7 Superposition states

Equation (7.51) states that there is a degeneracy of product solutions for each index value $n$. Equations (7.51), (7.53), and (7.55) give a solution

$$q_n(\mathbf{x}) = e^{-|\mathbf{x}|^2/(4a_0^2)} 2^{-n/2} \sum_{\substack{ijk \\ i+j+k=n}} a_{nijk} H_i(x/a_0\sqrt{2}) H_j(y/a_0\sqrt{2}) H_k(z/a_0\sqrt{2}), \tag{7.56}$$

$$a_{nijk} = const.$$

The Hermite polynomials are defined as (Abramowitz and Stegun, 1965)

$$H_n(x) = n! \sum_{m=0}^{[n/2]} (-1)^m \frac{(2x)^{n-2m}}{m!(n-2m)!}, \tag{7.57}$$

where the notation $[b]$ means the largest integer not exceeding $b$. The lowest-order polynomials are

$$H_0(x) = 1, \quad H_1(x) = 2x, \quad H_2(x) = 4x^2 - 2. \tag{7.58}$$

Using Eqs. (2.23) and (7.56), the PDF on momentum fluctuations is

$$p(\mathbf{x}) = p_0 e^{-|\mathbf{x}|^2/(2a_0^2)} \times \left\{ 1 + \sum_{n=1}^{N} 2^{-n} \left[ \sum_{\substack{ijk \\ i+j+k=n}} b_{nijk} H_i(x/a_0\sqrt{2}) H_j(y/a_0\sqrt{2}) H_k(z/a_0\sqrt{2}) \right]^2 \right\},$$

$$p_0 = a_{0000}^2/N. \quad (7.59)$$

The '1' is from the sum evaluated at $n = 0$. The constants $b_{nijk}$ are proportional to the $a_{nijk}$ in (7.56).

*Exercise*: Consider a scenario where there are no potentials, so the mean kinetic energy equals the mean total energy, $\langle E_{\text{kin}} \rangle = \langle E \rangle = 3kT/2$ (the latter by Eqs. (7.31) and (7.33)). On this basis the particular quadratic form (7.49) chosen for functions $f_n(\mathbf{x})$ leads ultimately to a requirement that the mean kinetic energy obey $\langle E_{\text{kin}} \rangle \leqslant 3kT$. However, this is no restriction since $\langle E_{\text{kin}} \rangle = 3kT/2$ from the above, and of course $3kT/2 \leqslant 3kT$ as required. Central to this result was showing that $\langle E_{\text{kin}} \rangle \leqslant 3kT$. Do this for a case $N = n = 0$.

*Hint*: Since the Fisher coordinates for this problem are real, it follows that $I \geqslant 0$. Also, since $\kappa = 1$, it is required that likewise the computed $J \geqslant 0$. When $J$ is computed for $N = 0$ using Eqs. (7.40), (7.47) and (7.49), in view of the fact that by Eqs. (7.53) and (7.54) both $A_0 > 0$ and $B < 0$, our requirement $J \geqslant 0$ leads to a requirement that $A_0 \geqslant |B| \langle x^2 \rangle$. The usual normalization relation and second moment $\langle x^2 \rangle$ relation for a PDF $q_0^2(x)$ are used for this purpose. Then, the use of $x = c\mu$, relation $\alpha_0^2 = mc^2 kT$ from Eq. (7.63), the definition $E_{\text{kin}} = \mu^2/(2m) = E$ (for this case), and $\langle E \rangle = 3kT/2$ from Eqs. (7.31) and (7.33) give the desired requirement $\langle E_{\text{kin}} \rangle \leqslant 3kT$.

### 7.4.8 Values of N; equilibrium and non-equilibrium solutions

Because of the free parameters $b_{nijk}$ in Eq. (7.59), the PDF on momentum obeys a multiplicity of solutions. Any particular one is defined by a set of the $b_{nijk}$. What can this mean? Since EPI Eq. (3.16) seeks a stationary solution, the result (7.59) indicates that there is a multiplicity of such solutions to this problem. Of these, one must represent the *equilibrium* solution, approached as time $t \to \infty$, with the others representing stationary solutions *en route* to equilibrium. Which one is the equilibrium solution?

The number $N$ of degrees of freedom is at this point unknown. Since it is also a unitless constant of the theory, we can use the *game corollary* of Sec. 3.4.14 to fix it. In Eq. (7.5) each term in the sum over $n$ contributes independently and positively toward the information $I$. Hence $I$ grows monotonically with $N$. Therefore the game corollary implies that $N$ should tend toward a minimum value. That is, the solution for $n = N = 0$ should be preferred, occurring far more often than higher-$N$ cases. From Eq. (7.59) the case $N = 0$ is that of a pure Gaussian law, i.e., an ordinary Maxwell–Boltzmann equilibrium distribution. Thus, the game corollary predicts that the Maxwell–Boltzmann distribution will occur far more often than higher-$N$ solutions (7.59).

By definition, the equilibrium solution for $p(\mathbf{x})$ is tended toward regardless of initial conditions. Thus, it is the physically preferred solution among the possibilities (7.59). Since, from the preceding paragraph, the game corollary solution is preferred on the basis of frequency of occurrence, it coincides with the equilibrium solution. That solution was the Maxwell–Boltzmann law. In this way the Maxwell–Boltzmann law is predicted to describe the equilibrium distribution of momentum in the ideal gas.

The remaining solutions for $N \geqslant 1$ therefore represent stationary solutions *en route* to the Maxwell–Boltzmann equilibrium solution. That is, they are examples of non-equilibrium solutions. In fact, Rumer and Ryvkin (1980) previously found these Hermite–Gauss functions to be non-equilibrium solutions that follow from the *Boltzmann transport equation*. The well-known "method of moments" was used. Our particular Hermite–Gauss solution (7.59) corresponds to a particular choice of these authors' expansion coefficients. We find it rather remarkable that the EPI approach, which avoids any use of the Boltzmann equation, attained the same class of solution.

By the form of Eq. (7.59) the coefficients $b_{nijk}$ must be unitless. Therefore, as with $N$ in the above, they should be computable by use of the game corollary. To do this requires expressing the information $I$ as a function of the $b_{nijk}$, then minimizing $|I|$ through their variation and subject to appropriate constraints on them. This is yet to be done.

The fact that the non-equilibrium solutions of (7.59) are stationary (obeying Eq. (7.41)) appears to indicate that they are, in some sense, more probable or more frequent in occurrence than other types of non-equilibrium solutions. This conjecture ought to be testable by experimental observation or by Monte Carlo computer simulation.

The preceding suggests that Fisher information might be used to generate both equilibrium and *non-equilibrium* thermodynamics, completely without recourse to the usual entropy measure. An important property of thermodynamics is its "Legendre transform structure" (Duering *et al.*, 1985). This is ordinarily stated using entropy as the measure of disorder. However, it was recently shown that the same kind of Legendre transform structure ensues for a thermodynamics that utilizes, instead, Fisher information as the measure of disorder (Frieden *et al.*, 1999).

Equation (7.48) has the form of a Schrödinger wave equation with a "potential" function $f_n(\mathbf{x})$ given by Eq. (7.49). For the case of a dilute, viscous gas, the coefficients $A_n$, $B_n$ in Eq. (7.49) can be generalized to depend upon the time. Then the potential becomes quadratic in the velocity (Frieden *et al.*, 2002), and consequently the equilibrium solution becomes directly the ground-state solution of a one-dimensional simple harmonic oscillator. Furthermore, *non-equilibrium thermodynamic* solutions then correspond to *quantum mechanical* admixtures of excited states. Just as in quantum mechanics, these thermodynamic solutions are obtainable by perturbing the ground state "wave function" $q_0(\mathbf{x})$ with linear terms. The resulting non-equilibrium distributions agree with standard non-equilibrium distributions based upon the use of entropy (Frieden *et al.*, 2002a, b). Most generally, the series (7.49) can be extended to include weighted higher-order powers of the velocity $\mathbf{x}$, where the time-dependences of the weights give rise to more general answers. Also see in this regard Sec. 7.4.13.

A useful further application of Eq. (7.48) is to econometrics, in particular, the valuation of financial securities. See Chapter 13. The tenor of the results there suggests that the laws of valuation follow those of generally non-equilibrium statistical mechanics.

### 7.4.9 Correspondence between derived $P(E_{\text{kin}\,j})$ and $p(\boldsymbol{\mu})$ functions

Equations (7.36c) and (7.38b) predict that, in the non-relativistic limit,

$$P(E_{\text{kin}\,j}) \to P(\boldsymbol{\mu}_j) \sim e^{-\mu^2/(2m\langle E\rangle)}, \tag{7.60}$$

whereas the equilibrium case of solution (7.59) predicts that

$$p(\boldsymbol{\mu}) \sim p_0 e^{-\mu^2 c^2/(2a_0^2)} \tag{7.61}$$

for some choice of $a_0$. Both expressions were derived for the non-relativistic case and, therefore, should agree. The fact that they do agree is a verification of the overall theory. This also confirms the argumentation of the previous section.

### 7.4.10 A retrospective on the $p(E)$ derivation

We assumed, with some loss of generality, that each $J_n = J_n(q_n)$ alone in Eq. (7.11). A possible $x$-dependence, as $J_n(q_n, x)$, was left out. Now we can see why it was. As in Eq. (7.56), a superposition of stationary solutions would result, only one of which is the required equilibrium solution $p(E)$. Again, the lowest-order one would represent the equilibrium solution, and this is the simple exponential form (7.25) as previously derived! The remaining solutions would represent non-equilibrium laws, as in Sec. 7.4.8. This work has not yet been carried out.

### *7.4.11 Equivalent constrained I problem*

We remarked in Sec. 1.8.8 that a constrained minimization (of $I$) called MFI superficially resembles the EPI approach. Using the quadratic form (7.49) in Eq. (7.47) leads to a minimization problem (7.41) with a "constraint" term

$$-4\int d\mathbf{x} \sum_n q_n^2(\mathbf{x})(A_n + Bx^2) = -4\left[\sum_n A_n \int d\mathbf{x}\, q_n^2(\mathbf{x}) + B\int d\mathbf{x}\, x^2 \sum_n q_n^2(\mathbf{x})\right]$$

$$= \lambda_1 \int d\mathbf{x}\, p(\mathbf{x}) + \lambda_2 \int d\mathbf{x}\, x^2 p(\mathbf{x}),$$

$$\lambda_1, \lambda_2 = const. \tag{7.62}$$

Normalization of $p(\mathbf{x})$ and of each PDF $q_n^2(\mathbf{x})$ is used, along with Eq. (2.23). Hence, in this case the EPI approach is *mathematically* equivalent to the "constrained minimization" approach MFI, where the constraints are those of normalization and a fixed second moment.

The correspondence is interesting but, in fact, largely coincidental. Two factors should be considered.

First, in contrast to MFI, EPI has a definite mechanism for finding the effective constraint terms. These arise out of the bound information $J$ which, by Eq. (3.18), *must be proportional to I.*

Second, to obtain the correct result by EPI, a minimal number of effective constraints must be used. That is, nature imposes a minimal set of constraint conditions upon the Fisher information, rather than a maximal set. As empirical evidence for this, in all of the derivations up to this point only one or, at most, two constraints have effectively been imposed via the information $J$. In fact, adding the most trivial of constraints – normalization – to, say, the electromagnetic derivation of Chapter 5 gives as the output wave equation the Proca equation (Jackson, 1975), not the electromagnetic Eq. (5.51). This may be easily verified. Or, adding a constraint on mean energy to the single normalization constraint in Sec. 7.3 would lead, not to the Boltzmann law, but to the square of an Airy function $Ai(E)$ (Fig. 13.2) as the predicted $p(E)$ (Frieden, 1988).

Thus, EPI is not an *ad hoc* approach wherein all constraints known to affect

$p(\mathbf{x})$ are tacked onto the minimization of $I$. It would give approximate results if used in this way. Instead, principle (3.18) of proportionality between $I$ and $J$ *must be used* to form the effective constraints through the action of $J$. MFI, by contrast, offers no such systematic method of finding its constraints.

### 7.4.12 Equilibrium law on magnitude of velocity

Equation (7.61) may be evaluated in the particular case of zero potential energy. Then Eq. (7.33) holds, and, since $\langle E_t \rangle = M\langle E_{\text{kin}} \rangle = M\langle \boldsymbol{\mu}^2 \rangle/(2m)$, it gives $\langle \boldsymbol{\mu}^2 \rangle = 3mkT$. The constants $p_0$, $a_0$ in (7.61) that satisfy normalization and the given moment $\langle \boldsymbol{\mu}^2 \rangle$ then define a

$$p(\boldsymbol{\mu}) = \frac{c^3}{(2\pi)^{3/2} a_0^3} e^{-\mu^2 c^2/(2a_0^2)}, \quad a_0^2 = mc^2 kT \tag{7.63}$$

for our particles with three degrees of freedom. This equation is of the separated form $p(\boldsymbol{\mu}) = p(\mu_1)p(\mu_2)p(\mu_3)$, where each component probability has the same variance $mc^2kT$. Here we seek the PDF on the magnitude $\mu$. This may readily (if tediously) be found by first transforming coordinates from the given rectangular coordinates $(\mu_1, \mu_2, \mu_3)$ to spherical coordinates $(\mu, \theta, \phi)$. The formal solution is

$$p(\mu, \theta, \phi) = |J(\mu_1, \mu_2, \mu_3/\mu, \theta, \phi)| p(\mu_1, \mu_2, \mu_3) \tag{7.64}$$

where $J$ is the Jacobian of the transformation (Frieden, 2001). After evaluating the Jacobian and integrating out over $\theta$ and $\phi$, the solution is

$$p(\mu) = A\mu^2 e^{-\mu^2/(2mkT)}, \quad A = \sqrt{2/\pi}\,(mkT)^{-3/2}. \tag{7.65}$$

This is the Maxwell–Boltzmann law on the magnitude of the momentum. Transformation of this law to $p(v)$ via relation $\mu = mv$ then gives the usual Maxwell–Boltzmann law on the velocity.

### 7.4.13 Anharmonic oscillators; solitons

An important prediction of statistical mechanics is that of *solitons* (Scott, 1999). These are solitary (thus the name) waves that travel with unchanged shape and undiminished energy over large distances. A diverse number of phenomena can be described by soliton theory. Among these are quantum lattice soliton theory, certain oceanic wave patterns and atmospheric phenomena, aspects of general relativity, Davydov solitons in nerve impulse propagation, Bose–Einstein condensates, heat conductivity in solids, and soliton-pulse optical communication. As it turns out, the previous derivation of the Maxwell–Boltzmann and related laws allows a generalization to include the existence of solitons. (We thank the economist Les Siegel for the observation.)

Comparing Eq. (7.50) with Eq. (D9) of Appendix D shows that (7.50) has the form of a Schrödinger wave equation in momentum coordinates $\mathbf{x}$, where the potential function $V_n(\mathbf{x})$ is effectively the term $B_n x^2$. This one-term potential arises from the assumption (above Eq. (7.49)) that the momentum $\mathbf{x}$ is small, so small that fourth- and higher-order terms can be neglected. Now let us segue to slightly higher momenta, and *include* a fourth-order term in the potential, as

$$V_n(\mathbf{x}) = B_n x^2 + C_n x^4, \qquad B_n,\ C_n = const.$$

The revised differential Eq. (7.50) will now likewise incorporate the term $C_n x^4$ as an added multiplier of $q_n(\mathbf{x})$. The differential equation now takes the mathematical form of the Schrödinger wave equation for an anharmonic oscillator, as in Scott (1999, p. 364). The correspondence is, however, purely formal since our theory is purely classical (no Planck's constant $\hbar$ enters in). Approximate solutions to this nonlinear wave equation therefore lead to the prediction of classical solitons (Scott, 1999), here in momentum $\mathbf{x}$ space.

In summary, whereas the *second*-order EPI analysis of Sec. 7.4.5 gave the Maxwell–Boltzmann law and non-equilibrium statistical mechanics, this fourth-order EPI analysis leads naturally to the prediction of classical solitons in momentum space.

## 7.5 Fisher information as a bound to entropy increase

Like the $I$-theorem (1.30), the following is a new finding that arises out of the use of classical Fisher information. As with the $I$-theorem, the output takes the form of an inequality. This is in contrast with the preceding EPI derivations which produce equality outputs. Thus, the derivation will not be a direct use of the EPI principle, but (as with EPI) will use the concepts of Fisher information and an invariance principle to form a fundamental physical law.

### *7.5.1 Scenario*

A system consists of one or more particles moving randomly within an enclosure. Denote the probability density for finding a particle at position $\boldsymbol{r} = (x, y, z)$ within the enclosure at the known time $t$ as $p(\boldsymbol{r}|t)$. A particle measurement $\boldsymbol{r}$ is made.

The enclosure is isolated. Hence, no particles leave the box and no new particles enter at its boundary $\boldsymbol{r} = \boldsymbol{B}$. Then the measurement $\boldsymbol{r}$ must lie within the enclosure, or

$$\int d\boldsymbol{r}\, p(\boldsymbol{r}|t) = 1, \tag{7.66}$$

a condition of normalization.

### 7.5.2 Shannon entropy

Denote by $H(t)$ the Shannon entropy of the system as evaluated at time $t$. This has the form Eq. (1.13),

$$H(t) = -\int d\boldsymbol{r}\, p(\boldsymbol{r}|t) \ln p(\boldsymbol{r}|t). \tag{7.67}$$

Suppose that the Shannon entropy represents the Boltzmann entropy as well. Then, by the second law of thermodynamics,

$$H_t \equiv \frac{dH}{dt} \geqslant 0. \tag{7.68}$$

(Through Sec. 7.5.5, we denote derivatives by subscripts without commas.) This establishes a definite lower bound to the change in entropy, but is there an upper bound as well? If so, what is it?

### 7.5.3 Invariance condition

Since no particles either enter or leave the enclosure, the particles obey an equation of conservation of flow,

$$p_t(\boldsymbol{r}|t) + \nabla \cdot \boldsymbol{P}(\boldsymbol{r}, t) = 0, \tag{7.69}$$

where $\boldsymbol{P}$ is a measure of flow whose exact nature depends upon the system. Denote the components of $\boldsymbol{P}$ as $(P_1, P_2, P_3)$. Thus, numbered subscripts denote vector components, whereas (from before) letter subscripts denote derivatives.

### 7.5.4 Dirichlet boundary conditions

Assume that there is no net flow of particles across the boundaries of the enclosure. That is,

$$\boldsymbol{P}(\boldsymbol{r}, t)\Big|_{\mathbf{B}} = 0. \tag{7.70}$$

Hence, $\boldsymbol{P}$ obeys Dirichlet boundary conditions (Eq. (5.55)). Also, assume that, if the boundary is at infinity, then

$$\lim_{r \to \infty} \boldsymbol{P}(\boldsymbol{r}, t) \to 0 \tag{7.71}$$

faster than $1/r^2$.

Since the enclosure is isolated, there must be vanishing probability that a particle is on the boundary,

$$p(\boldsymbol{r}|t)\Big|_{\mathbf{B}} = 0. \tag{7.72}$$

Hence $p$ also obeys Dirichlet boundary conditions. Also, for a boundary at infinity let $p$ obey

$$\lim_{r \to \infty} p(\boldsymbol{r}|t) \to 0 \tag{7.73}$$

faster than $1/r^3$. The latter is required by the condition (7.66) of normalization.

We will have need to evaluate the quantity $\boldsymbol{P} \ln p$ at the boundary. By conditions (7.70) and (7.72), this product is of the indeterminate form $-0 \cdot \infty$. Assume that the logarithmic operation "weakens" the $\infty$ so that condition (7.70) dominates the product,

$$\boldsymbol{P} \ln p\Big|_{\mathbf{B}} = 0. \tag{7.74}$$

### 7.5.5 Derivation

The partial derivative $\partial/\partial t$ of Eq. (7.67) gives

$$H_t = -\frac{\partial}{\partial t}\int d\boldsymbol{r}\, p \ln p = -\int d\boldsymbol{r}\, p_t \ln p - \int d\boldsymbol{r}\, p(1/p)p_t \tag{7.75}$$

after differentiating under the integral sign. The second right-hand integral gives

$$\int d\boldsymbol{r}\, p(1/p)p_t = \frac{\partial}{\partial t}\int d\boldsymbol{r}\, p = 0 \tag{7.76}$$

by normalization (7.66).

Next, use the flow Eq. (7.69) in the first right-hand integral of Eq. (7.75). This gives

$$H_t = \int d\boldsymbol{r} \nabla \cdot \boldsymbol{P} \ln p \equiv \iiint dz\, dy\, dx \left[\frac{\partial}{\partial x} P_1 + \frac{\partial}{\partial y} P_2 + \frac{\partial}{\partial z} P_3\right] \ln p. \tag{7.77}$$

Consider the first right-hand term. The innermost integral is, after integrating by parts,

$$\int dx \frac{\partial P_1}{\partial x} \ln p = P_1 \ln p\Big|_{\mathbf{B}} - \int dx\, (P_1/p)p_x = 0 - \int dx\, (P_1/p)p_x \tag{7.78}$$

by Dirichlet condition (7.74).

Analogous results follow for the second and third right-hand terms in Eq. (7.77). The result is that

$$H_t = -\int d\boldsymbol{r}\, \boldsymbol{P} \cdot \nabla p/p. \tag{7.79}$$

Squaring the latter and factoring the integrand gives

$$H_t^2 = \left[\int d\boldsymbol{r}(\boldsymbol{P}/\sqrt{p}) \cdot (\sqrt{p}\nabla p/p)\right]^2. \tag{7.80}$$

Temporarily replace the integral $d\boldsymbol{r}$ by a very fine sum over index $m$, and also execute the dot product as a sum over components $n$. This gives

$$H_t^2 = \left[\sum_{nm}\left(\frac{P_{nm}}{\sqrt{p_m}}\right)\left(\frac{\sqrt{p_m}}{p_m}\nabla_n p_m\right)\right]^2. \tag{7.81}$$

Now, the *Schwarz inequality* states that, for any two quantities $A_{nm}$, $B_{nm}$,

$$\left[\sum_{nm} A_{nm}B_{nm}\right]^2 \leqslant \sum_{nm} A_{nm}^2 \sum_{nm} B_{nm}^2. \tag{7.82}$$

Comparing Eqs. (7.81) and (7.82) suggests that we identify

$$A_{nm} = \frac{P_{nm}}{\sqrt{p_m}}, \qquad B_{nm} = \frac{\sqrt{p_m}}{p_m}\nabla_n p_m. \tag{7.83}$$

Then the two equations show that

$$H_t^2 \leqslant \sum_{nm}\frac{P_{nm}^2}{p_m}\sum_{nm}\frac{(\nabla_n p_m)^2}{p_m}. \tag{7.84}$$

Going back from the fine sum over index $m$ to the original integral $d\boldsymbol{r}$ gives

$$H_t^2 \leqslant \sum_n \int d\boldsymbol{r}\frac{P_n^2(\boldsymbol{r}, t)}{p(\boldsymbol{r}|t)}\sum_n \int d\boldsymbol{r}\frac{[\nabla_n p(\boldsymbol{r}|t)]^2}{p(\boldsymbol{r}|t)}. \tag{7.85}$$

Replacing the sums over components $n$ by dot product notation gives

$$H_t^2 \leqslant \int d\boldsymbol{r}\frac{\boldsymbol{P}\cdot\boldsymbol{P}}{p}\int d\boldsymbol{r}\frac{\nabla p \cdot \nabla p}{p}. \tag{7.86}$$

As in Sec. 2.4.2, assume that the PDF $p(\boldsymbol{r}|t)$ obeys shift invariance. Since one measurement $\boldsymbol{r}$ has been made (Sec. 7.5.1), the Fisher information obeys Eq. (2.17) with *its* index $n = 1$. This tempts us to associate the second integral in Eq. (7.86) with $I$. However, note the following potential complication: the coordinates in Eq. (2.17) are *fluctuations* from the ideal value, i.e., noise values, whereas our coordinates $\boldsymbol{r}$ in Eq. (7.86) are data component values. Can the second integral in Eq. (7.86) therefore still represent the information? The answer is yes, since there is only a constant shift between the two sets of coordinates. (The proof is left to the interested reader.) Hence, Eq. (7.86) becomes

$$H_t^2 \leq I \int d\boldsymbol{r}\, \frac{\boldsymbol{P} \cdot \boldsymbol{P}}{p}, \qquad I \equiv I(t) = \int d\boldsymbol{r}\, \frac{\nabla p(\boldsymbol{r}|t) \cdot \nabla p(\boldsymbol{r}|t)}{p(\boldsymbol{r}|t)}. \tag{7.87}$$

This shows that the entropy change during a small time interval is bounded from above. The bound is proportional to the square root of the Fisher information capacity for a position measurement. This upper bound is what we set out to find.

### *7.5.6 Entropy bound for classical particle flow*

Let the system consist, now, of many material particles. The particles interact under the influence of any potential, and they also collide with each other and with the boundary walls located at position $\boldsymbol{r} = \boldsymbol{B}$. In such a scenario, the flow vector is

$$\boldsymbol{P}(\boldsymbol{r}, t) = p(\boldsymbol{r}|t)\mathbf{v}(\boldsymbol{r}, t), \tag{7.88}$$

where $\mathbf{v}$ is the particle velocity (Lindsay, 1951, p. 284). Inequality (7.87) will hold for such a system, provided that conditions (7.69)–(7.74) hold. We show that this is the case.

Since the system is isolated, there is no net flow of particles in or out so that condition (7.69) holds by definition.

Condition (7.70) is now

$$p(\boldsymbol{r}|t)\mathbf{v}(\boldsymbol{r}, t)\Big|_{\mathbf{B}} = 0. \tag{7.89}$$

In order to satisfy condition (7.70), one or the other of the two factors must be zero on the boundary. If $\mathbf{v} = 0$ but $p \neq 0$ on the boundary that implies that, once a particle is on the boundary it cannot move away from it. Since $p \neq 0$ there, ultimately every particle will be on the boundary! The system degenerates into a collapsed state. We will not consider this kind of specialized solution. Hence the solution to (7.89) is taken to be

$$p(\boldsymbol{r}|t)\Big|_{\mathbf{B}} = 0. \tag{7.90}$$

By Eq. (7.88), this satisfies (7.70), stating that no particles cross the boundary.

Since no particles cross the boundary, normalization condition (7.66) must hold. But for normalization to hold, it must be that

$$\lim_{r \to \infty} p(\boldsymbol{r}|t) \to 0 \tag{7.91}$$

faster than $1/r^3$. Hence condition (7.73) holds.

By Eqs. (7.88) and (7.91), condition (7.71) holds. Also, condition (7.72) is now satisfied by Eq. (7.90).

Finally, condition (7.74) holds because, by Eq. (7.88),

$$\boldsymbol{P} \ln p = \mathbf{v} p \ln p \tag{7.92}$$

and of course

$$\lim_{p \to 0} p \ln p = 0. \tag{7.93}$$

The zero limit for $p$ is taken because of Eq. (7.90).

Since all of the requirements (7.70)–(7.74) hold, the inequality (7.87) holds for this problem.

### 7.5.7 Exercise

Prove the assertion (7.93) using l'Hôpital's rule.

### 7.5.8 Reduction of problem

Combining Eqs. (7.87) and (7.88) gives

$$\frac{dH(t)}{dt} \leqslant \sqrt{I \int d\boldsymbol{r}\, v^2 p}$$

or

$$\frac{dH(t)}{dt} \leqslant \sqrt{I \langle v^2 \rangle} \tag{7.94}$$

by the definition of the expectation $\langle \cdot \rangle$. The positive sign for the square-root is chosen because, by the second law, the change in $H$ must be positive (or zero). Equation (7.94) says that the rate of change of $H$ is bounded jointly by the Fisher information $I$ in a measurement $\boldsymbol{r}$ and the root-mean-square velocity. In some cases, the latter is a constant so that there is a direct proportionality $dH/dt \leqslant C\sqrt{I}$, $C = const$. This occurs, for example, for the random scenario of Sec. 7.2.1 where, in addition, there are no forces on the particles. Then Eq. (7.33) holds, and, with no potentials present, $\langle E_{\mathrm{t}} \rangle = \langle E_{\mathrm{kin}} \rangle = Mm\langle v^2 \rangle/2 = 3MkT/2$, so that

$$\langle v^2 \rangle \equiv C^2 = 3kT/m. \tag{7.95}$$

The result is the interesting expression

$$\frac{dH(t)}{dt} \leqslant I(t)^{1/2} \left( \frac{3kT}{m} \right)^{1/2}. \tag{7.96}$$

Classical particle flow describes the motion of an ideal fluid. Hence, for such a medium the rate of change of entropy is bounded above by the square root of the Fisher information.

### *7.5.9 Entropy bounds for electromagnetic flow phenomena*

The derivation in Sec. 7.5.5 shows that any flow phenomenon will obey the basic inequality (7.87) if Dirichlet boundary conditions (7.70)–(7.74) are obeyed. The phenomenon does not have to describe classical fluid flow specifically, as in the preceding sections. Quite a wide range of other phenomena obey the required boundary conditions, or can be restricted to cases that *do* obey the boundary conditions.

For example, the charge density $\rho$ and current $\boldsymbol{j}$ obey the conservation of flow Eq. (5.7). Then, if we assume Dirichlet boundary conditions for $\rho$ and $\boldsymbol{j}$, and that $\rho \geqslant 0$ obeys normalization, we have a mathematical correspondence $\rho(\boldsymbol{r}, t) \to p(\boldsymbol{r}|t)$, $\boldsymbol{j}(\boldsymbol{r}, t) \to P(\boldsymbol{r}, t)$ between electromagnetic quantities and quantities of the derivation in Sec. 7.5.5. Thus, for an entropy of charge density

$$H_\rho(t) \equiv -\int d\boldsymbol{r}\, \rho(\boldsymbol{r}, t) \ln \rho(\boldsymbol{r}, t), \qquad \rho \geqslant 0 \tag{7.97}$$

and an information quantity

$$I_\rho(t) \equiv \int d\boldsymbol{r}\, \frac{\nabla\rho \cdot \nabla\rho}{\rho}, \tag{7.98}$$

result (7.87) becomes

$$\left(\frac{dH_\rho}{dt}\right)^2 \leqslant I_\rho \int d\boldsymbol{r} \left(\frac{\boldsymbol{j}\cdot\boldsymbol{j}}{\rho}\right). \tag{7.99}$$

For a single moving charge, where $\boldsymbol{j} = \rho\mathbf{v}$, this simplifies to

$$\left(\frac{dH_\rho}{dt}\right)^2 \leqslant I_\rho \int d\boldsymbol{r}\, \rho v^2. \tag{7.100}$$

Similar results grow out of flow Eq. (5.8). With now required correspondences $q_4 \equiv \phi \to p$, $q_n \equiv A_n \to P_n$, the electromagnetic potentials take on the roles of $p$ and $\boldsymbol{P}$. Result (7.87) becomes

$$\left(\frac{dH_\phi}{dt}\right)^2 \leqslant c^2 I_\phi \int d\boldsymbol{r}\, \frac{\boldsymbol{A}\cdot\boldsymbol{A}}{\phi}, \qquad I_\phi = \int d\boldsymbol{r}\, \frac{\nabla\phi\cdot\nabla\phi}{\phi}. \tag{7.101}$$

The entropy is, here, that of the scalar potential $\phi$,

$$H_\phi(t) \equiv -\int d\boldsymbol{r}\, \phi(\boldsymbol{r}, t) \ln \phi(\boldsymbol{r}, t), \qquad \phi \geqslant 0. \tag{7.102}$$

This is a new concept. Note that it is mathematically well defined if one adds enough of a constant to the potential to keep it positive during the "ln" operation. As with the Fisher information $I_\phi$, it measures the "spread-out-edness" of the function $\phi$ over space $\boldsymbol{r}$.

### 7.5.10 Entropy bounds for gravitational flow phenomena

The analysis of Sec. 7.5.9 can apply in the same way to the gravitational flow phenomena (6.27) for the stress-energy tensor and (6.28) for the weak-field tensor. The entropy inequality (7.87) would hold for these tensors as well. However, the entropy of stress-energy and the entropy of the weak field would be new concepts that need interpretation.

### 7.5.11 Entropy bound for quantum electron phenomena

A flow equation for the quantum electron may be obtained as follows. Multiply the Dirac Eq. (4.42) on the left by the Hermitian adjoint $\psi^\dagger$ (transpose of $\psi^*$); multiply the Hermitian adjoint of (4.42) on the right by $\psi$, and subtract one of the two results from the other. This gives a flow equation

$$\frac{\partial}{\partial t} p(\boldsymbol{r}, t) + \nabla \cdot \boldsymbol{P}(\boldsymbol{r}, t) = 0, \tag{7.103a}$$

where

$$p(\boldsymbol{r}, t) = \psi^\dagger \psi, \quad \psi \equiv \psi(\boldsymbol{r}, t), \tag{7.103b}$$

and

$$\boldsymbol{P}(\boldsymbol{r}, t) = -c\psi^\dagger [\alpha] \psi. \tag{7.103c}$$

Quantity $[\alpha]$ is the Dirac vector of matrices $[\alpha_x, \alpha_y, \alpha_z]$ defined by Eqs. (4.43).

We show, next, that conditions (7.70) and (7.72) are obeyed. The wave function $\psi$ must continuously approach zero as $\boldsymbol{r}$ approaches any boundary $\mathbf{B}$ to measurement space (Schiff, 1955, pp. 29, 30). Since, by Eqs. (7.103b, c), both $p$ and $\boldsymbol{P}$ increase quadratically as $\psi$, necessarily

$$p(\boldsymbol{r}, t)\Big|_{\mathbf{B}} = 0, \tag{7.104a}$$

$$\boldsymbol{P}(\boldsymbol{r}, t)\Big|_{\mathbf{B}} = 0. \tag{7.104b}$$

Since the electron is always present somewhere in the enclosure, $p(t) = 1$. Then by Eq. (4.30b) $\psi$ obeys normalization requirement (7.66). Hence, if the boundary is at infinity, $p(\boldsymbol{r}|t)$ must fall off with $r$ faster than $1/r^2$. This satisfies requirement (7.73).

We turn to property (7.71), which requires that the boundary $\mathbf{B}$ be at infinity. Because property (7.70) holds, by Eq. (7.103b) $\psi$ must fall off with $r$ as $1/r$ or faster. Hence, by Eq. (7.103c), $\boldsymbol{P}$ must fall off with $r$ as $1/r^2$ or faster. Hence, requirement (7.71) is obeyed.

Finally, we consider requirement (7.74). By Eq. (7.103c) the $X$-component $P_1$ of $\boldsymbol{P}$ obeys

$$P_1 = -c(\psi_1 \psi_2 \psi_3 \psi_4)^* \begin{bmatrix} 0 & 0 & 0 & 1 \\ 0 & 0 & 1 & 0 \\ 0 & 1 & 0 & 0 \\ 1 & 0 & 0 & 0 \end{bmatrix} \begin{bmatrix} \psi_1 \\ \psi_2 \\ \psi_3 \\ \psi_4 \end{bmatrix}, \tag{7.105}$$

where we used the matrices (4.43) and (4.44) defining $[\alpha_x]$. After the matrix products in Eq. (7.105) are carried out, the result is

$$P_1 = -c(\psi_1^* \psi_4 + \psi_2^* \psi_3 + \psi_3^* \psi_2 + \psi_4^* \psi_1). \tag{7.106}$$

Then, by Eqs. (7.103b, c) and (7.106), we have

$$P_1 \ln p = -c(\psi_1^* \psi_4 + \psi_2^* \psi_3 + \psi_3^* \psi_2 + \psi_4^* \psi_1) \times \ln(|\psi_1|^2 + |\psi_2|^2 + |\psi_3|^2 + |\psi_4|^2). \tag{7.107}$$

Requirement (7.74) addresses the limiting form of this expression as $\boldsymbol{r} \to \mathbf{B}$ where, by Eqs. (7.103b) and (7.104a), all components $\psi_i = 0$, $i = 1$–$4$. There is arbitrariness in how we choose the components to approach the boundary. We first evaluate $\psi_2$ and $\psi_3$ on the boundary, letting $\psi_2 = \psi_3 = 0$ in (7.107). This gives

$$P_1 \ln p \Big|_{\mathbf{B}} = -2c \operatorname{Re}(\psi_1^* \psi_4) \ln(|\psi_1|^2 + |\psi_4|^2). \tag{7.108}$$

The right-hand side is of the form $0 \ln 0$, and so has to be evaluated in a limiting process. Now we will have $\psi_1$ and $\psi_4$ approach the boundary.

Denote a given boundary point by $\boldsymbol{R}$. Expand each of $\psi_1$ and $\psi_4$ in Taylor series about the point $\boldsymbol{R}$, dropping all quadratic and higher-power terms since the limit $\boldsymbol{r} \to \boldsymbol{R}$ will be taken:

$$\psi_i(\boldsymbol{r}, t) = \psi_i(\boldsymbol{R}, t) + d\boldsymbol{r} \cdot \nabla \psi_i(\boldsymbol{R}, t), \quad \text{where} \quad d\boldsymbol{r} = \boldsymbol{R} - \boldsymbol{r}, \quad i = 1, 4 \tag{7.109}$$

and

$$\lim d\boldsymbol{r} \to 0 \tag{7.110}$$

now defines the boundary.

The first right-hand term in (7.109) is zero, since all $\psi_i = 0$ on the boundary. Then, substituting Eq. (7.109) into Eq. (7.108) gives

$$P_1 \ln p \Big|_{\mathbf{B}} = -2c \operatorname{Re}[(d\boldsymbol{r} \cdot \nabla \psi_1^*)(d\boldsymbol{r} \cdot \nabla \psi_4)] \ln[|d\boldsymbol{r} \cdot \nabla \psi_1|^2 + |d\boldsymbol{r} \cdot \nabla \psi_4|^2] \Big|_{dr=0}. \tag{7.111}$$

Taking $dy = dz = 0$ eliminates terms in these differentials,

$$P_1 \ln p\bigg|_{\mathbf{B}} = -2c(dx)^2 \operatorname{Re}[(\psi_{1x}^*)(\psi_{4x})] \ln[(dx)^2|\psi_{1x}|^2 + (dx)^2|\psi_{4x}|^2]\bigg|_{dx=0}. \tag{7.112}$$

(Note that $\psi_{1x} \equiv \partial\psi_1/\partial x$, etc., is our derivative notation.) This is of the form $Au\ln(Bu)$, $u = (dx)^2$. In the limit $dx \to 0$ it gives 0, as in Eq. (7.93).

Retracing the steps (7.105)–(7.112) for the other components $P_2$, $P_3$ gives the same result. Hence requirement (7.74) is satisfied.

Hence, we have shown that the equation of flow (7.103a) and all required boundary value conditions (7.70)–(7.74) hold. This means that the inequality (7.87) follows. The inner product $\boldsymbol{P} \cdot \boldsymbol{P}$ in (7.87) is, by Eq. (7.103c),

$$\boldsymbol{P} \cdot \boldsymbol{P} \equiv \boldsymbol{P}^\dagger \boldsymbol{P} \equiv |\boldsymbol{P}|^2 = (-c\boldsymbol{\psi}^\dagger[\alpha]\boldsymbol{\psi})^\dagger(-c\boldsymbol{\psi}^\dagger[\alpha]\boldsymbol{\psi}) = |\boldsymbol{\psi}^\dagger c[\alpha]\boldsymbol{\psi}|^2 \tag{7.113}$$

by definition of the absolute value. Then inequality (7.87) becomes directly the inequality

$$\frac{1}{I}\left(\frac{\partial H}{\partial t}\right)^2 \leqslant \int d\boldsymbol{r}\, \frac{|\boldsymbol{\psi}^\dagger c[\alpha]\boldsymbol{\psi}|^2}{|\boldsymbol{\psi}|^2} = \int d\boldsymbol{r}\, \frac{v^2|\boldsymbol{\psi}|^4}{|\boldsymbol{\psi}|^2} = \int d\boldsymbol{r}\, v^2|\boldsymbol{\psi}|^2 = \langle v^2\rangle \leqslant c^2. \tag{7.114}$$

The first equality follows from a relation (Eisele, 1969, p. 236; Schiff, 1955, p. 328)

$$v\boldsymbol{\psi} = \pm c[\alpha]\boldsymbol{\psi} \tag{7.115}$$

expressing $c[\alpha]$ as an operator for the particle velocity component $v$. (Note that the choice of sign $\pm$ depends upon the particular component of the velocity, but does not matter because of the inner-product operation in the first integral (7.114).) The second equality is obtained by a cancellation. The third equality is by definition of the expectation, and the final inequality is the usual statement of special relativity (Sec. 3.5.5).

The outside inequality (7.114) may be expressed as

$$\left(\frac{\partial H}{\partial t}\right) \leqslant c\sqrt{I}. \tag{7.116}$$

This shows that the rate of gain of Shannon information, in bits/time, that an observer can acquire about the space-time coordinates of an electron is limited by the finiteness of the speed of light, and also by the size of the Fisher information capacity $I$ about those coordinates. Another way of looking at (7.116) is to notice that it provides a definition of the speed of light $c$: as the upper bound to the ratio of the rate of change of the entropy to the square-root of the Fisher information. Interestingly, this defines $c$ directly in terms of informations (albeit about an electromagnetic entity). Viewed abstractly, it sets an upper limit to the ratio of how rapidly we can learn to how much we already know.

### 7.5.12 *Historical notes*

The derivation Eqs. (7.66)–(7.87) was first given by Nikolov (1992) in a personal correspondence with the author. The derivation was later independently discovered by Plastino and Plastino (1995). Brody and Meister (1995) show many other important applications of the key result Eq. (7.87). The quantum mechanical application in Sec. 7.5.11 was published in Nikolov and Frieden (1994), except for the specific answer Eq. (7.116).

## 7.6 Overview

The EPI approach to statistical physics is, as usual, four-dimensional. The Fisher measurement coordinates are $(ix_E, c\boldsymbol{\mu})$ in energy-momentum space. By contrast, space-time coordinates $(i\boldsymbol{r}, ct)$ were used in derivation of quantum mechanics in Chapter 4. Hence, relativistic quantum mechanics arises out of EPI as applied to *space-time* coordinates, whereas statistical mechanics arises out of EPI as applied to *energy-momentum* coordinates. Of course, according to quantum mechanics (Eq. (4.4)) the two coordinate choices are *complementary*. Hence, EPI unifies quantum mechanics and statistical mechanics as resulting from the choice of one or the other of a pair of complementary coordinate spaces. It is interesting that in both derivations the use of mixed imaginary and real coordinates is essential.

Our chosen coordinates $(iE, c\boldsymbol{\mu})$ constitute a *four-dimensional* "phase space," in the usual language of statistical mechanics. This is in contrast with the standard approach, which takes place in a higher, six-dimensional $(\boldsymbol{\mu}, \mathbf{x})$ momentum-position phase space (Rumer and Ryvkin, 1980). That the Boltzmann and Maxwell–Boltzmann distribution laws can arise out of a lower-dimensioned analysis is of interest.

However, as we found in Sec. 3.5.5 of Chapter 3, *joint* energy-momentum coordinates are not appropriate Fisher four-coordinates, since they are they are *a priori dependent*, by Eq. (4.17). Hence we must be content to seek a lower level of knowledge, the *marginal* laws on *each* of $E$ and $\boldsymbol{\mu}$. These projections of the joint law are found, using EPI, in Secs. 7.3 and 7.4, respectively.

Both the Boltzmann probability law on energy and the Maxwell–Boltzmann law on momentum are seen to arise out of a common value $\kappa$ for the information efficiency. That value is $\kappa = 1$, indicating complete efficiency in the transfer of information from the physical phenomenon to the data (see also material below Eq. (3.18)).

We note that the Fisher coordinates $\mathbf{x}_\mu$ for the momentum problem are purely real. This means that the "game" (Sec. 3.4.12) between the observer and the "demon" is physical here. Both the general EPI solution (7.59) and the Maxwell–Boltzmann Eq. (7.65) follow as payoffs of the contest for maximum information.

A serendipitous result is the prediction (7.59) of multiple solutions for the PDF on momentum. These correspond to (a) the equilibrium Maxwell–Boltzmann PDF on momentum fluctuations, and (b) other stationary PDFs *en route* to equilibrium. The latter are non-equilibrium solutions, and fall into a category of Hermite–Gauss solutions previously found by Rumer and Ryvkin (1980) as solutions to the Boltzmann transport equation.

These non-equilibrium PDFs on momentum should have as counterparts non-equilibrium PDFs on the energy. The latter should be obtainable by replacement of the dependence $J(\mathbf{q})$ for the energy in Eq. (7.8) by a series of the form $J(\mathbf{q})(A + Bx + \cdots)$ as in Eqs. (7.47) and (7.49) for the momentum. See Sec. 7.4.10 for further discussion.

The general solution to the momentum-distribution problem is the differential Eq. (7.48). This has the general form of a Schrödinger wave equation (SWE). This SWE is a key result of generally non-equilibrium statistical mechanics (Frieden *et al.*, 2002a, b). By its use, *non-equilibrium thermodynamic* solutions correspond to analogous *quantum mechanical* admixtures of excited states. Just as in quantum mechanics, these thermodynamic solutions are obtainable by perturbing the ground state "wave function" $q_0(\mathbf{x})$ with linear terms. The resulting non-equilibrium distributions agree with standard non-equilibrium distributions based upon the use of entropy. The SWE (7.48) is found to have further applications to the anharmonic oscillator, soliton theory (Sec. 7.4.13) and (even) to econophysics (Chapter 13).

An important property of thermodynamics is its "Legendre transform structure" (Duering *et al.*, 1985). This is ordinarily stated using entropy as the measure of disorder. However, it was recently shown that the same kind of Legendre transform structure ensues for a thermodynamics that utilizes, instead, Fisher information as the measure of disorder (Frieden *et al.*, 1999).

The game corollary is used to predict the $N = 1$ probability law $p(E)$ for representing the distribution of energy values, and also to predict the prevalence of low-$n$ solutions $p_n(\mu)$ for the probability on momentum values. On this basis, the *lowest*-order solution $n = 0$ represents the equilibrium distribution, with higher orders corresponding to decreasingly likely distributions.

It is found that, if a PDF obeys an equation of continuity of flow (7.69), then its entropy increase must be limited by the Fisher information in a system measurement. The exact relation is the inequality (7.87). One might regard this result as an addendum to the second law of thermodynamics. That is, entropy shall increase, but (now) by not too much! This result has wide applicability. As we showed, it applies to classical particle flow (7.96), electromagnetic flow (7.99) and (7.101), gravitational flow (Sec. 7.5.10), and Dirac electron flow (7.116). The latter is a particularly interesting expression, bringing in the speed of light in a new way. *The speed fixes the ratio of how rapidly we can learn to how much we already know.* The general result (7.87) should have many other applications as well.

Further problems in statistical physics that are analyzed by EPI are: turbulent flow in Chapter 11; economic valuation of securities in Chapter 13; and growth and transport processes in Chapter 14.

# 8

# Power spectral $1/f$ noise

## 8.1 The persistence of $1/f$ noise

Consider a real, temporal signal $X(t)$ defined over a time interval $0 \leqslant t \leqslant T$. It has an associated (complex) Fourier spectrum

$$Z_T(\omega) \equiv T^{-1/2} \int_0^T dt\, X(t) e^{-i\omega t} \equiv (Z_r(\omega),\ Z_i(\omega)) \tag{8.1}$$

and an associated "periodogram"

$$I_T(\omega) = |Z_T(\omega)|^2 = Z_{\mathrm{r}}^2(\omega) + Z_{\mathrm{i}}^2(\omega). \tag{8.2}$$

Functions $Z_{\mathrm{r}}(\omega)$, $Z_{\mathrm{i}}(\omega)$ are, respectively, the real and imaginary parts of $Z_T(\omega)$. Define a power spectrum

$$S(\omega) = \lim_{T \to \infty} \langle I_T(\omega) \rangle. \tag{8.3}$$

The brackets $\langle \cdot \rangle$ denote an average over an ensemble of signals.

If $S(\omega) \approx const.$ then the signal is said to be "white noise." Of course, most physical phenomena exhibit a varying power spectrum. The most common of these is of the form

$$S(\omega) = A\omega^{-\alpha}, \quad A = const., \quad \alpha \approx 1. \tag{8.4}$$

This is usually called a "$1/f$ noise power spectrum" or, simply, "$1/f$ noise." Typical white-noise and $1/f$-noise signal traces are shown in Fig. 8.1.

A tremendously diverse range of phenomena obey $1/f$ noise. Just a partial list includes: voltage fluctuations in resistors, semiconductors, vacuum tubes, and biological cell membranes; traffic density on a highway; economic time series; musical pitch and volume; sunspot activity; flood levels on the river Nile; and the rate of insulin uptake by diabetics. See respective references: Handel (1971), Hooge (1976), Bell (1980), Johnson (1925), Holden (1976), Musha and Higuchi (1976), Granger (1966), Voss and Clarke (1978), Mandelbrot and Wallis (1969), and Campbell and Jones (1972).

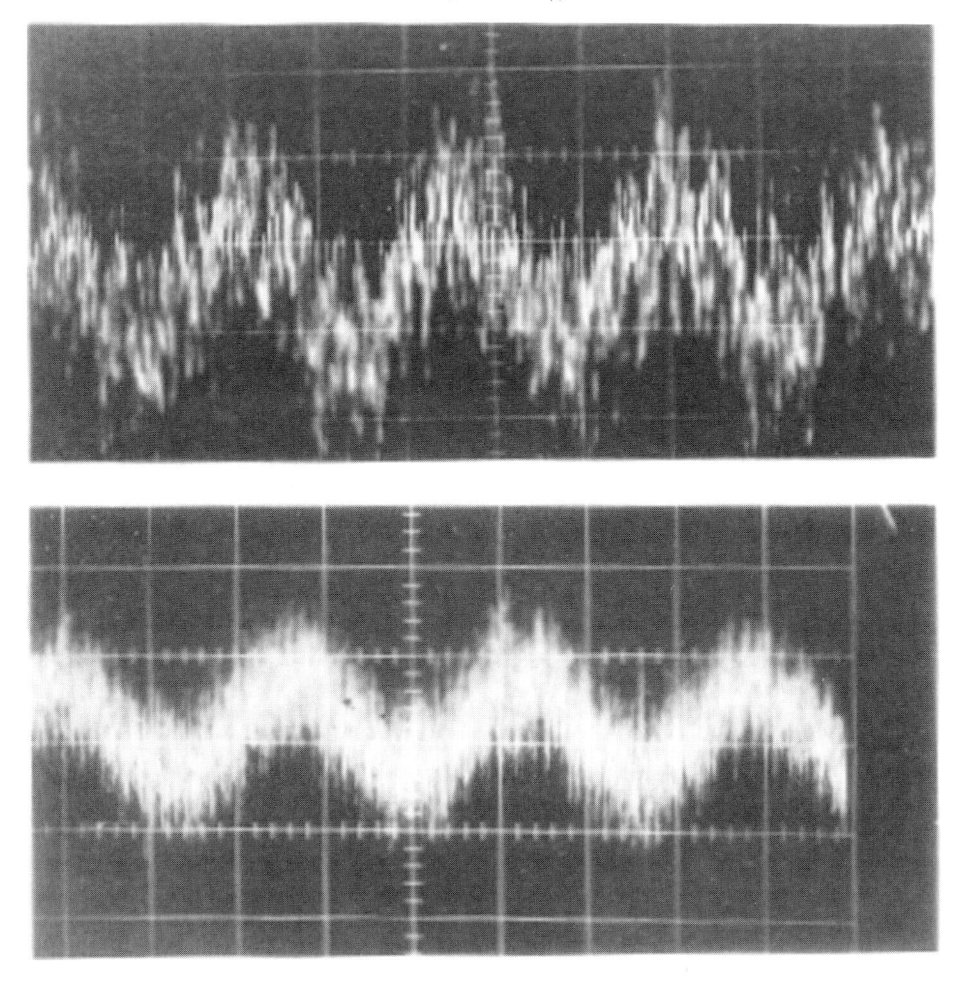

Fig. 8.1. A sinusoidal signal with superimposed noise: (top) $1/f$ noise; (bottom) white noise. (Reprinted from Motchenbacher and Connelly, copyright 1993, by permission of John Wiley & Sons, Inc.)

Numerous mathematical models have been advanced for achieving $1/f$ noise under differing conditions. Examples are models of fractal shot noise (Lowen and Teich, 1989), filtered white Gaussian noise (Takayasu, 1987), fractionally integrated white noise (Barnes and Allan, 1966), fractal Brownian motion (Mandelbrot, 1983), and a diffusion process driven by a white noise boundary condition (Jensen, 1991). Most of these models are based upon a situation of underlying white noise, which undergoes a modification to become $1/f$ noise.

However, the last mentioned model (Jensen, 1991) is of particular interest since a diffusion process obeys increasing entropy (Wyss, 1986), which implies increasing disorder. This reminds us of the $H$-theorem Eq. (1.29) and, more to the point, the $I$-theorem Eq. (1.30). This suggests that we attack the problem of deriving $1/f$ noise for $S(\omega)$ by the use of EPI. Some further justification for the use of this approach follows. (Of course the ultimate justification is that it works.)

## 8.2 Temporal evolution of tone amplitude

It is instructive to follow the evolution of a typical time signal $X(t)$ in terms of the Fisher information in a measurement. It will be shown that, as $T \to \infty$, the disorder of $X(t)$ increases and consequently $I \to$ a minimum value. Then, as we reasoned in Chapter 1 and Sec. 3.1.1, this suggests the use of EPI.

Consider the gedanken measurement experiment of Fig. 8.2. Time signal $X(t)$ is a musical composition, say, a randomly selected violin sonata. As time progresses the signal $X(t)$ is, of course, produced over increasing time intervals $(0, T_0)$, $(0, T_1)$, $(0, T_2)$, ..., where $T_0 < T_1 < T_2 \ldots$. The "ideal" parameter $\theta$ of the EPI approach is as follows.

Suppose that a note $\omega$ occurs in the zeroth interval $(0, T_0)$ with the complex amplitude

$$Z_0(\omega) \equiv \theta(\omega) \equiv (\theta_r(\omega), \theta_i(\omega)) \tag{8.5}$$

in terms of its real and imaginary parts. This is the ideal complex parameter value.

However, the observer is not necessarily listening during the zeroth interval. Instead, he hears during the $n$th interval the complex spectral amplitude

$$Z_n(\omega) = y_n \tag{8.6}$$

in our generic data notation (2.1). From such observation he is to best estimate $\theta(\omega)$. How should the mean-square error $e^2$ in such an estimate vary with the chosen interval number $n$ or (equivalently) time duration $T_n$?

For the interval number $n = 0$ the acquired data would be $Z_n(\omega) \equiv \theta(\omega)$, so of course $e^2$ would be zero. Suppose that the next interval, $(0, T_1)$, includes the ideal interval $(0, T_0)$ plus a small interval. See Fig. 8.2. Then, by Eq. (8.1), its Fourier transform $Z_1(\omega)$ "sees" the ideal interval plus a small tail region. Hence, $Z_1(\omega)$ will depart from $\theta(\omega)$ by only a small amount. Likewise, an optimum estimate of $\theta(\omega)$ made on this basis should incur small error $e^2$.

Next, a measurement $Z_2(\omega)$ based upon observation of $X(t)$ over time interval $(0, T_2)$ is made. It should incur a slightly larger error, since interval $(0, T_2)$ incurs more "tail" of $X(t)$ than its predecessor. Hence, the error after an estimate is made will likewise go up.

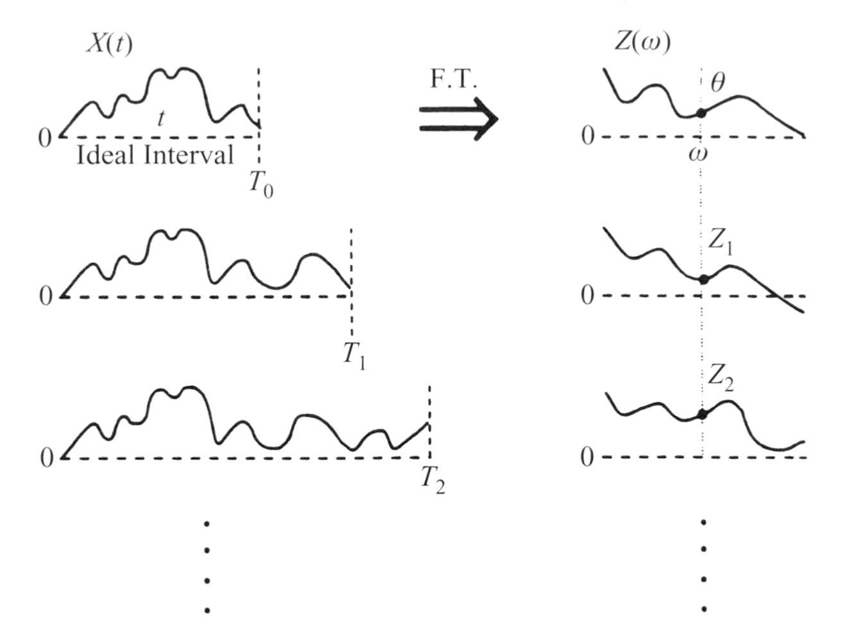


Fig. 8.2. Gedanken measurement–estimation experiment. The unknown tone amplitude $\theta(\omega)$ is caused by signal $X(t)$ over ideal interval $(0, T_0)$. Subsequent tone amplitudes $Z_1(\omega)$, $Z_2(\omega)$, ... are due to listening at the fixed $\omega$ over ever-longer time intervals. (Reprinted from Frieden and Hughes, 1994.)

It is obvious that this trend continues indefinitely. The error $e^2 \to \infty$ with time.

It will turn out that the underlying PDF is Gaussian, so that the optimum estimate achieves *efficiency* (Exercise 1.6.1). Then the equality is achieved in Eq. (1.1), and the Fisher information $I$ obeys

$$I = \frac{1}{e^2}, \tag{8.7}$$

in terms of the error $e^2$. Then, from the preceding paragraph, information $I \to 0$ with time, its absolute minimum value. By the reasoning in Sec. 3.1.1, this is its *equilibrium* value as well, so that the EPI principle should resultingly give the PDF attained at equilibrium. The power spectrum $S(\omega)$ attained at equilibrium will also be so determined since, for this problem, $I$ relates uniquely to the power spectrum, as found next.

## 8.3 Use of EPI principle

Since our measurement is of the (complex) spectral amplitude $Z = (Z_r, Z_i)$, the Fisher information (2.15) obeys (with $N = 1$)

$$I = \iint dZ_r\, dZ_i\, \frac{(\partial p/\partial\theta_r)^2 + (\partial p/\partial\theta_i)^2}{p} \tag{8.8}$$

in terms of the joint likelihood $p(Z_r, Z_i|\theta_r, \theta_i)$ of the data. As usual, we assume that an observation $(Z_r, Z_i)$ using a real instrument causes perturbation of the system PDF $p(Z_r, Z_i|\theta_r, \theta_i)$. This initiates the EPI process.

### *8.3.1 Information I in terms of power spectrum S(ω)*

For a certain class of time signals $X(t)$ the information expression (8.8) may be directly related to the power spectrum.

By the form of Eq. (8.4), $1/f$ noise exhibits long-term memory. Recalling that a frequency $\omega \sim 1/t$, where $t$ is a time lapse, by (8.4) a $1/f$-noise process has very large power at very long time lapses (very small $\omega$). This means that the joint statistics of $(Z(t), Z(t'))$ at two different times $t$, $t'$ do not simply depend upon the relative time difference $(t - t')$. The joint statistics must depend upon the absolute time values $t$, $t'$. A process $X(t)$ with this property is called non-stationary (Frieden, 2001).

A wide class of non-stationary signals $X(t)$ was recently defined and analyzed (Solo, 1992). This is called the class of intrinsic random functions ($IRF_0$) of order zero (Matheron, 1973). Such a class of signals is not stationary in its values $X(t)$ but *is stationary* in its changes $X(t + \tau) - X(t)$, $\tau \geqslant 0$. There is evidence that $1/f$-noise processes obey just such a property of stationarity (Keshner, 1982). Hence, we assume the process $X(t)$ to be an $IRF_0$ process.

It is known that such a process obeys a central limit theorem. Thus, $p(Z_r, Z_i|\theta_r, \theta_i)$ turns out to be separable normal. The Fisher information for a *single* normal random variable is found next.

By Eq. (1.2)

$$I = \int dx\, \frac{p'^2}{p} \equiv \int dx\, p\left(\frac{p'}{p}\right)^2 \equiv \left\langle \left(\frac{\partial \ln p}{\partial x}\right)^2 \right\rangle. \tag{8.9}$$

Then, by $p(x) = A \exp[-(x - \theta)^2/(2\sigma^2)]$, $A$ constant, we get $\ln p = B - (x - \theta)^2/(2\sigma^2)$, so that $\partial \ln p/\partial x = -(x - \theta)/\sigma^2$. Squaring this and taking the mean gives, by Eq. (8.9),

$$I = \frac{1}{\sigma^2} \tag{8.10}$$

after using $\langle (x-\theta)^2 \rangle \equiv \sigma^2$. Solo (1992) showed further that, for the $\mathrm{IRF}_0$ process, $\sigma^2 = S(\omega)/2$ and each of the two random variables $(Z_{\mathrm{r}}, Z_{\mathrm{i}})$ has the same variance. Then, by the added nature of Eq. (8.8) the resulting $I$ is simply twice the value in Eq. (8.10),

$$I = \frac{2}{\sigma^2} = \frac{4}{S(\omega)}. \tag{8.11}$$

This is the information in observing $Z$ at a single frequency $\omega$. Let us now generalize the data scenario of Sec. 8.2 to include observations over any band of frequencies, denoted as $\Omega$, which *excludes* the pure d.c. "tone" $\omega = 0$; the latter has no physical reality. Band $\Omega$ can be narrow or wide. The $\mathrm{IRF}_0$ process is also independent over all frequencies. Then by the additivity of information (Sec. 2.4.1) the total information is the integral

$$I = 4\int_{\Omega} d\omega\, \frac{1}{S(\omega)}. \tag{8.12}$$

The information $I$ is, then, a known functional of the power spectrum $S(\omega)$. We needed such an expression in order to start the EPI calculation of the equilibrium $S(\omega)$. However, note that this EPI problem differs *conceptually* from all others in preceding chapters. Here, the unknowns of the problem are not probability amplitudes $\mathbf{q}$. Indeed, the form of the PDF $p(Z_{\mathrm{r}}, Z_{\mathrm{i}}|\theta_{\mathrm{r}}, \theta_{\mathrm{i}})$ (and, hence, amplitude) over all frequencies *is known* to be independent normal. Instead, the unknowns are the power spectrum values $S(\omega)$ over the bandpass interval $(0, \Omega)$. These are unknown *parameters* of the known PDF – essentially the variances (see below Eq. (8.10)).

Hence, we are using EPI here to find, not a continuous PDF, but the identifying parameters $S(\omega)$ of a PDF of known form. This might be considered a parameterized version of the EPI procedure.

### 8.3.2 Finding information J

The use of EPI principle (3.16) and (3.18) requires knowledge of informations $I$ and $J$. The former obeys Eq. (8.12). The bound information $J$ must be found.

Let the functional $J$ be represented by the general form

$$J = 4\int_{\Omega} d\omega\, F[S(\omega), \omega], \tag{8.13}$$

where $F$ is a general function of its indicated arguments. Then, by Eqs. (8.12) and (8.13), the physical information (3.16) is

$$K = 4\int d\omega \left[\frac{1}{S(\omega)} - F[S(\omega), \omega]\right] \equiv \mathit{extrem}. \tag{8.14}$$

As in Chaps. 5–7, we find $J$ by simultaneously solving EPI Eqs. (3.16) and (3.18) for a common solution, here $S(\omega)$.

The Euler–Lagrange Eq. (0.34) for the extremum in Eq. (8.14) is

$$\frac{d}{d\omega}\left(\frac{\partial \mathscr{L}}{\partial S'}\right) = \frac{\partial \mathscr{L}}{\partial S}, \quad S' \equiv \partial S/\partial\omega, \quad \text{where} \quad \mathscr{L} = 1/S - F. \tag{8.15}$$

The result is that $F$ satisfies

$$\frac{1}{S^2} + \frac{\partial F}{\partial S} = 0. \tag{8.16}$$

On the other hand, the EPI solution (3.18) requires that $I$, $J$, and $\kappa$ obey

$$I - \kappa J = 4\int d\omega \left[\frac{1}{S} - \kappa F\right] \equiv 0. \tag{8.17}$$

The microlevel solution Eq. (3.20) requires this to be true at each value of the integrand,

$$\frac{1}{S} = \kappa F. \tag{8.18}$$

Simultaneously solving Eqs. (8.16) and (8.18) requires that

$$\kappa F/S = -\partial F/\partial S. \tag{8.19}$$

This simple differential equation has a general solution

$$F[S(\omega), \omega] = G(\omega)S(\omega)^{-\kappa}, \tag{8.20}$$

where $G(\omega) \geqslant 0$ is some unknown function. Substituting form (8.20) for $F$ into Eq. (8.17) gives

$$I - \kappa J \equiv 0 = 4\int d\omega\, [S(\omega)^{-1} - \kappa G(\omega)S(\omega)^{-\kappa}]. \tag{8.21}$$

Function $G$ is found as follows.

### 8.3.3 Invariance principle

As usual we require an invariance principle (see Sec. 3.4.5). *Let the combined EPI principle (8.21) remain invariant in form under an arbitrary change of scale in $\omega$.* That is, let it change by at most a constant of proportionality. This is a 'self-similarity' requirement for the overall EPI approach to the problem. It also closely resembles the invariances Eq. (3.36) to Lorentz transformation – a linear rotation of coordinates – that were required of the informations. By comparison, here we have a *stretch* of the coordinate $\omega$.

First we need to find how a given $S(\omega)$ transforms under a change of scale. Define a new variable

$$\omega_1 = \alpha\omega, \quad \alpha > 0. \tag{8.22}$$

We suppose that $\int d\omega\, S(\omega)$ represents the total power in the process. Then this should be a constant independent of any change of coordinates. Let $S_1(\omega_1)$ represent the power spectrum in the new coordinate system. Then (Frieden, 2001) the two power spectra $S_1(\omega_1)$, $S(\omega)$ relate as

$$S_1(\omega_1)\, d\omega_1 = S(\omega)\, d\omega. \tag{8.23}$$

(Note that this is *not* a statement of self-similarity for a function $S(\omega_1)$, owing to the subscript 1 of $S_1(\omega_1)$. It is just a statement of transformation. See Exercise 8.3.5 below.) Combining Eqs. (8.22) and (8.23) yields

$$S_1(\omega_1) = (1/a)S(\omega_1/a). \tag{8.24}$$

Equation (8.21) is the EPI principle (3.18) as applied to this problem. Hence, suppose that $G(\omega)$ satisfies Eq. (8.21). In the new coordinate system, the information (8.21) is identically

$$I_1 - \kappa J_1 = 4\int d\omega_1 \left[\frac{1}{S_1(\omega_1)} - \kappa G(\omega_1)S_1(\omega_1)^{-\kappa}\right]. \tag{8.25}$$

We demand that $G(\omega)$ make this zero as well, in view of the invariance principle that was postulated. Now using identity (8.24) gives

$$I_1 - \kappa J_1 = 4\int d\omega_1 \left[\frac{a}{S(\omega_1/a)} - \kappa G(\omega_1)a^{\kappa}S(\omega_1/a)^{-\kappa}\right] \equiv 0. \tag{8.26}$$

Change variable back to $\omega$ via transformation (8.22). Equation (8.26) becomes

$$I_1 - \kappa J_1 = 4a^2\int d\omega \left[\frac{1}{S(\omega)} - \kappa G(a\omega)a^{\kappa-1}S(\omega)^{-\kappa}\right] \equiv 0. \tag{8.27}$$

Now compare this expression with the like one (8.21) in the original coordinate system. For Eq. (8.27) to attain the invariant value $I_1 - \kappa J_1 = 0$ function $G(\omega)$ must be such that the integrand of Eq. (8.27) is proportional to that of Eq. (8.21). This gives a requirement

$$\kappa G(\omega)S(\omega)^{-\kappa} = \kappa G(a\omega)a^{\kappa-1}S(\omega)^{-\kappa}. \tag{8.28}$$

After cancellations, this becomes

$$G(a\omega) = a^{1-\kappa}G(\omega). \tag{8.29}$$

The general solution to this is

$$G(\omega) = B\omega^{1-\kappa}, \quad B = const. \tag{8.30}$$

This is the function $G(\omega)$ we sought.

### 8.3.4 Power spectrum

From Eq. (8.20), we therefore have $F = B\omega^{1-\kappa}S^{-\kappa}$. Substituting this into Eq. (8.18) gives the solution

$$S(\omega) = A\omega^{-1}, \quad A = (\kappa B)^{1/(\kappa-1)} = const. \tag{8.31}$$

The EPI approach derives the $1/f$ noise power spectrum. This result follows for any value of $\kappa$.

As a minor point, the solution (8.31) seems ill-defined in the case $\kappa = 1$. However, defining the constant $B = \kappa^{-1}\exp[C(1-\kappa)]$, with $C > 0$ a constant, permits the solution.

### 8.3.5 Exercise

If Eq. (8.24) were replaced by a requirement $S(\omega_1) = (1/a)S(\omega_1/a)$, a self-similarity condition on $S(\omega_1)$, then this condition *by itself* has the solution (8.31). Show this.

### 8.3.6 Exercise – the size of N

In all the preceding we assumed the data $Z$ values to be observed over a continuous bandwidth $\Omega$ of frequencies. This amounts to an infinite number $N$ of data values and, hence, degrees of freedom! But, in fact, the analysis (8.21)–(8.31) could have been carried through, with the same result (8.31), at but $N = 1$ frequency. Show this. Hence, *any number* $N$ of degrees of freedom suffices to derive the $1/f$ law by the EPI approach.

### 8.3.7 Values of the efficiency parameter κ

The result (8.31) is seen to be independent of the choice of parameter $\kappa$. From Eq. (3.18), $\kappa$ represents the ratio of the total acquired Fisher information $I$ in the data to the information $J$ that is intrinsic to the phenomenon. Hence, $\kappa$ represents an efficiency parameter for the relay of information from phenomenon to data.

Until now, a given physical phenomenon has been characterized by a unique value for $\kappa$. As examples, we found value $\kappa = 1$ for quantum mechanics (Chapter 4) and for statistical physics (Chapter 7), and value $\kappa = 1/2$ for electromagnetic theory (Chapter 5) and gravitational theory (Chapter 6) Here, for the first time, we have a phenomenon that is characterized by *any level* of efficiency $\kappa$.

### 8.3.8 Exercise – checks on solution

Show that the solution (8.31) for $S(\omega)$ and $A$, when used with (8.30) in (8.20) to form $F$, satisfies both the EPI extremum condition (8.16) and zero-condition (8.18).

### *8.3.9 Assumption of an efficient estimator*

In Eq. (8.7) we assumed the existence of an efficient estimator. (*Note*: this sense of the word "efficiency" is independent of the concept of "information efficiency" $\kappa$ used in Sec. 8.3.6). In fact, the separable normal law $p(Z_r, Z_i|\theta_r, \theta_i)$ assumed in Sec. 8.3.1 for the data values gives rise to an efficient estimator; see Sec. 1.6.1.

### *8.3.10 Reference*

The idea of using EPI to derive the $1/f$-noise law was published in Frieden and Hughes (1994). However, many aspects of the derivation in this chapter differ from those in the reference. The EPI approach has evolved since then.

## 8.4 Overview

The EPI principle regards the $1/f$ power spectral law (8.31) as an equilibrium state of the signal and its spectrum $Z(\omega)$. The assumption is that, as time $t \to \infty$, the PDF $p(Z_r, Z_i|t)$ on the data $Z$ at one (or more) frequencies approaches an equilibrium form $p(Z_r, Z_i)$ whose power spectrum has the required $1/f$ form. (*Note*: For simplicity, in this section we suppress the fixed parameters $\theta_r$, $\theta_i$ from the notation.) The equilibrium state is to be achieved by extremizing the physical information $K$ growing out of the PDF $p(Z_r, Z_i)$.

The requisite invariance principle (see Sec. 3.4.5) is the invariance of the form of the combined EPI principle (8.21) to a linear change of scale (Sec. 8.3.3). A physical assumption is that the time signal $X(t)$ be an $\text{IRF}_0$ random process. This assumption does not, by itself, imply $1/f$ noise (Solo, 1992).

We note that the Fisher coordinates for the problem, $Z_r$ and $Z_i$, are purely real. Then it follows (Sec. 3.4.12) that the information transfer game is played here. The payoff of the contest is the $1/f$ law.

The $1/f$-noise solution (8.31) follows under uniquely general circumstances. There is complete freedom in the value of the information efficiency $\kappa$. See Sec. 9.3.8 for more on this.

In that a value $\kappa = 1$ characterizes quantum phenomena (Chaps. 4 and 11), and $\kappa = 1/2$ characterizes classical phenomena (Chaps. 5 and 6), the unrestricted nature of $\kappa$ here implies that $1/f$ noise describes a wide range of quantum and classical phenomena. Perhaps this accounts for the ubiquitous nature of the $1/f$ law (see the list in Sec. 8.1). Adding to the generality is the arbitrary number $N$ of degrees of freedom for the problem (Sec. 8.3.6).

The result (8.31) is useful in predicting "pure" $1/f$ noise, i.e., noise with an exponent $\alpha = 1$ in definition (8.4). However, this leaves out physical phenomena for which $\alpha \neq 1$. The assumption that $X(t)$ is an $\mathrm{IRF}_0$ random process may have been decisive in this regard, only permitting derivation of the $\alpha = 1$ result. A more general class of random functions might exist for deriving Eq. (8.4) with a general value of $\alpha$.

# 9

# Physical constants and the $1/x$ probability law

## 9.1 Introduction

The universal physical constants $c$, $e$, $\hbar$, etc., are the cornerstones of physical theory. They are, by definition, the fundamental numbers that define the different phenomena that make up physics. The size of each sets the scale for a given field of physics. They are by their nature as unrelated as any physical numbers can be.

Naturally, their magnitudes depend upon the system of units chosen, i.e., whether c.g.s, m.k.s, f.p.s. or whatever. However, regardless of units, they cover a tremendous range of magnitudes (see Table 9.1 below, from Allen (1973)). There is great mystery surrounding these constants, particularly their values. By their definition as fundamentally unrelated numbers, they must be independent. Hence, no analytical combination of them should equal rational fractions, for example. To get around the fact that their values change with choice of units, many physicists advocate using unitless combinations of the constants as new, more "fundamental" constants. Examples are

$$\alpha_{\mathrm{e}} \equiv \frac{\mathrm{e}^2}{\hbar c} = \frac{1}{137.0360\ldots}, \quad \frac{m_{\mathrm{n}}}{m_{\mathrm{p}}} = 1.001\,38\ldots, \quad \text{and} \quad \frac{m_{\mathrm{e}}}{m_{\mathrm{p}}} = \frac{1}{1836.12\ldots}, \tag{9.1}$$

where $\alpha_{\mathrm{e}}$ is the electromagnetic fine structure constant and $m_{\mathrm{n}}$, $m_{\mathrm{p}}$, and $m_{\mathrm{e}}$ are, respectively, the masses of the neutron, proton, and electron.

We have found in preceding chapters that the fixed nature of the constants is, in fact, demanded by EPI theory. However, their *magnitudes* have not so far been fixed by the theory, nor by any accepted theory to date. (The game corollary of Sec. 3.4.14 might ultimately permit unitless constants of the type (9.1) to be computed analytically, however.) It has been conjectured that the constants in Table 9.1 have these values because, if they were different, we

Table 9.1. *The fundamental constants*

| Quantity | Magnitude | $\log_{10}$ | (−30, −18) | (−18, −6) | (−6, 6) | (6, 18) | (18, 30) |
|---|---|---|---|---|---|---|---|
| Velocity of light $c$ | $2.99 \times 10^{10}$ | +10 | | | | X | |
| Gravitational constant $G$ | $6.67 \times 10^{-8}$ | −7 | | X | | | |
| Planck constant $h$ | $6.63 \times 10^{-27}$ | −26 | X | | | | |
| Electronic charge $e$ | $4.80 \times 10^{-10}$ | −10 | | X | | | |
| Mass of electron $m$ | $9.11 \times 10^{-28}$ | −27 | X | | | | |
| Mass of 1 amu | $1.66 \times 10^{-24}$ | −24 | X | | | | |
| Boltzmann constant $k$ | $1.38 \times 10^{-16}$ | −16 | | X | | | |
| Gas constant $R$ | $8.31 \times 10^{+7}$ | +8 | | | | X | |
| Joule equivalent $J$ | $4.19 \times 10^{0}$ | 0 | | | X | | |
| Avogadro number $N_A$ | $6.02 \times 10^{23}$ | +24 | | | | | X |
| Loschmidt number $n_0$ | $2.69 \times 10^{19}$ | +19 | | | | | X |
| Volume gram-molecule | $2.24 \times 10^{+4}$ | +4 | ↓ | ↓ | X | | ↓ |
| Standard atmosphere | $1.01 \times 10^{+6}$ | +6 | | | | X | |
| Ice point | $2.73 \times 10^{+2}$ | +2 | | | X | | |
| Faraday $N_A e/c$ | $9.65 \times 10^{+3}$ | +4 | | | X | | |
| Totals | | | 3 | 3 | 4 | 3 | 2 |

would not be here to observe them. This is an example of use of the "anthropic principle" (Dicke, 1961; Carter, 1974). Thus, the values of the constants are, somehow, scaled to accommodate the presence of human beings. This would be an ultimate form of the "observer participancy" postulated by Wheeler in Sec. 0.1.

## 9.2 Can the constants be viewed as random numbers?

There are no analytical relationships that allow us to know one fundamental physical constant uniquely in terms of the others. If there were, the constants would not be fundamental. The absence of *analytical* relationships suggests *statistical* relationships: i.e., that the constants are a random sample of independent numbers from some "master" probability law $p(y)$, where $y$ is the magnitude of any physical constant.

Another argument for randomness (Barrow, 1991) is as follows. The constants that we deem to be "universal" arise out of a limited, four-dimensional worldview. But if, as proponents of "string theory" suggest, there are additional dimensions present as well, then our observed universal constants might derive from more fundamental constants that exist in other dimensions. And the process of projection into our limited dimensionality might have a *random component*, e.g., due to unknown quantum gravitational fluctuations.

Admitting, then, that the universal constants might have been randomly generated, what form of probability law should they obey?

The EPI principle has been formulated to derive such laws. The EPI view is that all physical phenomena – even apparently deterministic ones – arise as random effects. This includes, then, the physical constants. Let us see what probability law EPI can reasonably come up with for the physical constants.

## 9.3 Use of EPI to find the PDF on the constants

We next proceed through EPI in the usual way, identifying the appropriate Fisher variables, defining an appropriate invariance principle for finding the bound information, and then enforcing the two EPI conditions (3.16) and (3.18) in order to get solutions. The game corollary will be used once again to fix certain constants of the theory.

### *9.3.1 Fisher variables: logarithms of the constants*

The numbers in Table 9.1 identify the appropriate Fisher variable for the problem. Let $x$ denote the absolute value of a randomly chosen physical constant, $x \geqslant 0$. The number totals along the bottom row of Table 9.1 indicate approximately the same number (3, 3, 4, 3, and 2, respectively) of constants $x$ lying within each logarithmic interval. The intervals are at a constant subdivision, each containing 12 powers of 10. As each power of 10 is the *logarithm* of a physical constant, this suggests that *the logarithms of the constants obey close to a uniform probability law.* The latter is widely regarded as the simplest probability law. On the grounds that nature ought to be described as simply as possible, the appropriate *Fisher variable* is therefore the *logarithm* of a physical constant $x$,

$$z \equiv \ln(x), \quad x \geqslant 0, \tag{9.1}$$

rather than the constant $x$ itself. We call $z$ a "log constant." Hence we seek the probability law $p_Z(z)$ on the log-constants. Once this law is known, Jacobian theory will be used to convert it to the required law $p_X(x)$ in $x$, as follows.

Equation (9.1) states that each event interval $(z, z+|dz|)$ occurs as often as a corresponding event in $(x, x+|dx|)$. Thus,

$$p_Z(z)|dz| = p_X(x)|dx|. \tag{9.2}$$

This allows us to compute the required $p_X(x)$, as

$$p_X(x) = \left[ p_Z(z) \left| \frac{dz}{dx} \right| \right]_{z=\ln(x)} = p_Z(\ln(x))\frac{1}{x}, \tag{9.3}$$

the latter by differentiating (9.1).

As an example of the use of (9.3), suppose that the PDF on random variables $z$ *really is* uniform, $p_Z(z) = C$, $C = const.$, with $z$ confined to a symmetric interval of values $-\ln b \leqslant z \leqslant \ln b$. (A reasonable value for $b$ is $b = 10^{30}$ according to Table 9.1.) Then, by Eq. (9.3), $p_X(x) = C/x$. Also, by Eq. (9.1), the constants $x$ lie on the interval $(1/b, b)$. Then, requiring normalization for $p_X(x)$ gives $C = (2\ln b)^{-1}$. Hence

$$p_X(x) = \frac{1}{(2\ln b)x} \quad \text{for} \quad 1/b \leqslant x \leqslant b. \tag{9.4}$$

Thus, if the logarithms of random numbers obey a uniform probability law, the numbers themselves obey a $1/x$ law.

As a check on the answer (9.4) for the PDF, it has a normalization integral

$$\int_{1/b}^{b} dx\, p_X(x) = \frac{1}{2\ln b}\int_{1/b}^{b} dx\, \frac{1}{x} = \frac{2\ln b}{2\ln b} = 1 \tag{9.5}$$

as required.

### 9.3.2 Invariance principle: consequences of a zero-information vacuum

Vilenkin (1982; 1983) and others suggest that the universe arose out of "nothingness." That is, everything observable is a fluctuation from some unknown vacuum state. The universal constants $x$ or $z$ are the building blocks of physics. They are, by definition, independent of one another and impossible to determine by any prior knowledge. This suggests that they are "completely" free to vary, i.e., have more randomness in some sense than any other quantities in physics. The framework of EPI permits the expression of such maximal randomness: as arising from *zero* source information,

$$J = 0. \tag{9.6}$$

Zero source information characterizes other effects as well: for example, the "macroscopic" prior knowledge taken in both statistical mechanics and economic valuation theory (Chapter 13). Also, by the proportionality between $J$ and the squared rest mass in Eq. (4.18), $J$ is zero when a *massless particle* such as a gluon is the information source for a measured particle. Equation (9.6) expresses an absolute and, as such, constitutes our invariance principle.

A level of *zero* information (9.6) for the physical constants has another ramification: It suggests that the "ideal" value of every such constant is the *same number* $\theta$. We anticipate that the form of the PDF $p_X(x)$ will be invariant to the choice of units for the constants $x$. Therefore, the constant $\theta$ should likewise be invariant to a change of units. A change of units involves multiplications. The only finite constant that remains the same under arbitrary multiplications is the value zero, so necessarily

$$\theta = 0. \tag{9.7}$$

An immediate consequence of Eq. (9.7) is that the *observed* value $y$ of a physical log-constant $z$ is pure fluctuation, i.e.,

$$y \equiv \theta + z = z. \tag{9.8}$$

This allows us to regard $z$ *per se* as the observed value of a log-physical constant and, consequently, to regard the PDF $p_Z(z)$ as the ultimate law to be found.

### 9.3.3 Fisher information I and physical information

Temporarily allow for the presence of multiple PDFs $p_Z(z)$ denoted as $p_n(z)$, $n = 1, \ldots, N$. Higher-order $p_n(z)$ would correspond to "excited" states of the physical constants (if such states existed). By Eqs. (2.18) and (9.6), the EPI extremum problem for of the single, real coordinate $z$ is then

$$K \equiv I - J = I = 4 \sum_{n=1}^{N} \int_{-\ln b}^{\ln b} dz\, q_n'^2 \equiv min., \quad q_n^2(z) \equiv p_n(z), \tag{9.9}$$

$$q_n' \equiv dq_n(z)/dz.$$

(*Note*: Although each $p_n(z)$ must obey normalization, there is no add-on normalization constraint term present in (9.9) since, by hypothesis, EPI extremizes purely the physical information $I - J$.) The finite integration limits on $z$ are due to Eqs. (9.1) and (9.4).

The Euler–Lagrange solution Eq. (0.27) to the extremization problem (9.9) is directly

$$q_n'' = 0, \quad \text{with solution} \quad q_n = a(c_n z + d_n), \quad a,\ c_n,\ d_n = \text{real } consts. \tag{9.10}$$

(also previously found in Sec. 1.2.6.) The question of units enters in here (see also Sec. 9.4.1). By the form of the normalization integral for each PDF $q_n^2(z)$, the unit of $q_n$ must be that of $1/\sqrt{z_n}$. Let the constant $a$ in the second Eq. (9.10) carry the unit $1/z_n^{3/2}$. Then, to balance units in the equation, *the constants* $c_n$ *must be unitless*. (Such unitless constants can later be fixed by the game corollary.) Back substituting the solution (9.10) for $q_n$ into the integral in Eq. (9.9) gives

$$I = 8a^2(\ln b) \sum_{n=1}^{N} c_n^2. \tag{9.11}$$

### 9.3.4 Value of N

But, are there really "excited" states $q_n(z)$, $n = 2, 3, \ldots$? This would require a number of degrees of freedom $N > 1$. By the game corollary (Sec. 3.4.14), $N$ is found as the value that minimizes $||I||$ at solution, Eq. (9.11). Since each term in the sum (9.11) is positive or zero, $||I||$ is minimized for

$$N = 1 \tag{9.12}$$

degree of freedom. Thus $n = 1$ only, and *we may drop subscripts n*. There are no excited states, only the single state $p(z) \equiv p_Z(z)$. Also, the information (9.11) is now

$$I = 8(\ln b)a^2 c^2. \tag{9.13}$$

### 9.3.5 Further use of the game corollary

The value of $c$ is still to be found. Since $c$ is unitless, we can again apply the game corollary to the solution (9.13). This shows an absolute minimum value of $I = 0$ attained for

$$c \equiv c_n = 0. \tag{9.14}$$

Using this in Eq. (9.10) gives

$$q(z) = ad = const. \tag{9.15}$$

or, equivalently, a single PDF $p(z) \equiv p_Z(z) = a^2 d^2 = const.$ Since $p_Z(z)$ must obey normalization over the interval $-\ln b \leqslant z \leqslant \ln b,$ it follows that

$$p_Z(z) = \frac{1}{2 \ln b} = const., \qquad -\ln b \leqslant z \leqslant \ln b. \tag{9.16}$$

Thus, the intuition voiced in Sec. 9.3.1 is justified: According to the EPI extremum principle, including the game corollary, the PDF on the log-constants really is a flat one.

### 9.3.6 Probability law for the constants x

We now convert back to $x$-space, as in transition from Eq. (9.3) to Eq. (9.4),

$$p_X(x) = \frac{1}{(2 \ln b)x}, \qquad 1/b \leqslant x \leqslant b. \tag{9.17}$$

Thus, the constants *per se* obey a $1/x$ law. This indicates a high state of randomness for the constants. In fact, Jeffreys (1961) regards the $1/x$ probability law as describing a random variable *about which nothing is known* except its positivity. This qualitatively agrees with the invariance principle (9.6) of *zero* source information in the values of $x$. Apparently this is at the root of the quantitative agreement.

### 9.3.7 EPI zero solution

We found that a $1/x$ law (9.17) follows from use of the EPI extremum solution. Does the $1/x$ law also satisfy the EPI zero-solution? By Eq. (9.9), the zero-solution (3.18) is to obey

$$I - \kappa J = I = 4 \sum_{n=1}^{N} \int_{-\ln b}^{\ln b} dz \, q_n'^2 \equiv 0 \tag{9.18}$$

through choice of the $q_n(z)$. By the additivity of all terms in the sum, the unique solution to this problem is $q_n'(z) = 0$, or, equivalently, $q_n(z) = const.$ This is the same elemental solution (9.15) that satisfied the EPI extremum principle. Results (9.16) and (9.17) follow again as well. Hence both EPI principles give the same solution. This gives added strength to the result.

### 9.3.8 Value of $\kappa$

The solution (9.15) satisfies the condition (9.18) irrespective of the size of $\kappa$. Hence, the probability law on the physical constants holds for any value of the information efficiency $\kappa$. This property is likewise obeyed by $1/f$ power law phenomena (Chapter 8), and laws governing economic valuation and statistical mechanics (Chapter 13). See further discussion in Sec. 9.6.

### 9.3.9 Significance of the median value of 1

The only real checks we have on the prediction (9.17) are the data in Table 9.1. How can they be used for this purpose? (See also Sec. 9.5.3.)

The median $x_M$ of a general PDF $p(x)$ is defined by the property that *half* the number of random samples $x$ from $p(x)$ will obey $x \leqslant x_M$, that is $P(x \leqslant x_M) = 1/2$. Applying this requirement to the PDF (9.17) gives

$$P(x \leqslant x_M) \equiv \frac{1}{2} \equiv \int_{1/b}^{x_M} \frac{dx}{(2 \ln b)x} = \frac{1}{2 \ln b}(\ln x_M + \ln b). \qquad (9.19)$$

The solution, as can be readily verified, is $x_M = 1$. Notice that this median value is independent of the size of $b$. This is worthy of discussion.

Noise outlier events $x$ of a PDF $p(x)$ are in general values of $x$ that are large and improbable but, still, possible. For example, the PDF (9.17) allows noise outliers that are close to size $b$ in magnitude. The median is famous for being *insensitive* to noise outliers. For example, the PDF (9.17) has as a median the value $x_M = 1$, and we found it to be independent of the size of $b$. By comparison, the size of the *mean* usually *depends* upon the noise outliers. For example, for the PDF (9.17) it is directly

$$\langle x \rangle = (b - 1/b)/(2 \ln b). \qquad (9.20)$$

These results clearly point toward one test of the prediction (9.17) by use of the data in Table 9.1. Suppose that the mean $\langle x \rangle$ were to be used for this purpose. The test would be a comparison of the *theoretical* mean (9.20) with the *sample* mean (the ordinary average) of the data in Table 9.1. Notice that the theoretical mean (9.20) depends upon knowledge of the theoretical value of $b$. However, the value of $b$ is certainly unknown at present, and probably can never be known *a priori* by any theory. Moreover, the sample mean of the data in Table 9.1 is dominated by just the largest number in the list, i.e., one datum.

Such domination makes it highly unstable, and therefore highly inaccurate. For these reasons, testing the theory by use of the mean is untenable.

Consider instead use of the median. We found in Eq. (9.19) that its theoretical value is $x_M = 1$, *independent* of the size of the unknown parameter $b$. Hence, in comparison with $\langle x \rangle$, $x_M$ is known. The second requirement from the preceding was that the test statistic, here the median, be amenable to accurate estimation from the data. This is the case.

Given discrete data from a PDF, a widely accepted estimate of the median is the "sample median" of the data (Frieden, 2001). The sample median of 15 numbers, as in Table 9.1, is defined as the number that exceeds in size exactly seven other data values. The list shows that the theoretical median value of unity in fact exceeds six numbers in the list. Hence, the existing data on universal constants, as in Table 9.1, confirm pretty well one aspect of the result (9.17). The agreement between the numbers six and seven is in fact surprising considering the limited number of data that are in Table 9.1 and their huge range of magnitudes.

In general a prediction of a mean value or a median is for a population of values expressed in a *particular set of units*. Thus the prediction would seem to change with the choice of units. But, if so, how could we know that the numbers in Table 9.1 were in the appropriate units to have a median value of close to unity? In fact, the prediction of a median value of unity is *independent* of the choice of units: By Sec. 9.4.1 to follow, the PDF (9.17) is invariant to the choice of units for the physical constants. This adds further weight to the prediction of a median value of unity.

### *9.3.10 Checks on the solution*

In the preceding we found a self-consistent EPI solution without the need for combining the two EPI conditions. Instead, we solved each condition separately in, respectively, Secs. 9.3.3 and 9.3.7, and found that each condition was satisfied by the same solution (9.15).

## 9.4 Statistical properties of the $1/x$ law

The reciprocal law (9.17) has further statistical properties that give insight into the nature of the physical constants. As will be seen, there is something to be gained by viewing them as statistical entities.

### 9.4.1 Invariance to change of units

We show next that the reciprocal law maintains its form under a change of units. Subdivide the constants into classes $i$ according to units. For example, constants in class $i = 1$ might all have the units of mass, those in class $i = 2$ might be lengths, etc. Let a fraction $r_i$ of the constants be in class $i$, $i = 1, \ldots, n$. Of course

$$\sum_{i=1}^{n} r_i = 1, \tag{9.21}$$

a constant must be in one of the classes. Denote the PDF for the $x$ of a specific class $i$ as $p_i(x)$. Then, by the law of total probability (Frieden, 2001), the net probability density for a value $x$, regardless of class, obeys

$$p(x) = \sum_{i=1}^{n} r_i p_i(x). \tag{9.22}$$

Suppose that the units are now changed. This means that, for class $i$, all numbers $x$ are multiplied by a fixed factor $b_i$ (the change of units factor). Denote the resulting PDF for class $i$ as $P_i(x)$. It is elementary that

$$P_i(x) = \frac{1}{b_i} p_i\left(\frac{x}{b_i}\right). \tag{9.23}$$

As with Eq. (9.22), the *net* PDF after the change of units obeys

$$P(x) = \sum_{i=1}^{n} r_i P_i(x). \tag{9.24}$$

Then combining Eqs. (9.23) and (9.24) gives

$$P(x) = \sum_{i=1}^{n} \frac{r_i}{b_i} p_i\left(\frac{x}{b_i}\right). \tag{9.25}$$

It is reasonable to assume that the $x$ are identically distributed (independent of class $i$), i.e.,

$$p_i(x) = p_X(x). \tag{9.26}$$

Combining the last two equations gives

$$P(x) = \sum_{i=1}^{n} \frac{r_i}{b_i} p_X\left(\frac{x}{b_i}\right). \tag{9.27}$$

This gives the net new law $P(x)$ in terms of the old net law $p_X(x)$ (before the change of units). In the particular case (9.17), where the old law is the reciprocal law, Eq. (9.25) gives

$$P(x) = A^2 \sum_{i=1}^{n} \frac{r_i}{b_i} \frac{b_i}{x} = \frac{A^2}{x} \sum_{i=1}^{n} r_i = \frac{A^2}{x}, \qquad A^2 \equiv (2 \ln b)^{-1} \tag{9.28}$$

using Eq. (9.21), or

$$P(x) = p_X(x) \tag{9.29}$$

by Eq. (9.17). That is, the PDF for the constants in the new units is the same as in the old units. The reciprocal law is invariant to choice of units.

This result is actually indifferent to the assumption (9.26). Even if the individual classes had different PDF laws, when the net PDF $p(x)$ is a reciprocal law the net output law after change of units remains a reciprocal law (Frieden, 1986).

### 9.4.2 *Invariance to inversion*

Most of the constants are defined in equations that are of a multiplicative form. For example, the usual definition of Planck's constant $h$ is in the relation

$$E = h\nu, \tag{9.30}$$

where $E$ is the energy of a photon and $\nu$ is its frequency. This relation could instead have been written

$$E = \nu/h, \tag{9.31}$$

in which case $h$ would have the reciprocal value to that defined in Eq. (9.30). Of course neither definition is more "correct" than the other.

This illustrates an arbitrary aspect of the definition of the physical constants. But then, in the face of such arbitrariness of inversion, *does the concept of a PDF p(x) for the constants make sense*? Clearly it would only if the *same* law $p(x)$ resulted *independently of any choice of the fraction r* of inverted constants in a population of constants.

Let us, then, test our candidate law (9.17) for this property. Suppose that the reciprocated constants obey a PDF $p_r(x)$. Then, by the law of total probability (Frieden, 2001), the PDF for the entire population obeys

$$p(x) = rp_r(x) + (1 - r)p_X(x), \quad 0 \leqslant r \leqslant 1. \tag{9.32}$$

We can also relate the PDF $p_r(x_r)$ to the PDF $p_X(x)$. A number $x_r$ is formed from a number $x$ by the simple transformation

$$x_r = 1/x. \tag{9.33}$$

By transformation theory (Frieden, 2001)

$$p_r(x_r)|dx_r| = p_X(x)|dx|, \quad \text{so that} \quad p_r(x_r) = p_X\left(\frac{1}{x_r}\right)\frac{1}{x_r^2} \tag{9.34}$$

by Eq. (9.33). Substituting this into Eq. (9.32) gives

$$p(x) = rp_X(1/x)x^{-2} + (1-r)p_X(x) \tag{9.35}$$

as the relationship to be obeyed by a candidate law $p_X(x)$.

With our law $p_X(x) = A^2/x$, the right-hand side of Eq. (9.35) becomes

$$rA^2xx^{-2} + (1-r)A^2x^{-1} = A^2x^{-1}[r + (1-r)] = A^2/x. \tag{9.36}$$

Thus Eq. (9.35) gives $p(x) = A^2/x$, which was the law without reciprocations. Or, the $1/x$ law obeys invariance to reciprocation. Moreover it obeys it independently of the value of $r$, as was required.

### *9.4.3 Invariance to combination*

Combinations of constants, such as the quantity $e/m$ for the electron, are often regarded as fundamental constants in their own right. But if they are fundamental, they should obey the same law $p_X(x)$ that is obeyed by the fundamental constants $x$. Now, a general PDF does not have this property. For example, if numbers $x$ and $y$ are independently sampled from a Gaussian PDF, then the quotient $x/y = z$ is no longer Gaussian but, rather, Cauchy, with a PDF of the form $(1 + z^2)^{-1}$.

Nevertheless, our law $p_X(x) = A^2/x$ obeys the property of invariance to combination, provided that the range parameter $b$ is large (which it certainly is – see Sec. 9.3.1). The proof is in Frieden (1986).

### *9.4.4 Needed corrections to the estimated physical constants*

The $1/x$ law is a *prior* law, indicating which constants $x$ are expected prior to measurement. In particular, the law shows that small constants $x$ are more likely to occur than are large constants. The result is that, if an unknown constant is observed through some measurement technique, its maximum probable value is *less than* the observed value. Furthermore, if many repetitions of the experiment are performed, the maximum probable value of the constant is *less than* the arithmetic mean of the experimental outcomes. See Frieden (1986). An example follows.

Suppose that a single universal constant is measured $N$ times, and that each measurement $x_n$ is corrupted by added Gaussian noise of standard deviation $\sigma$. The conventional answer for a "good" estimate of the constant is the sample mean

$$\overline{x} = N^{-1}\sum_{n=1}^{N} x_n \tag{9.37}$$

of the measurements. By contrast, the *maximum probable* value $x_{MP}$ of the constant is (Frieden, 1986)

$$x_{MP} = \tfrac{1}{2}(\bar{x} + \sqrt{\bar{x}^2 - 4\sigma^2/N}). \tag{9.38}$$

Hence, if $N$ is of the order of 1, and/or the noise $\sigma$ is appreciable, the maximum probable value of the constant is considerably less than $\bar{x}$. The sample mean *would not* be a good estimate.

## 9.5 What histogram of numbers do the constants actually obey?

### 9.5.1 Log-uniform hypothesis

As mentioned in Sec. 9.3.1, the 15 physical constants listed in Table 9.1 seem to be log-uniform distributed. Let us now test this hypothesis. Taking the logarithm of each item $x$ in the table, we have essentially the powers of 10 as data: as listed, these are $+10$, $-7$, $-26$, $-10$, etc. These numbers range from $-28$ to $+24$. In view of the symmetric limits to the range of $y = \ln x$ in the theoretical model Eq. (9.17), we symmetrize the estimated limits to a range of $-30$ to $+30$. Thus the hypothesis is that

$$p_{\log x}(x) = \frac{1}{60} \quad \text{for} \quad -30 \leqslant x \leqslant +30 \tag{9.39}$$

and 0 otherwise. (Notice that this obeys normalization.)

To test this hypothesis, divide the total range $(-30, +30)$ into five equal subranges of length $60/5 = 12$. These are the subranges $(-30, -18)$, $(-18, -6)$, etc. as indicated in Table 9.1. The uniform law (9.39) predicts, then, that *the same number of events* $(15/5 = 3)$ should lie in each subrange. The actual numbers (see Table 9.1) are 3, 3, 4, 3, and 2, not bad approximations to the 3! We can quantify this agreement as follows.

### 9.5.2 Chi-square test – background

Given a die that is rolled $N$ times, the number of occurrences of roll outcome 1 is denoted as $m_1$, etc., for roll outcomes 2, ..., 6. As an example, for $N = 10$ rolls, a typical set of roll occurrences is $(m_i, \ldots, m_6) = (2, 1, 3, 0, 1, 3)$. As a check, notice that the sum of the occurrences is $N = 10$.

The chi-square test is used to measure the degree to which the *observed* occurrence numbers $m_i$ for a set of events $i$ agree with the *expected* occurrence numbers on the basis of a hypothesis $P_1, \ldots, P_M$ for the probabilities of the events. Let the probability of an event $i$ be $P_i$ and suppose that there are $N$ trials. The number $m_i$ of events $i$ in the $N$ trials will tend to be exactly $NP_i$. However, the *actual* number of times event $i$ occurs will randomly differ from this value. A measure of the disparity between an observation $m_i$ and a corresponding theoretical probability $P_i$ is, then, $(m_i - NP_i)^2$. If the numbers used for the $P_i$ are *correct* then, from the preceding, this disparity will tend to be small.

However, if the $P_i$ are actually *unknown*, and instead an estimated set of them (called a "hypothesis" set) is used, then the above disparity will tend to be larger as the hypothesis becomes more erroneous. Hence, a total disparity, weighted appropriately, is used to test the probable correctness of hypothetical $P_i$ values:

$$\sum_{i=1}^{M} \frac{(m_i - NP_i)^2}{NP_i} \equiv \chi^2. \tag{9.40}$$

This is called the "chi-square" statistic.

Since the $m_i$ are random, so must be the value of $\chi^2$. The PDF for $\chi^2$ is known. It obeys the (what else?) chi-square distribution. The number of event types $M$ is a free parameter of the chi-square distribution. Quantity $M - 1$ is called the "number of degrees of freedom" of the distribution. The chi-square distribution is tabulated (Frieden, 2001) for different values of the number of degrees of freedom.

On this basis, an observed value of $\chi^2$, call it $\chi_0^2$, has a known probability. For example, if $\chi_0^2$ is very large (by Eq. (9.40), indicating strong disagreement between the observations $m_i$ and the hypothetical $P_i$), the computed probability is very low. On the other hand, the maximum likelihood hypothesis is that events $m_i$ that occur were *a priori* highly probable (that's why they occurred). Or, as a corollary, *unlikely events don't happen*. Hence, the low probability associated with a high value of $\chi_0^2$ indicates a conflict: on the basis of the hypothetical probabilities $P_i$ the value $\chi_0^2$ was just too high. It shouldn't have occurred. The conclusion is, then, that either the observations $m_i$ are wrong or the hypothetical $P_i$ are wrong. Since the observations are presumed known to be correct, the onus is placed upon the hypothesis $P_i$. It is rejected.

The degree of rejection is quantified as follows. Denote the known PDF for $\chi^2$ as $p_{\chi^2}(x)$. The probability $\alpha$ that a value of $\chi^2$ at least as large as the observed value $\chi_0^2$ will occur is

$$\int_{\chi_0^2}^{\infty} dx\, p_{\chi^2}(x) \equiv \alpha. \tag{9.41}$$

This can be computed. It is also tabulated (see, e.g., Frieden, 2001, Appendix B). Obviously, if the lower limit $\chi_0^2$ in Eq. (9.41) is large, then $\alpha$ is small. And, from the preceding, a large $\chi_0^2$ connotes *rejection* of the hypothesis $P_i$. If $\alpha$ is so small as to be less than (say) value 0.05, then the hypothesis $P_i$ is rejected "on the level of significance" $\alpha = 0.05$.

Or, by contrast, if $\alpha$ is found to be fairly close to 1, the hypothesis $P_i$ may be accepted on that level of significance.

In this manner, the chi-square test may be used to accept or reject a hypothesis for an underlying probability law.

### 9.5.3 *Chi-square test – performed*

In Table 9.1 there are $M = 5$ possible occurrence intervals for the logarithm of a given constant to occur within. These are the intervals $(-30, -18)$, $(-18, -6)$, etc., indicated along the top row. Our hypothesis is the uniform one $P_i = 1/5$, $i = 1, \ldots, 5$ for the logged constants. Since there are $N = 15$ of them, quantities $NP_i = 3$ for all $i$. The observed values of $m_i$ are (3, 3, 4, 3, 2), respectively (see bottom row in Table 9.1). These numbers as used in Eq. (9.40) give a computed

$$\chi_0^2 = \frac{(3-3)^2 + (3-3)^2 + (4-3)^2 + (3-3)^2 + (2-3)^2}{3} = 0.667. \quad (9.42)$$

This value of $\chi_0^2$ coupled with the $M - 1 = 4$ degrees of freedom gives, using any table of the $\chi^2$ distribution, a value $\alpha = 0.95$.

Hence, the disparity between the data $m_i$ and the log-uniform hypothesis is so *small* that, if many sets of 15 data were randomly generated from the law, *for 95% of such data sets* the disparity $\chi_0^2$ would *exceed* the observed disparity of 0.667. The hypothesis is well confirmed by the data.

It should be noted, however, that, because of the small population $N = 15$ of constants, the test-interval size was necessarily quite coarse – 12 orders of magnitude. Hence, the hypothesis of a log-uniform law was only confirmed on a coarse subdivision of points. It would require more constants to test on a finer scale. However, it is doubtful that significant departures from the flat curve would result, since 15 widely separated points already fall on that curve.

## 9.6 Overview

The preceding has shown that *the empirical data at hand* – the actual values of the physical constants in Table 9.1 – obey the hypothesis that their logarithms are uniformly distributed. As we showed in Sec. 9.3.1, this is equivalent to a $1/x$ probability law for the numbers themselves. Also, as was shown in Sec. 9.4.1, this result would hold *regardless of the set of units* assumed for the numbers. Hence, the result is not a "fluke" of the particular (c.g.s.) units employed in Table 9.1.

The $1/x$ law is also invariant to arbitrary inversion of any number of the constants (Sec. 9.4.2), and to forming new constants by taking quotients of arbitrary pairs of the constants (Sec. 9.4.3).

The fact that the $1/x$ law holds independently of the choices of units, inversion, and quotients means that it would have been observed regardless of human history. It is a truly *physical* effect.

Knowledge of such a law allows us to achieve improved estimates of physical constants. See Sec. 9.4.4. Given multiple measurements of a constant, its maximum-probable estimate is *not* the sample mean of the measurements. It is *less than* the sample mean, and obeys Eq. (9.38). This follows from the bias of a $1/x$ PDF toward small numbers: *a priori*, a number $x$ is more likely to be small than to be large.

A fascinating property of the $1/x$ law Eq. (9.17) is that its median value is 1 (Sec. 9.3.9). Essentially, the value 1 lies halfway on the range of physical phenomena. It is noteworthy that this result holds independently of units, inversion, and combination, since the $1/x$ law itself is invariant under these choices. Therefore, the median value of 1 is a *physical effect*. On the other hand, we reasoned (Sec. 9.3.2) that the "ideal" value of any physical constant is 0. Hence, the numbers 0 and 1 are cardinal points in the distribution of the physical constants.

The fact that every subrange of exponents given in Table 9.1 should contain, theoretically, the same number of constants, can be used to crudely predict the values of future constants. For example, there should be additional constants in the subrange (18, 30) of exponents (in the c.g.s. units of the table, of course).

As we saw, the law that is predicted by EPI is validated by the data. This, in turn, tends to validate the model assumptions that went into the EPI derivation. Among these are: (a) the argument of Barrow (1991) in favor of randomness among the physical constants (Sec. 9.2); and (b) the argument by Vilenkin (1982, 1983) that the Universe – and its constants – arose out of nothingness (Sec. 9.3.2.).

The EPI derivation (Sec. 9.3) is based upon (i) the use as Fisher variables of the logged constants; and (ii) the invariance principle (9.6) that the source information for the constants has the invariant value $J = 0$. Property (i) is strongly implied by the actual occurrences of the known data in Table 9.1. Property (ii) follows because the constants are, by definition, independent of one another and impossible to determine by any prior knowledge. Therefore they effectively arise out of *zero* prior knowledge $J$.

The $1/x$ law for the constants and the $1/f$ power spectral law of Chapter 8 share many mathematical properties. Both obey an invariance to change of scale (cf. Exercise 8.3.5 and Sec. 9.4.1). This is sometimes called a property of "self-similarity." As an alternative to the use of EPI, both effects can be derived purely from a requirement of self-similarity. Finally, both EPI derivations are valid for any value of the efficiency constant $\kappa$. Thus, both apply to quantum ($\kappa = 1$) and classical ($\kappa < 1$) effects. Note that, to be consistent, the EPI output $1/x$ law had to have this property since by hypothesis it describes the constants from *all* physical effects.

What we have effectively found is that the PDF for any population of *basically unrelated* numbers (of apples, oranges, street addresses, etc.) should be of the $1/x$ form. The fundamental physical constants are unrelated numbers of this type. But, so also are other populations. An example is all the numbers that an individual encounters during a typical year. Perhaps a more testable example is all the numbers that are output from the central processor of a mainframe computer at a general-purpose facility such as a university. It would be interesting to check whether the histogram of these numbers follows a $1/x$ form.

# 10

# Constrained-likelihood quantum measurement theory

## 10.1 Introduction

In preceding chapters, EPI has been used as a *computational procedure* for establishing the physical laws governing various measurement scenarios. We showed, by means of the optical measurement model of Sec. 3.8, that EPI is, as well, a *physical process* that is initiated by a measurement. Specifically, it arises out of the interaction of the measuring instrument's probe particle with the object under measurement. This perturbs the system probability amplitudes, which perturbs the informations $I$ and $J$, etc., as indicated in Figs. 3.3 and 3.4. The result is that EPI derives the phenomenon's physics as it exists at the input space to the measuring device.

We also found, in Sec. 3.8, the form of the phenomenon's physics at the output to the measuring instrument. This was given by Eq. (3.51) for the output probability amplitude function.

The analysis in Sec. 3.8 was, however, severely limited in dimensionality. A one-dimensional analysis of the measurement phenomenon was given. A full, covariant treatment would be preferable, i.e., where the space-time behaviors of all probability amplitudes were determined.

Such an analysis will be given next. It constitutes a covariant *quantum theory of measurement*. This covariant theory will be developed from an entirely different viewpoint than that in Sec. 3.8. The latter was an analysis that focussed attention upon the probe–particle interaction and the resulting wave propagation through the instrument. The covariant approach will concentrate, instead, on *the meaning* of the acquired data to the observer as it reflects upon the quantum state of the measured particle. Hence, whereas the previous analysis was purely physical in nature, this one will be "knowledge-based" as well as physical.

Such a knowledge-based orientation is consistent with our thesis (and

Wheeler's) that the measurer is part of the measured phenomenon. For such an observer, *acquired knowledge reflects physical state as well*. We return to this thesis in the Overview section.

This may be quantified as follows. Suppose that the observer measures the three-position $\bar{\boldsymbol{r}}$ of a particle. (By convention, barred quantities are data.) This single piece of information affects the observer's state of knowledge of the particle's position and, hence, the physical state of the wave function (as above). This is a kind of physical manifestation of the principle of maximum likelihood. For example, if $\bar{\boldsymbol{r}} = 0$ then $\psi(\bar{\boldsymbol{r}})$ for the particle should, by the principle of maximum likelihood, *tend to be* high near $\boldsymbol{r} = 0$. Such a result is ordinarily taken to provide a mere prediction, or estimate, of an unknown quantity. Now it graduates into a statement of physical fact. Thus, the "prediction" takes the physical form of the "reduction of the wave function," an effect that is derived in the analysis below (this was also derived in one dimension in Sec. 3.8).

The theory shares many similarities of form with earlier work of Feynman (1948); Mensky (1979; 1993); Caves (1986); and Caves and Milburn (1987). These investigators used a path integral-based theory of continuous quantum measurements. For brevity, we will call this overall approach "Feynman–Mensky theory." Valuable background material on alternative approaches to measurement may also be found in books edited by Pike and Sarkar (1987), and by Cvitanovic, Percival, and Wirzba (1992).

Our development parallels Feynman–Mensky theory, but with important *caveats*. It is both (i) a natural extension (literally an "add-on") to the preceding EPI theory, and (ii) a further use of the simple notions of classical statistics and information. In particular, the classical notion of the *log-likelihood* (Sec. 1.6.2) of a *fixed* set of measurements is used. We develop this theory next. Key results that parallel those of Mensky and other workers in the field will be pointed out as they arise in the development.

## 10.2 Measured coordinates

The measurements that were used in Chapter 4 to derive relativistic quantum mechanics were the space-time coordinates of a particle. As we found, this gives rise to the physical information $K \equiv K_0$ obeying Eq. (4.27). In that treatment, as here, time was assumed to be as unknown, and random, as the space coordinates of the particle. We now modify $K_0$ with a constraint to the effect that three-position measurements have been made at a sequence of *imperfectly measured* time values. Each has a random component obeying the marginal law Eq. (4.30b),

$$p_T(t) = \int d\boldsymbol{r}\, \psi^*(\boldsymbol{r}, t)\psi(\boldsymbol{r}, t), \tag{10.1}$$

where $\psi(\boldsymbol{r}, t)$ is the EPI four-dimensional wave function. PDF $p_T(t)$ defines the experimental fluctuations $t$ in time from the "true" laboratory clock value $\theta_0$. (*Caveat*: By contrast, Feynman–Mensky theory presumes the time values to be perfectly known.)

Consistently with the knowledge-based approach, each data value will contribute a constraint term to supplement (in fact, to be added to) the information $K_0$. That is, knowledge of the measurements will be used to mathematically *constrain* the information. This is a standard, and convenient, step of the Lagrange approach (see Sec. 0.3.8).

Hence, assume some measured time values $\tau_m$ and intervals,

$$\tau_m = \theta_{0m} + t_m \quad \text{and} \quad (t_m, t_m + dt) \equiv t_m, \qquad m = 1, \ldots, M, \tag{10.2}$$

respectively, during which particle *three-position* measurements

$$\bar{\mathbf{y}}(t_1), \ldots, \bar{\mathbf{y}}(t_M) \equiv \bar{\mathbf{y}}_1, \ldots, \bar{\mathbf{y}}_M \equiv \bar{\mathbf{y}} \tag{10.3}$$

are collected. For simplicity of notation, we have suppressed the deterministic numbers $\theta_{0m}$ from the arguments of the $\bar{\mathbf{y}}$ values. The data $\bar{\mathbf{y}}$ are taken to be fixed, non-random numbers, which is the usual stance of classical likelihood theory (Sec. 1.6; Van Trees, 1968).

*Caveat*: Many versions of Feynman–Mensky theory instead treat the data $\bar{\mathbf{y}}$ as *random*, to be integrated over during normalization. As in the preceding, we treat them as *fixed* data – the usual orientation of classical likelihood theory.

For simplicity of language, we sometimes say that the measurements take place *at* a time $t_m$, which really means *during* the infinitesimal time interval $(t_m, t_m + dt)$.

The data $\bar{\mathbf{y}}$ are acquired with an instrument that realistically adds random noise to each reading. If the noise-free measurements are $\mathbf{y}_1, \ldots, \mathbf{y}_M \equiv \mathbf{y}$ then the data obey

$$\bar{\mathbf{y}} = \mathbf{y} + \boldsymbol{\gamma} \tag{10.4}$$

(cf. Eq. (3.45)), where $\boldsymbol{\gamma}_1, \ldots, \boldsymbol{\gamma}_M \equiv \boldsymbol{\gamma}$ denotes the random noise values.

As with Eq. (10.2), the space coordinates are themselves stochastic, obeying

$$\mathbf{y} = \boldsymbol{\theta} + \boldsymbol{r} \tag{10.5}$$

(cf. Eq. (3.43)), where $\boldsymbol{\theta} \equiv \boldsymbol{\theta}_1, \ldots, \boldsymbol{\theta}_M$ denotes the ideal (classical mechanics) space positions of the particle. Measurements $\bar{\mathbf{y}}$ are taken so as to know the values $\mathbf{y}$.

Substituting Eq. (10.5) into Eq. (10.4) shows that the measurements $\bar{\mathbf{y}}$ are degraded from the ideal space position values $\boldsymbol{\theta}$ by *two* sources of "noise," the intrinsic quantum mechanical fluctuations $\boldsymbol{r}$ and the instrument noise values $\boldsymbol{\gamma}$.

The aim of making the measurements is to observe the quantum fluctuations $\boldsymbol{\gamma}$. Toward this end, if we define a quantity

$$\overline{\boldsymbol{r}} \equiv \mathbf{y} - \boldsymbol{\theta} \tag{10.6}$$

as *associated* data, then these relate to the quantum fluctuations $\boldsymbol{r}$ as

$$\overline{\boldsymbol{r}} = \boldsymbol{r} + \boldsymbol{\gamma}. \tag{10.7}$$

From Eq. (10.6), the associated data are fixed; they are the departures of the (fixed) data from the (fixed) ideal three-positions $\boldsymbol{\theta}$. Equation (10.7) states that the associated data are degraded from the ideal fluctuations $\boldsymbol{r}$ by the added noise $\boldsymbol{\gamma}$ due to the instrument. For simplicity, in the following we will often simply call the "associated data" the "data."

Our aim is to see how the acquisition of fixed data modify the output wave function $\psi(\boldsymbol{r}, t)$ as predicted by EPI. It is intuitive that such an effect must exist. If, for example, the data value $\overline{\boldsymbol{r}}_m$ is zero then, by the classical *principle of maximum likelihood*, there must have been *a forteriori* a relatively high probability for the true quantum position $\boldsymbol{r}_m$ to be zero. Hence, the probability *amplitude* $\psi(\boldsymbol{r}_m, t_m)$ should be relatively high at $\boldsymbol{r}_m = 0$. The corresponding output law $\psi(\boldsymbol{r}, t)$ from EPI should, therefore, show this effect as well. We now proceed to quantify the effect.

## 10.3 Likelihood law

Any model of measurement has to address the question of how the measuring instrument affects the ideal parameter values under measurement. According to classical statistics, that specifier is the *likelihood law* (Sec. 1.2.3). This is the probability of a fixed data value $\overline{\boldsymbol{r}}_m$ conditional upon its ideal value $\boldsymbol{r}_m$, denoted as $p(\overline{\boldsymbol{r}}|\boldsymbol{r}_m)$. This probability is a single, scalar number that is often used as a quality measure of the instrument (O'Neill, 1963). Consequently, it is also a quality measure of the data. Hence, if this quality measure is large, the data are confirmed; if it is small, they should be ignored.

In fact, rather than the likelihood law *per se*, in classical statistics it is the *logarithm* of the likelihood law that has real import. For example, the maximum likelihood principle is conventionally expressed as $\ln p = max.$ (Van Trees, 1968). Also see Sec. 1.6.2. In statistical mechanics it is the *average of the log-likelihood*, i.e., the *entropy*, which has fundamental significance; likewise with standard communication theory (Shannon, 1948). For these reasons, and because "it works," we adopt the log-likelihood as well.

In the preceding examples, the average arose out of the stochastic nature of each likelihood term defining the system. Likewise, in our case there are many data values present, and these are taken over a sequence of time values, some of which are *a priori* more probable than others; see Eq. (10.1). It is logical to give less weight to log-likelihood terms that are less likely to occur, and conversely.

The considerations of the two previous paragraphs suggest that we use, as our scalar measure of data quality,

$$\langle \ln[C^2 p(\bar{\boldsymbol{r}}|\boldsymbol{r})] \rangle_{\text{time}} \equiv \sum_{m=1}^{M} p_T(t_m) \ln[C^2 p(\bar{\boldsymbol{r}}_m|\boldsymbol{r}_m)] \equiv L. \tag{10.8}$$

This is the time-averaged log-likelihood, aside from the presence of an arbitrary constant parameter $C$, which merely adds in a constant value $\ln(C^2)$. The negative of the remaining sum can be considered an "entropy" of the measurements. The arbitrary additive constant value $\ln(C^2)$ is permitted since (like optical phase) only *changes* in the entropy are physically observable.

## 10.4 Instrument noise properties

Equation (10.8) simplifies in the following way. Assume that the noise $\boldsymbol{\gamma}$ is "additive," i.e., independent of the ideal values $\mathbf{y}$ in Eq. (10.4) and, consequently, independent of the ideal values $\boldsymbol{r}$ in Eq. (10.7). Then from (10.7) fluctuations $\bar{\boldsymbol{r}}_m$ simply follow those of the noise $\boldsymbol{\gamma}_m$. If $p_\Gamma(\boldsymbol{\gamma})$ denotes the PDF on the noise, and if the noise is statistically shift invariant, then

$$p(\bar{\boldsymbol{r}}_m|\boldsymbol{r}_m) = p_\Gamma(\bar{\boldsymbol{r}}_m - \boldsymbol{r}_m). \tag{10.9}$$

The noise PDF $p_\Gamma$ arises as the squared modulus of a generally complex probability amplitude $w$ called the "instrument function,"

$$p_\Gamma(\boldsymbol{\gamma}_m) \equiv w^*(\boldsymbol{\gamma}_m) w(\boldsymbol{\gamma}_m), \quad m = 1, \ldots, M. \tag{10.10}$$

## 10.5 Final log-likelihood form

Combining Eqs. (10.8)–(10.10) gives

$$L = \sum_{m=1}^{M} p_T(t_m)\{\ln[w(\bar{\boldsymbol{r}}_m - \boldsymbol{r}_m)] + \text{c.c.}\}, \tag{10.11a}$$

where the notation c.c. denotes the complex conjugate of the preceding term. *Note*: For purposes of simplicity, we ignore the constant $C$ that actually multiplies amplitude function $w$ in this equation and in equations through Eq. (10.21). $C$ is regained in Eq. (10.22).

Combining Eqs. (10.1) and (10.11) gives our final likelihood form

$$L = L_0[\psi, \psi^*, w] + L_0[\psi, \psi^*, w^*], \tag{10.11b}$$

where functional

$$L_0[\psi, \psi^*, w] \equiv \int d\boldsymbol{r}\, dt\, \psi^*(\boldsymbol{r}, t)\psi(\boldsymbol{r}, t) \sum_{m=1}^{M} \delta(t - t_m) \ln[w(\bar{\boldsymbol{r}} - \boldsymbol{r})]. \quad (10.12)$$

Note that the sifting of the Dirac delta functions $\delta(t - t_m)$ collapses the $\int dt$ back into the sum over values $t_m$ in Eq. (10.11). Thus, by Eqs. (10.11b) and (10.12) the likelihood $L$ has two distinct contributions, one due to the instrument amplitude function $w$ and the other due to its complex conjugate $w^*$.

## 10.6 EPI variational principle with measurements

We are now ready to find the effects of the measurements upon the EPI output wave functions. Previously, we found that the log-likelihood function measures the quality of the data. From our knowledge-based viewpoint, *the higher the quality of the data, the stronger should be its effect upon the EPI principle*. The question is, how do we implement this viewpoint?

Equation (0.39) shows how to constrain an extremum principle with knowledge of data. Accordingly, we weight and add the log-likelihood terms $L_0[\psi, \psi^*, w]$ and $L_0[\psi, \psi^*, w^*]$ to the information $K_0$ given by Eq. (4.27). The resulting EPI variational principle will show a tradeoff between extremizing the original information $K_0$ and the new, add-on log-likelihood terms. The larger (stronger) are the log-likelihood values the more will they be extremized, *at the expense of* $K_0$; and vice versa.

Accordingly, the EPI variational principle Eq. (3.16) becomes

$$K_0[\psi, \psi^*] + \alpha_1 L_0[\psi, \psi^*, w] + \alpha_2 L_0[\psi, \psi^*, w^*] = extrem. \quad (10.13)$$

The square brackets identify the amplitude functions that functionals $K_0$ and $L_0$ depend upon. Parameters $\alpha_1$ and $\alpha_2$ are Lagrange multipliers that are to be fixed so as to agree with the form of Feynman–Mensky theory, wherever possible.

## 10.7 Klein–Gordon equation with measurements

The Lagrangian (or information density) for the problem is the total integrand of Eq. (10.13), obtained by the use of Eqs. (4.27) and (10.12). Substituting this Lagrangian into the Euler–Lagrange Eq. (4.24) gives

$$-c^2\hbar^2\left(\nabla - \frac{ie\boldsymbol{A}}{c\hbar}\right)\cdot\left(\nabla - \frac{ie\boldsymbol{A}}{c\hbar}\right)\psi + \hbar^2\left(\frac{\partial}{\partial t} + \frac{ie\phi}{\hbar}\right)^2 \psi + m^2c^4\psi$$
$$+ \psi \sum_{m=1}^{M} \delta(t - t_m)\{\beta_1 \ln[w(\bar{\boldsymbol{r}} - \boldsymbol{r})] + \beta_2 \ln[w^*(\bar{\boldsymbol{r}} - \boldsymbol{r})]\} = 0. \quad (10.14)$$

This is the Klein–Gordon Eq. (4.28) plus two measurement terms. The latter terms have new constants $\beta_1$, $\beta_2$, linearly related to $\alpha_1$, $\alpha_2$. These need to be determined. Subscript $n$, present in Eq. (4.28), is dropped here since there is only one complex component $\psi_1 \equiv \psi$ (Sec. 4.1.22) in this scalar amplitude case.

Equation (10.14) is extremely non-linear in time. It shows that *between* measurements, i.e., for times $t \neq (t_m, t_m + dt)$, the ordinary Klein–Gordon Eq. (4.28) re-emerges. Hence, usual EPI output equations describe the phenomenon between, rather than during, measurements. This is consistent with the analysis in Sec. 3.8, where it was found that EPI determines the phenomenon at the input space to the instrument, i.e., just as the measurement is initiated (but before the data value is acquired). The generalization that Eq. (10.14) gives is that the EPI output must define the phenomenon, as well, *after* the measurement; that is, after the output convolution Eq. (3.51) has been attained.

Equation (10.14) also shows that *at* the measurements, i.e., for values of $t = (t_m, t_m + dt)$, the constraint terms infinitely dominate the expression. This is very non-linear behaviour, and true of other theories of measurement as well (see the edited books cited as references at the beginning of the chapter). We defer evaluating this case until taking its non-relativistic limit in Sec. 10.9.

The foregoing results are obtained by use of the first EPI principle, Eq. (3.16). We now proceed to the second half of EPI, Eq. (3.18). Unfortunately, this development has not yet been carried through, and can only be sketched. In analogy with the development in the preceding section, this should lead to properties of the *Dirac equation with measurements*.

## 10.8 On the Dirac equation with measurements

In Sec. 4.2.5, the total Klein–Gordon Lagrangian was factored to produce the "pure" Dirac equation (the equation between measurements). Here, by comparison, the Lagrangian is the total Lagrangian of Eq. (10.13); *this, then, requires factoring*. As in Chapter 4, replace scalar functions $\psi$ and $w$ by vector functions $\boldsymbol{\psi}$ and $\mathbf{w}$. Factorization approaches are given in Matveev and Salle (1991, pp. 26, 27). We expect one factor to contain amplitudes $\boldsymbol{\psi}$ and $\mathbf{w}$, and the other $\boldsymbol{\psi}^*$ and $\mathbf{w}^*$. Equate, in turn, each factor to zero (by EPI principle Eq. (3.18)). The first factor will yield a first-order differential wave equation for the particle (involving wave function $\boldsymbol{\psi}$ and instrument function $\mathbf{w}$), while the second will yield a first-order differential equation for the anti-particle (in terms of wave function $\boldsymbol{\psi}^*$ and instrument function $\mathbf{w}^*$). The latter effects are analogous to those of Sec. 4.2.6.

Also, in analogy with Eq. (4.38), the wave function for the anti-particle will be that for the direct particle as evaluated at $-t$, and the anti-particle will be measured by the instrument function as evaluated at $-t$. The anti-particle both travels backward in time *and is detected* backward in time. The latter effect may be clarified as follows.

The physical behavior of the anti-particle completely mirrors that of the particle (Feynman and Weinberg, 1993), so the anti-particle is often described as a "virtual" image of the particle. This means that, when the particle is measured at time $t$, *automatically* the anti-particle is measured at time $-t$; no new measurement is actually made. The anti-particle measurement is as virtual as the particle it measures.

However, the "virtual" particle and its measurement are as "real" physically as are the "actual" particle and its measurement. One man's particle is another's anti-particle. This gives us a formal way of reconstructing the past, albeit relative to an arbitrary laboratory time origin and with no real gain in information: the past *measured* history of an anti-particle is redundant with the future measured history of its particle mate.

## 10.9 Schrödinger wave equation with measurements

For simplicity, consider the case of a general scalar potential $\phi(\boldsymbol{r}, t)$ but a vector potential $\boldsymbol{A}(\boldsymbol{r}, t) = 0$. The non-relativistic limit of the Klein–Gordon part of Eq. (10.14) can be taken as before (see Appendix G). The result is a SWE (G8) (or (10.30)) plus measurement terms,

$$i\hbar \frac{\partial \psi}{\partial t} = -\frac{\hbar^2}{2m} \nabla^2 \psi + e\phi\psi$$

$$+ \psi \sum_{m=1}^{M} \delta(t - t_m)\{\varepsilon_1 \ln[w(\bar{\boldsymbol{r}} - \boldsymbol{r})] + \varepsilon_2 \ln[w^*(\bar{\boldsymbol{r}} - \boldsymbol{r})]\}. \quad (10.15)$$

The constants $\varepsilon_1$, $\varepsilon_2$ are the new, adjustable Lagrange multipliers.

### *10.9.1 Transition to Feynman–Mensky theory*

The result (10.15) is analogous to that of Mensky (1993), provided that the constants $\varepsilon_1$, $\varepsilon_2$ take the values

$$\varepsilon_1 = i\hbar, \quad \varepsilon_2 = 0. \quad (10.16)$$

Setting $\varepsilon_2 = 0$, in particular, eliminates from Eq. (10.15) the only complex-conjugated term in the equation, which seems correct on the basis of simplicity alone. However, this term should remain present in the measurement-inclusive Dirac equation of Sec. (10.8) since there both $\psi$ and $\psi^*$ are present. The values Eq. (10.16) bring Eq. (10.15) into the Feynman–Mensky form

$$i\hbar\frac{\partial\psi}{\partial t} = -\frac{\hbar^2}{2m}\nabla^2\psi + e\phi\psi + i\hbar\psi\sum_{m=1}^{M}\delta(t-t_m)\ln[w(\bar{\boldsymbol{r}}-\boldsymbol{r})]. \quad (10.17)$$

We analyze it next.

### 10.9.2 Analysis

As with the Klein–Gordon measurement Eq. (10.14), Eq. (10.17) is highly non-linear in its time-dependence. At times $t \neq t_m$ between measurements, because $\delta(t-t_m) = 0$, (10.17) becomes the ordinary Schrödinger wave Eq. (10.30). However, *at* a time $t$ within a measurement interval $(t_m, t_m + dt)$ the sum in Eq. (10.17) collapses to its $m$th term, and dominates the right-hand side, so that the equation becomes

$$i\hbar\frac{\partial\psi}{\partial t} = i\hbar\psi\delta(t-t_m)\ln[w(\bar{\boldsymbol{r}}(t)-\boldsymbol{r}(t))]. \quad (10.18)$$

We inserted back the time-dependences within the argument of $w$ for clarity. Now multiply both sides of Eq. (10.18) by $dt/(i\hbar\psi)$ and integrate $dt$ over the measurement interval $(t_m, t_m + dt)$, $dt \geqslant 0$.

The left-hand side becomes

$$\int_{t_m}^{t_m+dt} dt\,\frac{1}{\psi}\frac{\partial\psi}{\partial t} = \int_{\psi_0}^{\psi_+}\frac{d\psi}{\psi} = \ln\left(\frac{\psi_+(\boldsymbol{r}_m)}{\psi_0(\boldsymbol{r}_m)}\right), \quad (10.19a)$$

$$\psi_+(\boldsymbol{r}_m) \equiv \psi(\boldsymbol{r}_m, \bar{\boldsymbol{r}}_m, t_m + dt) \equiv \psi(\boldsymbol{r}_m, t_m + dt), \quad (10.19b)$$

$$\psi_0(\boldsymbol{r}_m) \equiv \psi(\boldsymbol{r}_m, t_m). \quad (10.19c)$$

The notation $\psi_+(\boldsymbol{r}_m)$ denotes the wave function as evaluated within the measurement interval. The joint event $(\boldsymbol{r}_m, \bar{\boldsymbol{r}}_m)$ shown in the argument of $\psi$ comes about because, by hypothesis, the event $\boldsymbol{r}_m$ occurs within the measurement interval, i.e., co-jointly with the occurrence of the measurement $\bar{\boldsymbol{r}}_m$. The single event $\boldsymbol{r}_m$ is equivalent to the joint event $(\boldsymbol{r}_m, \bar{\boldsymbol{r}}_m)$.

The right-hand side of Eq. (10.18) becomes, because of the sifting property of the delta function,

$$\ln w(\bar{\boldsymbol{r}}_m - \boldsymbol{r}_m). \quad (10.20)$$

### 10.9.3 Bohm–Von Neumann measurement theory

The net result is that

$$\ln\left(\frac{\psi_+(\boldsymbol{r}_m)}{\psi_0(\boldsymbol{r}_m)}\right) = \ln[w(\bar{\boldsymbol{r}}_m - \boldsymbol{r}_m)] \quad \text{or} \quad \psi_+(\boldsymbol{r}_m) = \psi_0(\boldsymbol{r}_m)w(\bar{\boldsymbol{r}}_m - \boldsymbol{r}_m). \quad (10.21)$$

With the understanding that all coordinates are evaluated at time $t = t_m$, the latter simplifies in notation to

$$\psi_+(\boldsymbol{r}) \equiv \psi(\boldsymbol{r}, \overline{\boldsymbol{r}}) = C\psi_0(\boldsymbol{r})w(\overline{\boldsymbol{r}} - \boldsymbol{r}) \tag{10.22}$$

(cf. Eq. (3.53)). Here we have regained the joint-event notation $(\boldsymbol{r}, \overline{\boldsymbol{r}})$ of Eq. (10.19b), as well as the constant $C$ multiplier of $w$ that was suppressed after Eq. (10.10).

Equation (10.22) agrees with the well-known measurement theory of Bohm (1952a, b), Von Neumann (1955), and others. It shows that the wave function $\psi(\boldsymbol{r}, \overline{\boldsymbol{r}})$ at the measurement is essentially the wave function $\psi_0(\boldsymbol{r})$ just before the measurement times the instrument amplitude function. Or, the effect of a measurement is a *separation* of the system wave function from the instrument function. The effect is commonly called the "reduction of the wave function" by the measurement, since the product should ordinarily be narrower than the input amplitude function $\psi_0$ (see Secs. 10.9.6 and 10.9.7).

A similar wave function separation, Eq. (3.53), was found in the analysis of the optical measuring device. In fact Eq. (10.22) agrees with Eq. (3.53). The right-hand sides agree identically, after the use of Eq. (10.6). The left-hand sides also agree, when one recalls that $\boldsymbol{r}$ and $\overline{\boldsymbol{r}}$ are, here, three-dimensional generalizations of $x$ and $\overline{y}$.

### 10.9.4 Value of constant C

We assume that the measurement procedure is passive: the inferred PDF $p_+(\boldsymbol{r}) \equiv |\psi_+(\boldsymbol{r})|^2$ at a measurement cannot exceed the PDF $p_0(\boldsymbol{r}) \equiv |\psi_0(\boldsymbol{r})|^2$ prior to the measurement,

$$|\psi_+(\boldsymbol{r})|^2 \leqslant |\psi_0(\boldsymbol{r})|^2. \tag{10.23}$$

Then, by Eq. (10.22), the constant $C$ must have a value such that

$$|Cw(\gamma)| \leqslant 1, \quad \text{all } \gamma, \quad \text{or } C|w_{\max}| = 1, \tag{10.24}$$

where $w_{\max}$ is the maximum value of $w$ over its argument.

### 10.9.5 Initialization effect

In Sec. 10.9.2 we found that, between measurements, the wave function (10.17) is the simple Schrödinger equation, whereas *at* each measurement the wave function is "instantaneously" perturbed into a form obeying Eq. (10.22). The upshot is that *each measurement merely re-initializes* the wave function, after which it freely "evolves" according to the ordinary Schrödinger wave equation. Hence, with the understanding that the notation $\psi$ means the output $\psi_+$ of the *previous* or $(m-1)$th measurement, we may rewrite Eq. (10.17) at the $m$th measurement as

$$i\hbar\frac{\partial\psi}{\partial t}=-\frac{\hbar^2}{2m}\nabla^2\psi+e\phi\psi+i\hbar\psi\delta(t-t_m)\ln[w(\bar{\boldsymbol{r}}-\boldsymbol{r})]. \tag{10.25}$$

### 10.9.6 A one-dimensional example

In Fig. 10.1 we sketch a one-dimensional example $\boldsymbol{r}=x$. Figure 10.1(a) shows the system probability amplitude curve $\psi_0(x)$ just prior to the measurement, and the instrument amplitude curve $w(x)$ centered upon the measurement $\bar{x}$. According to the effect Eq. (10.22), the result of the measurement is a new probability amplitude curve $\psi_+(x)$ obeying

$$\psi_+(x)\equiv\psi(x,\bar{x})=C\psi_0(x)w(\bar{x}-x) \tag{10.26a}$$

(cf. Eq. (3.53)). This is shown in Fig. 10.1(b). The new curve $\psi_+(x)$ is narrower than the input $\psi_0(x)$, showing that the measurement has narrowed the realm of possibilities for coordinate $x$ of the particle. An increased amount of position information has been acquired by the measurer. (Note that this is in the specific sense of *Fisher information* about the unknown particle position; Shannon information does not measure the information about an unknown *parameter*.) The *narrowed* product form (10.26a) and the *gain* in Fisher information are fundamental effects of the measurement.

The probability amplitude $\psi(\bar{x})$ for the measurement value $\bar{x}$ of an ideal value $x$ obeys

$$\psi(\bar{x})\equiv\int dx\,\psi(x,\bar{x})=C\int dx\,\psi_0(x)w(\bar{x}-x) \tag{10.26b}$$

by Eq. (10.26a). This has the form of a convolution, and checks with the optical instrument result Eq. (3.51).

### 10.9.7 Exercises

One observes that the output curve has not generally been reduced to a single point of complete "collapse." That would have required a narrower measurement function $w(x)$, in fact to the limit of a delta function. Even the maximum probable position value after the measurement, indicated as coordinate $\hat{x}$ in Fig. 10.1(b), is not quite the same as the measurement value $\bar{x}$. Why?

In general, the output curve $\psi_+(x)$ in Fig. 10.1(b) is narrower than *either* curve $\psi_0(x)$ or $w(x)$ of Fig. 10.1(a). Prove that this is the case if both of the latter curves have the normal form (Gaussian with finite mean). (*Hint*: Form the product and complete the square in the exponent.)

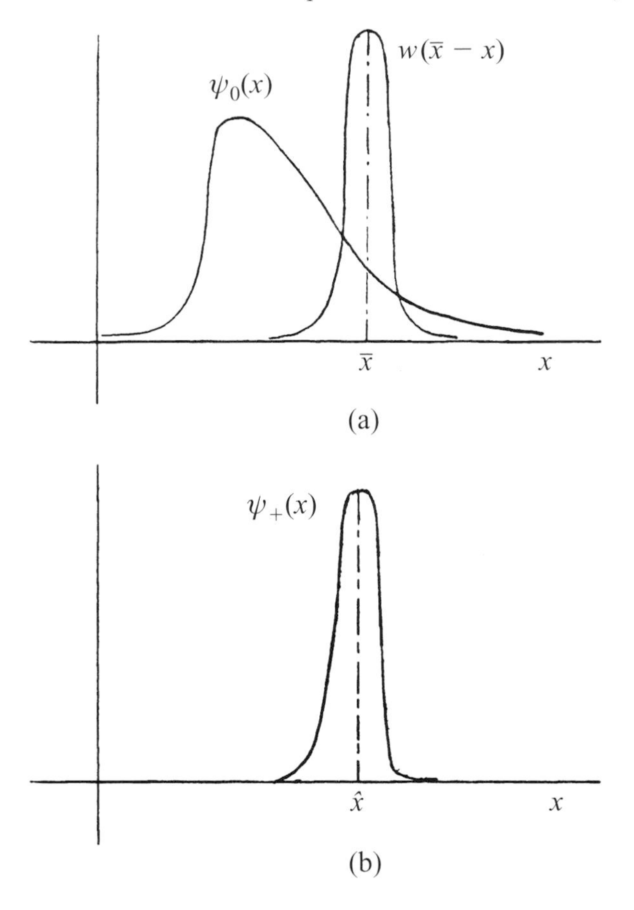


Fig. 10.1 (a) Wave function prior to measurement, and instrument measurement function; (b) wave function at measurement.

### *10.9.8 Passive Gaussian case*

We continue next with the specifically Gaussian case

$$w(x) = \frac{1}{(2\pi\sigma^2)^{1/4}} e^{-x^2/(4\sigma^2)}, \quad C = (2\pi\sigma^2)^{1/4}, \tag{10.27}$$

where we used Eq. (10.24) to find $C$. Gaussian noise is, of course, pervasive in nature because of the central limit theorem. The one-dimensional version of Eq. (10.25), including the multiplier $C$ of $w$ given by Eq. (10.27), becomes

$$i\hbar \frac{\partial \psi}{\partial t} = -\frac{\hbar^2}{2m}\psi'' + e\phi\psi - i\hbar\psi\delta(t - t_m)\frac{(\bar{x} - x)^2}{4\sigma^2}. \tag{10.28}$$

This shows that, in the case of a very poor, passive measuring instrument where the error variance obeys $\sigma^2 \to \infty$, the between-measurements Schrödinger wave equation essentially results once again. Making a measurement at

$t = (t_m, t_m + dt)$ with such an instrument is equivalent to making almost no measurement at all. This is a reasonable result, and again shows the effect of knowledge (here a lack thereof) upon physics.

This result is confirmed, as well, by Eq. (10.26a), which becomes

$$\psi_+(x) = \psi_0(x)e^{-(\bar{x}-x)^2/(4\sigma^2)} \approx \psi_0(x)\left[1 - \frac{(\bar{x}-x)^2}{4\sigma^2}\right] \tag{10.29}$$

in the limit $\sigma^2 \to \infty$ of a very poor instrument. Here $\psi_+$ is barely changed from $\psi_0$ by knowledge of the measurement $\bar{x}$. It suffers an *infinitesimal change* from the input state.

### 10.9.9 Time-reversal effects

(This section may be skipped in a first reading.) Between measurement times $t_m$, Eq. (10.17) becomes the ordinary Schrödinger wave equation

$$i\hbar \frac{\partial \psi}{\partial t} = -\frac{\hbar^2}{2m}\nabla^2\psi + e\phi\psi, \quad \psi = \psi(\boldsymbol{r}, t). \tag{10.30}$$

Under reversal of the time coordinate, i.e.,

$$t \to -t, \quad dt \to -dt, \tag{10.31}$$

and with complex conjugation, Eq. (10.30) becomes

$$i\hbar \frac{\partial \psi^*}{\partial t} = -\frac{\hbar^2}{2m}\nabla^2\psi^* + e\phi\psi^*, \quad \psi^* = \psi^*(\boldsymbol{r}, -t). \tag{10.32}$$

This assumes a real potential function $\phi$ as well. Comparing Eqs. (10.30) and (10.32) shows that Eq. (10.32) is again the Schrödinger wave equation, now in the wave function $\psi^*(\boldsymbol{r}, -t)$. Hence, if $\psi(\boldsymbol{r}, t)$ is a solution to the Schrödinger wave equation, so is $\psi^*(\boldsymbol{r}, -t)$, the "time-reversed" solution.

One might think that, since measured lab time is usually regarded as moving forward, the time-reversed solution has no physical reality. However, it does indeed have some observed effects, one of which is the so-called "Kramers degeneracy" of energy eigenfunctions (Merzbacher, 1970, p. 408). Another, as we will find next, is that of a measured time-reversed wave.

The Feynman–Mensky Eq. (10.17) has been derived as the non-relativistic limit of Eq. (10.14), which was derived by EPI after varying the *complex conjugate* $\psi^*$ of the wave function in Eq. (10.13): see Eqs. (4.24) and (4.25). But the Euler–Lagrange approach for a complex function $\psi$ should give valid results irrespective of whether it is $\psi$ or $\psi^*$ that is varied. Accordingly, let us now vary $\psi$ in Eq. (10.13). Carrying through the analogous steps to those that gave rise to Eq. (10.17) now gives, after a complex conjugation,

$$i\hbar \frac{\partial \psi}{\partial t} = -\frac{\hbar^2}{2m} \nabla^2 \psi + e\phi\psi - i\hbar\psi \sum_{m=1}^{M} \delta(t - t_m) \ln[w^*(\bar{\boldsymbol{r}} - \boldsymbol{r})]. \tag{10.33}$$

For times $t \neq (t_m, t_m + dt)$ between measurements, this is the same as Eq. (10.17). However, at a measurement time things are slightly different. Repeating the analogous steps to Eqs. (10.18)–(10.22) now gives

$$\psi(\boldsymbol{r}, t + dt) = C\psi(\boldsymbol{r}, t)/[w^*(\bar{\boldsymbol{r}} - \boldsymbol{r})]. \tag{10.34}$$

Cross-multiplying Eq. (10.34) shows that, by the reasoning of Secs. 10.9.3, 10.9.6, and 10.9.7, $\psi(\boldsymbol{r}, t)$ is now narrower than $\psi(\boldsymbol{r}, t + dt)$. Or, the unmeasured wave is *narrower than* the measured wave, where we assumed that $dt > 0$. Consequently there is more Fisher information about particle position in the *unmeasured* state than in the *measured* state. By the reasoning in Sec. 10.9.6, this is unacceptable physically.

We noticed that the preceding result assumed a positive time differential $dt$. Other time values were also assumed to be positive. To try to obtain reasonable results, we now evaluate Eq. (10.34) for *negative* times and time intervals, i.e.,

$$t \to -t, \quad dt \to -dt, \quad t_m \to -t_m, \quad (t_m, t_m + dt) \to (-t_m - dt, -t_m). \tag{10.35}$$

(See, by comparison, Eqs. (10.31) for the time-reversed Schrödinger wave Eq. (10.32).) Since $-t_m - dt \leqslant -t_m$, $\psi(\boldsymbol{r}_m, -t_m - dt)$ and $\psi(\boldsymbol{r}_m, -t_m)$ now represent the wave functions just prior to and at the measurement, respectively. Equation (10.34) becomes, with the replacements (10.35) and after a cross-multiplication and complex conjugation,

$$\psi^*(\boldsymbol{r}, -t) = B\psi^*(\boldsymbol{r}, -t - dt) w(\bar{\boldsymbol{r}} - \boldsymbol{r}) \tag{10.36}$$

(subscripts $m$ suppressed as before). $B$ is a constant. This, now, shows the desired effect: the wave function $\psi^*(\boldsymbol{r}, -t)$ at the measurement is *narrower than* the wave function $\psi^*(\boldsymbol{r}, -t - dt)$ before measurement. Once again the Fisher information on position has been increased.

Also, Eq. (10.36) shows that, for negative times, the measurement function $w(\bar{\boldsymbol{r}} - \boldsymbol{r})$ acts as a "transfer function" of measurement for the system wave function $\psi^*$. That is, when the transfer function operates upon the system wave function in its unmeasured state, the result is the system wave function in its measured state. By symmetry one would have expected the complex conjugated quantity $w^*(\bar{\boldsymbol{r}} - \boldsymbol{r})$ to have filled that role here.

These results are for negative times. It is interesting that, even for positive times, Eq. (10.22) exhibits the same transfer function $w(\bar{\boldsymbol{r}} - \boldsymbol{r})$. This seems to be consistent with the Feynman–Mensky Eq. (10.17), the basis for these effects, which likewise contains $w(\bar{\boldsymbol{r}} - \boldsymbol{r})$ rather than its complex conjugate.

Equation (10.36) may be considered to be a time-reversed measurement of the wave function $\psi^*$ that mirrors a corresponding measurement at *positive* time of the wave function $\psi$ in Eq. (10.34).

## 10.10 Overview

The primary aim of this chapter was to develop a covariant, four-dimensional theory of measurement. In contrast with other theories of measurement, and consistently with the covariant approach of EPI, we allowed the measurement times themselves to suffer from random noise. A secondary aim was to confirm the one-dimensional measurement effects that were found in Sec. 3.8 from a different viewpoint.

The culmination of this measurement theory, Eq. (10.14) or Eq. (10.17), confirms that, in the input space (see Fig. 3.3) of a measuring instrument (i.e., for time $t < t_m$), the laws of physics are exactly those that EPI predicts. In this case it is the Klein–Gordon law, as derived in Chapter 4, or the Schrödinger wave equation, as derived in Appendix D. The physical law *that contains* the measurement information is obtained *at* the measurement time $t = t_m$, with the result Eq. (10.22). This amounts to simply *re-initializing the wave function* (Sec. 10.9.5).

The result Eq. (10.22) shows *how* mathematically the EPI amplitude law is changed by the measurement, namely through simple multiplication by an instrument amplitude function. As we found, this confirms Eq. (3.53), the corresponding result for the optical measuring device of Sec. 3.8. Equation (10.22) has been known for over 40 years, of course, independently of the measurement theory of this book (Bohm, 1952a; Von Neumann 1955). The multiplication amounts to a transfer of the system statistics from an unmeasured to a measured state, where the instrument amplitude function $w$ acts as the transfer function.

This view of measurement, supplemented by that in Sec. 3.8, defines EPI as a physical process. Its progression is as follows.

(1) The measuring instrument's probe particle perturbs the input, or object, amplitude function $\psi_0$ (see Sec. 3.8.1).
(2) Perturbing the complex wave function $\psi_0$ means, by Eq. (2.24), perturbing the real amplitude functions $q_n$, $n = 1, 2$, by amounts $\delta q_n(\mathrm{x})$.
(3) These cause a perturbation $\delta I$ in the intrinsic information, through the defining Eq. (2.19) for $I$.
(4) This, in turn, causes an equal perturbation $\delta J$ in the bound information through the conservation law Eq. (3.13). As we saw in Sec. 3.8, *this law is identically obeyed in the presence of a unitary transformation* between conjugate spaces such as $(\boldsymbol{r}, ct)$ and $(\boldsymbol{\mu}, E/c)$.

(5) The invariance principle (Sec. 3.4.5) causes the information densities $i_n$, $j_n$ to obey proportionality, so that their sum total (integrated) values obey

$$I = \kappa J. \tag{10.37}$$

Alternatively, this equation follows identically, with $\kappa = 1$, when the invariance principle is the unitary transformation mentioned in item (4).

(6) Since the perturbations $\delta I$ and $\delta J$ are equal (point (4) above), they must obey the EPI extremum principle Eq. (3.16),

$$I - J \equiv K = extrem. \tag{10.38}$$

(see Sec. 3.4.1). Together, Eqs. (10.37) and (10.38) comprise the EPI principle.

(7) The EPI principle is physically enacted. The "knowledge game" of Secs. 3.4.11–3.4.13 is played if the coordinates are real, or if the time coordinate separates out due to energy conservation (see exercise in Sec. 3.4.12).

(8) The outcome of the game is the physical occurrence of the law that defines the $q_n(\mathbf{x})$ or (by Eq. (2.24)) the complex $\psi_n(\mathbf{x})$. Any unitless constants of the law are fixed by the game corollary (Sec. 3.4.14).

(9) This probability law is randomly sampled by the measuring instrument, giving the data value, the acquisition of which was the aim of the whole procedure. This means that the act of seeking data is physically self-realizing; it elicits from nature the very probability law that randomly forms the data. One might say that EPI is an "epistem*active*" process, in contrast with the more passive connotations of the word "epistem*ology*." That is, a question is asked, and the response is not only the requested answer, but the mechanism for determining the answer as well. A kind of local reality is so formed (Sec. 3.10).

(10) The solution $\psi_n(\mathbf{x})$ and resulting information densities $i_n$, $j_n$ define the future evolution and *dynamics* (Sec. 3.7) of the system until a new measurement is made.

These results imply two general modes of operation of the EPI principle: (a) a "passive" mode, whereby EPI is used as a computational tool (as in other chapters); and (b) an "active" mode, whereby real data are sought and EPI executes as a *physical process*. Mode (a) is utilized in a quest for *knowledge* of the physical law governing a measurement scenario. Mode (b) occurs during a quest for real data. As we saw, either quest elicits the law that generates the requested data.

We saw that an act of measurement elicits a physical law. The measurement was of a well-defined particle in the presence of a well-defined potential function. Does EPI yield a physical law under less definite measurement conditions? How does a reduced level of knowledge affect the law?

For simplicity, specify the particle position by a single coordinate $x$. We found before (Secs. 3.8.5 and 10.1) that the physical law will follow from any perturbation of the intrinsic wave function of the particle as, for example, occurs in the input port of the measuring instrument. However, some "measurements" are of a less precise kind than this. Rather than measuring the position of a particle in the presence of a definite potential function $\phi(x, t)$, the measurement might be so poor, or indirect, as to effectively occur in the presence of any one randomly chosen potential over a class of possible potentials $\phi_i(x, t)$, $i = 1, \ldots, n$. In other words, any one of a class of possible particle interactions randomly took place. This occurs, e.g., if the "measurement" determines that the particle lies *somewhere* within a large container, different parts of which are subjected to different potentials $\phi_i(x, t)$. This "measurement" implies the knowledge that there is a range of possible *ideal* positions $\boldsymbol{\theta}_i$, $i = 1, \ldots, n$ present for the particle. This is a mode of operation of EPI that is, experimentally, less well defined than mode (b) above, which has a single ideal position $\boldsymbol{\theta}$. What will be the effective output amplitude law $\psi_{\text{eff}}(x, t)$?

We work in the non-relativistic limit. In the case of a definite measurement of position $\boldsymbol{\theta}_i$ in the presence of a definite potential $\phi_i(x, t)$, the output law will be the ordinary SWE (10.30) for the wave function $\psi(x|i)$.

But here the potential $\phi_i(x, t)$ arises randomly. Suppose that it occurs according to a probability amplitude $c_i$. The $c_i$, $i = 1, \ldots, n$, define the degree of vagueness of the measurement. If all $c_i$ are equal, then the measurement is maximally vague; a case of maximum ignorance. At the opposite extreme, if all $c_i = 0$ except for one $c_j = 1$, then we are back to definite measurement, mode (b).

Since the SWE is linear in $\psi(x|i)$, multiplying the equation by $c_i$ and summing over $i$ gives a net SWE (not shown) in an effective probability amplitude $\psi_{\text{eff}}(x, t)$ obeying

$$\psi_{\text{eff}}(x) = \sum_{i=1}^{n} c_i \psi(x|i). \tag{10.39}$$

This is due to an effective potential $\phi_{\text{eff}}(x, t)$ obeying

$$\phi_{\text{eff}}(x, t) = \frac{\sum_i \phi_i(x, t) c_i \psi(x|i)}{\sum_i c_i \psi(x|i)}. \tag{10.40}$$

Equation (10.39) has the form of a "coherent," weighted superposition of amplitudes. Notice that, in the case of all $c_i \geqslant 0$, it represents a broader PDF on $x$ than in the case of a single definite measurement. Also, the effective potential (10.40) is widened, or blurred, by the superposition of individual potentials. These effects reflect an increased state of ignorance of the measurer due to the vagueness of the measurement scenario. Hence, under vague measurement conditions the output amplitude law is vague as well, although the output physical law is still the well-defined SWE.

Interestingly, there are cases where the coherence effect (10.39) is lost, and recourse must be made to a "density function" type of average (Louisell, 1970),

$$\rho(x, x') = \sum_{i=1}^{n} P_i \psi^*(x|i)\psi(x'|i), \quad P_i = |c_i|^2. \tag{10.41}$$

We see that, in the limit $x \to x'$, Eq. (10.41) becomes

$$p_{\text{eff}}(x) = \sum_i P_i p(x|i), \quad p(x|i) \equiv |\psi(x|i)|^2. \tag{10.42}$$

This is an "incoherent" or classical-statistics superposition of states. Comparison with Eq. (10.39) is germane.

If we let the difference $\Delta x \equiv x' - x \to 0$, as previously, then, from Sec. 1.4.3 we suspect that yet another transition to Fisher information is at hand; in this case, the average over all interaction states $i$. The specific relation is

$$\int dx \lim_{x' \to x} \frac{\partial^2 \rho(x, x')}{\partial x\, \partial x'} = \frac{1}{4} \sum_{i=1}^{n} P_i I_i \equiv \frac{\langle I \rangle}{4}. \tag{10.43}$$

This may easily be verified by operating $\partial^2/\partial x\, \partial x'$ on Eq. (10.41), multiplying by the Dirac $\delta(x - x')$, integrating both sides $dx'$ and using the sifting property Eq. (0.63b) of the delta function, then integrating both sides $dx$. The representation Eq. (3.38) for information $I$ must also be used. Note that, if all $P_i = const. = 1/n$, indicating maximum ignorance as to system state, $\langle I \rangle = I$ in Eq. (10.43).

We conclude that knowledge of the density matrix for a measurement scenario implies knowledge of the average Fisher information over the possible measurement states. This is another intriguing connection between quantum mechanics and classical measurement theory. See also in this regard Frieden (2002).

An underlying premise of EPI is that *the observer is part of the observed phenomenon* (see Sec. 10.1). Thus, a full quantum treatment of the data collection process should include the observer's internal interactions with the observed data values. This would require at least a five-dimensional analysis of the measuring apparatus and the observer as they jointly interact: four-dimensional space-time, plus one "reading" internal to the observer. It is interesting that we did not have to take such an approach in the preceding – the simple use of Lagrange constraint terms served to avoid it. In effect, inputs of knowledge replaced detailed biological analysis.

However, a one-dimensional analysis that directly *includes* the observer may be performed using the coherent optical model of measurement given in Sec. 3.8. Simply regard the coherent optical device of Fig. 3.3 as a model for the visual workings of the eye. That is, the particle is in the object space of the eye, the lens L is the lens of the eye, the measurement space M is the retina, the data value $\bar{y}$ corresponds to a retinal excitation at that coordinate, etc. This provides a model for the eye–particle measurement interaction. Then, as found in Sec. 3.8, visual observation of the particle at $\bar{y}$ elicits its physics (the appropriate wave equation), both in the input space to the eye and at the retina (the latter by Eq. (3.51)). These effects serve to confirm the hypothesis made by Wheeler at the beginning of the book.

The development of this chapter did not achieve all our aims. We analyzed the EPI measurement procedure in the particular scenario of quantum mechanics, the subject of Chapter 4. However, other chapters used EPI to derive physical laws arising out of "non-quantum" phenomena, such as classical electromagnetic theory, gravitational theory, etc. Would these be amenable to the same kind of measurement theory as developed here?

Also, there is an unresolved issue of dimensionality. Neither active nor passive observers can know what ultimate dimension to use in characterizing each measurement $\mathbf{x}$ of a particle or field. Coordinate $\mathbf{x}$ in definition (2.19) of $I$ was made to be four-dimensional, by the dictates of covariance (Sec. 3.5.8). However, it might prove more productive to use a generally $M$-dimensional version of (2.19). In fact, *EPI operates on any level of dimensionality, giving the projection of the "true" law into the M dimensions of the measurement space*. For example, use of $M = 1$ for a particle gives rise to the one-dimensional Schrödinger wave equation (see Appendix D), whereas use of $M = 4$ gives rise to either the Dirac equation or the Klein–Gordon equation (Chapter 4). Use of higher values of $M$, such as values of 10 or more, may give rise to string theory or a variant of it. Thus, EPI gives rise to a projection of the true law into the measurement space assumed. As with the famous shadows on the walls of Plato's cave, the ultimate, "true" dimensionality of the phenomenon tends to be an unknown. However, as we have seen, the game corollary (Sec. 3.4.14) can provide reasonable answers.

# 11

# Research topics

## 11.1 Scope

Preceding chapters have dealt with relatively elementary applications of the EPI principle. These are problems whose answers are well established within known experimental limits. Historically, such problems constituted "first tests" of the principle. As we found, EPI agreed with the established answers, with several extensions and refinements. The next step in the evolution of the theory is to apply it "in the large," i.e., to more difficult problem areas, those of active, current research by the physics and engineering communities.

Three such are the phenomena of (i) quantum gravity, (ii) turbulence at low Mach number, and (iii) elementary particle theory. We show, next, the current status of EPI work on these problems. Aspects of the approaches that are, as yet, uncertain are pointed out in the developments. It is expected that final EPI versions of the theory will closely resemble the first "tries" given below.

## 11.2 Quantum gravity

### *11.2.1 Introduction*

The central concept of gravitational theory is the metric tensor $g_{\mu\nu}(\mathbf{x})$. (For a review, see Exercise 6.2.5 and the material preceding it.) This defines the local distortion of space at a continuous four-coordinate $\mathbf{x}$. The Einstein field Eq. (6.68) permits the metric tensor to be computed from knowledge of the gravitational source – the stress-energy tensor $T_{\mu\nu}$ (Sec. 6.3.4). This is a deterministic view of gravity. That is, for a given stress-energy tensor and initial conditions, a given metric tensor results. This view holds over macroscopic distances $\mathbf{x}$.

However, at very small distances the determinism is lost. It is believed that the metric tensor fluctuates *randomly* on the scale of the Planck length, Eq. (6.22), the order of $10^{-33}$ cm. This corresponds to a local curvature of $10^{33}$ cm$^{-1}$.

If the metric tensor fluctuates, what amplitude functions $\mathbf{q}[g_{\mu\nu}(\mathbf{x})]$ govern the joint fluctuations of all $g_{\mu\nu}$ at all $\mathbf{x}$? This is a probability amplitude with massively joint statistics. The determination of this amplitude function is a major aim of the field of study called quantum gravity. Many astronomers believe the wave function to obey a differential equation called the "Wheeler–DeWitt" equation. We will attempt to show how this equation follows from the EPI procedure. As mentioned above, the derivation is not yet complete but should strongly indicate the direction to take.

### 11.2.2 Analogies with quantum mechanics

Assume, at first, a scalar quantum theory. It seems reasonable to model it after the one taken in Chapter 4 for deriving the scalar Klein–Gordon equation of relativistic quantum mechanics. Thus, there is a measurement space and its Fourier transform, a "momentum" space. Also, Parseval's theorem is again used to compute the bound information $J$ in momentum space as the equivalent amount of Fisher information $I$ in measurement space. To keep the derivation brief, we freely use results from seminal papers by DeWitt (1967) and by Hartle and Hawking (1983).

### 11.2.3 Measurements and perturbations

For simplicity we restrict our attention to spatially closed universes. This permits use of a three-metric $g_{ij}$ in place of the general four-metric $g_{\mu\nu}$. (Recall from Sec. 6.2.1 that Latin indices go from 1 to 3 whereas Greek indices go from 0 to 3.) Relations connecting $g_{ij}$ and $g_{\mu\nu}$ are given in DeWitt (1967).

As in the quantum mechanical development (Sec. 4.1.4), we pack the real probability amplitudes $\mathbf{q}$ as the real and imaginary parts of a *complex* scalar amplitude $\psi$. See Eq. (2.24). Presuming a scalar case, we use $N = 2$ functions $\mathbf{q}$ so that $\psi \equiv q_1 + iq_2$ (an irrelevant factor $1/\sqrt{2}$ is ignored). Thus we want to determine the wave function $\psi[g_{ij}(\mathbf{x})]$. It should be noted that each component $g_{ij}$ is defined at every possible space-time coordinate $\mathbf{x}$ throughout the Universe. Since coordinate $\mathbf{x}$ is continuous, $\psi[g_{ij}(\mathbf{x})]$ is actually a *functional* (Sec. 0.2). It represents the wave function for every possible universe of metric tensor values; indeed, an all-inclusive wave function!

The observer measures the $g_{ij}$ at a sequence of closely spaced points $\mathbf{x}$ throughout all of four-space. Thus, essentially an infinity of real measurements are made. Each measurement locally perturbs the overall wave function $\psi[g_{ij}(\mathbf{x})]$ with the result that it is perturbed everywhere, as $\delta\psi[g_{ij}(\mathbf{x})]$. As we saw (Chaps. 3 and 10), this excites the EPI process. Since a unitary transformation (Eq. (11.18)) will be found to connect the space of values $g_{ij}(\mathbf{x})$ with a physically meaningful momentum space, EPI again holds rigorously (See Secs. 3.8.5 and 3.8.7).

It is noted that the positions $\mathbf{x}$ are regarded as deterministic, in this problem.

### 11.2.4 Discrete aspect of the problem

Since the EPI approach is based upon a series of *discrete* measurements, we must replace the continuous position coordinate $\mathbf{x}$ by a fine subdivision $\mathbf{x}_1, \ldots, \mathbf{x}_N$ of such values. (A return to the continuous limit is, however, taken at the end of the derivation.) Thus, denote the value of the metric tensor $g_{ij}$ at the four-position $\mathbf{x}_n$ as the discrete quantity $g_{ijn}$. Also, then

$$\psi \equiv \psi[g_{ij}(\mathbf{x})] = \psi(g_{ijn}). \tag{11.1}$$

The wave function is now no longer a functional, simplifying matters. It is an ordinary function of the random variables $g_{ijn}$, as EPI requires.

### 11.2.5 Fisher coordinates for the problem

The first step of the EPI approach is to identify the physical coordinates of the problem. These correspond to coordinates $\mathbf{y}$, $\boldsymbol{\theta}$, and $\mathbf{x}$ in the basic data Eq. (2.1). In particular, what is the "ideal" value $\theta_{ijn}$ of $g_{ijn}$? Taking the cue from the quantum development in Sec. 4.1.2, we regard the ideal to be the classical, deterministic value $\overline{g}_{ijn}$ of the metric. This is the one that obeys the Einstein field Eq. (6.68) at a position $\mathbf{x}_n$. Then, since we want to find $\psi(g_{ijn})$, we must regard the $g_{ijn}$ as the fluctuation values $\mathbf{x}$ in Eq. (2.1). Hence, our intrinsic data are the $y_{ij}(\mathbf{x}_n)$, where

$$y_{ij}(\mathbf{x}_n) = \overline{g}_{ij}(\mathbf{x}_n) + g_{ij}(\mathbf{x}_n) \quad \text{or} \quad y_{ijn} = \overline{g}_{ijn} + g_{ijn} \tag{11.2}$$

for all $i$, $j$, $n$, and where the $\overline{g}_{ij}(\mathbf{x})$ obey the deterministic field Eq. (6.68).

### 11.2.6 The Fisher information in "Riem" space

Equation (2.19) expresses $I$ in flat, rectangular coordinate space $\mathbf{x}$. The implied metric in (2.19) is (1, 1, 1, 1). We now want to express $I$ in $g_{ijn}$ space.

Temporarily revert to the case of a *continuous* four-position $\mathbf{x}$, at which the Fisher coordinates are $g_{ij}$. This is a generally curved space of coordinates.

Curved space has a Riemannian metric and is called "superspace" or, more briefly, "Riem" space. Coefficients $G_{ijkl}$ obeying

$$G_{ijkl}(\mathbf{x}) \equiv \frac{1}{2\sqrt{g}}(g_{ik}g_{jl} + g_{il}g_{jk} - g_{ij}g_{kl}), \qquad g \equiv \det[g_{ij}] \tag{11.3}$$

may be regarded (DeWitt, 1967, Appendix A) as the metric of Riem space. We adopt this metric as well.

What is a length in the Riem space defined by this metric? Consider a tensor function $V^{ij}$ of generally all metric components $g_{kl}$ evaluated at a continuous position $\mathbf{x}$. The squared magnitude of this tensor then obeys (Lawrie, 1990, p. 38) a Pythagorean theorem

$$|V(\mathbf{x})|^2 \equiv \sum_{ijkl} G_{ijkl} V^{ij} V^{*kl}, \qquad G_{ijkl} \equiv G_{ijkl}(\mathbf{x}), \qquad V^{ij} \equiv V^{ij}(\mathbf{x}). \tag{11.4}$$

Next, return to the *discrete* position case, where coordinates $g_{ijn}$ are metric components $g_{ij}$ evaluated at discrete coordinates $\mathbf{x}_n$. A function $V$ in this space has the discrete representation $V = V^{ij}(\mathbf{x}_n) \equiv V^{ijn}$, so the squared length $|V_n|^2$ is now

$$|V_n|^2 = \sum_{ijkl} G_{ijkln} V^{ijn} V^{*kln}, \qquad G_{ijkln} \equiv G_{ijkl}(\mathbf{x}_n). \tag{11.5}$$

Take the particular case of

$$V^{ijn} \equiv \frac{\partial \psi^*}{\partial g_{ijn}}. \tag{11.6}$$

This denotes the gradient with respect to Riem space coordinates $g_{ij}$ of $\psi^*$ as evaluated at the four-position $\mathbf{x}_n$. Then, by Eq. (11.5),

$$|V_n|^2 \equiv |\nabla \psi(g_{ijn})|^2 = \sum_{ijkl} G_{ijkln} \frac{\partial \psi^*}{\partial g_{ijn}} \frac{\partial \psi}{\partial g_{kln}}. \tag{11.7}$$

Now, our random Fisher variables are the $g_{ijn}$ (see Eq. (11.2)). Then, by Eqs. (2.19) and (2.24), we get

$$I = 4\int d\mathbf{g} \sum_{n=1}^{N} |\nabla\psi(g_{ijn})|^2, \qquad d\mathbf{g} \equiv dg_{111} \cdots dg_{33N}. \tag{11.8}$$

The summation over $n$ is required by the added dimensionality of each new discrete position. This result is analogous to the quantum mechanical result Eq. (4.2), with coordinates $\mathbf{g}$ replacing $(\boldsymbol{r}, t)$ of the quantum mechanical problem.

Combining Eqs. (11.7) and (11.8) gives the target information expression,

$$I = 4\int d\mathbf{g} \sum_{ijkln} G_{ijkln} \frac{\partial \psi^*}{\partial g_{ijn}} \frac{\partial \psi}{\partial g_{kln}}. \tag{11.9}$$

### 11.2.7 Free-wave EPI solution

An interesting special case is that of nearly flat space and the absence of sources. As will be seen, this is analogous to the case of a Klein–Gordon particle with zero mass and charge.

We first seek the EPI extremum solution to Eq. (3.16),

$$K \equiv I - J = extrem. \tag{11.10}$$

The bound information $J$ is a function of the sources that are present (Sec. 3.4.5). Hence, in the absence of sources,

$$J = 0. \tag{11.11}$$

The Euler–Lagrange Eq. (0.34) is here

$$\sum_{ijn} \frac{d}{dg_{ijn}} \left( \frac{\partial \mathscr{L}}{\partial \psi^*_{,ijn}} \right) = \frac{\partial \mathscr{L}}{\partial \psi^*}, \qquad \psi_{,ijn} \equiv \frac{\partial \psi}{\partial g_{ijn}}. \tag{11.12}$$

Because of Eq. (11.11), the Lagrangian $\mathscr{L}$ is simply the integrand of Eq. (11.9). From the latter we see that

$$\frac{\partial \mathscr{L}}{\partial \psi^*_{,ijn}} = 4 \sum_{kl} G_{ijkln} \frac{\partial \psi}{\partial g_{kln}}, \quad \text{and} \quad \frac{\partial \mathscr{L}}{\partial \psi^*} = 0. \tag{11.13}$$

The zero follows because information $I$ of course does not depend explicitly upon $\psi^*$ while $J$, which would have, is zero here.

Substituting results (11.13) into Eq. (11.12) gives a solution

$$\sum_{ijkl} \sum_{\mathbf{x}_n} G_{ijkl}(\mathbf{x}_n) \frac{\partial^2 \psi}{\partial g_{ij}(\mathbf{x}_n)\, \partial g_{kl}(\mathbf{x}_n)} = 0, \tag{11.14}$$

where we show the explicit $\mathbf{x}$-dependence of all quantities. A "Ricci theorem" (DeWitt, 1967) $\partial G_{ijkl}(\mathbf{x}_n)/\partial g_{i'j'}(\mathbf{x}_n) = 0$, all $i, j, k, l, i', j'$, was used. Equation (11.14) is our free-wave solution according to EPI theory.

### 11.2.8 Free-wave Wheeler–DeWitt solution

In Eq. (11.14), the sum on $n$ represents the contribution from all discrete four-positions. As was noted (Sec. 11.2.4), measurements can be made only at discrete positions. Nevertheless, let us take a quasi-continuous limit as such positions become ever more finely spaced. Denote the constant spacing between $\mathbf{x}$ values as $\Delta\mathbf{x}$. Multiply and divide Eq. (11.14) by $\Delta\mathbf{x}^2$. In the limit of small $\Delta\mathbf{x}$ it becomes

$$\Delta\mathbf{x} \int d\mathbf{x} \sum_{ijkl} G_{ijkl}(\mathbf{x}) \frac{\delta^2 \psi}{\delta g_{ij}(\mathbf{x})\, \delta g_{kl}(\mathbf{x})} = 0. \tag{11.15}$$

We used the correspondence Eq. (0.58) between partial derivatives $\partial/\partial g$ and *functional* derivatives $\delta/\delta g$.

With $\Delta x$ small but finite, the integral in (11.15) must be zero. As in Eq. (3.20), we demand that the zero be attained at each value $\mathbf{x}$ of the integrand:

$$\sum_{ijkl} G_{ijkl}(\mathbf{x}) \frac{\delta^2 \psi}{\delta g_{ij}(\mathbf{x})\, \delta g_{kl}(\mathbf{x})} = 0. \tag{11.16}$$

This is the Wheeler–DeWitt equation in the absence of sources and in negligibly curved space. See Kuchar (1973) for details. The notation $\delta$ in (11.16) denotes covariant (Sec. 3.6.2), functional (Eq. (0.58)) derivatives.

### 11.2.9 Klein–Gordon type solution

The answer Eq. (11.14) or Eq. (11.16) has the mathematical form of a Klein–Gordon equation for the scalar wave $\psi$ in a curved space, i.e., Riem space $g_{ij}(\mathbf{x})$. The mixed partial derivatives allow for the presence of such curvature. By comparison, the Klein–Gordon equation was found, in Eq. (4.28), to be obeyed by a particle of zero spin in $\mathbf{x}$-space. If we further specialize that scenario to one of a massless and charge-free particle, then Eq. (4.28) resembles Eqs. (11.14) and (11.16) even more closely. Thus, relativistic quantum mechanics and quantum gravity obey parallel theories even though their spaces are vastly different.

### 11.2.10 EPI solution for a pure radiation universe

Although mass–energy sources are still assumed to be absent, we now allow for a non-negligible curvature of space. This means the presence of a generally non-zero Ricci curvature scalar, ${}^3R = {}^3R(g_{11}, \ldots, g_{33})$. Such a situation defines a "pure radiation universe." This explicit dependence upon the Fisher coordinates $g_{ij}$ will now lead to a non-zero bound information $J$. As before, we develop the theory in analogy to that of the quantum development in Sec. 4.1.

### 11.2.11 Momentum conjugate space

For convenience, we stay with the case of the continuum of points $\mathbf{x}$. In this limit, by definition (0.58) the information expression (11.9) becomes

$$I = 4\int d\mathbf{x} \int Dg \sum_{ijkl} G_{ijkl} \frac{\delta\psi^*}{\delta g_{ij}(\mathbf{x})} \frac{\delta\psi}{\delta g_{kl}(\mathbf{x})},$$

$$Dg \equiv \lim_{N\to\infty} \prod_{ij} \prod_{n=1}^{N} dg_{ij}(\mathbf{x}_n). \tag{11.17}$$

(An irrelevant constant multiplier $\Delta\mathbf{x}$ has been ignored; see Eq. (11.15)). The partial derivatives become functional derivatives. The inner integral is called a "functional integral" (Ryder, 1987, pp. 162–4, 176–9). In effect, it is integrated over all possible functions $g_{ij}$ at all possible argument values $\mathbf{x}$. For fixed $i$, $j$ this represents a kind of a metric "path" through $\mathbf{x}$ space, analogous to the paths through temporal space of the Feynman path integral.

We now define a *conjugate momentum space* that is analogous with that in Chapter 4 describing conventional quantum mechanics. The Fisher coordinates for the problem are $g_{ij}(\mathbf{x})$ for all $i$, $j$. Denote these as $\mathbf{g}(\mathbf{x})$. The new momentum space will connect with the space of coordinates $\mathbf{g}(\mathbf{x})$ by a unitary transformation, in particular, a Fourier transformation. This unitary transformation validates the EPI approach for this problem (see Secs. 3.8.5 and 3.8.7).

Our unknown functional $\psi[\mathbf{g}(\mathbf{x})]$ represents the probability amplitude for all metrics evaluated at all points of four-space. Then, likewise define a conjugate functional $\phi[\boldsymbol{\mu}(\mathbf{x})]$ representing the probability amplitude for all *momentum values* $\mu^{ij}$ as evaluated over all points of four-space. In analogy with conventional quantum mechanics, a Fourier relation is to connect the two,

$$\psi[\mathbf{g}(\mathbf{x})] = \int D\mu\, \phi[\boldsymbol{\mu}(\mathbf{x})] e^{-i\int d\mathbf{x}\, \mathbf{g}\cdot\boldsymbol{\mu}}, \tag{11.18}$$

where

$$D\mu \equiv \lim_{N\to\infty} \prod_{ij} \prod_{n=1}^{N} (2\pi)^{-1}\, d\mu^{ij}(\mathbf{x}_n), \qquad \mathbf{g}\cdot\boldsymbol{\mu} \equiv \sum_{ij} g_{ij}(\mathbf{x})\mu^{ij}(\mathbf{x}).$$

(Compare with Eq. (4.4).) This is a *functional* Fourier relation. It is mathematically similar in form to the well-known path integrals used by Feynman to reformulate quantum mechanics (Ryder, 1987, p. 178). Although Eq. (11.18) seems a reasonable relation on these grounds, and will lead to the desired end Eq. (11.32), we have no other evidence for its validity at this time.

Regarding units in the Fourier Eq. (11.18), the exponent as it stands has units of *length*$^4$–*momentum* (note that the metric $g_{ij}$ is unitless). But mathematically it must be unitless. This implies that it should be divided by a constant that carries the units of *length*$^4$–*momentum*. It seems reasonable to use for this purpose $\hbar L^3$, where $L$ is the Planck distance given by Eq. (6.64). Hence, momentum $\boldsymbol{\mu}$ is assumed to be expressed in units of $\hbar L^3$.

### 11.2.12 Finding bound information J

As in Sec. 4.1.11, we find $J$ by representing the Fisher information (11.17) in momentum space. This entails use of Parseval's theorem, as follows.

A functional derivative (see Eq. (0.59)) of Eq. (11.18) gives

$$\frac{\delta\psi}{\delta g_{kl}(\mathbf{x})} = -i\int D\mu\, \phi[\boldsymbol{\mu}(\mathbf{x})]\mu^{kl}(\mathbf{x})\, e^{-i\int d\mathbf{x}\, \mathbf{g}\cdot\boldsymbol{\mu}}. \tag{11.19}$$

A similar expression holds for $\delta\psi^*/\delta g_{ij}$. We will also have need for a *functional* representation of the Dirac delta function:

$$\int Dg\, e^{i\int d\mathbf{x}\, \mathbf{g}\cdot(\boldsymbol{\mu}'-\boldsymbol{\mu})} = (2\pi)^{6N}\delta[\boldsymbol{\mu}'(\mathbf{x}) - \boldsymbol{\mu}(\mathbf{x})]. \tag{11.20}$$

(Compare with Eq. (0.70).) This expression is easily derived by evaluating the $6N$ separated integrals of the left-hand side that follow the use of the identity (11.17) for $Dg$.

Substituting Eq. (11.19) and its complex conjugate into Eq. (11.17), switching orders of integration, and using the Dirac delta function representation (11.20) gives

$$I = 4(2\pi)^{-6N}\int d\mathbf{x}\sum_{ijkl} G_{ijkl}\int D\mu\, \phi(\boldsymbol{\mu})\mu^{kl}\int D\boldsymbol{\mu}'\, \phi^*(\boldsymbol{\mu}')\mu'^{ij}(2\pi)^{6N}\delta(\boldsymbol{\mu}' - \boldsymbol{\mu}). \tag{11.21}$$

In analogy with quantum mechanical Eq. (4.9), this contracts to

$$I = 4\int d\mathbf{x}\sum_{ijkl} G_{ijkl}\int D\mu\, \phi[\boldsymbol{\mu}(\mathbf{x})]\phi^*[\boldsymbol{\mu}(\mathbf{x})]\mu^{ij}\mu^{kl} \tag{11.22}$$

after use of the "sifting" property (Eq. (0.69)) of the Dirac delta function. The right-hand sides of Eqs. (11.17) and (11.22) are equal, since they both equal $I$, which shows that the transformation Eq. (11.18) is unitary (see definition Eq. (4.5)). Then EPI holds identically for this measurement scenario (Secs. 3.8.5 and 3.8.7).

Since $\phi(\boldsymbol{\mu})$ is, by definition, a probability amplitude, its squared amplitude $\phi\phi^*$ is a probability density. Then the inner integral in Eq. (11.22) becomes an expectation $\langle\ \rangle$ over all momenta at a given $\mathbf{x}$, and

$$I = 4\int d\mathbf{x}\sum_{ijkl} G_{ijkl}\langle\mu^{ij}\mu^{kl}\rangle. \tag{11.23}$$

As a check on this calculation, we note that in quantum mechanics (Eqs. (4.12) and (4.13)) the Fisher information $I$ was, as here, proportional to the squared momentum. Hence, quantum mechanics and quantum gravity continue to develop along parallel lines. Also, the formulation of quantum gravity by DeWitt (1967, p. 1118) is through a Hamiltonian that, likewise, contains squared momentum terms. A Lagrangian approach (as here) should, then, also contain squared momentum terms.

By Eq. (11.4), Eq. (11.23) further collapses to

$$I = 4\int d\mathbf{x}\,\langle|\boldsymbol{\mu}|^2\rangle, \tag{11.24}$$

the mean-squared momentum over all space.

As is usual in quantum gravity, the squared momentum $|\boldsymbol{\mu}|^2$ is associated with the kinetic energy. We are assuming a pure radiation universe, i.e., no matter–energy inputs to the system. The total energy of such a universe should be zero (Hawking, 1988), so that the kinetic energy equals the negative of the potential energy. The potential energy is associated with minus the quantity $g^{1/2}\,{}^3R$ (DeWitt, 1967, p. 1117), a quantity that Wheeler (1962, pp. 41, 60) calls the "intrinsic curvature invariant" of energy. Thus, Eq. (11.24) becomes

$$I = 4\int d\mathbf{x}\,\langle g^{1/2}\,{}^3R\rangle \equiv J. \tag{11.25}$$

By Eq. (3.18), the information transfer efficiency is now $\kappa = 1$, exactly as it was for quantum mechanics (Sec. 4.1.11).

All quantities within the $\langle\,\rangle$ signs of Eq. (11.25) are functions of $\mathbf{g}$. Therefore, the averaging is appropriately done in $\mathbf{g}$-space,

$$\langle g^{1/2}\,{}^3R\rangle = \int Dg\,\psi^*(\mathbf{g})\psi(\mathbf{g})g^{1/2}\,{}^3R. \tag{11.26}$$

Using this in Eq. (11.25) gives

$$J = 4\int d\mathbf{x}\int Dg\,\psi^*\psi g^{1/2}\,{}^3R. \tag{11.27}$$

### *11.2.13 Physical information K*

Use of Eqs. (11.17) and (11.27) gives

$$K \equiv I - J = 4\int d\mathbf{x}\int Dg\left(\sum_{ijkl} G_{ijkl}\frac{\delta\psi^*}{\delta g_{ij}}\frac{\delta\psi}{\delta g_{kl}} - \psi^*\psi g^{1/2}\,{}^3R\right). \tag{11.28}$$

### *11.2.14 Wheeler–DeWitt equation*

The Lagrangian in Eq. (11.28) uses functional derivatives. To find its extremum solution we need an Euler–Lagrange equation of the same type,

$$\sum_{ij} \frac{\delta}{\delta g_{ij}} \left( \frac{\delta \mathscr{L}}{\delta \psi^*_{,ij}} \right) = \frac{\delta \mathscr{L}}{\delta \psi^*}, \tag{11.29}$$

$$\psi_{,ij} \equiv \frac{\delta \psi}{\delta g_{ij}}. \tag{11.30}$$

Using as $\mathscr{L}$ the integrand of Eq. (11.28), we have

$$\frac{\delta \mathscr{L}}{\delta \psi^*_{,ij}} = 4 \sum_{kl} G_{ijkl} \frac{\delta \psi}{\delta g_{kl}}, \quad \frac{\delta \mathscr{L}}{\delta \psi^*} = -4\psi g^{1/2}\, {}^3R. \tag{11.31}$$

Then the Euler–Lagrange solution Eq. (11.29) is

$$\sum_{ijkl} G_{ijkl} \frac{\delta^2 \psi}{\delta g_{ij} \delta g_{kl}} + g^{1/2}\, {}^3R\psi = 0. \tag{11.32}$$

This is the Wheeler–DeWitt equation for a pure radiation universe. It is essentially a Klein–Gordon equation, with the form of Eq. (4.26). The functional derivatives $\delta$ should be regarded as *covariant* functional derivatives throughout the derivation (Sec. 3.6.2).

Hence, ordinary quantum mechanics and quantum gravity continue to develop along parallel lines. In particular, the unitary transformation is vital to both derivations.

It is interesting to compare this approach with the path-integral approach taken by Hawking (1979) and by Hartle and Hawking (1983). These authors *postulate* the use of a path-integral approach, in analogy to the use of path integrals by Feynman and Hibbs (1965) in ordinary quantum mechanics. Moreover, the integrand is of a postulated *form*, involving exponentiation of the classical gravitational action.

By comparison, the EPI use of path integrals appears to arise a little more naturally. There are two path integrals in the EPI approach. The path integral Eq. (11.17) arises as the continuous limit of Fisher information over Riem space. The form of that integral is fixed as the usual Fisher form. The second path integral, Eq. (11.18), arises as the continuous limit over momentum space of a Fourier relation between the metric and its associated momentum. The form of that integral is the usual Fourier one.

### *11.2.15 Need for the inclusion of spin 1/2 in formalism*

In the preceding, the wave function $\psi$ was a scalar quantity. As we saw in Sec. 4.2.1, scalar quantum mechanics describes the kinematics of a spin-0 particle. Particles with integral values of the spin are called bosons. Correspondingly, the Wheeler–DeWitt equation (11.32) describes the kinematics of a spin-2 particle, the graviton. This is, then, another boson. On the other hand, Wheeler (1962) has emphasized that quantum gravity must allow for the inclusion of spin-1/2 particles as well, such as the neutrino. Particles having half-integer spin, like the neutrino, are called fermions. How, then, can a gravitational theory of fermions be formulated?

Recall that in Sec. 4.2 we were able to develop the Dirac equation, which describes the quantum mechanics of a particle of half-integral spin, by using a *vector* wave function $\boldsymbol{\psi}$. Since quantum mechanics and quantum gravity have, to this point, been developed by analogous steps, it should follow that a quantum gravity for particles of half-integral spin can be developed analogously. This entails using the zero-aspect Eq. (3.18) of EPI to form a solution. The following steps are proposed.

Factor the integrand of Eq. (11.28), just as the integrand of Eq. (4.32) was factored. See also Matveev and Salle (1991, pp. 26, 27). This requires abandonment of the scalar nature of $\psi$ for a vector $\boldsymbol{\psi}$. Next, matrices analogous to the Dirac matrices Eq. (4.34) need to be found. The results will be replacement of the second-order differential Lagrangian in Eq. (11.28) with a product of two first-order factors. Setting either to zero (via EPI principle (3.18)) will give the vector gravitational wave equation. It will be interesting to compare the answer with other vector wave equations of quantum gravity.

### *11.2.16 The original deus ex machina?*

It was mentioned in Sec. 3.10, and confirmed by the behavior of Eq. (10.17), that measurements act, to the object under measurement, as unpredictable, discontinuous, irreversible, instantaneous operations, somewhat like so many *deus ex machina* activities. We now view the current measurement problem from this viewpoint.

This approach to quantum gravity agrees with the approach taken in Chapter 4 to derive the Klein–Gordon equation, except for one important factor: the number of real measurements that are required to be taken. We assume, here, essentially an infinity of measurements of the metric tensor to be made, i.e., throughout all of four-space. By comparison, in Chapter 4 a *single* measurement of a joint wave function $\boldsymbol{\psi}(\mathbf{x})$ was required. In principle, a single measurement should work here as well since we have assumed likewise that the wave function $\psi[\mathbf{g}(\mathbf{x})]$ *connects jointly* the behavior of all metric values over all space. The measurement of the metric would then suffice to provide the perturbation $\delta\psi[\mathbf{g}(\mathbf{x})]$ over all space that EPI needs in order to execute.

On the other hand, we did achieve the "correct" answer (Eq. (11.32)) by the use of an infinity of measured points. Therefore this approach is correct as well.

(*Caveat*: This assumes that the Wheeler–DeWitt equation is fully correct. Not everyone agrees with this, however, notably DeWitt (DeWitt, 1997); Wheeler seems to still support the equation (Wheeler, 1994). We will proceed on the assumption that it is at least approximately valid. The "end" that we are working toward, in this section, might justify the "means.")

But, if both approaches are correct, this implies that, somehow, a single measurement is equivalent to an infinity of measurements over all space. How could this occur?

An obvious answer is that the single measurement takes place when all of space is squeezed into a single point. This, presumably, is at the Big Bang. The implication of this is that a single observation of the metric occurred at the onset of the Universe, and this generated the Wheeler–DeWitt equation for the pure radiation universe which existed then. In other words, the gravitational structure of the Universe was created out of a single, primordial quest for knowledge.

## 11.3 Nearly incompressible turbulence

### *11.3.1 Introduction*

Turbulence is one of the last great mysteries of classical physics. It describes the seemingly random, or quasi-random, flow of gases and liquids that occurs under normal operating conditions. A good general reference is Bradshaw (1978).

A qualitative description of turbulence is easy to give. Turbulence is a three-dimensional non-stationary motion of the medium. The motion is characterized by a continuous range of fluctuations of velocity, with wavelengths ranging from (i) minimum lengths that are determined by viscous forces, to (ii) maximum lengths that are determined by the boundary conditions of the flow. The turbulence process is *initiated* at the maximal lengths. An example is a body of warm air just beginning to rise from a parked car on a windless day. The maximal length is the length of the automobile. At this stage, the flow is smooth, laminar, and deterministic. But, because of frictional and viscous interaction with the surrounding air, as this body of air rises it breaks down into many smaller lengths of rising air, called "eddies," with conservation of energy transmitting the internal energy of the original mass into these eddies. The process continues in this manner, to ever smaller eddies. At the smallest eddies the energy is transformed into heat due to the action of the viscous forces, which is then radiated away.

The quantitative picture is, however, less precise. Turbulent flow obeys classical physics, whose laws are well understood. However, the process is what is called mathematically "ill-posed," in that the slightest imprecision in the knowledge of initial conditions propagates into huge errors in any subsequent physical (deterministic) analysis after but a finite amount of time. This is the so-called "butterfly effect" that makes weather forecasting so tenuous an activity. Interestingly, there is an intermediary state between the initial, deterministic flow and complete randomness of subsequent small scale states. The situation at this intermediary state is only quasi-random. This state is the subject of "chaos" theory; see, e.g., Sagdeev *et al.* (1988).

For these reasons, turbulence is often analyzed statistically. See, e.g., Ehlers *et al.* (1972). That is, probability laws are sought in order to model the joint fluctuations in velocity, pressure, temperature, etc., that define a turbulence process. It is known that the key such PDF is the one on joint velocity at all points in space (Batchelor, 1956). But this PDF is, so far, too complicated to find or to use. It is also known that lower-order joint PDFs or single-point PDFs on velocity are close to Gaussian in form, with any departures arising out of the essentially non-linear behavior of the turbulence (Bradshaw, 1978, p. 15). The theory below verifies this effect.

Another statistical quantity of interest is the joint PDF on density and velocity in a fluid. This PDF was recently determined by Cocke (1996) using an approach that is very close to that of EPI. We present it next.

### *11.3.2 Fisher coordinates*

All physical quantities are assumed to be evaluated at a given position and time in the fluid. We do not consider the correlation between such quantities at different positions or at different times.

The Fisher coordinates of the problem should comprise a four-vector (see Sec. 3.5). Mass flux

$$w^{\mu} \equiv \rho v^{\mu}, \quad \mu = 0, 1, 2, 3 \tag{11.33}$$

with $\rho$ the local density and $v^{\mu}$ the four-velocity, is one such quantity. (We initially use tensor notation; see Sec. 6.2.1.) Here,

$$v^{\mu} \equiv \frac{dx^{\mu}}{d\tau} \tag{11.34}$$

with $x$ the position and $\tau$ the proper time. As is usual with a four-vector, $w$ obeys a conservation (of mass) equation

$$\frac{\partial w^{\mu}}{\partial x^{\mu}} = 0, \tag{11.35}$$

using summation notation (see Sec. 6.2.2).

We restrict our attention to non-relativistic velocities $v^{\mu}$. This is on the grounds of simplicity, and also because data which one would need to confirm the theory are lacking for relativistic turbulence. In the non-relativistic limit, by Eq. (3.33) the proper time $\tau = t$, the laboratory time, so that $dx^0/d\tau = d(ct)/dt = c$. It follows from Eqs. (11.33) and (11.34) that

$$w^{\mu} = \rho(c, \mathbf{v}), \quad \mathbf{v} \equiv \frac{d\mathbf{x}}{dt} = (v_x, v_y, v_z), \tag{11.36}$$

the ordinary Newtonian fluid velocity. Hence, the Fisher coordinates become

$$\rho c \text{ and } \mathbf{w}, \quad \mathbf{w} \equiv \rho \mathbf{v}. \tag{11.37}$$

The variables $(\rho c, \mathbf{w})$ are fluctuations from ideal (here, mean) values $\theta_\rho$, $\theta_w$ that need not be specified. Thus, we seek the PDF $p(\rho c, \mathbf{w})$. As usual, we work with the corresponding amplitude function $q(\rho c, \mathbf{w}) \equiv q(\rho, \mathbf{w})$ in simpler notation.

### 11.3.3 Fisher information I

Given this choice of coordinates, the information $I$ is, from Eq. (2.19),

$$I = 4\int d\mathbf{w}\, d\rho \left[\nabla_w q \cdot \nabla_w q + \frac{1}{c^2}\left(\frac{\partial q}{\partial \rho}\right)^2\right]. \tag{11.38}$$

The notation $\nabla_w$ signifies the gradient in three-dimensional $\mathbf{w}$-space.

### 11.3.4 Energy effects

The kinetic energy density is defined as

$$E_{\text{kin}} \equiv \tfrac{1}{2}\rho v^2 = \frac{w^2}{2\rho} \tag{11.39}$$

by Eq. (11.37). The medium also has an internal energy density $\epsilon(\rho)$, which can, it turns out, be left unspecified.

### 11.3.5 Skewness effects

The PDF on pressure is known to be highly skewed. It is an exponential for negative pressure and nearly a Gaussian for positive pressure (Pumir, 1994). Since pressure and density $\rho$ often vary nearly linearly with one another (see Sec. 11.3.11), one would expect the PDF on $\rho$ to likewise exhibit strong skewness. Given these facts, it is reasonable to build the possibility of skewness into the one input to EPI that brings prior physical knowledge into the theory: the bound information $J$.

### 11.3.6 Information J

A turbulent medium may be regarded as a non-ideal gas. We found the PDF $p(\mathbf{v})$ for an ideal gas in Chapter 7. This suggest that we represent the bound information $J$ for a turbulent medium as that for an ideal gas, except for extra terms expressing skewness effects as mentioned above. The tactic turns out to be valid in that it leads to empirically correct results.

From Eqs (7.40), (7.47), and (7.49) the $J$ for an ideal gas was equivalent to a normalization constraint plus a constraint on mean-squared velocity. The latter is equivalent to mean kinetic energy. The turbulent medium is characterized by the internal energy $\epsilon(\rho)$ as well. Hence, we express $J$ here as a sum of constraints on normalization, kinetic energy, and internal energy, suitably skewed:

$$J = 4\int d\mathbf{w}\, d\rho \left[\frac{\lambda_1 w^2}{2\rho} + H(\rho - \rho_1)(\lambda_2 + \lambda_3\epsilon(\rho))\right] q^2. \tag{11.40}$$

Function $H(\rho)$ is the step function $H(\rho) = 1$ for $\rho \geqslant 0$, $H(\rho) = 0$ for $\rho < 0$. The constant $\rho_1$ allows for the skewness effects mentioned above. Using the step function in Eq. (11.40), these are

$$\int_{\rho_1}^{\infty} d\mathbf{w}\, d\rho\, q^2 = const., \qquad \int_{\rho_1}^{\infty} d\mathbf{w}\, d\rho\, \epsilon(\rho) q^2 = const. \tag{11.41}$$

Constants $\lambda_1, \lambda_2, \lambda_3$ are found by the theory.

It is to be noted that we have, here, departed from the usual procedure for finding $J$, which is to *solve for it* via knowledge of an invariance principle. Rather, it was formed by analogy to the $J$ for an ideal gas, as discussed.

### 11.3.7 Net EPI variational principle

By the use of Eqs. (11.38) and (11.40), the EPI principle Eq. (3.16) becomes

$$\begin{aligned} K &\equiv I - J \\ &= 4\int d\mathbf{w}\, d\rho \left[\nabla_w q \cdot \nabla_w q + \frac{1}{c^2}\left(\frac{\partial q}{\partial \rho}\right)^2 - \frac{\lambda_1 w^2}{2\rho} q^2 \right. \\ &\qquad \left. - H(\rho - \rho_1)(\lambda_2 + \lambda_3 \epsilon(\rho)) q^2 \right] \\ &= extrem. \end{aligned} \tag{11.42}$$

This is to be solved for the amplitude function $q(\rho, \mathbf{w})$ under the boundary conditions that $q^2$ be integrable and continuous, with continuous first derivatives, and that $q^2(0, \mathbf{w}) = 0$.

### 11.3.8 Euler–Lagrange solution

The Euler–Lagrange Eq. (0.34) for this problem is

$$\sum_{i=1}^{3} \frac{d}{dw_i}\left(\frac{\partial \mathscr{L}}{\partial q_{,i}}\right) + \frac{d}{d\rho}\left(\frac{\partial \mathscr{L}}{\partial q_{,\rho}}\right) = \frac{\partial \mathscr{L}}{\partial q},$$
$$q_{,i} \equiv \frac{\partial q}{\partial w_i}, \quad i = 1, 2, 3, \quad q_{,\rho} \equiv \frac{\partial q}{\partial \rho}. \tag{11.43}$$

The Lagrangian $\mathscr{L}$ is the integrand of Eq. (11.42). Then the Euler–Lagrange solution Eq. (11.43) is

$$\nabla_w^2 q + \frac{1}{c^2}\frac{\partial^2 q}{\partial \rho^2} + \left[\frac{\lambda_1}{2\rho} w^2 + H(\rho - \rho_1)(\lambda_2 + \lambda_3 \epsilon(\rho))\right] q = 0. \tag{11.44}$$

Some simplification arises out of an assumption of isotropy for the turbulence. This allows us to use

$$q(\rho, \mathbf{w}) = q(\rho, w) \quad \text{and} \quad \nabla_w^2 q = \frac{1}{w^2}\frac{\partial}{\partial w}\left(w^2 \frac{\partial q}{\partial w}\right) \tag{11.45}$$

in Eq. (11.44).

### 11.3.9 Numerical solutions

The problem (11.44) and (11.45) is very difficult to solve, even numerically. However, an approximate solution may be found, for cases of low Mach number $M \equiv v_{\mathrm{rms}}/v_0$, where $v_{\mathrm{rms}}$ is the root mean-square velocity in the medium and $v_0$ is the speed of sound. Then any velocity $v$ obeys

$$v^2 \ll v_0^2. \tag{11.46}$$

We will sketch the steps of the derivation. Details can be found in Cocke (1996).

Solutions are found for two different regions of density, $\rho < \rho_1$ and $\rho > \rho_1$. The solution in region 1 is

$$q_1(\rho, w) = R(\rho)W(w), \quad R(\rho) = Ae^{-B\rho},$$
$$W(w) = Ce^{-Dw^2}, \quad A, B, C, D = const. \tag{11.47}$$

Thus, $q$ (and the PDF $p$) obey an exponential falloff in density and a Gaussian dependence upon velocity.

It is well known that turbulent velocity does approximately obey Gaussian statistics. However, there are more zeros and more strong gusts than would be the case for a pure Gaussian dependence. The low Mach number approximation Eq. (11.46) may be responsible for the Gaussian result. If so, then relaxing this condition will produce better agreement with the known facts. This is left for future research.

The solution in region 2 is

$$q_2(\rho, w) = S(\rho)W(w), \qquad S(\rho) \approx \mathrm{Ai}(a\rho + b), \qquad a, b = const. \tag{11.48}$$

Here it was assumed that the fluctuations in density are small enough that the internal energy function $\epsilon(\rho)$ is well approximated by a linear function of $\rho$. Function $\mathrm{Ai}(\rho)$ is Airy's function. An asymptotic formula is

$$\mathrm{Ai}(x) \approx \frac{e^{-2x^{3/2}/3}}{2\sqrt{\pi}x^{1/4}}. \tag{11.49}$$

The PDF on density is, therefore, proportional to the square of Airy's function.

The two solutions $q_1(\rho)$ and $q_2(\rho)$ are made to be continuous in their values and in their derivatives $\partial q/\partial\rho$ at the connecting point $\rho = \rho_1$. This is accomplished through adjustment of the constants of the problem (the $\lambda_i$, $a$, $b$, etc.). It is interesting that EPI predicts analogous results (Eq. (13.24) *et vecin.*) in econophysics.

### *11.3.10 EPI zero solution*

The preceding was the solution to EPI principle Eq. (3.16). This is the extremization half of the principle. But the other half, the zero-condition Eq. (3.18), must also be satisfied. The premise of EPI is that all solutions to *both* the extremization and the zero-condition must have physical significance.

If we take the information efficiency constant $\kappa = 1$ here, then Eq. (11.42) is now to be zeroed. Hence, what is the solution $q$ to this problem? In fact, the solutions we obtained above for extremizing Eq. (11.42) also must zero it! This follows essentially because the constraint terms in Eq. (11.42) are proportional to factor $q^2$. For a proof, see Frieden and Cocke (1996).

Hence, the EPI *zero-solution* for this problem does not elicit any additional physical knowledge. This is for a *scalar* amplitude $q$. By contrast, the use of vector amplitudes $\mathbf{q}$, as in Sec. 4.2, might lead to useful new results.

### *11.3.11 Discussion*

It was found that, in the limit of low Mach number, velocity fluctuations obey Gaussian statistics. Also, for $\rho < \rho_1$ the density fluctuations obey exponential statistics, whereas for $\rho > \rho_1$ the PDF for density fluctuations is the square of an Airy function.

It is possible to compare these density fluctuation results with some of the simulation results of Pumir (1994). Pumir computed PDF curves of *pressure* fluctuation, based upon simulations using the Navier–Stokes equations with a viscosity term. These pressure fluctuation curves may be interpreted as *density* fluctuation curves as well since, in our low-Mach limit, the pressure is a linear function of the density $\rho$.

One such simulation curve is shown in Fig. 11.1 (dotted line). The

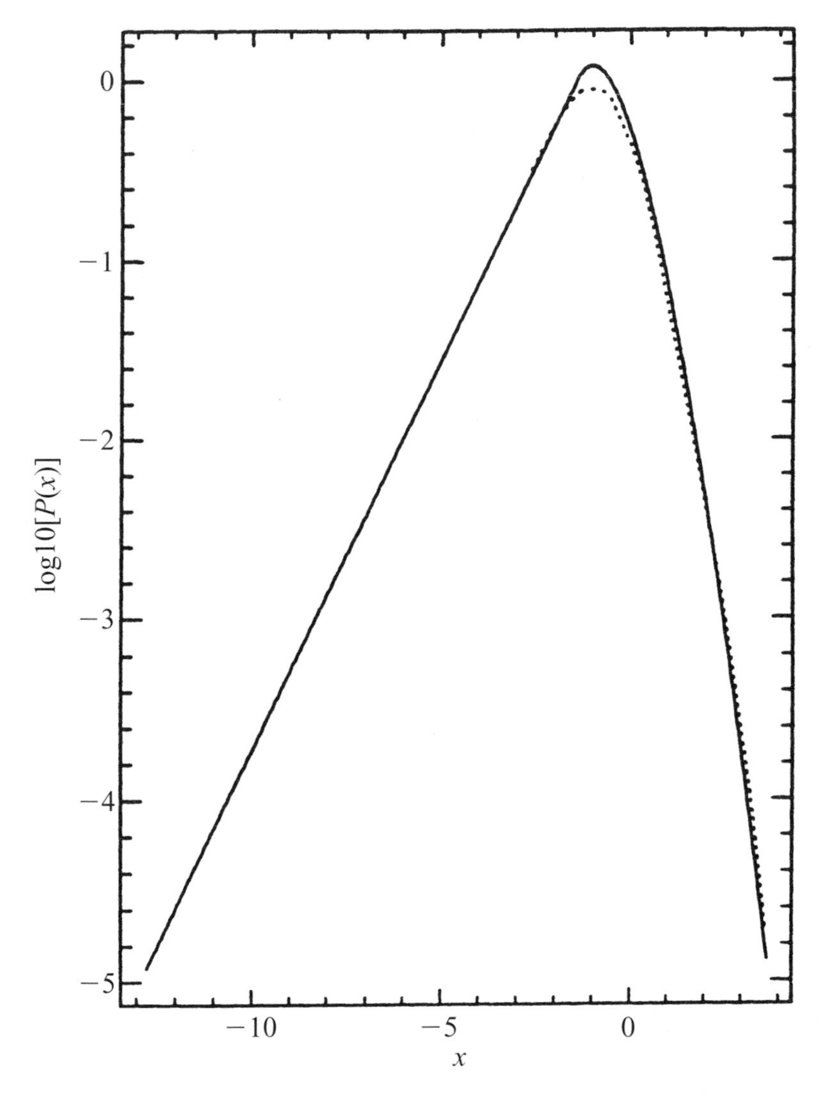


Fig. 11.1. The solid curve is the logarithm of our unnormalized model PDF $P(\rho) \equiv q^2(\rho)$ as a function of the dimensionless variable $x$, with $x \propto \rho$. Pumir's simulations are represented by the dotted curve. (Reprinted with permission from *Physics of Fluids* **8**, 1609 (1996). Copyright 1996, American Institute of Physics.)

$x$-coordinate (horizontal) is linear in the density $\rho$. The peak probability value corresponds to density value $\rho_1$. To its left is an exponential solution and to its right is a near-Gaussian. These can be considered as "ground truth," since they are based upon physical simulation of the turbulence.

By comparison, adjustment of the constants in the above EPI approach gives the solid curve. The agreement with Pumir's simulation curve is seen to be very good over four decades of density. The only significant disagreement occurs near the maximum, where Pumir's curve lies below the EPI curve. However, Pumir's approach included physical effects which we ignored – the Navier–Stokes equations with a viscosity term. It may be that the inclusion of such effects into the EPI principle would give better agreement in this region. Rigorous use of the EPI principle, in place of the *ad hoc* approach taken here, might improve things as well.

As a followup to this work, Cocke and Spector (1998) used Fisher information to compute the correlation between velocity and pressure of the turbulent medium.

## 11.4 Topics in particle physics

The EPI approach has not yet been used to derive the full theory of particle physics. However, progress has been made on selected topics of the subject. See Frieden and Plastino (2001a) and the material in the next two sections.

### *11.4.1 Empirical formulae for the Weinberg and Cabibbo angles*

The Higgs mass phenomenon of elementary particles provides a mechanism (maybe, *the* mechanism) by which mass is created. This is out of vector bosons, e.g., the so-called Z-particle. The Weinberg angle $\theta_{\rm W}$ is a key parameter of the Higgs mass phenomenon. It is defined as

$$\tan^2\theta_{\rm W} \equiv \frac{g'^2}{g^2}, \tag{11.50}$$

where $g'$ and $g$ are two fundamental coupling constants. It is well known that the standard model of elementary particles does not provide a numerical value for the angle; it must instead be found experimentally (Martin and Shaw, 1992). At a Z-particle energy level of 93 GeV/$c^2$, the value for the Weinberg angle (Groom *et al.*, 2000) is currently *measured* as

$$\theta_{\rm W} = 28.737\,75 \pm 0.010\,88^\circ, \quad \sin^2\theta_{\rm W} = 0.231\,17 \pm 0.000\,16. \tag{11.51}$$

The Weinberg angle is one characteristic constant of electroweak interactions. Another is the *Cabibbo* angle $\theta_{\rm C}$ (Martin and Shaw, 1992). As with the

Weinberg angle, the standard model of elementary particles does not provide an *a priori* numerical value for the angle. It must instead be found experimentally, the current state of the art value being

$$\theta_C = 12.685\,54 \pm 0.135\,04^\circ \qquad (11.52)$$

(Groom *et al.*, 2000). An analytical expression for the angle has long been sought.

We now turn to analytical formulae for the two angles. Beyond the context of the standard model, so-called *grand unified theories* exist that do give analytical values for the Weinberg angle. One such theory (Lawrie, 1990), based upon so-called SU(5) symmetry, arrives at a figure of

$$\sin^2 \theta_W = 0.375. \qquad (11.53)$$

However, this value is very far from the experimental result (11.51), and is conjectured by Aitchison and Hey (1982) to hold true at higher interaction energy values than are currently attainable. Another SU(5) theory (Lawrie, 1990) gives for the Weinberg angle $\sin^2 \theta_W \approx 0.21$ at energies comparable to the mass $M_W$. This is pretty close to the experimental value (11.51).

Empirical formulae for $\theta_W$ and $\theta_C$ are suggested by the theory in Chaps. 14 and 15. It is found in Chapter 14 that both biological growth and physical growth (in the form of transport effects) follow a common law – the Lotka–Volterra equation. Also, in Chapter 15 it is found that two basic types of biological growth – ideal growth and *in situ* cancer growth – depend fundamentally upon the *Fibonacci constant* $\gamma$; see Eqs. (15.36) and (15.40).

The Fibonacci constant $\gamma$ (also called the "golden mean") specifically arises as the *ratio* of two numbers: in an ideally growing population, the total population at the $n$th and $(n - 1)$st generations in the limit as $n \to \infty$ (see Chapter 15). Likewise, the Weinberg angle arises analytically out of Eq. (11.50) as the *ratio* of two coefficients, $g'^2$ and $g^2$. By the Higgs mass effect, mass is created whenever a boson interacts with a Higgs particle. The result is a temporal Higgs mass growth process.

If the Higgs mass growth with time can likewise be considered "ideal," we can expect the growth to depend in one or more ways upon the Fibonacci constant. In fact, Higgs mass growth *is* related to ideal Fibonacci growth (Rosen, 1996). One property of the mass growth is the Weinberg angle. By this reasoning the Weinberg angle might depend upon $\gamma$. We sought such a relation.

In that nature tends to follow simple laws, we sought a *simple* relation connecting $\theta_W$ with $\gamma$. At the same time, it had to agree well with the empirical value (11.51) of the angle. One such law is

$$\tan^2 \theta_W = \tfrac{1}{2}(\gamma - 1) = \tfrac{1}{4}(\sqrt{5} - 1) = 0.309\,02 \qquad (11.54a)$$

to five decimal places. Equation (15.45) was used to form the second equality. The dependence of (11.54a) upon $\gamma$ is linear and indeed simple. From (11.54a),

$$\sin^2\theta_W = 0.236\,068, \qquad \theta_W = 29.069\,38^\circ. \tag{11.54b}$$

This value for $\theta_W$ is about $0.3^\circ$ larger than the mid-experimental value (11.51). This is pretty good agreement. The difference might be due to the particular Z-particle energy used in the experiment.

Next we turn to finding an empirical formula relating the Cabibbo angle $\theta_C$ to $\gamma$. Now a simple formula that agrees well with the empirical value (11.52) of the angle is sought. The simplest we could so form is

$$\tan^2(4\theta_C) = \gamma. \tag{11.55a}$$

Using it with the value Eq. (15.45) for $\gamma$ gives

$$\theta_C = 12.956\,82^\circ. \tag{11.55b}$$

This is larger than the mid-experimental value (11.52) by about $0.3^\circ$. Again, the disagreement might be due to the particular Z-particle energy used in the experiment. Interestingly, this departure of about $0.3^\circ$ for the Cabibbo angle is *the same*, in magnitude and sign, as was found between the experimental and theoretical values of the *Weinberg* angle. See below Eq. (11.54b).

Thus, the two empirical formulae (11.54a) and (11.55a) give consistent results to within a constant bias (of $0.3^\circ$) away from the current experimental values. A constant bias suggests that the departures are of a *systematic*, rather than random, nature. This also is consistent with the possibility that the two formulae (11.54a) and (11.55a) are *correct*, up to an additive bias due to a currently unknown systematic effect such as energy range.

Finally, eliminating $\gamma$ between the two Eqs. (11.54a) and (11.55a)) relates the two angles,

$$\tan^2(4\theta_C) = 1 + 2\tan^2\theta_W. \tag{11.56}$$

The standard model does not predict a relation between the two angles.

I thank physics doctoral candidate Gary P. Weber for help on this problem.

### *11.4.2 EPI prediction of mesons*

Quarks are spin-1/2 fermions and, as such, obey the Dirac Eq. (4.41) when in a free field. Quarks are thought to be the fundamental building blocks of composite hadrons, forming them by being bound together with anti-quarks in pairs, or as quark trios, anti-quark trios, etc. A quark $q$–anti-quark $\overline{q}$ pair is called a "meson," and denoted as $q\overline{q}$. We show here that EPI predicts the existence of mesons. A critical role is played by the property of "asymptotic freedom" of the constituent particles.

Assume the measurement of the space-time of a composite particle – a hadron. The theory now follows that of Chapter 4. The EPI zero-solution $\psi$ to the problem was required to satisfy

$$k[\psi] = \mathbf{v}_1[\psi] \cdot \mathbf{v}_2[\psi] = 0 \qquad (11.57)$$

by Eq. (4.40) and Appendix E. Here, by Eqs. (4.37a, b) and (4.41),

$$\mathbf{v}_1[\psi] = i[\boldsymbol{\alpha}] \cdot \nabla\psi - \eta[\boldsymbol{\beta}]\psi + i\lambda\,\partial\psi/\partial t,$$

$$\mathbf{v}_2[\psi] = i[\boldsymbol{\alpha}^*] \cdot \nabla\psi^* + \eta[\boldsymbol{\beta}^*]\psi^* - i\lambda\,\partial\psi^*/\partial t. \qquad (11.58)$$

These equations define general fermions. See Sec. 4.2.11. We specialize to the case where the fermions are quarks.

Then Eq. (11.57) is more specifically a statement

$$\mathbf{v}_1[\psi^q] \cdot \mathbf{v}_2[\psi^{\bar{q}}] = 0, \qquad (11.59)$$

where $\psi^q$ is the amplitude function for a quark and $\psi^{\bar{q}}$ is that for an anti-quark (Secs. 4.2.6 and 4.2.7). Notice that Eq. (11.59) may be satisfied by either condition,

$$\mathbf{v}_1[\psi^q] = 0 \qquad (11.60a)$$

or

$$\mathbf{v}_1[\psi^{\bar{q}}] = 0, \qquad (11.60b)$$

or *both*

$$\mathbf{v}_1[\psi^q] = 0 \quad \text{and} \quad \mathbf{v}_2[\psi^{\bar{q}}] = 0. \qquad (11.60c)$$

We next find a meaning, in the context of measurement, for each of these three alternatives.

Condition (11.60a) states that the *single* measured particle was a quark. This is physically possible. It had long been conjectured that the quark and anti-quark could exist as free particles in a deconfined quark–gluon plasma (Bhaduri, 1992). In fact this effect was recently observed (Heinz and Jacob, 2000).

Alternatively, condition (11.60b) states that the single measured particle was an *anti*-quark. This is again possible, as in the preceding.

However, the condition (11.60c) at first seems implausible: it states that the measured particle was *both* a quark and an anti-quark. How can one particle be two? The only possibility is that the quark and anti-quark are bound *together* – a pair $q\bar{q}$ defining a meson – as the measured particle. In this way, *EPI predicts the existence of bound particle pairs*.

Further evidence for the $q\bar{q}$ assertion is the so-called "asymptotic freedom" – or independent behavior – (Yndurain, 1999) of the quarks at small separations. By this property, $q$ and $\bar{q}$ behave independently except when they are relatively far apart. Now, the validity of Eqs. (11.60c) was essential to the $q\overline{q}$ assertion. These require the quark and anti-quark wave functions $\psi^q$ and $\psi^{\overline{q}}$ to obey *distinct* wave equations, i.e., to act as independent wave functions. We see, then, that this behavior is allowed by the property of asymptotic freedom.

The picture of this measurement problem that is provided by EPI is, then, the following. The original, unfactored, information Lagrangian for an observed nucleon defines its wave Eq. (4.26). This equation is obeyed by the amplitude function of the *composite* nucleon. However, when the Lagrangian is factored as in Eq. (11.57), equating each factor to zero gives the wave equation that is obeyed by *each* quark (or anti-quark) of the composite nucleon. Thus, the mathematical factorization of an information Lagrangian plays a key *physical* role: *each factor identifies a fundamental constituent particle of the measured composite particle.*

We have examined only the simplest such measurement problem, where a *single meson* is measured. The EPI measurement of *composite mesons* and/or other multiple-quark combinations, complemented by analysis analogous to that preceding, should lead to predictions of the existence of these particles as well. See in this regard Frieden and Plastino (2000).

## 11.5 On field operators in general

Probability amplitudes $\psi$ have, so far, been treated by EPI as ordinary mathematical *functions*, not field operators. However, a field operator approach is useful in (a) solving *many-body problems*, such as in the case of rich particle jets in nuclear reactions, and (b) describing processes that involve the creation or destruction of particles, such as in radioactive beta decay and meson–nucleon interaction. In fact, *the use of field operators is mathematically equivalent to a many-body formulation of non-operator amplitude functions* (Robertson, 1973). Such non-operator amplitude functions characterize most of the preceding EPI applications, for example. However, in view of the benefits (a) and (b) to be gained by an operator approach, it is instructive to consider whether the amplitude functions in EPI output wave equations for many-body problems may, in fact, be *regarded* as operators. The answer is yes, as discussed next.

Suppose that, in applying EPI, we take a complete operator approach: all amplitude functions $\psi$ are now regarded as operators $\psi^{\text{op}}$. As usual we seek the wave equations to be obeyed by them. Integrals of operators are operators. Then, in Eq. (4.2), information $I$ is an operator $I^{\text{op}}$. Also, any integral form defining $J$, such as Eq. (4.9), casts $J$ as an operator $J^{\text{op}}$ as well.

Does it make sense to apply EPI, that is,

$$I^{\text{op}} - J^{\text{op}} = extrem., \tag{11.61}$$

$$I^{\text{op}} - \kappa J^{\text{op}} = 0, \tag{11.62}$$

to the operators $I^{\text{op}}$, $J^{\text{op}}$? If so, what are the general operator solutions $\psi^{\text{op}}$ to the resulting EPI equations? It is known that the operator solution $\psi^{\text{op}}$ to extremum condition (11.61) satisfies *the same* Euler–Lagrange equation as does the *non*-operator (ordinary) solution $\psi$ to the corresponding *non*-operator problem $I - J = extrem.$ (see, e.g., Schiff, 1955, p. 343).

In fact, a previous such solution, $\psi\mu, \mu_2 \ldots \mu_{\text{k}}$ in Eq. (4.46), is indeed an operator since it is a multi-subscripted tensor. Likewise, the second EPI Eq. (11.62) represents the same condition on an operator $\psi^{\text{op}}$ as on a non-operator $\psi$.

Thus any EPI solution $\psi$ either an ordinary function or an operator function.

$$\psi \leftrightarrow \psi^{\text{op}} \tag{11.63}$$

as determined by tensor rank.

The derived wave equation for operator function $\psi$ is often called the "classical" wave equation of the scenario. A classical wave equation is the point of usual starting for developing each field operator (second-quantized) theory. Subsequent steps involve defining an appropriate gauge, defining an associated momentum operator field, finding appropriate commutation relations for $\psi^{\text{op}}$ with itself and with the momentum operator field, etc. The second-quantization problem is not at all trivial, with whole books devoted to accomplishing it. See, for example, Yndurain (1999) or Muta (1998) for the particular case of quantum chromodynamics. Consequently, accomplishing second quantization is regarded as beyond the scope of this book.

# 12

# EPI and entangled realities: the EPR–Bohm experiment

Electron spin was derived by the use of EPI in Chapter 4, in the context of the Dirac equation. This was for the scenario of a single particle. Here, by comparison, we treat spin for a *two-particle* scenario, that of the well-known EPR–Bohm experiment. A general two-particle PDF can exhibit correlation. We show that the well-known probabilities of spin-pair combinations in the EPR–Bohm experiment follow simply from EPI; and that these do obey certain well-known correlation effects. See also a related derivation by S. Luo (2002).

Information $J$ was previously regarded as the "bound" information (Chapter 3), in the sense of the amount of information that is bound to the unobserved source phenomenon. Here we find that it can also represent a degree of information *entanglement* between *two particles*: At the EPI solution, the information $J$ for the observed particle *equals* the information $I$ for the *un*observed one. Or, the source of information for the observed particle is the unobserved one.

A final point of interest will be the use, in Sec. 12.4, of the game corollary $|I| = min.$ (Eq. (3.21)) to determine a vital free parameter of the problem. This approach is also used in Chapter 4 to establish the appropriate number $N$ of amplitude functions to describe various quantum particles, and in Chapter 15 to determine the unknown exponent of a power law for cancer growth. All parameters in the preceding are unitless, as is required for the use of Eq. (3.21). See further discussion in Sec. 12.10.

First we outline the EPR–Bohm experiment. Then we give its predicted spin combinations on the basis of quantum mechanics. Following that, we carry through an EPI derivation of the spin combinations. Finally, we reapply the EPI derivation to an optically analogous polarizer–analyzer experiment, with analogous results.

## 12.1 EPR–Bohm experiment

In everyday activities, we are capable of operating independently of other people. This is essentially because people are very complex, decoherent organisms. However, one of the stranger aspects of quantum mechanics is its prediction of *non-separate* or *entangled* realities. This is illustrated by the EPR–Bohm experiment (Einstein *et al.*, 1935; Bohm, 1951), as sketched in Fig. 12.1.

The source in Fig. 12.1 is a diatomic molecule that decays into a pair of identical spin-1/2 particles, numbered (1) and (2), that head off in opposite directions. The analyzers **a** and **b** consist of Stern–Gerlach magnets that can measure the spins of the two particles.

Specifically **a** measures $S_a = \mathbf{a} \cdot \mathbf{S}_1$, the component of the spin of particle 1 along **a**; and analogously for **b** and particle 2. Let the planes of **a** and **b** be rotated with respect to one another by an angle $x$, $0 \leqslant x \leqslant 2\pi$. Let + designate the event of an observed spin value of $S_a = +\hbar/2$, and − designate an observed spin event $S_a = -\hbar/2$. Also, let

$$S_a S_b \equiv S_{ab} = (++), (--), (+-), (-+) \tag{12.1}$$

describe the event space for *joint* (simultaneous ) spin values $(+\hbar/2, +\hbar/2)$, $(-\hbar/2, -\hbar/2)$, etc.

According to quantum mechanics (Meystre, 1984), there is *strong correlation* between the observed values $S_a$ and $S_b$ of the two spins. In fact, the four possible *joint* spin probabilities $P(S_{ab}|x)$ obey

$$\begin{aligned} P(++|x) &= P(--|x) = \tfrac{1}{2}\sin^2(x/2), \\ P(+-|x) &= P(-+|x) = \tfrac{1}{2}\cos^2(x/2). \end{aligned} \tag{12.2}$$

(Notice that then $\sum_{ab} P(S_{ab}) = 1$, normalization is trivially obeyed at each angle $x$). The aim of this chapter is to show that Eqs. (12.2) for $P(S_{ab}|x)$ follow from use of EPI.

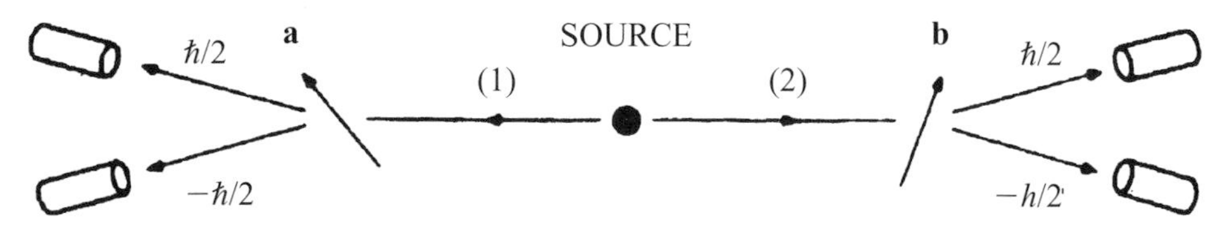


Fig. 12.1. Experimental arrangement of the EPR–Bohm experiment. The angle between the analyzers **a** and **b** is $x$ (from Meystre, 1984, with kind permission of Kluwer Academic Publishers).

## 12.2 Invariance principles

The role of invariance principles in an EPI calculation is to provide a link between the observations and the physics *defining* the observations. Here the physics is that of quantum spin correlation. Interestingly, for this EPI problem the invariance principles that will be imposed are simply *boundary conditions* that follow as particular cases of Eqs. (12.2). The invariance principles are Eqs. (12.3a)–(12.3e) below.

First, take the simple case $x = 0$ where the two analyzer planes have the same orientation. Then (12.2) gives

$$P(+ + |0) = P(- - |0) = 0, \tag{12.3a}$$

meaning that two spins up, or two spins down, are never seen; and $P(+ - |0) = P(- + |0) = 1/2$, meaning that *one* spin *up* and the other *down* or vice versa are *always* seen.

Analogously, if the angle $x = \pi$, Eqs. (12.2) show that one spin up and the other down are *never* seen, $P(+ - |\pi) = P(- + |\pi) = 0$.

Consequently, if either $x = 0$ or $x = \pi$, observing one particle's spin *gives absolute knowledge* of the other's. Clearly, in these cases the spins cannot be viewed as being formed independently or, consequently, obeying separate realities. Instead they obey *entangled realities*.

Also, for *any* angle $x$, Eqs. (12.2) show that either spin value $S_b$ occurs with equal probability. By the law of total probability (Frieden, 2001)

$$P(S_b|x) = P(S_b + |x) + P(S_b - |x) = \tfrac{1}{2} = P(S_b) \quad \text{for } S_b = +, -. \tag{12.3b}$$

The last equality is by

$$P(S_b) = \int dx\, P(S_b|x)p(x) = \int dx\, \frac{1}{2}\frac{1}{2\pi} = \frac{1}{2},$$

using Eq. (12.3d) below.

Another useful property of Eqs. (12.2) is that they show no preference of spin-up over spin-down events, i.e.,

$$P(S_{+-}|x) = P(S_{-+}|x) \quad \text{and} \quad P(S_{--}|x) = P(S_{++}|x). \tag{12.3c}$$

That is, if the experiment were viewed by an upside-down observer, the statistical results would look the same.

It is reasonable to assume that all angles $x$ are *a priori* equally probable over the fundamental interval,

$$p(x) = 1/(2\pi), \qquad 0 \leqslant x \leqslant 2\pi. \tag{12.3d}$$

This also defines a situation of maximum prior ignorance in $x$.

Finally, by elementary considerations

$$P(S_a|S_b) \equiv P(S_{ab})/P(S_b) = \int dx\, P(S_{ab}|x)p(x)/P(S_b).$$

Evaluating the integral by the use of Eqs. (12.2) and (12.3d), and using Eq. (12.3b), gives

$$P(S_a|S_b) = \tfrac{1}{2}, \quad \text{all } S_a,\, S_b. \tag{12.3e}$$

## 12.3 EPI problem

We want to show that Eqs. (12.2) for $P(S_{ab}|x)$ are particular predictions of an appropriate EPI problem. The problem is as follows. *One particle* (say particle (1) in Fig. 12.1) *is observed* for its spin value $S_a$. The spin value $S_b$ of the other particle is *not* observed. It is regarded as unknown but fixed. Also, $S_b$ is not to be estimated from the observation $S_a$. Instead, the rotation angle $x$ between the two planes of Fig. 12.1 is regarded as the unknown parameter of the problem. It is to be estimated from the datum $S_a$.

Since $S_a$ is the datum and $S_b$, $x$ are fixed unknowns, the *likelihood law* for the problem is $P(S_a|S_b, x)$. This probability form is to be found by EPI. Once it is known, $P(S_{ab}|x)$ is readily found from it by the use of identity (12.20) below.

Surprisingly, the mathematics of this EPI problem follows analogous steps to that on deriving the Boltzmann law of statistical mechanics (Sec. 7.3). For the reader's convenience, corresponding equation numbers are shown.

### 12.3.1 Fisher information

The Fisher information that measures the accuracy in estimating the angle $x$ on the basis of a measurement $S_a$ is then

$$\begin{gathered} I(S_b, x) = 4\sum_a q'^2_{ab}(x), \qquad S_b = (+, -), \\ q'_{ab} \equiv dq_{ab}/dx, \qquad P(S_a|S_b, x) \equiv q^2_{ab}(x), \end{gathered} \tag{12.4}$$

in terms of probability *amplitudes* $q_{ab}$. Equation (12.4) represents *two* equations, one for each of the two indicated values of $S_b$. Also the sum is over the *two* possible spin values $S_a = +, -$. The notation $I(S_b, x)$ indicates the information in observation of particle (1) about the unknown angle $x$, in the presence of some fixed (and unknown) value $S_b$ of the spin in particle (2).

### 12.3.2 Expression for Fisher information capacity

Since the angle $x$ is continuously any value on the interval $(0, 2\pi)$, Eq. (12.4) represents an infinity of Fisher information values. On the other hand, EPI requires the use of a *single*, scalar information *capacity* value $I_1$. We get such a value from two different points of view.

First, proceed as in Secs. 2.2.2–2.3.2, forming the information capacity as *the sum* of the informations over the appropriate prior measurement space. Note in this regard that Eq. (2.7) actually represents a *doubly* subscripted information $I_\nu \equiv I_{n\nu}$, the subscript $n$ having been suppressed for simplicity (at the beginning of Sec. 2.2.2). The correspondence is then that $I_{n\nu} \equiv I_{b\nu} = I(S_b, x_\nu)$ here. The sum is therefore to be taken over all possible values of the unknown spin $S_b$ of particle 2 and angle $x_\nu$. The result is an information *capacity*

$$I_1 \equiv \sum_{b\nu} I(S_b, x_\nu) \to \sum_b \int dx\, I(S_b, x) = 4 \sum_{ab} \int dx\, q'^2_{ab}(x) \tag{12.5}$$

since the $x_\nu \to x$ are *continuous*. Eq. (12.4) was used. The limits on all integrals in this EPR–Bohm problem are from 0 to $2\pi$ owing to Eq. (12.3d). The indicated sum $ab$ is over all joint spin values $S_{ab}$.

Alternatively, by Exercise 12.3.7 below, $q_{ab}(x)$ – a probability amplitude *on random events* $S_a$ conditional upon fixed events $S_b, x$ – is proportional to $\psi_{ab}(x)$, the probability amplitude *on random* $x$, conditional upon fixed events $S_{ab}$. The information inequality (2.47) holds for such probability amplitudes $\psi_{ab}(x)$, with the channel capacity proportional to the right-hand side of Eq. (12.5). The presence of a proportionality factor does not affect the derivation to follow.

### 12.3.3 Bound information $J_1$

Represent the bound information $J_1$ for the observed particle as

$$J_1 \equiv 4 \sum_{ab} \int dx\, j_{ab}[q_{ab}] \tag{12.6}$$

(cf. Eq. (7.11)). The bound information is defined, as usual, to represent the unseen *source* of the observation $S_a$. We return to this important point below.

### 12.3.4 EPI conditions

The two EPI conditions are the extremum condition (3.16)

$$K \equiv I_1 - J_1 = 4 \sum_{ab} \int dx\, (q'^2_{ab} - j_{ab}[q_{ab}]) = extrem. \tag{12.7}$$

(cf. Eq. (7.12)), and the zero-condition (3.18)

$$I_1 - \kappa J_1 \equiv I_1 - J_1 = 4\sum_{ab} \int dx\, (q'^2_{ab} - j_{ab}[q_{ab}]) = 0. \qquad (12.8)$$

We took

$$\kappa = 1, \quad \text{or} \quad I_1 = J_1, \qquad (12.9)$$

as is usual for a quantum problem (see also Chaps. 4, 10, and 11). That $I_1 = J_1$ will agree with the concept of entangled realities for the two particles: see Sec. 12.7.

### 12.3.5 Self-consistent solution

We look for a *simultaneous* solution $q_{ab}$, $j_{ab}$ to Eqs. (12.7) and (12.8). This is the self-consistent EPI option (a) described in Sec. 3.4.6. The *extremum* solution to (12.7) obeys, by Eq. (0.34),

$$q''_{ab} = -\frac{1}{2}\frac{dj_{ab}}{dq_{ab}} \qquad (12.10)$$

(cf. Eq. (7.15)). The full derivative $dj_{ab}/dq_{ab}$ replaces the corresponding partial derivative because, by (12.6), $j_{ab} \equiv j_{ab}[q_{ab}]$ alone.

We next turn to the *zero*-condition (12.8). It is useful to first integrate Eq. (12.5) by parts, giving

$$I_1 = 4\sum_{ab} \int dx\, (C_{ab} - q_{ab}q''_{ab}),$$
$$C_{ab} = \frac{1}{2\pi}[q_{ab}(2\pi)q'_{ab}(2\pi) - q_{ab}(0)q'_{ab}(0)] \qquad (12.11)$$

(cf. Eq. (7.16)). The EPI solutions for the $q_{ab}(x)$ turn out to obey $C_{ab} = 0$, but at this point the $C_{ab}$ must be regarded as unknown finite constants. The zero-condition (12.8) is then, by Eqs. (12.6) and (12.11),

$$I_1 - J_1 = 4\sum_{ab} \int dx\, [C_{ab} - q_{ab}q''_{ab} - j_{ab}] = 0 \qquad (12.12)$$

(cf. Eq. (7.17)). The microscale solution (3.20) is then

$$C_{ab} - q_{ab}q''_{ab} - j_{ab} = 0 \qquad (12.13)$$

(cf. Eq. (7.18)) for each joint particle spin state $S_{ab}$.

According to plan we now combine the extremum solution and zero solution (12.10) and (12.13). Multiplying Eq. (12.10) by $q_{ab}$ and using the resulting product $q_{ab}q''_{ab}$ in Eq. (12.13) gives

$$\frac{1}{2} q_{ab} \frac{dj_{ab}}{dq_{ab}} = j_{ab} - C_{ab} \tag{12.14}$$

(cf. Eq. (7.19)). This may be placed in the convenient form

$$\frac{2\, dq_{ab}}{q_{ab}} = \frac{dj_{ab}}{j_{ab} - C_{ab}}$$

for integration, giving

$$j_{ab} \equiv j_{ab}[q_{ab}] = \frac{q_{ab}^2}{A_{ab}^2} + C_{ab}, \qquad A_{ab} = const. \tag{12.15}$$

(cf. Eq. (7.20)). The constants of integration $A_{ab}$ are at this point generally complex.

With $j_{ab}$ thereby known, we substitute it into either Eq. (12.10) or Eq. (12.13) to give a differential equation in $q_{ab}$ of

$$q''_{ab} = -\frac{q_{ab}}{A_{ab}^2} \tag{12.16}$$

(cf. Eq. (7.24a)). Since the amplitude functions $q_{ab}$ of EPI are generally real, so likewise must be the $q''_{ab}$, so that, by (12.16), the constants $A_{ab}^2$ must be real.

The solutions $q_{ab}(x)$ to (12.16) are *generally complex exponentials* $B_{ab}\exp(ix/A_{ab}) + C_{ab}\exp(-ix/A_{ab})$, $B_{ab}$, $C_{ab}$ = *consts*. Because the $A_{ab}^2$ are real (see preceding), it must be that either $A_{ab} = a_{ab}$ or $A_{ab} = ia_{ab}$ for some real constants $a_{ab}$. By these respective choices the solution is either a pure trigonometric $B_{ab}\exp(ix/a_{ab}) + C_{ab}\exp(-ix/a_{ab})$ or a pure exponential $B_{ab}\exp(x/a_{ab}) + C_{ab}\exp(-x/a_{ab})$. The trigonometric law can be alternatively expressed as $B_{ab}\sin(x/a_{ab}) + C_{ab}\cos(x/a_{ab})$. Or, since we had $A_{ab} = a_{ab}$ for that solution, it becomes $B_{ab}\sin(x/A_{ab}) + C_{ab}\cos(x/A_{ab})$. Here the $B_{ab}$, $C_{ab}$ are new constants, which are real since, again, the $q_{ab}(x)$ must be real.

Which of the two choices of solution should be chosen here? At this point we diverge from the derivation of Sec. 7.3. There the interval for $x$ was *semi-infinite* since $x$ was an energy. Then a trigonometric choice of solution could not obey normalization. For this reason the (decreasing) exponential choice (7.25b) was made. By comparison, Eq. (12.3d) states that the range of $x$ is *finite*: limited to the interval $(0, 2\pi)$. Also, since $x$ is an *angle* in the probability law $P(S_a|S_b, x)$, a periodic dependence upon $x$ is appropriate. Its corresponding amplitude law $q_{ab}(x)$ is then likewise periodic. Hence the *trigonometric choice* is the physically meaningful one here. As mentioned above, this has the general form

$$q_{ab}(x) = B_{ab}\sin(x/A_{ab}) + C_{ab}\cos(x/A_{ab}), \qquad A_{ab}, B_{ab}, C_{ab} = \textit{real consts}. \tag{12.17a}$$

### 12.3.6 Choices for $A_{ab}$ in view of orthogonality

The sine and cosine functions in (12.17a) form basis functions for the amplitude functions $q_{ab}(x)$. It will be useful for us to enforce orthogonality upon these sine and cosine functions, specifically over the continuous space of $x$. The basis functions for quantum amplitude functions *obey* orthogonality. Thus, we must first show that the $q_{ab}(x)$ are proportional to quantum amplitude functions $\psi_{ab}(x)$ over the space of $x$.

The complication is that the $q_{ab}(x)$ describe the random behavior of events $S_a$, not events $x$. Therefore they are not yet quantum amplitude functions $\psi_{ab}(x)$. However, the $q_{ab}(x)$ may readily be shown to be *proportional to* the quantum solutions $\psi_{ab}(x)$.

### 12.3.7 Exercise: establishing that $\psi_{ab}(x) \propto q_{ab}(x)$

Show the proportionality. *Hint*: By Bayes' rule and the definition of conditional probabilities,

$$p(x|S_{ab}) \equiv \psi^2_{ab}(x|S_{ab}) \equiv \psi^2_{ab}(x) = \frac{P(S_a|S_b, x)P(S_b)p(x)}{P(S_a|S_b)P(S_b)}.$$

Then use Eq. (12.3d), the last Eq. (12.4), and (12.3e).

Therefore the sine and cosine basis functions for the $q_{ab}(x)$ in (12.17a) can now be required to be mutually *orthogonal* over the space of $x$. By (12.3d) the latter is interval $x = (0, 2\pi)$. Thus we require

$$\int_0^{2\pi} dx \sin(x/A_{ab}) \cos(x/A_{ab}) = 0. \tag{12.17b}$$

Evaluating the integral gives the requirement that

$$A_{ab} = \frac{2}{n_{ab}}, \qquad n_{ab} = 1, 2, \ldots. \tag{12.17c}$$

Hence, each of the $A_{ab}$ must be some one of these values. It is found below that, in fact, all the $A_{ab}$ are equal – the value (12.17c) where all $n_{ab} = 1$. This will follow from a requirement that the $A_{ab}$ obey the game corollary.

## 12.4 Use of game corollary to fix the constants $A_{ab}$

Although (12.17c) is a condition on the possible values of the $A_{ab}$, it does not uniquely fix them. We need another condition. Note in this regard that the angles $x$ are unitless, so that the unknown $A_{ab}$ are also unitless (see Eq. (12.17a)). Since the unknowns are unitless, they may be found by the use of Eq. (3.21) – the *game corollary* of the EPI procedure.

To carry through, we first need to compute the $C_{ab}$. Using Eqs. (12.17a) and (12.17c) in the definition (12.11) of $C_{ab}$ gives directly

$$C_{ab} = B_{ac}C_{ac}\frac{n_{ab}}{2}[\cos^2(n_{ab}\pi) - 1] = 0. \tag{12.18a}$$

### 12.4.1 Exercise: values of $C_{ab}$

Carry through the derivation.

A second, necessary step is to compute the integral

$$\int dx\, q_{ab}^2(x) \equiv \int dx\, P(S_a|S_b, x) = \int dx\, \frac{p(x, S_a|S_b)P(S_b)}{p(x, S_b)}$$

$$= \int dx \frac{p(x, S_a|S_b)\frac{1}{2}}{P(S_b|x)p(x)} = \int dx\, \frac{p(x, S_a|S_b)\frac{1}{2}}{\frac{1}{2}\frac{1}{2\pi}}$$

$$= 2\pi \int dx\, p(x, S_a|S_b) = 2\pi P(S_a|S_b) = 2\pi\frac{1}{2} = \pi. \tag{12.18b}$$

The first step is by definition (12.4) of $q_{ab}^2(x)$, the second by Bayes' rule, the third by Eq. (12.3b) and the definition of a conditional probability, the fourth by Eqs. (12.3b) and (12.3d), the fifth by cancellations, the sixth by the summation law for probabilities, and the seventh by Eq. (12.3e).

It is now possible to express the information $I_1$ *purely* in terms of the $n_{ab}$. Using Eqs. (12.6), (12.9), (12.15), (12.17c), (12.18a) and (12.18b) gives

$$I_1 = J_1 = 4\sum_{ab}\frac{1}{A_{ab}^2}\int_0^{2\pi} dx\, q_{ab}^2 = 4\pi\sum_{ab}\frac{1}{A_{ab}^2} = \pi\sum_{ab} n_{ab}^2. \tag{12.18c}$$

This will give us a handle on finding the $n_{ab}$.

The game corollary of the EPI approach, Eq. (3.21), is here a statement

$$I_1 = min. \tag{12.18d}$$

through choice of the $A_{ab}$. Condition (12.18d) is also consistent with the "information hoarding game" described in Sec. 3.4.12, by which the aim of the "demon" is to attain a *minimum* payout of information $I_1$.

As discussed in Sec. 3.4.12, the minimum is generally attained by *maximal blur* of the amplitude functions, here the $q_{ab}(x)$. This is verified by the forms of Eqs. (12.17a) and (12.18c): By Eq. (12.17a), maximum blur in each $q_{ab}(x)$ is attained by a *maximum* value for $A_{ab}$; and by Eq. (12.18c), the *information payout $I_1$ is minimized* by, again, a *maximum* value for each $A_{ab}$.

Now, the values of the $A_{ab}$ are limited to the choices (12.17c). Clearly the maximum possible value for each $A_{ab}$ is then

$$A_{ab} = 2, \quad \text{all } S_a \text{ and } S_b, \tag{12.19}$$

corresponding to the choice $n_{ab} = 1$. This attains the requirement (12.18d).

## 12.5 Getting the constants $B_{ab}$, $C_{ab}$ by orthogonality

There are eight constants $B_{ab}$, $C_{ab}$ in (12.17a) that need to be found. We want to find them next by use of the invariance principles in Sec. 12.2. However, the latter are largely properties of the *joint* probabilities $P(S_a S_b|x)$, whereas (by the last Eq. (12.4)) our $q_{ab}(x)$ are probability amplitudes for values of $S_a$ *conditional upon* values of $S_b$. The connection between the two is, from elementary probability theory,

$$P(S_{ab}|x) \equiv P(S_a S_b|x) \equiv P(S_a|S_b, x)P(S_b|x) = q_{ab}^2(x)P(S_b|x) = \tfrac{1}{2}q_{ab}^2(x) \tag{12.20}$$

by Eq. (12.3b).

Equation (12.20) shows that $q_{ab}(x) = 0$ if $P(S_a S_b|x) = 0$. Then evaluating Eq. (12.17a) at $x = 0$ and using Eq. (12.3a) gives

$$C_{++} = C_{--} = 0. \tag{12.21}$$

We next turn to finding $B_{+-}$ and $B_{-+}$.

We found in Exercise 12.3.7 that the $q_{ab}(x)$ are proportional to quantum amplitude functions $\psi_{ab}(x)$. Therefore $q_{++}(x)$ must be orthogonal to $q_{+-}(x)$. Thus, by Eqs. (12.17a), (12.19), and (12.21),

$$\int dx \sin(x/2)[B_{+-} \sin(x/2) + C_{+-} \cos(x/2)] = 0. \tag{12.22}$$

Notice that the integrand term $\sin^2(x) \geqslant 0$ at all $x$. Therefore it contributes a positive value to the integral. The other term $\sin(x/2)\cos(x/2)$ contributes zero, since its integral obeyed our orthogonality condition (12.17b). Therefore, the required zero for the integral (12.22) can be attained only if

$$B_{+-} = 0 = B_{-+}, \tag{12.23}$$

the latter because of the up–down symmetry (12.3c).

By the results (12.19), (12.21), and (12.23), Eq. (12.17a) becomes

$$q_{++}(x) = B\sin(x/2), \quad q_{--}(x) = B\sin(x/2), \quad B \equiv B_{++},$$

$$q_{-+}(x) = C\cos(x/2), \quad q_{+-}(x) = C\cos(x/2), \quad C \equiv C_{-+}. \tag{12.24}$$

The identities $B_{++} = B_{--}$ and $C_{-+} = C_{+-}$ were also used via the up–down symmetry condition (12.3c). There are now but two constants to evaluate, $B$ and $C$.

In view of Eqs. (12.20), the normalization condition on probabilities $P(S_a S_b|x)$ amounts to requiring

$$2 = q^2_{++}(x) + q^2_{--}(x) + q^2_{+-}(x) + q^2_{-+}(x). \quad (12.25)$$

Then the use of Eqs. (12.24) requires

$$2 = 2B^2 \sin^2(x/2) + 2C^2 \cos^2(x/2) = 2(B^2 - C^2)\sin^2(x/2) + 2C^2 \quad (12.26)$$

by elementary trigonometry. In the outer equality the right-hand side is a function of $x$ whereas the left is not. The only way the two can balance for arbitrary $x$ is for

$$B^2 = C^2 = 1. \quad (12.27)$$

Use of Eqs. (12.27) in (12.24) gives

$$q_{++}(x) = q_{--}(x) = \sin(x/2), \quad q_{+-}(x) = q_{-+}(x) = \cos(x/2). \quad (12.28)$$

(We arbitrarily chose positive signs; this will not matter since only their squares are of import; see next.) Using these in Eqs. (12.20) gives the final joint probabilities

$$P(+\,+|x) = P(-\,-|x) = (1/2)\sin^2(x/2),$$
$$P(+\,-|x) = P(-\,+|x) = (1/2)\cos^2(x/2). \quad (12.29)$$

These verify the EPR–Bohm spin Eqs. (12.2).

### 12.5.1 Exercise: self-consistency of the solution

Check that the solutions (12.28) do simultaneously satisfy the two EPI conditions (12.10) and (12.13). *Hint*: This is readily done using Eqs. (12.15), (12.19), and (12.28), duly keeping track of minus signs.

### 12.5.2 Exercise: showing that $P(S_{ab}) = 1/4$

(a) Show that the joint probability of spins $S_a$ and $S_b$ obeys $P(S_{ab}) = 1/4$. *Hint*: Use the law of total probability to express $P(S_{ab})$ in terms of $P(S_{ab}|x)$ and $p(x)$, then use Eqs. (12.20) and (12.28) to evaluate the required integral.

(b) By summing out this result over spins $S_b$ show that $P(S_a) = 1/2$. Show similarly that $P(S_a) = 1/2$. Then, combining this result with that of (a), we see that

$$P(S_{ab}) = P(S_a)P(S_b). \quad (12.30)$$

This describes a situation of independent spins and, therefore, *separate realities*. Doesn't this violate the spirit of the EPR–Bohm experiment?

The probability $P(S_{ab})$ in (12.30) represents, by definition, that of a joint spin occurrence $S_{ab}$ in the *absence* of information about the value of $x$. Consequently, this event arises as the outcome of an EPR–Bohm experiment in which the angle $x$ is *not known*, and maximally so by (12.3d). Now, as Eqs. (12.28) indicate, the spin correlation effects *vary* with knowledge of $x$, and strongly so. Hence, one might expect the correlations to be reduced (at least) by the averaging over $x$ that occurs when $x$ is not known. In fact Eq. (12.30) states that they are not only reduced, they disappear! That is, under these conditions the spins $S_a$ and $S_b$ are independent.

This has an interesting physical interpretation. Ehrenfest's theorem states that the average value of a quantum mechanical parameter equals its value from *classical* mechanics. In this light, Eq. (12.30) states that, when averaged over angle $x$, the quantum EPR–Bohm spin entanglement reverts to *classical* separation of variables and realities.

## 12.6 Uncertainty in angle

The aim of the EPI *measurement* experiment (not the EPI derivation) was to estimate the angle $x$ based upon observation of a spin value $S_a$. We can now compute how good the estimate will be.

The value of the information capacity $I_1$ at the solution is computed next. It was shown in Eqs. (12.18c) that

$$\int dx\, q_{ab}^2(x) = \pi, \qquad \text{all } S_a,\, S_b. \tag{12.31a}$$

Note that this is not necessarily unity by definition (12.4) of $q_{ab}^2(x)$. Then Eqs. (12.6), (12.9), (12.15), (12.18a), (12.19), and (12.31a) give

$$I_1 = 4\sum_{ab}\int dx\, j_{ab}[q_{ab}] = 4\sum_{ab}\int dx\, \frac{q_{ab}^2(x)}{4} = 4\cdot 4\cdot \pi/4 = 4\pi. \tag{12.31b}$$

Using this value of $I_1$ in the Cramer–Rao inequality Eq. (1.1) gives a mean-square error $e^2$ in the estimated angle obeying

$$e^2 \geqslant \frac{1}{4\pi}, \qquad \text{or} \qquad e \geqslant 0.28 \text{ rad}. \tag{12.32}$$

*Note*: Actually the Cramer–Rao inequality states that $e^2 \geqslant 1/I$, where $I$ is the *Fisher* information. By comparison, our information $I_1$ is a Fisher information *capacity* (12.5), obeying $I_1 \geqslant I$. But then identically $1/I_1 \leqslant 1/I \leqslant e^2$ by Eq. (1.1). Hence $e^2 \geqslant 1/I_1$ as well. This is the statement (12.32).

Equation (12.32) states that the observation of a single spin value in the face of *maximum* ignorance (12.3d) gives small, but finite, information about the angle. The error of 0.28 rad is rather large. However, with more precise prior information about angle, i.e., a narrower $p(x)$ than as in (12.3d), the recalculated error $e^2$ would, of course, be smaller.

### 12.7 Information $J$ as a measure of entanglement

In Eq. (12.9) we used the hypothesis that $J_1 = I_1$. This hypothesis is used in all applications of EPI to quantum problems. What does this equality mean in the context of the current problem?

In the current problem $I_1$ is the information about angle that is contained in an observation of the spin of particle (1). The spin state of particle (2) remains *unobserved*. (This is why its spin $S_b$ could have either value + or − throughout the analysis.) Likewise, the bound information $J_1$ for particle (1) is, by Eq. (12.9), the information at the unobserved *source* of the data. These imply that *the bound information for particle* (1) *originates at particle* (2). It follows that the data space, i.e., the observed spin reality of particle (1), is equivalent, on the level of information, to the source space of the unobserved particle. This equivalence of the two spaces is ultimately why the inferences about the spin of the unobserved particle (2) in Sec. 12.3 can be made. The inferences define a situation of entangled realities for the two particles. Note that the entanglement is generally complete (see Sec. 12.8).

These results lead us to propose that a general EPI statement $I = J$, *regardless of application*, implies an entanglement of the data space of the observed particle with the space of the unobserved particle. In this way, EPI is found to have a built in mechanism for predicting entangled realities. In these scenarios, $J$ represents not only a "bound" information but, also, an "entangled" information. Quantum applications always obey $I = J$. Therefore, in quantum applications, $J$ may always be regarded as an entangled information.

It appears that there are two types of such EPI entanglement. In this chapter the entanglement is that of an observed particle with an unobserved particle. This is entanglement in the true sense. By contrast, in Chapter 4 it is the entanglement of the state $\psi_n$ of the observed particle in space-time, with *its own state* $\phi_n$ in momentum-energy space. The latter is, then, a trivial case of *self*-entanglement.

Given this significance for the condition $I = J$, it is important to trace its origin. In Sec. 3.4.17 we found that the condition $I = J$ is merely an expression of the fact that *any $L^2$ norm, such as I, is invariant to a unitary transformation*. On this basis we propose that the underlying basis for *all* *quantum entanglement* is the peculiar $L^2$ form of the Fisher information $I$. But one step further back, Fisher information has the $L^2$ form because it obeys the *Cramer–Rao inequality* (1.1). Then quantum entanglement follows, most fundamentally, as the result of a quest for finding the true value of a parameter, i.e., a quest for knowledge. But, of course, the same can be said of all results of EPI applications in this book.

## 12.8 Information game

Since the coordinates of this problem are purely real, the solution to the EPI Eqs. (12.7) and (12.8) for the problem obeys a *game of information hoarding* (Sec. 3.4.12). The game is between the observer of spin $S_a$ and the "information demon" (Sec. 3.4.13). The demon personifies the decay process of the diatomic molecule in Fig. 12.1. By Secs. 2.1.3, 2.6, and 3.4.12, the observer seeks to *maximize* his received information $I_1$ by making spin values $S_a$, $S_b$ occur as *independently* as possible, while from the form of Eqs. (12.18c) the demon seeks to *minimize* his expenditure of information $I_1$ by maximizing the values of the constants $A_{ab}$. This is equivalent to *maximal blurring* in $x$, by the form of Eq. (12.17a). The payoff of the game is therefore at a saddle point solution (Sec. 3.4.11), and this is satisfied by the output probability laws (12.29) of the EPI calculation.

Thus, by Eq. (12.3a) *et seq.*, the payoff of *the game* predicts *complete entanglement* of realities when angle $x = 0$ or $\pi$. This is the entanglement phenomenon couched in terms of game theory.

A practical aspect of the game is that it allowed the constants $A_{ab}$ of the problem to be found in Sec. 12.4.

## 12.9 Analogous polarization experiment

There is a well-known optical analog to the EPR–Bohm spin-pair experiment of Fig. 12.1. This is a polarization experiment. In place of the Stern–Gerlach apparatus on the far right there is a polarizer, e.g., a Nicol prism that polarizes incident light from the right such that its direction of vibration is (say) purely vertical. The light proceeds to the far-left space, where it encounters an analyzer whose direction of preferred vibration is rotated out of the page by a fixed angle $x$. Hence $x$ is the angle between the direction of vibration of the polarized light and the preferred direction of the analyzer. Let the angle $x$ be *unknown* to the observer.

Let a random variable $a = +, -$ define the two alternatives for a photon that transits the apparatus: $a = +$ is the event that the photon is *observed to pass through* the analyzer and $a = -$ is the event that it is *not observed to pass through*. The EPI measurement experiment is the observation of a value of $a$. Since $x$ is unknown, this measurement follows a likelihood law $P(a|x)$. This is the physical law governing the observation. It is to be found using EPI.

The problem may be attacked by EPI as in the preceding spin-pair experiment. Therefore we can be brief. Corresponding equations of the spin-pair analysis are indicated parenthetically, as following Eq. (12.33a).

It is an experimental fact that if $x = 0$, all incident photons pass through the analyzer, whereas if $x = \pi/2$, none do; that is,

$$P(+|0) = 1, \quad P(-|0) = 0, \quad P(+|\pi/2) = 0, \quad P(-|\pi/2) = 1 \qquad (12.33a)$$

(cf. Eq. (12.3a) *et seq*.). The law $P(a|x)$ should depend upon the magnitude of the angle $x$, not its sign. Therefore, the effective range of $x$ is now *half* the fundamental interval

$$x = (0, \pi) \qquad (12.33b)$$

(cf. Eq. (12.3d)). This is compared with the *full* interval $(0, 2\pi)$ used in the EPR–Bohm case. The boundary conditions (12.33a, b) are regarded as *invariance principles* for generating the solution below.

The channel capacity information in the observation $a$ is

$$I_1 = 4\sum_a \int dx\, q_a'^2(x), \quad P(a|x) \equiv q_a^2(x), \quad a = -, + \qquad (12.34)$$

(cf. Eq. (12.5)). All integrals are here from 0 to $\pi$, in view of (12.33b).

The bound information is now

$$J_1 \equiv 4\sum_a \int dx\, j_a[q_a] \qquad (12.35)$$

(cf. Eq. (12.6)). Complete entanglement of informations is again assumed,

$$\kappa = 1 \text{ or } I_1 = J_1 \qquad (12.36)$$

(cf. Eq. (12.9)).

Steps analogous to Eqs. (12.7)–(12.16) are taken. Results are as follows.

The new constants $A_a \equiv A_+, A_-$ replace the $A_{ab}$ of the EPR analysis. The requirement $I =$ *minimum* of the *game corollary* now gives

$$A_+ = A_- = 1 \qquad (12.37)$$

(cf. Eq. (12.19)). The result (12.37) differs from the EPR–Bohm result (12.19) effectively because the range for $x$ is now the half interval (12.33b) as opposed to the full interval (12.3d).

The amplitudes obey

$$q_+(x) = B_1 \cos(x), \quad q_-(x) = B_2 \sin(x), \quad B = const.,$$

$$\text{i.e.,} \quad P(+|x) = B_1^2 \cos^2(x), \quad P(-|x) = B_2^2 \sin^2(x) \tag{12.38}$$

(cf. Eqs. (12.24)). The second Eq. (12.34) and the second and third boundary conditions (12.33a) were used. Finally, imposing the normalization property $P(+|x) + P(-|x) = 1$ upon solutions (12.38) gives

$$B_1^2 = B_2^2 = 1. \tag{12.39}$$

Then back substitution of (12.39) into (12.38) gives

$$P(+|x) = \cos^2(x), \quad P(-|x) = \sin^2(x). \tag{12.40}$$

This is equivalent to Malus' law in electromagnetic theory.

Here the entanglement is between the data space of the observed photon and its source space at the polarizer. Specifically, the state $a$ of the particle as to whether it is emerging or not emerging from the analyzer *is entangled with* the unknown polarization state (direction of vibration) of the polarized photons. According to Eq. (12.40), the degree of entanglement increases as the angle $x$ approaches either zero or $\pi/2$. The limiting values are *unit* probability either for the event $a =+$ (emergence) for the *zero* angle, or for the event $a = -$ (non-emergence) for the angle $\pi/2$.

Thus, *if* in repetitions of the experiment the observer sees the photon as always emerging, or as never emerging, he knows that the angle angle $x = 0$ in the first instance or $x = \pi/2$ in the second. Equivalently, in the former case he knows that the polarizer and the analyzer are parallel, whereas in the second case he knows that they are perpendicular. This describes *complete* entanglement of the realities of the spaces of the polarizer and the analyzer.

As in Sec. (12.8) for the EPR–Bohm problem, this polarization problem obeys the game aspect of EPI. The game is between the observer at the analyzer and a demon at the polarizer. Each tries to maximize its level of Fisher information through adjustment of the curve $q_a(x)$ governing the effect. The observer tries to *maximize* his gain of information through adjustment of the dependence of $q_a(x)$ upon the two values of $a$, and the demon tries to *minimize* his expenditure of information through adjustment of the dependence of $q_a(x)$ upon $x$. (*Note*: These adjustments are obviously not independent.) The payoff of the game is the solution "point" (12.40), defining the probability laws for the problem. Equation (12.40) show complete entanglement at angles $x = 0$ or $\pi/2$.

## 12.10 Discussion: Can EPI determine all unitless constants?

Note the use of the game corollary $|I| = min.$ to determine the vital free parameters $A_{ab}$ and $A_a$ of these two problems. This approach is used throughout the book to determine unknown unitless constants; e.g., in Chapter 7 to fix the Boltzmann energy distribution as a unique law; and in Chapter 15 to determine the unknown exponent of a power law for cancer growth. In all cases the parameters are determined *analytically*.

That all such constants are *unitless* is of interest. Generalizing from this experience with EPI, the implication is that it can be used to determine the values of *all* unitless universal physical constants. Examples are the various *fine structure constants*.

Of course, the fine structure constants are, themselves, defined in terms of the ordinary physical constants such as $c$, $e$, $\hbar$, etc. Therefore knowledge of the former should result in new, *analytical relations linking* the latter. These relations have not yet been found.

The existence of these relations would imply that the physical constants are not all independent; or equivalently, only a subset of them is truly fundamental.

## 12.11 "Active" information and $J$

The *physical* information $K = I - J$ of Chapter 3 was specifically formed so as to apply to *physical processes*. This is to be compared, e.g., with the intended applications of the Shannon form of information, which are to communication networks; or the original intended area of application of the Fisher form, to estimation theory in general.

An alternative approach to forming a physical information was taken by Bohm and Hiley (1993) (referred to as B–H in the following), who qualitatively defined something called the *active information*. (Note that EPI is not synonymous with the Bohm intepretation of quantum mechanics. See below.)

Although its precise mathematical form is left undefined by B–H, active information is qualitatively defined as having the physical ability "to in-form, which is actively to put form into something or to imbue something with form" (all quoted phrases in this section are from B–H). An example of the latter is a radio wave, which is too weak an energy to be detected unaided by us, but which takes the form of audible sound once it enters the radio. This form has been generated in some way by the active information interacting with the incoming radio wave.

"Active information is ... capable of participating in the thing to which it refers." Thus, active information has a participatory aspect, suggesting Wheeler's principle of Sec. 0.1. As an example on the microlevel, the pilot wave of de Broglie and Bohm somehow 'acts' to guide the quantum particle. Thus, in the famous two-slit experiment, the motion of the particle is guided by active information provided by the wave function $\psi$. The latter is (according to B–H) merely a kind of field function, and acts to guide the particle through one or the other of the two slits by means of an active information of some kind.

Although the mathematical form of the active information is left unspecified, "... all information is at least potentially active." However, entropy and Shannon information are regarded as, specifically, *passive* informations by B–H. What mathematical form, then, should the active information have?

From the EPI derivation of the EPR–Bohm spin results (12.29), information $J$ obeys many of the requirements of a B–H active information. First of all, it is an information. Second, it is an information of *entanglement*, providing all the information about the *un*observed particle that is present in the *observed* particle's data space (Sec. 12.7). This relay of information $J$ is not only to the observer, but (by the B–H picture) also to the observed particle itself. This elicits the usual dual effect of EPI.

(a) The relay of information to the *observer* gives rise to his *knowledge* of the output law (12.29) governing spins; this is via the EPI derivation (Secs. 12.3–12.5).
(b) The relay of information $J$ to the *observed particle* gives rise to a sampled datum from the output law, the spin value $S_a$. This implies that the information relay causes the particle to *obey* the output spin law (12.29). The cause is some unknown interactive mechanism, perhaps a "rich and complex inner structure which can respond to information and direct its self-motion accordingly." In this way, the relay of information $J$ ties together the spin states of the unobserved and observed particles. Thus, $J$ "gives form" (as required) to the spin properties of the *un*observed particle.

The information $J$ plays analogous roles to the above in other EPI applications of this book. On this basis the intended role of the B–H active information seems to be filled at least in part by the information $J$.

Aspects of Bohm mechanics that differ from those of EPI should also be mentioned. The two approaches may be compared as follows.

(1) Bohm mechanics regards the laws of mathematical physics as absolutes, existing independently of any observer. Hence it takes an ontological viewpoint. EPI regards such laws as resulting from a quest for knowledge of the value of a parameter $\theta$ (Chapter 3; Sec. 10.10). The quest is enacted via a measurement. By this view the laws are fundamentally observer-dependent and participatory. Whereas Bohm mechanics is ontological, EPI is epistemological.
(2) Bohm mechanics is fundamentally deterministic. It assumes the presence of definite trajectories $\boldsymbol{r}(t)$ for particles, and gives a prescription for computing them from a known field function $\psi(\boldsymbol{r}, t)$. EPI regards all physical effects as fundamentally statistical, principally because of their uniform tie-in to Fisher information, which measures statistical error.
(3) As a consequence of (2) preceding, Bohm mechanics assumes the existence of a

*definite value* for a parameter $\theta$. EPI assumes this as well, in that it is knowledge of the value of $\theta$ (see (1) preceding) that is being sought when making a measurement. A counterargument is that $\theta$ drops out of the Fisher $I$ expression (Sec. 1.2.4) under shift-invariant conditions, and therefore knowledge of specific values of $\theta$ is not crucial to resulting EPI applications. However, the derivation of $I$ (Sec. 1.2.3) is under the assumption that a definite value for $\theta$ *exists*, whatever this value might be. Note that this is also precisely the stance of *classical* estimation theory (Sec. 1.2.1), the progenitor of EPI (see, e.g., Sec. 1.8.13). Hence knowledge that $\theta$ *has* a definite value is essential to EPI, although its actual numerical size is not needed.

(4) Bohm mechanics interprets $\psi(\boldsymbol{r}, t)$ as being, *not* a probability amplitude function (in view of (2) preceding), but rather a kind of field function, whose form may be computed from the SWE. In turn, the trajectories $\boldsymbol{r}(t)$ are computed from $\psi(\boldsymbol{r}, t)$. (Note an alternative definition and route to the calculation of trajectories, in Frieden and Plastino (2001b).) By comparison, consistently with (2) preceding, EPI regards $\psi(\boldsymbol{r}, t)$ in all its applications as a *probability* amplitude function.

(5) Bohm mechanics *assumes* the form of the SWE in order to compute $\psi(\boldsymbol{r}, t)$. The emphasis in Bohm mechanics is upon computing particular trajectories and other "absolutes" of motion. By comparison, the aim of EPI is to derive physical laws, so it *derives* the SWE, and also its relativistic versions (Chapter 4, Appendix D).

# 13

# Econophysics

by B. Roy Frieden and Raymond J. Hawkins

(Specifically economic terms that may be unfamiliar to the reader are defined in the glossary, Sec. 13.10. These terms are identified by sans serif font in the text the first couple of times they are used.)

The overall aim of the discipline of econophysics is to apply the methods of mathematical physics to problems of economics. Probably the most basic phenomenon connecting economics and physics is *Brownian motion*, as analyzed by Louis Bachelier (1900) in his prescient Ph.D. thesis. Fast forwarding 100 years, a nice recent introduction to econophysics is the book by Mantegna and Stanley (2000). A classic book that brings together famous works by Bachelier, Mandelbrot, Osborne, and Cootner is one edited by Cootner (1964). Two recent books that (1) demonstrate how statistical physics techniques can be effectively applied to the economic problem and (2) extend notions that are developed in the Cootner book, are those of Bouchard and Potters (2000) and Voit (2001). Another physical effect used to advantage in economics is the *heat equation*. Using this in conjunction with the Ito calculus, Black and Scholes (1973) formed their famous valuation model. See also in this regard Merton (1974). Entropic uses in valuation have been advanced by Hawkins *et al.* (1996), and Hawkins (1997). A good example of the current state of econometrics, in general, in financial markets is Campbell *et al.* (1997). A wealth of further references on econophysics may be found in these books and on the World Wide Web.

This book has been concerned with generating wave equations, such as the SWE, that govern the fluctuations in *physical* data. These wave equations follow from a variational (or zero) principle – EPI – which is, in turn, an expression of Wheeler's principle of a "participatory universe" (Sec. 0.1). Financial economics has, of course, a strong participatory (human) component. Therefore it seems to be an eminent candidate for obeying Wheeler's principle and EPI. An important financial measure is the probability law $p(x)$ on fluctuations $x$ in the price of a security. Thus, such laws ought to be computable using EPI. See further in this regard Hawkins and Frieden (2004).

A variational principle, such as EPI, that can be used to derive general probability laws $p(x)$ provides an operational calculus for the incorporation of new knowledge. This is via Lagrange constraint terms (Sec. 0.3.8). This attribute of a variational principle has long been exploited in the physical sciences. It is also becoming increasingly popular in finance and economics, where, for example, a principle of maximum entropy has found application as a useful and practical computational approach to financial economics (Maasoumi, 1993; Sengupta, 1993; Golan *et al.*, 1996; Fomby and Hill, 1997). It is generally felt that a candidate probability law $p(x)$ should be minimally biased toward particular values of $x$, i.e., *maximally smooth* in some sense, while maintaining consistency with the known information about $x$.

While this criterion of smoothness has often been used to motivate the use of maximum entropy (Buck and Macaulay, 1990), other variational approaches provide similar – and potentially superior – degrees of smoothness to probability laws (Frieden, 1988; Edelman, 2001). It is the purpose of this chapter to show that EPI – which provides just such a variational approach – can be used to reconstruct probability densities of interest in financial economics. How, then, does the phenomenon of price fluctuation fit within the EPI framework of measurement?

## 13.1 A trade as a "measurement"

The trade (purchase, sale, etc.) price $y$ of a security is a direct measure of its "valuation." (The simpler word "value" would be preferred, but is too generic in this context.) Consider the trade, at a time $t$, of a security $A$, such as a stock, bond, option, etc. Denote by $\theta$ the "true" or ideal valuation $\theta$ of $A$ at the time $t$. There is no single, agreed-upon way of computing $\theta$, but for our purposes it can be taken to be simply the arithmetic average value of $y$ over all trades of $A$ worldwide at that time.)

*Regarding time dependence*: All quantities $\theta$, $p(x)$, $q(x)$ are assumed to be evaluated at the time $t$ of the trade. However, for brevity, $t$ is suppressed. Note that, since $t$ has a general value, we are not limiting attention to cases where $t$ is large and distribution functions approach equilibrium distributions. The analysis will hold for general $t$, i.e., for generally *non-equilibrium* distributions.

Basically, trading is a means of price discovery. Therefore a trade valuation $y$ generally differs from its "true valuation" $\theta$: The disparity is due to a wide range of causes, among them imperfect communication of all information relevant to valuation of the security, and the famous emotional effects of greed and fear in the minds of buyers and sellers. (This is a case where the phenomenon – the valuation – is affected not only by the *act* of observation, i.e., the trade, but also by the *emotions behind* the trade.) Even in cases of perfect communication, there is inherent heterogeneity in the mapping of such information to prices. The result is differing expectations as to whether the price of a security will increase or decrease. As Bachelier (1900) aptly noted, "Contradictory opinions concerning these changes diverge so much that at the same instant buyers believe in a price increase and sellers in a price decrease." Indeed, if this were not the case the trade would not take place.

Let $x$ denote the difference between the traded price $y$ and ideal valuation $\theta$, so that

$$y = \theta + x. \tag{13.1}$$

This is the basic data Eq. (1.0) of the EPI process. Up to now this equation has described a physical *measurement*. Now we see that it can represent a physical *trade*. In essence, a traded price is a sample, or "measurement," of the valuation of the security. The trader is also therefore an "observer," in the sense used in Chapter 1 *et seq*. Likewise the difference $x$ between the ideal and observed valuations is generally unpredictable and, hence, regarded as random "noise." It is a noise of valuation.

By definition, the PDF $p(x)$ on the valuation noise governs the ensemble of such trades of $A$ by the observer under repeated initial conditions. Thus, $p(x)$ is a well-defined PDF. Given the efficiency of today's communication networks, to a good approximation $p(x)$ should also represent the PDF on noise over all worldwide trades of $A$ at that time. That is, $p(x)$ should also obey ergodicity. But, aside from this matter, how can $p(x)$ be estimated?

There is nothing in the construction of EPI (Chapter 3) that limits its application to scenarios of physics. EPI is the outgrowth of a flow of Fisher information (Chapter 3). The manifestation of the Fisher information must be known (*mass* in Chapter 4, *current* in Chapter 5, *lactic acid* from cancer in Chapter 15), but is otherwise arbitrary. Here it is valuation $y$ in (say) *dollars*. Therefore we apply EPI to the given problem of estimating $p(x)$.

The approach also turns out to apply to the problem of estimating a PDF $p(x)$ of *classical statistical mechanics*, where $x$ is now some macroscopic observable such as energy, pressure, velocity, etc. (see also Chapter 7). Although the following development is in the specific context of the economic valuation problem, we point out along the way corresponding results for the statistical mechanics problem.

We emphasize that *estimates* of PDFs, rather than exact answers, are sought in these problems. By Sec. 0.3.8, the anwer to a variational problem depends intimately upon the nature of its constraints. However, the types of data that are used to define the PDFs for our problems will largely be selected *arbitrarily*, ultimately out of convenience (e.g., the number $M$ of call prices $c(k)$ used in Eq. (13.25)). It results that the constraints (13.2) upon the estimated PDF are likewise arbitrary, so that the output PDF is by and large a manmade construction. Hence it cannot be an absolute answer. By comparison, the bound informations $J$ utilized in Chapter 7, for example, act as the constraints imposed by *nature* (see Sec. 3.1.3). Thus, they give absolute ("nature's") results for the PDFs of statistical mechanics, in contrast to what is attained here.

These effects have direct counterparts in economics, giving rise to the famous dilemma of choosing between a "fundamentalist" and a "technical" approach to valuation (Sec. 13.4). The fundamentalist approach aims to base the valuation upon "nature's constraints"; the technical approach bases it entirely upon empirical data. Thus, by the preceding paragraph, the technical approach always gives an approximate answer; whereas the fundamentalist approach is *potentially* capable of providing an absolute answer. The proviso is that the "right" technical factors be factored in.

Since the output PDFs of valuation are but estimates, by what measure of quality can each be judged? As mentioned above, it is generally felt that a candidate PDF should be minimally biased, i.e., *maximally smooth*. Hence, degree of smoothness (by whatever measure) is taken to be the main measure of quality. Also, in some applications the *theoretical* PDF is known ground truth (Sec. 13.7.3), and therefore provides an absolute standard of comparison.

## 13.2 Intrinsic versus actual data values

As in previous chapters, $y$ denotes an *intrinsic* datum, i.e., serves to define the PDF $p(x)$ on intrinsic fluctuations $x$ (Chaps. 1 and 2), rather than representing actual data. The data are here valuations denoted as $d_m$. These depend upon $p(x)$ through relations

$$d_m = \int_a^b dx\, f_m(x) p(x), \quad m = 1, \ldots, M, \tag{13.2}$$

as given below. The limits $a$, $b$ and kernel functions $f_m(x)$ are known for each trade problem.

In all applications of EPI the activity of taking data plays a vital role in initiating the EPI process. This is the case here as well. The execution of the trade (and the execution of all other trades) of the given security during the time interval $(t, t + dt)$ necessarily affects the PDF $p(x)$ on its valuation.

Because the time interval is very short, the result is a perturbation of $p(x)$. Therefore informations $I$ and $J$ are perturbed, initiating the EPI extremum principle as described in Chapter 3.

Returning to the classical statistical mechanics problem, the observer ordinarily knows certain *macroscopic data* $d_m$, $m = 1, \ldots, M$ of the type (13.2), corresponding to values of temperature, pressure, etc. How should knowledge of these data effect an EPI solution $p(x)$ to the problem?

## 13.3 Incorporating data values into EPI

In nearly all previous applications of EPI the actual *values of the data* were *not* used to explicitly affect the output PDF or amplitude function $\psi$. The exception was in Chapter 10. (As discussed in Sec. 10.9.5, the data can be imposed upon an EPI output $\psi$ by requiring $\psi$ to obey known initial conditions.)

Alternatively, as discussed in Sec. 0.3.8, the data may be mathematically imposed upon the solution by the variational process itself. Data give information about the particular member of an ensemble $\psi(\boldsymbol{r}, t)$ of possible amplitude functions that are present. Thus, knowledge of, for example, a particular electron position gives information about the particular Feynman path $\boldsymbol{r}(t)$ it has taken. This should, for example, cause $\psi(\boldsymbol{r}, t)$ to have large values along that path and nearby ones, but small values elsewhere. How, then, can the variational EPI approach be modified to accommodate data?

This was taken up in Chapter 10, under slightly more general circumstances. There, the unknown PDF was on the intrinsic quantum fluctuations $\boldsymbol{r}$ in position. However, the data were *measured* positions $\overline{\boldsymbol{r}}$, and, by Eq. (10.7), these incorporated both intrinsic fluctuations $\boldsymbol{r}$ and *additional* random errors $\gamma$ due to instrument noise. Therefore, since $\boldsymbol{r}$ differed from $\overline{\boldsymbol{r}}$, it did not make sense to force $\boldsymbol{r} \equiv \overline{\boldsymbol{r}}$ at the data values. That is, the constraints were not made to be *absolute* constraints. The result was instead a *probabilistic* constraint Eq. (10.22) allowing $\boldsymbol{r}$ to randomly depart from $\overline{\boldsymbol{r}}$. In particular, by Eq. (10.22), the degree of allowed departure from each datum is governed by the width of the probability amplitude function $w(\overline{\boldsymbol{r}} - \boldsymbol{r})$ on random instrument noise values $\gamma$.

Conveniently, the economic problem at hand is simpler. Ordinarily, the numerical valuation data $d_m$ do not suffer from additive measurement "noise" due to a measuring instrument amplitude function $w(\overline{\boldsymbol{r}} - \boldsymbol{r})$. Therefore they represent exact prior knowledge about the unknown PDF $p(x)$, so that any estimate of the latter should be forced to obey the data *exactly*. How can this be done?

These data constraints are so-called "equality constraints." As discussed in Sec. 0.3.8, equality constraints may be imposed upon a Lagrangian extremum problem by simply adding them to the objective functional via undetermined multipliers $\lambda_m$. Here the objective functional is the physical information $I - J$. Thus, for our problem the EPI extremum principle becomes a constrained problem

$$I - J + \sum_{m=1}^{M} \lambda_m \left[ \int_a^b dx\, f_m(x) p(x) - d_m \right] = extrem. \tag{13.3}$$

(The EPI zero condition will not be used.)

Equation (13.3) is capable of giving exact answers $p(x)$, as were the corresponding quantum measurement Eqs. (10.13) and (10.15). As with any EPI problem, the validity of the output $p(x)$ depends upon the validity of the choice of the information $J$.

Continuing the parallel development for statistical mechanics, by analogous reasoning the distribution laws $p(x)$ for mechanics problems obey the same general principle (13.3).

## 13.4 Evaluating *J* from the standpoint of the "technical" approach to valuation

Two extreme strategies for valuating a security are termed (a) *fundamental* and (b) *technical*. A pure fundamentalist arrives at a valuation decision $y$ only after detailed examination of all attributes of the business – its yearly sales, debts, profits, etc. By comparison, a pure technician bases his/her pricing decision entirely upon knowledge of the sales history of prices $y$. (This is an extreme form of empiricism.) Most investors use some combination of the two approaches. Both approaches (a) and (b) to valuation seem to be equally successful. Seeking a simple analysis, we assume the simpler approach. This turns out to be the technical one (b). In a purely technical approach, the observer eschews all knowledge of the fundamentals of the security. Thus, the observer presumes there to be zero prior information about *valuation*, the quantity $\theta$. In EPI, prior information equates to bound information $J$, so that here

$$J = 0. \tag{13.4}$$

Since the information efficiency $\kappa$ occurs in EPI only in multiplication of $J$, the result (13.4) means that, for these valuation problems, the value of $\kappa$ is irrelevant.

The level of prior knowledge (13.4) also has a powerful effect upon the statistical mechanics problem of estimating $p(x)$. In limiting his prior knowledge to the observation of *macroscopic* variables, the observer rules out any knowledge about $p(x)$ that might follow from a theoretical analysis of *micro-*

*scopic* effects (as in Sec. 7.4.5). Hence, in effect the "technical approach" of economics corresponds to the "macroscopic observables approach" of classical mechanics. Both approaches take a stance of purposeful ignorance regarding the intricate details of the unknown processes. Of course, ignoring detailed effects is to ignore vital information. Hence, both applications give *approximate answers* $p(x)$.

## 13.5 Net principle

Representing $I$ by functional (1.24), reverting to the use of an amplitude function $q(x)$ via Eq. (1.23), and using Eq. (13.4), Eq. (13.3) becomes

$$4\int_a^b dx\, q'^2(x) + \sum_{m=1}^{M} \lambda_m \left[ \int_a^b dx\, f_m(x) q^2(x) - d_m \right] = extrem., \qquad p(x) \equiv q^2(x). \tag{13.5}$$

Since $x$ is real, the extremum solution is always a minimum (Sec. 0.3.4). Thus, for this problem the EPI solution is also the constrained MFI solution (minimum Fisher information) of Eq. (0.23).

As seen in examples below, minimizing the Fisher information has the benefit of yielding a probability law where smoothness is ensured across the range of support. This is in contrast to maximum entropy, where smoothness tends to be concentrated in regions where the probability density is very small.

A further interesting aspect of the problem (13.5) is that its solution will obey mathematically a Schrödinger wave equation.

The Lagrangian of Eq. (13.5) has been derived on the basis of information flow, and entirely without any explicit consideration of energy. The data $d_m$ are moments of $p(x)$, and are likewise ordinarily independent of energy. The conventional "action integral" approach, by which Lagrangians are regarded as energies, could not, then, produce this Lagrangian. EPI extends the Lagrangian approach into new problem areas. (See also Sec. 1.1.)

Since maximum entropy is comparatively well known in financial economics, and shares with Fisher information a common variational structure, we shall use it as a point of comparison for all EPI solutions. Maximum entropy solutions $p(x)$ from data $d_m$ obey, by Eq (1.13),

$$-\int_a^b dx\, p(x) \ln p(x) + \sum_{m=1}^{M} \lambda_m \left[ \int_a^b dx\, f_m(x) p(x) - d_m \right] = extrem. \tag{13.6}$$

The two principles (13.5) and (13.6) will be applied to corresponding data $d_m$ in three valuation problems below. To distinguish the resulting two solutions to each problem, Fisher information-based solutions to (13.5) will be denoted as $q_{\mathrm{FI}}$ or $p_{\mathrm{FI}}$, and maximum entropy solutions to (13.6) will be denoted as $p_{\mathrm{ME}}$.

Returning to the statistical mechanics problem, principle (13.6) is precisely that of E. T. Jaynes (1968) for estimating an unknown PDF $p(x)$.

## 13.6 Formal solutions

The two estimation principles (13.5) and (13.6) are readily solved in general.

The net Lagrangian of Eq. (13.5) is

$$\mathscr{L} = 4q'^2(x) + \sum_{m=1}^{M} \lambda_m f_m(x) q^2(x). \tag{13.7}$$

From this we see that

$$\frac{\partial \mathscr{L}}{\partial q'} = 8q'(x), \qquad \frac{\partial \mathscr{L}}{\partial q} = 2\sum_{m=1}^{M} \lambda_m f_m(x) q(x). \tag{13.8}$$

Using these in the Euler–Lagrange Eq. (0.13) gives

$$q''(x) = \frac{q(x)}{4} \sum_{m=1}^{M} \lambda_m f_m(x). \tag{13.9}$$

This is the wave equation on valuation fluctuation $x$ we sought at the outset. It holds at a general trade time $t$ (Sec. 13.1), so that in effect $q(x) = q(x|t)$ with the $t$-dependence suppressed for brevity. Particular solutions $q_{\mathrm{FI}}(x)$ of the wave equation are defined by particular values of the multipliers $\lambda_m$, in applications below.

As usual for EPI outputs, the solution (13.9) has the form of a differential equation. In fact it has the form of the energy-eigenvalue Schrödinger wave equation (Eq. (D9) of Appendix D). However, since Planck's constant $\hbar$ does not enter in, it is not descriptive of quantum mechanics in particular. Of course, depending upon application, $\hbar$ can enter *implicitly* through an appropriate constraint term. Likewise, energy *per se* does not enter into Eq. (13.9).

Since the economic problem and that of statistical mechanics have so far shared a common EPI approach, the statistical mechanics problem must likewise have the solution (13.9). See also in this regard Eq. (7.48) and the references Frieden *et al.* (2002a; 2002b). Moreover, as in these statistical mechanics applications, the time $t$ is arbitrary. That is, the output $q(x)$ of the EPI solution (13.9) represents generally a *non*-equilibrium solution to the valuation problem. In summary, EPI predicts the following: *Economic valuation problems and non-equilibrium statistical mechanics problems have a common general solution in the form of the energy eigenvalue SWE.*

The rest of this chapter is, in effect, devoted to applying the general solution (13.9) to specific economic problems. (Its application to particular statistical mechanics problems may be found in Frieden *et al.*, 2002a; 2002b.)

*Note on terminology*: The SWE (13.9) is also called in physics the "stationary SWE." Unfortunately, the term "stationary" when applied to economic problems suggests relaxation of the system to an equilibrium state over some large time $t$. But in fact we are permitting the time $t$ to be arbitrary (see above). Hence use of the word "stationary" would be misleading, and it is replaced with the alternative terminology "energy eigenvalue."

In practice, the form of the differential Eq. (13.9) as a SWE is fortuitous, since it permits past decades of the development of computational approaches to its solution to now be applied to economic problems. In particular, a number of analytical solutions are known, for which a substantial collection of numerical solutions exists (see below).

We next turn to the maximum entropy solution. The Lagrangian of Eq. (13.6) is

$$\mathscr{L} = -p(x)\ln p(x) + \sum_{m=1}^{M} \lambda_m f_m(x) p(x). \tag{13.10}$$

From this we see that

$$\frac{\partial \mathscr{L}}{\partial p'} = 0, \quad \frac{\partial \mathscr{L}}{\partial p} = -1 - \ln p(x). \tag{13.11}$$

Using these in the Euler–Lagrange Eq. (0.13) gives directly

$$p_{\mathrm{ME}}(x) = \exp\left[-1 - \sum_{m=1}^{M} \lambda_m f_m(x)\right]. \tag{13.12}$$

Since $p_{\mathrm{ME}}(x)$ must obey normalization, Eq. (13.12) is more conveniently cast in the pre-normalized form (Jaynes, 1968)

$$p_{\mathrm{ME}}(x) = \frac{1}{Z(\lambda_1, \ldots, \lambda_M)} \exp\left[-\sum_{m=1}^{M} \lambda_m f_m(x)\right], \tag{13.13a}$$

where

$$Z(\lambda_1, \ldots, \lambda_M) \equiv \int_a^b dx \exp\left[-\sum_{m=1}^{M} \lambda_m f_m(x)\right]. \tag{13.13b}$$

The reader can easily verify that $p_{\mathrm{ME}}(x)$ defined by Eq. (13.13a) explicitly obeys normalization. Given this, it should not be re-imposed by a multiplier $\lambda_m$ corresponding to a normalization datum $d_m = 1$.

From Eqs. (13.5) and (13.6), both the EPI and maximum entropy estimates depend upon empirical information in the form of equality constraints. In the following sections, increasingly complex forms of empirical information are given. The efficient generation of yield curves from observed bond prices, and of probability densities from observed option prices, will result. The calculations are based upon the mathematical correspondence between (i) the resulting differential equations for probability laws in these examples from financial economics and (ii) past work with the Schrödinger equation. In addition, we shall see that the probability densities generated using Fisher information are, in general, smoother than those obtained from maximum entropy. A brief summary of the results is given in Sec. 13.9. The two Appendices J and K show details of the calculations in somewhat expanded form.

## 13.7 Applications

In this section we use the EPI approach to motivate practical computational approaches to the extraction of probability densities and related quantities from financial observables. In each case we contrast the results obtained by this method with those of maximum entropy. We begin in Secs. 13.7.1 and 13.7.2 with fixed income problems, where the observables are expressible in terms of partial integrals of a probability density and of first moments of the density. In Secs. 13.7.3 and 13.8 we conclude the applications by examining the canonical problem of extracting probability risk-neutral densities from observed option prices, and extend this analysis by showing how Fisher information can be used to calculate a general measure of volatility for a security.

### *13.7.1 PDF on the term structure of interest rates: yield-curve construction*

A basis for the use of information theory in fixed-income analysis comes from the perspective developed by Brody and Hughston (2001). They observed that the price of a zero-coupon bond – also known as the discount factor $D(T)$ – can be viewed as a complementary PDF where the time $T$ to maturity (aka tenor) is taken to be an abstract random variable. The associated probability density $p(t)$ satisfies $p(t) > 0$ for all $t$, and

$$D(T) = \int_T^\infty p(t)\,dt$$

$$= \int_0^\infty \Theta(t - T)p(t)\,dt, \tag{13.14}$$

where $\Theta(x)$ is the Heaviside step function, defined as $\Theta(x) = 0$ if $x < 0$ and 1 otherwise.

To illustrate the differences between the Fisher information and maximum entropy approaches to term structure estimation, we consider the case of a single zero-coupon bond of tenor $T = 10$ years and a price of 28 cents on the dollar or, equivalently, a discount factor of 0.28. Comparing Eqs. (13.2) and (13.14), this is a case $M = 1$ with $f_1(t) = \Theta(t - T)$. Hence the maximum entropy solution (13.13a) is

$$p_{\mathrm{ME}}(t) = \frac{e^{-\lambda\Theta(t-T)}}{T + 1/\lambda}. \tag{13.15}$$

The corresponding discount factor is readily found from (13.14) to be $D(T) = 1/(\lambda T + 1)$, from which the Lagrange multiplier $\lambda$ is found to be 0.2571.

The Fisher information solution is obtained by solving Eq. (13.9) with the constraints of normalization and one discount factor:

$$\frac{d^2 q(t)}{dt^2} = \frac{q(t)}{4}[\lambda_0 + \lambda_1 \Theta(t - 10)]. \tag{13.16}$$

This problem is known (see, e.g., Landau and Lifshitz, 1977) to have the solution

$$q(t) = \begin{cases} A\cos(\alpha t) & \text{if } t \leqslant T \\ B\exp(-\beta t) & \text{if } t > T, \end{cases} \tag{13.17}$$

where the coefficients $A$ and $B$ are determined by requiring that $q(t)$ and $q'(t)$ (or the logarithmic derivative $q'(t)/q(t)$) match at $t = T$. Carrying through the requirement yields

$$\tan(\alpha T) = \beta/\alpha. \tag{13.18}$$

Choosing $\alpha T = \pi/4$ and matching amplitudes, we find that $B = A\exp(\pi/4)/\sqrt{2}$, so that

$$p_{\mathrm{FI}}(t) = \begin{cases} 8(\pi + 4)^{-1}\alpha\cos^2(\alpha t) & \text{if } t \leqslant T \\ 4(\pi + 4)^{-1}\alpha\exp(\pi/2 - 2\beta t) & \text{if } t > T. \end{cases} \tag{13.19}$$

Also $D(T) = 2/(\pi + 4) = 0.28$.

The maximum entropy PDF (13.15), the Fisher information PDF (13.19), and the corresponding term structure of interest rates are plotted in Fig. 13.1. In the uppermost panel we see both PDFs as functions of tenor with the Fisher information result denoted in all panels by the solid line. The maximum entropy result is uniform until the first (and in this case only) observation is encountered; beyond which a decaying exponential is observed. The discontinuity in slope traces from the *global* nature of the entropy measure (Sec. 1.3.1). The Fisher information result is smoother, reflecting the need to match both the amplitude and the derivative at the data point. This traces from the *local* nature of the Fisher measure (Sec. 1.3.1).

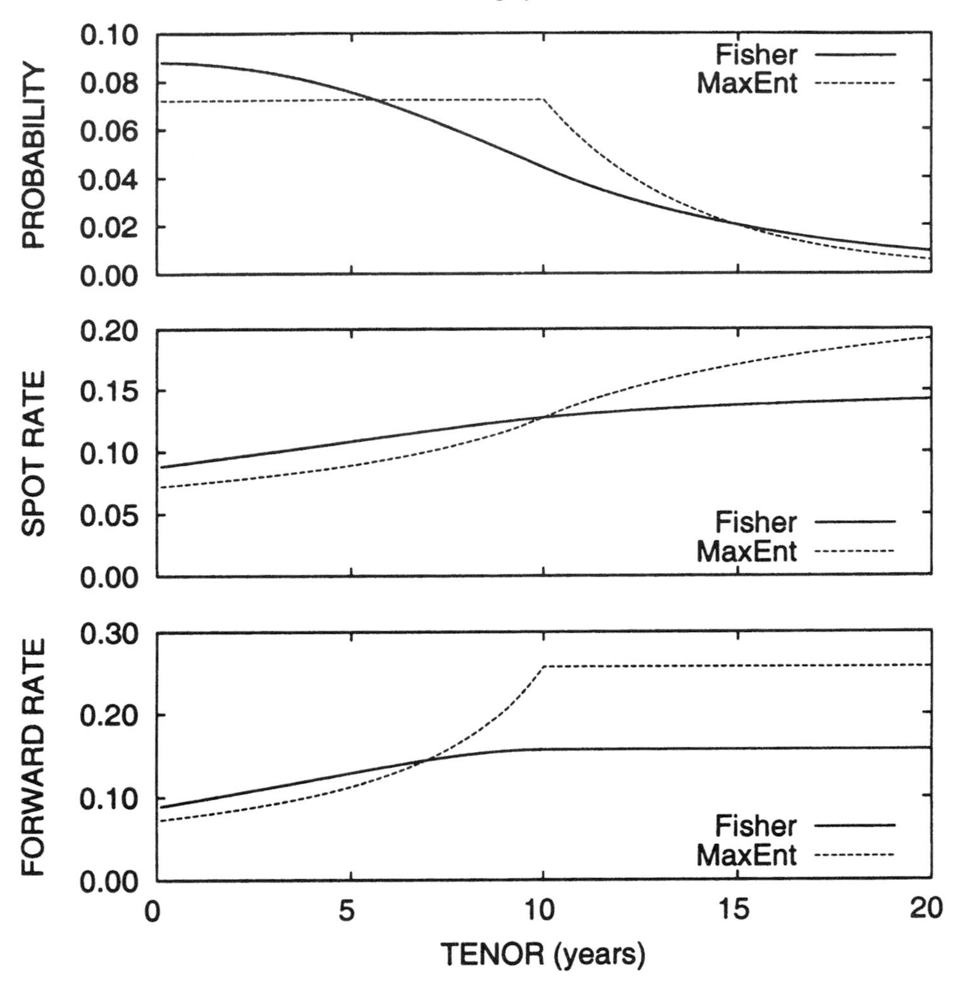


Fig. 13.1. The probability densities and derived interest rates for a single 10-year discount factor of 0.28.

It is difficult to make a real esthetic choice between the two curves in the upper panel. To aid in making such a choice, we turn to two important derived quantities: the spot rate $r(t)$ and the forward rate $f(t)$. These are related to the discount factor by

$$D(T) = e^{-r(T)T} \tag{13.20a}$$

$$= e^{-\int_0^T f(t)dt}. \tag{13.20b}$$

The spot rates are shown in the middle panel of Fig. 13.1. Both methods yield a smooth result, with the Fisher information solution showing less structure than the maximum entropy solution. A greater difference between the two methods is seen in the lowermost panel of Fig. 13.1, where the forward rate is shown. It is the structure of this function that is often looked to when assessing the relative merits of a particular representation of the discount factor. The forward rate reflects the structure of the probability density as expected from the relationship

$$f(t) = p(t)/D(t). \tag{13.21}$$

The maximum entropy result shows more structure than the Fisher information result, again, due to the continuity at the level of the first derivative imposed on $p(t)$ by the Fisher information approach.

It is a comparatively straightforward matter to extend this approach to the construction of a term structure of interest rates that is consistent with *any number* of arbitrarily spaced zero-coupon bonds. A particularly convenient computational approach based on transfer matrices is presented in Appendix J.

Previous work on inferring the term structure of interest rates from observed bond prices has often focussed on the somewhat *ad hoc* application of splines to the spot or forward rates (McCullough, 1975; Vasicek and Fong, 1982; Shea, 1985; Fisher *et al.*, 1995).

By comparison, the work of Frishling and Yamamura (1996) is similar in spirit to the EPI approach, in that it minimizes $df(t)/dt$. Their paper deals with the often unacceptable results that straightforward applications of splines to this problem of inference can produce. In a sense, EPI can be seen as an information-theoretic approach to imposing the structure sought by Frishling and Yamamura on the term structure of interest rates. A similar pairing of approaches is the minimization of $d^2f(t)/dt^2$ by Adams and Van Deventer (1994) and the recent entropy work of Edelman (2003).

### 13.7.2 The probability density for a perpetual annuity

Material differences between the probability densities generated by Fisher information and maximum entropy are also seen when the observed data are moments of the density – a common situation in financial applications. As an example, the value of a perpetual annuity $\xi$ due to all possible values $t$ of the tenor is given by (Brody and Hughston, 2001; 2002)

$$\xi = \int_0^{+\infty} tp(t)\, dt. \tag{13.22}$$

This first-moment constraint provides an interesting point of comparison for the maximum entropy approach employed by Brody and Hughston and our

Fisher information approach. Comparing Eqs. (13.2) and (13.22), here $M = 1$ and $f_1(t) = t$. Then the maximum entropy solution (13.13a) is directly

$$p_{\mathrm{ME}}(t) = \frac{1}{\xi} \exp(-t/\xi). \tag{13.23}$$

The EPI solution $q(t)$ to Eq. (13.9) was obtained by Frieden (1988) as

$$p_{\mathrm{FI}}(t) = c_1 \, \mathrm{Ai}^2(c_2 t). \tag{13.24}$$

$\mathrm{Ai}(x)$ is Airy's function, and the constants $c_i$ are determined uniquely by normalization and the constraint Eq. (13.22).

We show these two probability densities as a function of tenor $t$ for a perpetual annuity with a price of \$1.00 in Fig. 13.2. The two solutions are qualitatively similar in appearance: Both monotonically decrease with tenor and are quite smooth. However, since the Fisher information solution starts out at zero tenor with a lower value than the maximum entropy solution and then crosses the maximum entropy solution so as to fall off more slowly than maximum entropy solution in the mid-range region, the Fisher information solution is in appearance even smoother. This can be quantified as follows.

A conventional measure of smoothness of a PDF is the size $H$ of its entropy: The smoother the PDF is the larger is $H$. As we found in Chapter 1, a smooth

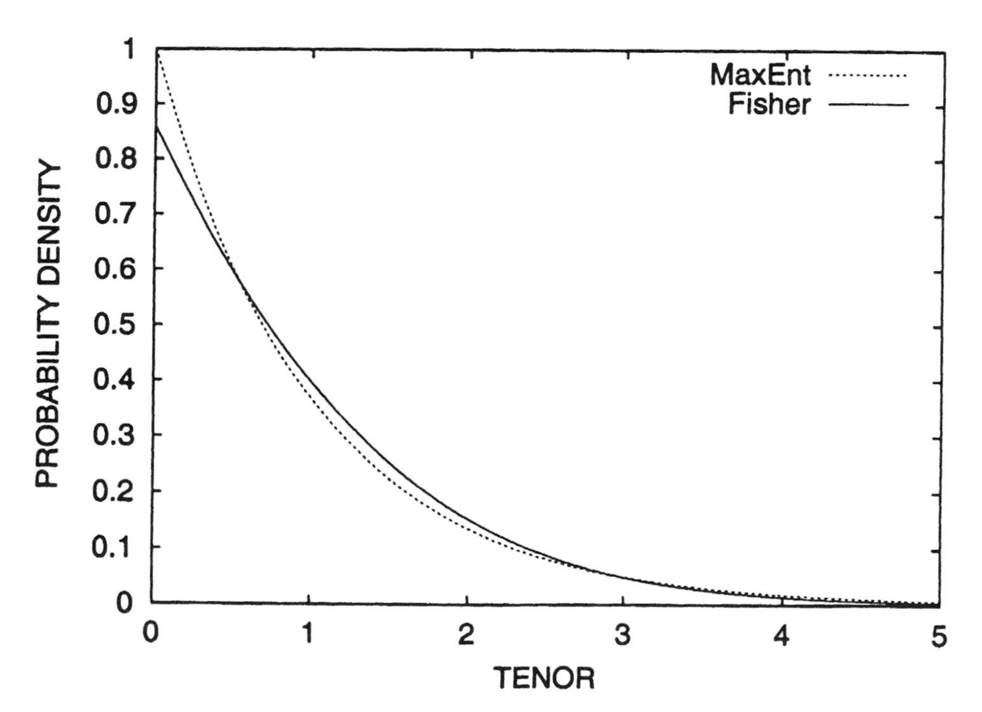


Fig. 13.2. The two PDFs associated with a perpetual annuity of price 1.0.

PDF also has a *small* value of the Fisher information $I$. We want to compare the smoothness of PDFs by the alternative measures $H$ or $I$. We therefore need to form out of $I$ a quantity that, like $H$, becomes *larger* as the PDF becomes smoother. Such a quantity is of course $1/I$ (Frieden, 1988). By Eq. (1.1), $1/I$ has the further significance of defining the Cramer–Rao error bound, although that property isn't used here.

The $1/I$ value of the Fisher information solution (13.24) in Fig. 13.2 is found to be 1.308 by numerically integrating Eq. (1.2). By comparison, $1/I$ for the maximum entropy solution (13.23) in Fig. 13.2 is found to be 1.000. Hence, the Fisher information solution is significantly smoother on the basis of $1/I$ as a criterion.

We also compared the relative smoothness of these solutions using Shannon's entropy $H$ (Eq. (1.13)) as a measure. The maximum entropy solution is found to have an $H$ of 1.0, whereas the Fisher information solution has an $H$ of 0.993. Hence the maximum entropy solution does indeed have a larger Shannon entropy than does the Fisher information solution, but it is certainly not much larger. Since the two solutions differ much more in their $1/I$ values than they do in their Shannon $H$ values, it appears that $1/I$ is a more sensitive measure of smoothness, and hence biasedness, than is $H$.

An interesting result emerges from an examination of the smoothness as a function of *range*. Since over the range $0 \leqslant t \leqslant 5$ shown in Fig. 13.2 it appears that the Fisher information is smoother than the maximum entropy solution by *any* quantitative criterion, we also computed the Shannon $H$ values over this limited interval. The results are $H = 0.959$ for the maximum entropy solution and $H = 0.978$ for the Fisher information solution. Thus the Fisher information solution has the larger entropy! This shows that the Fisher information solution is smoother than the maximum entropy solution, *even by the maximum entropy criterion*, over this *finite* interval. Of course the maximum entropy solution must (by definition) have the larger entropy over the *infinite* interval. It follows that the larger entropy value results from values of the PDF in the long *tail* region $t > 5$.

However, in this tail region the maximum entropy PDF has negligibly *small* values for most purposes. Hence, the indications are that the criterion of maximum entropy places an unduly strong weight on the behavior of the PDF in tail regions of the curve.

Note by comparison that, over the interval $0 \leqslant t \leqslant 5$, the $1/I$ values for the two solutions still strongly differ, with 1.317 for the Fisher information solution and 1.007 for the maximum entropy. Again, the Fisher $1/I$ measure is the more sensitive measure. Moreover, these are close to their values as previously computed over the entire range $0 \leqslant t \leqslant \infty$. Hence, Fisher information gives comparatively lower weight than does entropy to the tail regions of the probability densities.

### *13.7.3 The probability density in option prices*

The reconstruction of probability densities for option prices has been the topic of much research (see Jackwerth, 1999, and references therein). As mentioned before, smooth estimates are preferred. As with term-structure estimation, approaches to smoothing probability densities have often been constructed in an *ad hoc* manner. Most of these have been by the use of *global* smoothing measures such as maximum entropy. However, a satisfactory degree of smoothness has not been consistently achieved by these approaches (e.g. maximum entropy approaches taken by Stutzer, 1994; 1996; Hawkins *et al.*, 1996; Buchen and Kelley, 1996; Hawkins, 1997; Avellaneda, 1998).

By comparison, Fisher information naturally enforces *local* smoothness for use in reconstructing option-based densities. It has given visibly (and usually quantitatively) smoother density estimates than are obtained by the use of maximum entropy, in all test cases carried out by us. The following are typical examples.

We first consider the price $c(k)$ of a European call option. This obeys

$$c(k) = e^{-r(t)t} \int_0^\infty \max(x - k, 0)\, p(x)\, dx. \tag{13.25}$$

Here $k$ is the strike price, $t$ is the time to expiration, and $r(t)$ is the risk-free spot rate. We take $c(k)$, $k$, $r(t)$, and $t$ to be known and ask what $p(x)$ is associated with the observed $c(k)$. The observed call value $c(k)$ can also be viewed as the mean value of the function $e^{-r(t)t} \max(x - k, 0)$, and it is from this viewpoint that a natural connection between option pricing theory and information theory can be made.

Given a set of $M$ observed call prices $\{c(k_m)\}$ the EPI solution is obtained by solving Eq. (13.9) subject to these constraints and normalization,

$$\frac{d^2 q(x)}{dx^2} = \frac{q(x)}{4}\left[\lambda_0 + e^{-r(t)t} \sum_{m=1}^{M} \lambda_m \max(x - k_m, 0)\right], \qquad p(x) \equiv q^2(x). \tag{13.26}$$

The particular parameter $\lambda_0$ plays a special role in this equation. Rearranging (13.26) puts it into the "energy-eigenvalue form" of the SWE,

$$\left[\frac{d^2}{dx^2} - \frac{e^{-r(t)t}}{4} \sum_{m=1}^{M} \lambda_m \max(x - k_m, 0)\right] q(x) = \frac{\lambda_0}{4} q(x).$$

This shows that parameter $\lambda_0$ takes on the role of the energy eigenvalue for a corresponding quantum solution $q(x)$.

In general there is a family of solutions $(\lambda_0, q(x))$ to (13.26), with a "ground state" solution corresponding to the lowest value of $\lambda_0$. Larger values of $\lambda_0$ define higher-order solutions that we call "excited states" $q(x)$ in analogy with those in quantum mechanics. A solution $(\lambda_0, q(x))$ with $\lambda_0$ a *minimum* will be *smoothest* or least biased, since general-order solutions $q(x)$ to the SWE are expressed as polynomials (Hermite–Gaussian, Laguerre, etc.) of corresponding order, so that the higher the order the more oscillatory $q(x)$ is.

The ground state solution is also presumed to represent the *equilibrium solution*, with excited states corresponding to non-equilibrium solutions (as in Sec. 7.4.8). The "distance" from equilibrium (defined appropriately) is presumed to increase with order. In the examples below we show only equilibrium solutions.

What qualitative features should we expect to see in the ground state solution $q(x)$ of (13.26)? In this particular problem the solution should be something like a Gaussian centered near $x = 1$, since the log-normal distribution that the "observables" were calculated from has that feature. In fact $q(x)$ should, like a Gaussian, have a single dominant mode, since, as the strike price for a call (put) moves further above (below) the current level of the underlying asset, the price falls off.

Next we can ask, what does this imply for the summation on the right-hand side of Eq. (13.26)? The summation has precisely the role of the potential in the quantum mechanical SWE. A well-known class of quantum mechanical problems utilizes a potential well of some sort. Hence, if the summation amounts to a potential well, the nature of the solutions will be well understood. This will in fact be the case.

Two numerical approaches to solving Eq. (13.26) are as follows. First, the numerical method described in Appendix J can be applied. This requires either replacing the terms multiplying $q(t)$ on the right-hand side of (13.26) with a stepwise approximation, or treating the solution between strike prices as a linear combination of Airy functions. Alternatively, (13.26) can be integrated directly using the Numerov method (Appendix K). The latter was in fact our route to solution.

The maximum entropy solution can be obtained in a straightforward manner, substituting Eq. (13.13a) into Eq. (13.25) and varying the $\{\lambda_m\}$ to reproduce the observed option prices.

We applied the Fisher information approach to the example discussed by Edelman (2003) of $M = 3$ call options on a stock with an annualized volatility $\sigma$ of 20% and a time to expiration $t$ of 1 year. In this example the interest rate $r$

is set to zero, the stock price $S_t$ is \$1.00 and the three option prices with strike prices $k$ of \$0.95, \$1.00, and \$1.05 have Black–Scholes prices \$0.105, \$0.080, and \$0.059, respectively. We note that the stock price is included within the data of these examples as the price of a call option with a strike price of $k = \$0.00$. *This effectively adds another datum to each example.*

Given these "observations" together with the normalization requirement on $p(x)$ we generated the solutions shown in the two panels of Fig. 13.3. The smooth dash–dot curve is the log-normal distribution from which the "observed" option prices were sampled, and also is the target distribution we are trying to recover from the option prices. The Fisher information results are given by the two solid curves, and the maximum entropy results are the corresponding two dashed curves.

We begin by considering the case of the probability density implicit in a *single* at-the-money (i.e., $k = S_t = \$1.00$) option price. (This single-datum problem is effectively a case of $M = 2$ data; see above). The results are shown in the upper panel. The maximum entropy solution is a sharply peaked product of two exponentials. In contrast, the Fisher information solution has, by virtue of requiring continuity of the first derivative, a much smoother appearance. It also is much closer (by any reasonable measure) to the target log-normal density than is the maximum entropy solution. However, we do see that, with only a single option price, the *asymmetric* features of the log-normal density are not recovered by either reconstruction.

Adding the option prices at the $k = \$0.95$ and $k = \$1.05$ strikes (from the Edelman example) to the problem results in the densities shown in the lower panel of Fig. 13.3. With these added data the agreement between the Fisher information PDF and the log-normal PDF is improved substantially: the peaks of the two PDFs are nearly the same and the characteristic asymmetry of the log-normal density is now *seen* in the Fisher information result. The maximum entropy solution continues to be sharply peaked and to deviate substantially from the log-normal PDF.

Sharply peaked densities have appeared in previous work involving limited financial observables (see, for example, Kuwahara and Marsh, 1994), so this feature should not be interpreted as a specific indictment of the maximum entropy method. In fact, a simple generalization of maximum entropy, the Kullback–Leibler entropy (Eq. (1.16)) given by

$$G \equiv -\int dx\, p(x) \ln\left[\frac{p(x)}{r(x)}\right] dx, \tag{13.27}$$

works relatively well with use of an appropriate reference density $r(x)$. Hawkins *et al.* (1996) reconstructed in this way smooth implied densities from

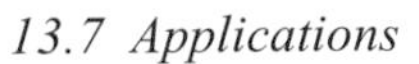


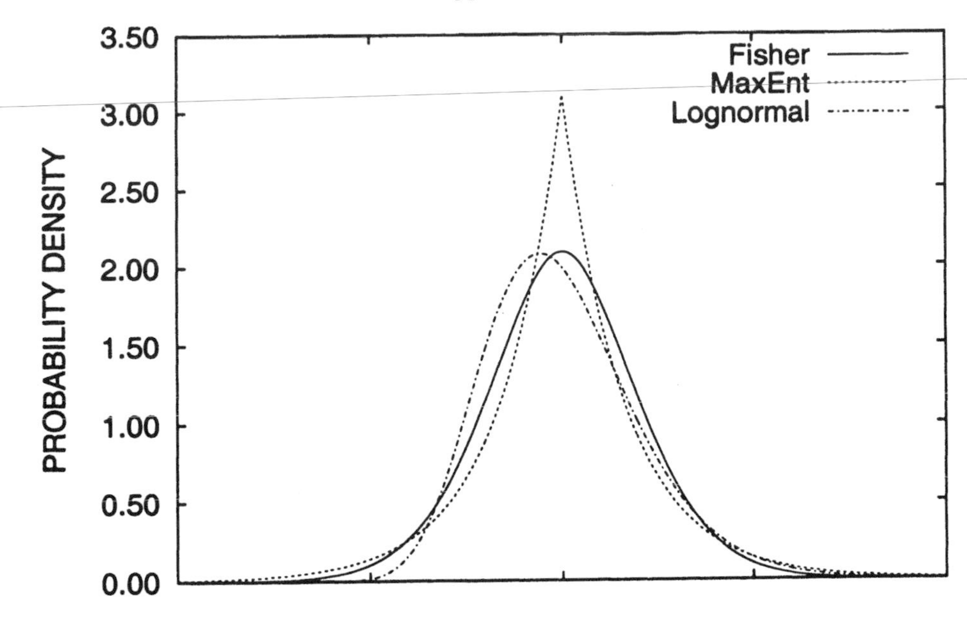


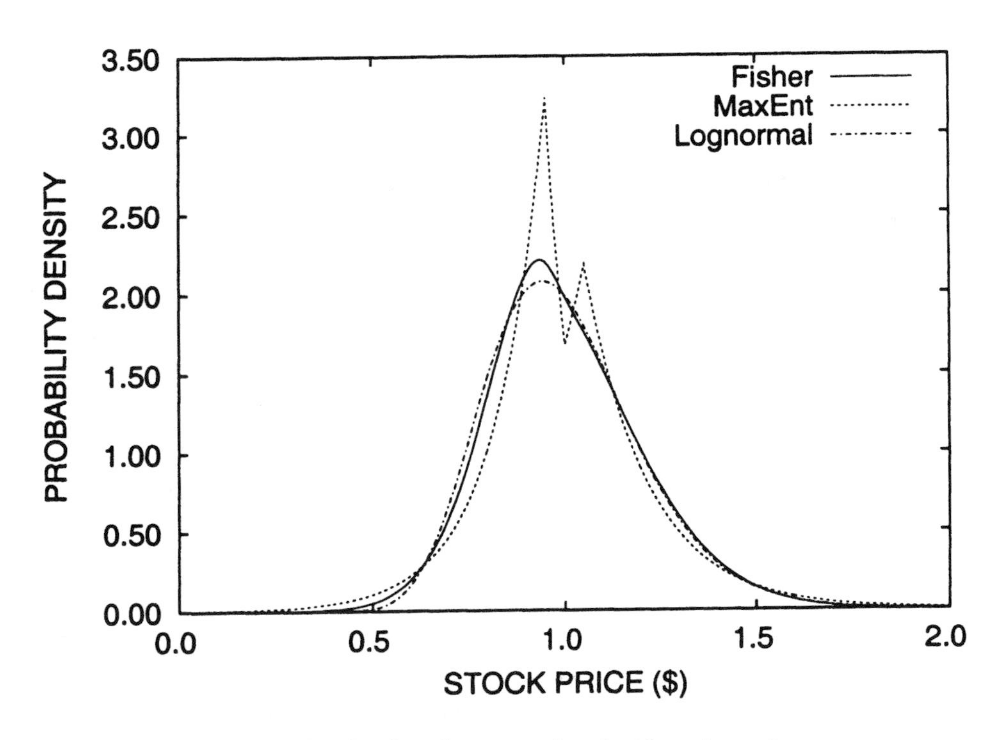


Figure 13.3. Probability density functions associated with option prices.

S&P-500 index options, using a log-normal reference density based upon the at-the-money option ($k = \$1.00$) price.

These examples also point out one of the key differences between the maximum entropy and Fisher information approaches: needed assumptions. When observations are limited – as they often are in financial economics applications – smooth maximum entropy densities often require strong prior knowledge, in the form of a critically chosen density function $r(x)$. Fisher information simply imposes a continuity constraint.

## 13.8 A measure of volatility

We showed that Fisher information, via the EPI principle, provides a way of calculating the densities implied by option prices. Fisher information also provides, via the Cramer–Rao inequality (1.1) of Chapter 1, a value

$$e^2_{\min} = 1/I \tag{13.28}$$

of the intrinsic uncertainty $e^2_{\min}$ in knowledge of the unknown ideal parameter $\theta$. In valuation problems $\theta$ is the unknown valuation (Sec. 13.1) of a security. Then the uncertainty $e^2_{\min}$ is that in knowledge of the valuation of the security. If we want the term "volatility" to measure the smallest expected level of uncertainty in knowledge of the valuation, then $e^2_{\min}$ is the volatility and it relates to $I$ by Eq. (13.28). In this application the value of $I$ to be used in Eq. (13.28) would be formed out of the EPI solution $q(x)$ to the valuation problem.

Equation (13.28) is also a *conservative* measure of the volatility, since it expresses the minimum expected level of uncertainty in the observer about the valuation of the security. The actual uncertainty is the volatility figure or higher.

As a trivial example, if $p(x)$ is a Gaussian, then the information $I = 1/\sigma^2$, with $\sigma^2$ the variance. Then, by Eq. (13.28), the volatility $e^2_{\min} = \sigma^2$. Of course, $p(x)$ does not have to be Gaussian. Equation (13.28) provides a natural generalization of the concept of implied volatility to *non*-Gaussian underlying-asset distributions.

## 13.9 Discussion and summary

We have shown how, in the presence of limited data, EPI can be used to solve some canonical inverse problems in financial economics. The *general solution* is the differential Eq. (13.9). The resulting probability densities are generally *smoother* or less biased than those obtained by other methods such as maximum entropy. They also often well approximate the underlying PDF that gave rise to the data. EPI also has the virtue of providing a natural measure of, and computational approach for determining, implied *volatility*, via the Cramer–Rao inequality.

EPI (or, equivalently in this problem, MFI) provides a variational approach to the estimation of probability laws that are consistent with financial observables. Using this approach, one can employ well-developed computational approaches from formally identical problems in the physical sciences. A key benefit of use of the Fisher measure is that the output PDFs are forced to have *continuous first derivatives*. This imparts to the estimated PDFs and resulting observables (integrals of these PDFs) a degree of smoothness that one expects in good estimates.

As a parallel development, we also attacked the problem of PDF estimation in *non-equilibrium classical mechanics*. Again, the lack of microscopic analysis and the limited amount of data do not allow an exact answer. The approximate answer to both the economic and mechanics problems is found to obey the differential Eq. (13.9). This has the mathematical form of the SWE. In fact, the entire *Legendre-transform* structure of thermodynamics can be expressed using Fisher information *in place of* the Shannon entropy – an essential ingredient for constructing a statistical mechanics (Frieden *et al.*, 1999).

On comparing the maximum entropy approach with that of EPI, it was seen that the former has the virtue of simplicity, always giving an exponential answer for the PDF. By comparison, the EPI solution is always in the form of a differential equation in the PDF. Although somewhat more complicated, this is, in some sense, a virtue because the probability densities of physics and, presumably of finance and economics, generally obey differential equations. Indeed, the latter applications offer great potential for *future applications* of EPI.

A theory of *macroeconomics* that is related to EPI is that of Aoki (1998; 2001). The aim of this approach is to place all or nearly all heterogeneous microeconomic agents within one stochastic and dynamical framework. The result is a description of economic dynamics that is provided by equations like the Chapman–Kolmogorov and Fokker–Planck equations – equations of non-equilibrium statistical mechanics. Since the Fokker–Planck equation and the SWE (13.9) share eigensolutions (Risken, 1996), Aoki's theory might often share solutions with those of EPI.

In summary, in cases of limited data, the elegant and powerful Lagrangian approach to physics, with a framework provided by EPI, should prove to be a very useful tool for estimating the distribution functions of finance, economics, and statistical mechanics.

## 13.10 Glossary

**Coupon bond:** A debt obligation that makes periodic interest payments during the life of the obligation and pays the principal amount together with the final coupon payment on the maturity date.

**Discount factor:** The fraction of (say) a dollar that an investor would need to put into a risk-free investment today so that he would have one dollar on a future date. It is equal to the price of a zero coupon bond maturing on that date divided by the principal amount of the zero coupon bond.

**European call option:** A call option (see below) that permits the holder to buy the corresponding underlying asset (e.g., stock) at, but not before, the expiration date of the option. By comparison, an "American" call option permits purchase of the underlying asset at any time up to expiration.

**Forward rate:** An interest rate to apply for a specified period beginning at a specified future date.

**Log-normal distribution:** The risk-neutral log-normal distribution is

$$p(S_T, T|S_t, t) = \frac{1}{S_T\sqrt{2\pi\sigma^2(T-t)}} \exp\left(-\frac{[\ln(S_T/S_t) - r(T-t)]^2}{2\sigma^2(T-t)}\right).$$

Substituting this into (13.25) yields the Black–Scholes call price

$$c(k) = S_t N(d_1) - ke^{-r(T-t)} N(d_2),$$

where $N(x)$ is the cumulative normal density function and where

$$d_1 \equiv \frac{\ln(S_t/k) + (r + \sigma^2/2)(T-t)}{\sigma\sqrt{T-t}},$$

$$d_2 \equiv d_1 - \sigma\sqrt{T-t}.$$

**Macroeconomic:** Regarding the overall state of an economy, as to its total output, employment, consumption, investment, price and wage levels, interest and growth rates, etc.

**Microeconomic:** Regarding the supply, demand, and prices of individual commodities and securities.

**Option:** Gives the buyer the right, but not the obligation, to buy or sell an asset at a set price on or before a given date. Investors, not companies, issue options. Investors who purchase call options bet the stock will be worth more than the price set by the option (the strike price), plus the price they paid for the option itself. Buyers of put options bet the stock's price will go down below the price set by the option. An option is part of a class of securities called derivatives, so named because these securities derive their value from the worth of an underlying investment.

**Perpetual annuity:** An annuity that pays interest forever.
**Risk-free spot rate:** The theoretical yield on a zero-coupon bond.
**Risk-neutral density:** The PDF $p(x)$ in the European call valuation Eq. (13.25).
**Spot rate:** The theoretical yield on a zero-coupon bond.
**Strike price:** The stated price for which an underlying asset may be purchased (in the case of a call) or sold (in the case of a put) by the option holder upon exercise of the option contract.
**Tenor:** The time remaining before maturity of a bond.
**Term structure of interest rates:** Relationship between interest rates on bonds of different maturities, usually depicted in the form of a graph often called a yield curve.
**Volatility:** A measure of risk based on the standard deviation of the asset return. Also, volatility is a variable that appears in option pricing formulae. In the option pricing formula, it denotes the volatility of the underlying asset return from now to the expiration of the option.
**Zero-coupon bond:** A bond that pays only the principal amount on the maturity date: there are no coupon payments (cf. definition of coupon bond above).

## 13.11 Acknowledgements

We thank Professor David Edelman for providing a preprint of his work and for stimulating discussions. We also thank Dana Hobson, Professor David A. B. Miller, Leif Wennerberg, and Professor Ewan Wright for helpful discussions and suggestions.

RAYMOND J. HAWKINS, PH. D. is a Managing Partner at Mulsanne Capital Management, 111 Chestnut St., Ste. 807, San Francisco, CA 94111; and also an Adjunct Professor at the Optical Sciences Ctr., University of Arizona, Tucson 85721. Email address: rhawkins@mulsanne.net.

# 14
# Growth and transport processes

## 14.1 Introduction

The *population components* of certain systems evolve according to what are called "growth processes" in biology and "transport processes" in physics. These systems are generally "open" to exterior effects such as, e.g., uniform or random energy inputs. Thus they do not necessarily follow the closed-system model of Sec. 1.2.1. However, as noted in Sec. 1.2.3, the Cramer–Rao inequality and the definition (1.9) of Fisher information hold for open systems as well. It will result that EPI can be applied to these problems as well, provided that the defining form (1.9) of Fisher information is used.

These growth processes are described by probability laws that obey first-order differential equations in the time. Examples are the *Boltzmann transport* equation in statistical mechanics, the rate equations describing the *populations of atomic energy levels* in the gas of a laser cavity, the *Lotka–Volterra* equations of *ecology*, the equation of *genetic change* in genetics, the equations of *molecular growth* in chemistry, the equations of *RNA cell replication* in biological cell growth, and the master equation of *macroeconomics*. These equations of growth describe diverse phenomena, and yet share a similar form. This leads one to suspect that they can be derived from a common viewpoint. The connection is provided by information (not energy; Sec. 1.1), and they are all derived by a single use of EPI.

### *14.1.1 Definitions*

Consider a generally open system containing $N$ kinds of "particles," at population levels $m_n$, $n = 1, \ldots, N$. The total number of particles in the system is

$$M \equiv \sum_n m_n.$$

(The upper limit on sums is understood to be $N$ unless otherwise indicated.) The particles can be *biological and/or physical* in nature, e.g., some mixture of living creatures, atoms, and/or molecules. In an ecological system the particles are living creatures and physical particles, such as of water, that are a resource for the creatures.

The situation is generally dynamical. As time passes, the particles interact. The interactions cause various population levels $m_n$ to grow and recede, each in turn achieving its "day in the sun." Hence, the population sizes $m_n$ vary with the time $t$. The total population $M$ can vary as well.

In order to ultimately bring in probabilities, we first define *relative* population sizes

$$m_n/M \equiv p_n, \qquad \text{where all } p_n,\ n = 1, \ldots, N \equiv \mathbf{p},$$

a vector. The relative sizes $p_n = p_n(t)$ then vary with the time as well.

Assume $M$ to be so large that the law of large numbers holds, i.e., for each relative population size

$$m_n/M \equiv p_n \equiv p(n|t) \equiv p_n(t), \tag{14.0}$$

the *probability* that the $n$th population type will be observed in the random drawing of a particle from the system at the time $t$. Thus, $n$ is taken to be random, and $t$ deterministic, in this problem. The $p_n(t)$ are called the temporal "growth functions" of the system. We often also call the $p_n(t)$ the "population levels" or "relative population levels," depending upon context.

The general conditional nature of the PDF $p(n|t)$ will be preserved throughout the calculation. In this class of problems shift invariance is *not* assumed to be present.

To clarify concepts, consider a system that is an aquarium. Suppose for simplicity that there are but $N = 4$ types $n$ of particle in the aquarium – snails, fish, plants, and water. A cubic millimeter (say) of *substance* is randomly sampled from the aquarium. This is the "particle." The particle is observed to be a piece of snail, fish, plant, or water, corresponding to values $n = 1$, 2, 3, or 4, respectively. The particle is presumed to be so small that there is vanishing probability that it will hold multiple substances such as *both* snail and fish simultaneously. Number $M$ is here the volume of the aquarium expressed in cubic millimeters. The volume $M$ will slowly deplete, by evaporation, but will ordinarily be quite large, so that the law of large numbers holds and the relative population sizes are again probabilities.

### 14.1.2 Normalization property

Since the $p_n$ are probabilities, they obey a normalization property

$$\sum_{n=1}^{N} p_n = 1 \quad \text{at each } t. \tag{14.1}$$

Differentiating $d/dt$ this equation $i = 1$ or 2 or ... times gives directly

$$\sum_{n=1}^{N} \frac{d^i p_n}{dt^i} = 0. \tag{14.2}$$

The case $i = 1$ will be particularly important to the EPI derivation.

### 14.1.3 The growth equations

The equations are inherently discrete in the index $n$ that identifies population type. Most of the equations are derivable from elementary considerations. The aim of this chapter is to show that all may be derived, alternatively, using EPI.

A few important ones follow. They contain coefficients $g_n$, $d_n$, etc., that are assumed to be *known functions* of the probabilities $\mathbf{p}$ and the time $t$. The coefficients are inputs that effectively drive the differential equations, acting as EPI *source terms* (Caveat 2 of Sec. 0.1).

(1) In *ecological systems* the constituent populations grow according to *Lotka–Volterra* equations (Lotka, 1956)

$$\frac{dp_n}{dt} \equiv \dot{p}_n = p_n(g_n + d_n), \quad \text{where} \quad g_n = \sum_{k=1}^{N} g_{nk} p_k, \tag{14.3}$$

$$\text{with} \quad g_{nk} \equiv g_{nk}(\mathbf{p}, t),$$

$$d_n = d_n(\mathbf{p}, t), \quad \text{and } p_n = p(n|t) \equiv p_n(t), \quad \text{for } n = 1, \ldots, N.$$

The $g_n$, $d_n$, and $g_{nk}$ are change coefficients that are defined in Sec. 14.5.2. Equations (14.3) were derived by Lotka (1956), and by Glansdorff and Prigogine (1971), by linearizing the hydrodynamic equations through an *approximate Taylor's expansion*. The equations have been used to model predator–prey interactions (Vincent and Grantham, 1997) and, even, entire ecological systems, including resources (Fath *et al.*, 2003).

(2) In *genetics*, the equations of *genetic change* are

$$\dot{p}_{nk} = p_{nk}(w_{nk} - \langle w \rangle), \quad \langle w \rangle \equiv \sum_{nk} w_{nk} p_{nk} \tag{14.4}$$

(Crow and Kimura, 1970). The double subscript $nk$ indicates a diploid mating of a gamete $n$ from one individual with a gamete $k$ from another. A corresponding genotype $nk$ results from the mating. The growth coefficients $w_{nk}$ are defined in Sec. 14.6.1. The Eqs. (14.4) of genetic change help to quantify Darwin's theory of evolution.

Crow and Kimura notably stated that it would be desirable to derive Eqs. (14.4) from a variational principle of the type used in physics. This comment provided, in fact, the original motivation for our paper (Frieden *et al.*, 2001) and, subsequently, this chapter.

(3) In a *laser resonator cavity* consisting of two-level atoms, by comparison with the preceding first-order quadratic differential equations, a system of first-order, *linear rate equations*

$$\begin{aligned} \dot{p}_1 &= R(p_2 - p_1) + R' p_2, \\ \dot{p}_2 &= -R(p_2 - p_1) - R' p_2. \end{aligned} \tag{14.5}$$

occurs. These describe the relative population levels $p_1(t)$, $p_2(t)$ of the two levels (Allen and Eberly, 1987; Milonni and Eberly, 1988). The growth coefficients $R$, $R'$ are defined in Sec. 14.7.1. Among other things, the Eqs. (14.5) predict the population inversions necessary to acquire lasing.

(4) Consider an *ideal, dilute gas*. This is a system of indistinguishable molecules that interact solely by elastic collisions. If the gas obeys the properties listed in Sec. 14.8, its evolution obeys the following rate equation:

$$\dot{p}_n = p_n(g_n^+ - g_n^-), \quad g_n^\pm \equiv g_n^\pm(\mathbf{p}, t) \tag{14.6}$$

as functions. Here $p_n(t) \equiv p((\boldsymbol{r}, \mathbf{v})_n, t)$, the probability of the discrete space–velocity values $(\boldsymbol{r}, \mathbf{v})_n$ at the time $t$ (see details in Sec. 14.8.2). In the continuous limit discussed in Sec. 14.8.3 this becomes

$$\dot{p} = p(g^+ - g^-). \tag{14.7}$$

This is the basic form of the Boltzmann transport equation (Reif, 1965). It says in particular that the probability law $p$ obeys a first- rather than a second-order differential equation. The right-hand side is calculated on the basis of specific properties of the gas. If the gas obeys the properties listed in Sec. 14.8, the result is Eq. (14.31). Replacing the right-hand side of (14.6) by this, and the left-hand side by identity (14.58), gives what is usually considered *the* Boltzmann transport equation

$$\frac{\partial p}{\partial t} + \mathbf{v} \cdot \frac{\partial p}{\partial \boldsymbol{r}} + \dot{\mathbf{v}} \cdot \frac{\partial p}{\partial \mathbf{v}} = \iint d\Omega' \, d^3\mathbf{v}_1 \, V\sigma(p'p_1' - pp_1). \tag{14.8}$$

The parameters in the right-hand side integral are given above Eq. (14.31). The

Boltzmann transport equation takes other, special forms, depending upon the particular form taken by the change coefficients $g^{\pm}$, including the Navier–Stokes equation (see Sec. 14.12).

The EPI approach below will derive the *discrete* Boltzmann transport Eq. (14.6). The latter states that the evolution obeys a first- rather than a second- or higher-order differential equation in the time. This is a key step toward defining the evolution of the system in complete detail.

Next, a transition (Eq. (14.54)) is made to the continuous Eqs. (14.7) and (14.8). This merits further discussion. Transport phenomena are initially defined over a discrete phase space (Sec. 14.8.1) consisting of contiguous space-momentum "boxes" of the finite size $\hbar$. The transition $\hbar \rightarrow 0$ to continuous phase space is only an idealization, and in fact "is really physically meaningless, [since it] would lead to a specification of the system more precise than is allowed by quantum theory" (Reif, 1965, p. 51). In fact, in agreement with this comment, the EPI derivation, below, of the discrete Boltzmann transport Eq. (14.6) does takes place within the discrete phase space. This is, then, a consistent aspect of the EPI approach. It is only *after* the EPI derivation of (14.6) that the usual transition $\hbar \rightarrow 0$ to continuous phase space is made, giving the desired end result (14.8).

By comparison, the standard derivation of the Boltzmann equation (Reif, 1965) is *ab initio* wthin *continuous* phase space. This seems to violate the spirit of Reif's statement above about meaningless phase space resolution. However, it usually has no serious consequences. By either the standard or the EPI approach, the problem of meaningless phase space resolution can be avoided by simply not evaluating the Boltzmann equation at a finer subdivision of space-momentum values than the limiting box size $\hbar$ allows.

(5) *Replicating molecules* of RNA (ribonucleic acid) are often thought to characterize the growth of the earliest forms of life (Bernstein *et al.*, 1983). These and other molecules of chemical growth processes grow according to a law

$$\dot{p}_n = p_n\left(\frac{b_n A}{A_n + A} - d_n\right). \tag{14.9}$$

Here $A_n$ is the level of the resource for replication $n$ that gives half its maximum rate of replication; $A$ is the total resource available; $b_n$ is the maximum rate of replication, from (14.9) when $A \gg A_n$; and $d_n$ is the mortality rate.

Other such growth laws are defined, in less detail, in Sec. 14.12.

## 14.2 General growth law

The discrete growth laws (14.3)–(14.6) and (14.9) can all be expressed simultaneously as one law

Table 14.1. *Identifying the change coefficients $g_n$, $d_n$*

| Application | Equation | Change coefficient identification |
|---|---|---|
| (1) | (14.3) | $g_n \to \sum_m g_{nm} p_m, \quad d_n \to d_n$ |
| (2) | (14.4) | $g_n \to w_{nk}, \quad d_n \to -\langle w \rangle$ |
| (3) | (14.5) | $g_1 \to -R, \quad d_1 \to (p_2/p_1)(R + R'),$ $g_2 \to -(R + R'), \quad d_2 \to (p_1/p_2)R$ |
| (4) | (14.6) | $g_n \to g_n^+, \quad d_n \to -g_n^-$ |
| (5) | (14.9) | $g_n \to \dfrac{b_n A}{A_n + A}, \quad d_n \to -d_n$ |

$$\dot{p}_n = p_n(g_n + d_n), \quad \text{where} \quad g_n \equiv g_n(\mathbf{p}, t) \quad \text{and} \tag{14.10}$$

$$d_n \equiv d_n(\mathbf{p}, t), \qquad n = 1, \ldots, N$$

are called "growth" and "depletion" coefficients, respectively. All coefficients together are called "change" coefficients. The change coefficients are the *effective sources* of the EPI problem, so it is assumed that these are known, at least numerically.

The aim is to show that the EPI principle may be used to derive this *first-order* differential equation. It is interesting to compare this with a *second-order* differential equation, as would be appropriate for describing position in Newtonian mechanics. In the first-order Eq. (14.10), the "velocities" $\dot{p}_n$ depend upon the time only through its current value $t$. Thus *they do not exhibit memory effects*. By comparison, in a second-order Newtonian equation the velocities depend upon knowledge of the force at *former* times as well as the present: the system does exhibit memory.

When Eq. (14.10) is integrated over time to form the instantaneous $p_n(t)$, past growth effects are incorporated into the current value. Thus, in comparison with its rate of change $\dot{p}_n(t)$, the *instantaneous population* level $p_n(t)$ does exhibit memory. This memory acts to stabilize, to a degree, the population curves $p_n(t)$, forcing them toward a continuous dependence upon the time.

### *14.2.1 Correspondences for the change coefficients*

The change coefficients $g_n$, $d_n$ in (14.10) identify with particular coefficients in Eqs. (14.3)–(14.6) and (14.9) as given in Table 14.1. The arrows make the identifications.

We emphasize that the laws (14.3)–(14.9) have been derived by other approaches. These are given in the references cited. What we want to show is that the laws of growth follow, as well, from a single EPI derivation.

As usual for an EPI derivation, the appropriate measurement must first be identified. This is done next for each of the applications.

### *14.2.2 Units and definition of the change coefficients*

From the form of Eq. (14.10), the change coefficients $g_n$, $d_n$ have units of $t^{-1}$. As the source functions for the problem, they are assumed to be *known* functions

$$g_n = g_n(\mathbf{p}, t), \quad d_n = d_n(\mathbf{p}, t) \tag{14.11}$$

of $\mathbf{p}$ and $t$. See Secs. 14.5–14.8 for particular change coefficient functions.

### *14.2.3 Plan*

The overall aim of the chapter is to derive the generic growth law (14.10) by means of EPI. After this derivation, in Sec. 14.10 substitution in of the *particular, known* source functions $g_n$ and $d_n$ given in Secs. 14.5–14.8 will yield the particular growth laws mentioned above. These cover a wide range of phenomena. The list is further broadened in Sec. 14.12.

Deriving the general law (14.10) will have some useful ramifications.

(i) It will permit us to conclude that EPI applies holistically to the observer, i.e., both to his external world and to his internal world:

  (a) *externally*, defining the microscopic and macroscopic phenomena he purposely observes, including the ecology surrounding him; and
  (b) *internally*, defining the growth of the competing populations (functioning cells, bacteria, possible cancer cells, etc.) that exist within his body. These govern his ability to live and, ultimately, *to observe*.

(ii) The EPI approach will give rise to a general uncertainty principle (14.67) of growth. This predicts that, under certain conditions, the evolutionary age $t$ of a system and its mean-square rate of change of populations at that time are complementary variables: The larger one is, the smaller is the other. This has an important consequence regarding ecological extinction effects.

(iii) The fact that the growth law (14.10) applies both to physical systems and to organic molecules, such as of RNA in application (5) above, verifies what has been called "The Darwinian dynamic" (Bernstein *et al.*, 1983). This is also consistent with a model of the origin of life, according to which life originated in imitation of an inorganic growth process that acted as a template for it (see Sec. 14.17).

As usual for an EPI derivation, the appropriate measurement must first be identified. This is done next for each of the applications.

## 14.3 Measured parameters

We assume that any particle of the system moves at non-relativistic speeds. This frees us from the requirement (Chapter 3, Sec. 3.5.8) of measuring *four*-vectors.

All systems under consideration consist of populations that *grow and regress in time*, that is, they evolve. Knowledge of the time is quite important to knowledge of the state of such a system. For example, in the ideal gas problem (4), the time value can indicate whether the system has reached its equilibrium state; or, in an evolving ecological system (1), the time value indicates approximately what the current state of evolution is. For example, if we know that the time is in the Mesozoic era, the state of evolution favors ferns and dinosaurs rather than flowering plants and apes. All these inferences assume that we know the growth law $p_n(t)$ for the system.

Assume, then, that the given system is turned "on" at some unknown time in the past, and *we require knowledge of the elapsed time* $t$ since its inception. Assume, for this purpose, that the growth law for the system is known, and a datum is observed whose value could be used, if desired, to imply the value of the time. The datum carries the Fisher information $I(t)$ *about the time*. The particular observed data are next listed for the system types (1)–(5) given in Sec. 14.1.

(1) and (2) *In an ecological system*, the observation is taken to be that of a certain species, either specified by a single "name" $n$ or as a gamete combination $nk$, in a randomly sampled volume, say a cubic millimeter, of the system. The type of species that is observed gives information about the time. As a simplistic example, if the *observed* species is a dinosaur then this indicates that the time is somewhere in the Mesozoic era.

(3) *In a system consisting of two-level atoms*, the observation is the random observation of an atom at the energy level $i$, where $i = 1$ or 2. Combining this with knowledge of the growth Eqs. (14.5) again permits an estimate of the time. For example, if they predict an *eventual* dominance of level 2, then the observation of that level implies that the time is relatively large.

(4) *In the ideal gas*, the datum is the observation of the $n$th discrete position–velocity pair $(\boldsymbol{r}, \mathbf{v})_n$ value of a randomly selected molecule. This observation gives *information about the time* because it is known that, with increasing time, (i) the particles must diffuse outward from their initial spatial configuration toward $r = \infty$, and (ii) the distribution of speeds should evolve toward the Maxwell–

Boltzmann equilibrium distribution (Chapter 7). Then, e.g., if $\boldsymbol{r}$ is observed to be very large, by the effect (i), the time ought to be relatively large.

(5) In *molecular replication* the datum $n$ is the observation of a replicator $n$, e.g., an RNA molecule.

In summary, the measured quantity is, depending upon the application,

$$\text{species } n, \quad \text{gamete } nk, \quad \text{atomic level } i,$$

$$\text{the joint } (\boldsymbol{r}, \mathbf{v})_n, \quad \text{or replicator } n. \tag{14.12}$$

We presumed in the preceding that the growth law $p_n(t)$ for the system was known. The EPI principle that is used below gives, in fact, a growth law that, if known, *would enable the time to be optimally inferred*. The reason is that the principle accomplishes $I - J = extrem.$, and the extremum is a *minimum* for this problem. Therefore $I$ is optimally close to its maximum possible value, $J$, so that, by Eq. (1.28), the error in estimation of the time is a minimum.

Previous EPI applications were to closed systems (Sec. 1.2.1). Here, by comparison, the system is generally open. During acquisition of a datum (14.12), the system is subject to possible outside influences, such as energy flux, radiation, etc. (depending upon the system). These necessarily influence the system growth, and therefore the growth law $p_n(t)$, in some manner. It is assumed that their effects upon the measured datum are known (at least) statistically. Thus, the aim of the EPI observer becomes that of finding the system PDF on the basis of the datum and this known random effect. The key, of course, is finding the appropriate invariance principle.

## 14.4 The invariance principle

The EPI growth equation solutions for problems (1)–(5) will be found to follow from a single invariance principle. This is established next.

### *14.4.1 Derivation of invariance principle*

There is no single *obvious* invariance principle that is common to biological, physical, and economic growth processes. However, by internal consistency of EPI, if its solution is correct, the solution must obey the sought invariance principle. Hence, with 20 : 20 hindsight of the solution, Eq. (14.10), we form the invariance principle from it, keeping in mind that, in a real situation, where the solution is unknown, this would of course be impossible to do. In these real applications the invariance principle would simply be regarded as an *ansatz* (working hypothesis). However, in fact most of the applications (1)–(5) are known to obey the principle, so that the *ansatz* has manifestly wide applicability. These cases will be pointed out.

Summing Eq. (14.10) over all $n$, and using Eq. (14.2) with $i = 1$, gives

$$\sum_n (g_n + d_n) p_n = 0, \qquad g_n = g_n(\mathbf{p}, t), \qquad d_n \equiv d_n(\mathbf{p}, t). \tag{14.13}$$

This is the invariance principle that will be used.

Note that the left-hand side of Eq. (14.13) is the mean overall change $\langle g + d \rangle$ at the given time. Since this time is arbitrary, the principle therefore states that the *mean overall change is zero* at all times,

$$\langle g + d \rangle = 0. \tag{14.14}$$

We regard this condition (14.13) or (14.14) of zero mean overall change as a *postulate* or *ansatz* of the growth processes we seek to describe. Its validity, in any application, must be verified before the resulting EPI prediction of the growth Eq. (14.10) is accepted.

### 14.4.2 *Implications of the invariance principle* per se

It is important to establish that we actually have a problem here. If the required growth law (14.10) is implied by (14.13) *itself*, there would be no need to use EPI. However, this is not the case. For example, consider the alternative growth law to (14.10)

$$\sum_{i=1}^{i_0} \alpha_i(t) \frac{d^i p_n}{dt^i} = p_n(g_n + d_n)\beta(t) + \dot{P}_n(t), \quad \text{where} \quad i_0 \geqslant 1 \tag{14.15a}$$

is an arbitrary upper limit to the sum. The $\alpha_i(t)$ and $\beta(t)$ are arbitrary functions, and $P_n(t)$ is an arbitrary probability law on $n$. Thus (14.15a) represents a *family* of possible growth laws, consisting of differential equations of *general order* $i_0$. By comparison, Eq. (14.10) is limited to being a *first-order* differential equation. Summing both sides of (14.15a) over $n$ and using identity (14.2) does give back Eq. (14.13), i.e., does satisfy the invariance principle. Hence the invariance principle generally gives a family of growth laws, not uniquely the required result (14.10).

By comparison, it is the nature of an extremum principle to select a *unique* solution from among many contenders. Thus, when the invariance principle (14.13) is used *with* EPI, this will pick the required form (14.10) out of the infinite family of possible growth law solutions.

### *14.4.3 On detailed balance*

We note in passing that invariance principle (14.13) is not the same as the so-called "principle of detailed balance" (Reif, 1965, p. 384). The latter is defined for a system at a time $t$ that is so large that *equilibrium* is attained. By comparison, Eq. (14.13) holds at all times. (See material preceding Eq. (14.14).)

### *14.4.4 On the change coefficients*

As was discussed in Secs. 14.1.3 and 14.2.2, the change coefficients $g_n$, $d_n$, etc., have the role of *known inputs* that effectively drive the growth equations. This is analogous to the role played by the charge-current sources of Chapter 5 in electromagnetic theory, i.e., *source* terms that are *specified* independently of EPI. These coefficients are addressed next, in separate sections for each of the applications (1)–(5). Depending upon application, they are either given or shown to follow from simple dynamical considerations.

## 14.5 Change coefficients for an ecological system

Consider a living biological system such as the aquarium, mentioned previously, containing snails, fish, plants, and water. An outside source of light illuminates the aquarium. Here there would be $N = 4$ populations under observation, of which three are alive and one is a non-living resource. Such a system is usually regarded as obeying the Lotka–Volterra Eqs. (14.3) (Crow and Kimura, 1970; Fath *et al.*, 2003). We identify next its change coefficients.

In this application the change coefficients $g_n$, $d_n$ are generally functions of all the population levels $p_n(t)$ and also quantities that cannot be represented as "growing populations," such as the temperature $T$, level of humidity, level of sunlight, etc. These are outside influences (Sec. 14.3) on the system. Their effects upon the system are *random*, but quantifiable through the change coefficients (see below). Their random influence causes the resulting Lotka–Volterra equations to become stochastic in nature (Frieden, 2001).

### *14.5.1 Spontaneous changes*

The form of the first Eq. (14.3) indicates that a coefficient $d_n$ is an absolute input to the problem, independently of other components $p_k$. (By comparison, the second equation shows that $g_n$ depends upon all $p_k$.) It therefore describes a *spontaneous* change. If $d_n > 0$ the change is one of growth; if $d_n < 0$ it is one of loss. The reason for the notation $d$ is that most often

$$d_n < 0, \tag{14.15b}$$

i.e., the spontaneous events are those of *d*epletion of a resource (say, water molecules) or the *d*eath of a creature.

### 14.5.2 Interactive changes

Aside from spontaneous changes, changes can be *interactive*. Thus, the second Eq. (14.3) indicates that a coefficient $g_{nk}$ represents a relative change in the population level $p_n$ due to the *interaction* of $p_n$ with a component $p_k$. The change could be positive, negative, or zero. Also, in general the diagonal coefficients

$$g_{nn} \neq 0 \tag{14.16}$$

since, by possible depletion effects, the current level $p_n$ of a population can affect its own rate $\dot{p}_n$. The $g_{nk}$ are in general functions of all the $p_n$, as indicated in Eqs. (14.3).

Equations (14.3) have conventionally been used to describe predator–prey relations among competing animals in an ecosystem (Vincent and Grantham, 1997). Use of the equations has even been extended to include mixed animate–*inanimate* systems, such as entire ecological communities (Fath *et al.*, 2003). Here, some of the components $n$ are not living creatures but, rather, resources that the animals need for survival, such as water. These can also be modelled as changing "populations," but now of relative *mass* rather than of relative number of individuals.

Denote as

$$\Delta p_{n|k} \equiv g_{nk}, \quad \text{where} \quad g_{nk} \equiv g_{nk}(\mathbf{p}, t), \tag{14.17}$$

the change in $p_n$ due to a single population component $k$, where $k = 1, 2, \ldots,$ or $N$. The gain coefficients $g_{nk}$ are assumed to be known.

Next, *multiply both sides of* (14.17) *by* $p_k$ *and sum over all* $k$. Results are as follows:

(i) The left-hand side of (14.17) gives

$$\sum_k \Delta p_{n|k} p_k \equiv g_n, \tag{14.18}$$

by definition the average change in $p_n$ due to interactions with all population components.

(ii) The right-hand side of (14.17) gives directly $\sum_k g_{nk} p_k$. Since this equals the left-hand side, Eq. (14.18), we get

$$g_n = \sum_k g_{nk} p_k. \tag{14.19}$$

This is the origin of the expression for $g_n$ in Eqs. (14.3).

In this ecological application it is not obvious that the invariance principle (14.13) is obeyed. It must remain a postulate of the EPI approach to the problem. Since EPI will derive the Lotka–Volterra Eqs. (14.3) on the basis of (14.13), the conclusion will be that the Lotka–Volterra equations apply to the extent that the invariance principle (14.13) holds, and this has to be justified on a case-by-case basis. A case for which it is readily justified is as follows.

## 14.6 Change coefficients of genetic growth

Here the growth coefficients are the $w_{nk}$. These are called "fitness coefficients."

### *14.6.1 Fitness coefficients defined*

A fitness coefficient $w_{nk}$ is defined as the relative number $\Delta p_n$ of offspring per mating *nk* that have survived to reproduce at the time $t$. These fitness coefficients are enforced by the surrounding environment and the instantaneous population levels $p_{nk}$, and can generally depend upon *the time*. Thus, they drive the system of Eqs. (14.4) and provide effective source functions for it. Notice from Eqs. (14.4) that $\langle w \rangle$ is the mean instantaneous fitness value.

### *14.6.2 Growth scenario*

The presumed conditions under which the genetic growth takes place are fairly general.

(a) There are multiple gene loci present on each chromosome.
(b) Individual births and deaths are so frequent as to occur nearly continuously in time. The population generations are generally overlapping.
(c) The fitness coefficients $w_{nk}$ generally vary with the time (i.e., from one generation to the next).
(d) The matings follow any selection rules, i.e., random, selective, or any combination thereof.

As with the equations of ecological growth, genetic growth is influenced by random outside factors. These are embodied in the fitness coefficients $w_{nk}$.

### 14.6.3 Fitness coefficients as interactive and spontaneous changes

Equations (14.4) are quite similar in mathematical form to the Lotka–Volterra Eqs. (14.3). Moreover they share a similar interpretation. Thus, the quantity

$$(w_{nk} - \langle w \rangle) \tag{14.20}$$

represents the amount by which the change $\Delta p_n$ in the population level $p_n$ exceeds the mean change. The sign of $(w_{nk} - \langle w \rangle)$ is crucial to whether growth or depletion is taking place.

In comparison with the scope of Eqs. (14.3), the Eqs. (14.4) of genetic change are limited to describing the populations of living creatures, *not* including inanimate populations. (Inanimate populations do not, of course, have gametes.) Equations. (14.4) are derivable from elementary considerations. See any book on genetics, e.g. by Crow and Kimura (1970). They also have been derived by EPI (Frieden *et al.*, 2001).

The mean fitness is simply

$$\langle w \rangle = \sum_{nk} w_{nk} p_{nk}, \qquad w_{nk} \equiv w_{nk}(t). \tag{14.21}$$

By *normalization* Eq. (14.1), of course also

$$\langle w \rangle = \sum_{nk} \langle w \rangle p_{nk}. \tag{14.22}$$

Then taking the difference of (14.21) and (14.22) gives

$$\sum_{nk} (w_{nk} - \langle w \rangle) p_{nk} = 0. \tag{14.23}$$

This has just the *form of the invariance law* (14.13), under the correspondences

$$g_{nk} = w_{nk} \quad \text{and} \quad d_n = -\langle w \rangle \tag{14.24}$$

for this particular case.

Hence, in this application, the invariance law (14.13) is easily *derived*, i.e., without recourse to the growth law solution (14.10).

## 14.7 Change coefficients in laser resonator

Consider an atomic two-level population of particles comprising a gas. Let energy level 1 be lower than level 2. Transitions $1 \rightleftarrows 2$ occur from one population level to the other. These transitions give rise to the rate Eqs. (14.5). The latter may be derived using basic principles of quantum electrodynamics (Allen and Eberly, 1987; Milonni and Eberly, 1988).

The population changes are, as in the *biological* case (1), both *spontaneous* and *interactive* in nature, the latter due to the presence of an imposed electromagnetic field. (The field is the exterior system influence for this problem alluded to in Sec. 14.3.) The spontaneous changes occur at the constant rate $R'$, and the interactive ones occur at the constant rate $R$. Dimensional analysis of Eqs. (14.5) shows that the rate $R$ is, as with any change coefficient in (1) or (2) above, the change in $p_1$ per unit $p_1$ per unit time. The rate $R'$ depends similarly upon $p_2$.

The aim of this section is to express the induced and spontaneous changes $g_n$ and $d_n$, $n = 1, 2$, in the two populations in terms of the other parameters $R$, $R'$, $p_1$, $p_2$ of the scenario. Since these expressions are rather easily derived, rather than merely stating them we show the derivation. The simple model provided by Allen and Eberly is used.

*Historical note*: The similarity in form between Eqs. (14.3) and (14.5) prompted my colleague, the late Fred A. Hopf, and his co-workers (Bernstein *et al.*, 1983) to investigate parallels between population growth in a two-level atomic gas and population growth in a biological system.

### *14.7.1 Spontaneous resonator changes*

The spontaneous changes (losses) are easiest to find. Denote by $d_1$ the spontaneous loss from the lower level 1. Level 1 may lose population only by induced emission, as there is no lower level to which spontaneous emission can go. Owing to the constant induced field, the loss is a constant amount $R$,

$$d_1 = -R. \tag{14.25}$$

The higher level 2 may lose population either by induced emission – at rate $R$ due to the external field – or by spontaneous emission at rate $R'$ to the lower level 1. The total loss from level 2 then obeys

$$d_2 = -(R + R'). \tag{14.26}$$

The *induced changes* $g_1$, $g_2$ remain to be found.

### *14.7.2 Induced resonator changes*

In this laser application the invariance principle (14.13) is known to hold (Allen and Eberly, 1987). This permits us to use Eq. (14.13) to generate the coefficients. The equation is here

$$p_1 g_1 + p_2 g_2 = -(p_1 d_1 + p_2 d_2) = p_1 R + p_2(R + R'), \tag{14.27}$$

the latter by (14.25) and (14.26). The *outside equation* may be solved for the remaining unknowns $g_1$ and $g_2$, provided that we have another equation in these quantities.

Since level 2 loses at the rate $(R + R')$, and this energy can only go to level 1, level 1 should likewise *gain* at a rate proportional to $R + R'$. That is,

$$g_1 = +K(R + R'), \tag{14.28}$$

with $K \geqslant 0$ some constant to be found. In the outside Eq. (14.27), quantities $R$ and $R + R'$ may be regarded as independent algebraic parameters. Therefore a solution should give equal coefficients of $R$ and of $R + R'$ on both sides. Substituting Eq. (14.28) into the outside Eq. (14.27) and matching coefficients of $R + R'$ on both sides gives $K = p_2/p_1$. Thus, from (14.28) directly

$$g_1 = \frac{p_2}{p_1}(R + R'). \tag{14.29}$$

Finally, substituting this into the outside Eq. (14.27) allows solution for $g_2$,

$$g_2 = \frac{p_1}{p_2} R. \tag{14.30a}$$

All the required induced and spontaneous changes $g_n$, $d_n$ for the problem have therefore been parametrized, as required.

In the above we used the fact that the change Eq. (14.27) is known to be obeyed. This is another case where the generic constraint Eq. (14.13) is known to the observer.

## 14.8 Change coefficients for ideal gas

The ideal gas is presumed to have the following properties (Reif, 1965).

(a) It is so dilute that only *two*-particle collisions need be considered.
(b) As with the other systems in this chapter, the gas is subjected to external influences. Here the influence is that of outside force fields of various types. Although the collision events of a real gas would be affected by the external fields, the ideal gas is presumed to be unaffected.
(c) Any possible correlations between the velocities of colliding molecules prior to their collision can be ignored. This is called a condition of "molecular chaos," and is only true as gas density approaches zero.
(d) The mean de Broglie wavelength of particles of the gas is small compared with the mean separation between particles. Then the motion of a particle between collisions can be described adequately as the motion of a wave packet or *classical* particle.

### *14.8.1 Discrete phase space*

Since molecules are indistinguishable, they cannot be specified as being of discrete "types" $n$ as in the sense of the above biological applications.

However, on the quantum level, the particle is pictured as moving through a phase space of *discrete* space-momentum cells. Because of the Heisenberg uncertainty principle (4.47), each cell has a linear size on the order of $\hbar$ and, therefore, an elemental volume of size $\hbar^3$. To prescribe the particle's trajectory on any finer scale than this would be meaningless (see also below Eq. (14.8)).

Hence, the particle's phase space values are in general discrete. For our purposes it is more convenient to use space-*velocity* values. Denote these as quantities $\mathbf{s}_n$. From the preceding, each

$$\mathbf{s} \equiv (\boldsymbol{r}, \mathbf{v})_n, \qquad n = 1, \ldots, N, \tag{14.30b}$$

where $m$ is the particle mass. The triple-equal sign indicates that each value $s_n$ identifies one possible space-velocity $(\boldsymbol{r}, \mathbf{v})_n$ value for the particle. Notice that the six-dimensional space of space-velocity values may be represented one-dimensionally in this way, via so-called "lexicographic mapping."

In this problem any trajectory taken by a particle through phase space will necessarily be abrupt and jumpy, because the phase space values $(\boldsymbol{r}, \mathbf{v})_n$ given in (14.30b) are discrete. This is taken up next.

### 14.8.2. Discrete trajectory

Let the probability $p(\mathbf{s}_n|t) \equiv p((\boldsymbol{r}, \mathbf{v})_n, t) \equiv p_n(t)$ define the frequency with which the particle occupies a random cell $n$ of phase space at a known (not random) time $t$. (With $t$ known, the notation $(p((\boldsymbol{rv})_n|t)$ is preferred, but we stick with the preceding because of its widespread use in physics texts. See also Sec. I.1 of Appendix I.) Thus, the growth problem is still a discrete one. As was discussed, this will enable a common EPI derivation to treat all the growth problems simultaneously.

As a distinction in nomenclature from previous applications, rather than describing "growth" the system function $p_n(t)$ now describes the *evolution* of the particle through phase space. Or more precisely, it describes our *knowledge* of this evolution.

In the ideal gas all changes are *interactive*, as the result of two-particle collisions (see item (a) preceding). Hence the changes are of the type $g_n$; there no longer are spontaneous changes $d_n$.

### 14.8.3 Transition to smooth trajectory

The discrete situation (14.30b) is maintained until after the EPI derivation (Sec. 14.10) of the growth equation (14.6) is obtained. It is then necessary to insert into this equation the particular change coefficients for the problem. The aim of this section is to calculate these changes.

To calculate these change coefficients, it is convenient as usual to analyze the gas particles as moving through a *continuous* phase space (subject to the proviso below Eq. (14.8)). Therefore we take the classical limit $\hbar \to 0$ of Eq. (14.30b). This causes the discrete phase space values

$$(\boldsymbol{r}, \mathbf{v})_n \to (\boldsymbol{r}, \mathbf{v}), \tag{14.30c}$$

continuous ones. Hence, the measurement is of a continuous pair value $(\boldsymbol{r}, \mathbf{v})$, and this takes place during a differential time interval $dt$. Also, the formerly discrete probabilities

$$p_n \to p(\boldsymbol{r}, \mathbf{v}, t), \tag{14.30d}$$

a continuous function. The final continuous limit

$$g_n^{\pm} \to g^{\pm}(\mathbf{p}, t) \equiv g^{\pm}(\boldsymbol{r}, \mathbf{v}, t), \tag{14.30e}$$

is from the formerly discrete changes to a continuous function. The argument $(\boldsymbol{r}, \mathbf{v}, t)$ arises from Eq. (14.30d).

The discrete Eq. (14.6) will be *derived* by EPI. Then transition to the continuous Eq. (14.7) is made via Eqs. (14.30c)–(14.30e). Equation (14.7) is thereby an EPI result. It states that $p$ evolves as, specifically, a *first*-order differential equation in the time. Equation (14.7) also states that the quantity $p(g^+ - g^-)$ is the net change in $p$ per unit time. We use the standard approach of Reif (1965) to compute changes $pg^+$ and $pg^-$ next.

### *14.8.4 Continuous molecular change coefficients*

Denote as "A molecules" those within a volume element $d^3\mathbf{r}$ with velocity near $\mathbf{v}$ which are scattered *out of* the velocity range $(\mathbf{v}, \mathbf{v} + d\mathbf{v})$ after a collision. Denote as "$A_1$ molecules" those within the same volume element $d^3\mathbf{r}$ that *do* the preceding scattering, and which have a common velocity value $\mathbf{v}_1$.

Primes indicate values after a collision. An A molecule changes its velocity from $\mathbf{v}$ to one near a value $\mathbf{v}'$, and an $A_1$ molecule changes its velocity from $\mathbf{v}_1$ to one near a value $\mathbf{v}_1'$.

Define the relative velocities before scattering and after scattering as $\mathbf{V} \equiv \mathbf{v} - \mathbf{v}_1$ and $\mathbf{V}' \equiv \mathbf{v}' - \mathbf{v}_1'$, respectively.

Denote as $\sigma\, d\Omega'$ the number of molecules per unit time that emerge after a collision with a final relative velocity $\mathbf{V}'$ in a direction within the solid angle range $d\Omega'$. Then the changes are

$$g^+ p = \iint d\Omega' d^3\mathbf{v}_1 V \sigma p' p_1' \quad \text{and} \quad g^- p = \iint d\Omega' d^3\mathbf{v}_1 V \sigma p p_1$$

(Reif, 1965), so that their difference obeys

$$(g^+ - g^-)p = \iint d\Omega'\, d^3\mathbf{v}_1\, V\sigma(p'p'_1 - pp_1). \tag{14.31}$$

Here, probabilities $p \equiv p(\boldsymbol{r}, \mathbf{v}, t)$, $p' \equiv p(\boldsymbol{r}, \mathbf{v}', t)$, $p_1 \equiv p(\boldsymbol{r}, \mathbf{v}_1, t)$ and $p'_1 \equiv p(\boldsymbol{r}, \mathbf{v}'_1, t)$.

The derivation of Eq. (14.31) requires use of the principles from classical mechanics of conservation of energy and mass during collisions. Classical mechanics derives, in turn, as the extreme weak-field limit of the Einstein field Eq. (6.68), and the latter were derived by EPI. Then Eq. (14.31) follows from the overall EPI approach taken in the book.

## 14.9 Change coefficients for replicating molecules

In application of the replication Eq. (14.9) to RNA replication, $A_n$ may be computed from basic laws of chemical physics, involving factors such as free energies of association (Bernstein *et al.*, 1983). The depletion coefficients obey $d_n \geqslant 0$ and, hence, represent a death process. See also below Eq. (14.9).

With all the change coefficients now known, we may finally proceed with the general EPI derivation.

## 14.10 EPI derivation of general growth law

All the problems (1)–(5) will now be solved *simultaneously*. These are *discrete* problems, in that the unknown probabilities $p_n$ are discrete. Thus, the continuous limit approximation for (4) that was taken in Secs. 14.8.3 and 14.8.4 will *not* be used here. That continuous limit will be taken again, but *after* the EPI derivation.

We now seek the *self–consistent* EPI solution $p_n(t)$ (Sec. 3.4.6) to the general problem. Here, the common solution $p_n(t)$ to the two EPI requirements

$$I - J = extrem., \tag{14.32}$$

$$I - \kappa J = 0 \tag{14.33}$$

is sought.

Use of Eqs. (14.32) and (14.33) requires knowledge of the coefficient $\kappa$, and of the functionals $I$ and $J$. We get these, in turn, next.

*Fisher information level I*: For this discrete problem, the Fisher information at the time $t$ is the discrete average indicated in Eq. (1.9),

$$I(t) = \sum_n \frac{z_n^2}{p_n}, \quad p_n \equiv p(n|t), \quad z_n \equiv dp_n/dt. \tag{14.34}$$

The quantities $\mathbf{y}$, $\theta$ in Eq. (1.9) correspond respectively to $n$, $t$ here. The notation $z_n$ is used for brevity. The upper limit on sums is $N$ unless otherwise specified.

It is interesting that this derivation uses the direct Fisher information, rather than the Fisher channel capacity derived in Sec. 2.3.2. This is because the system is described by a *single*, one-dimensional probability law $p(n|t)$, and under these conditions the channel capacity reverts to the Fisher information.

*Coefficient* $\kappa$: The transport and growth effects (14.10) to be derived are described by probabilities *per se*, rather than probability *amplitudes*. This is a clue that these effects are effectively, if not fundamentally, classical in nature. The applications (1), (2) and (4) are *fundamentally* classical (see, e.g., property (d) in Sec. 14.8). However, even the transport effect (14.5), describing laser gas energy levels, is a simple rate equation and, hence, *effectively* classical. The reasons are as follows.

The gas consists of colliding atoms that obey a density matrix whose diagonal elements $p_1$, $p_2$ and *off*-diagonal elements obey a quantum rate equation (shown in Milonni and Eberly, 1988). However, the off-diagonal elements relax temporally at very different rates. This permits their values to be expressed in terms of the values of the slowly changing *diagonal* elements $p_1$, $p_2$ (Milonni and Eberly). Substituting these expressions into the quantum rate equation then gives the effectively classical rate Eq. (14.5).

Of course, these physical details are not presumed to be known by the observer: Otherwise there would be no need for deriving Eqs. (14.5) using EPI. Rather, the observer is assumed to know only that (a) the invariance principle (14.13) is obeyed; and (b) this transport effect and all others in this chapter are effectively classical.

Classical effects require the use of a coefficient

$$\kappa = 1/2 \tag{14.35}$$

(see Chaps. 5 and 6). Hence that is the value assumed here.

*Fisher bound information level* $J$: With complete generality, represent $J$ by a sum involving all parameters of the problem,

$$J \equiv \sum_n J_n(\mathbf{g}, \mathbf{p}, \mathbf{z}, t), \quad \mathbf{g} \equiv (g_n, d_n, n = 1, \ldots, N), \tag{14.36}$$

$$g_n \equiv g_n(\mathbf{p}, t), \quad d_n \equiv d_n(\mathbf{p}, t),$$

$$\mathbf{p} \equiv (p_1, \ldots, p_N), \quad \mathbf{z} \equiv (z_1, \ldots, z_N),\ z_n \equiv dp_n/dt.$$

The $J_n$ are the indicated functions of their parameters. Note that the vector $\mathbf{g}$ depends upon both sets of coefficients $g_n$ and $d_n$.

*Extremum solution*: Using Eqs. (14.34) and (14.36) in Eq. (14.32) gives a problem

$$K \equiv \sum_m \left[ \frac{z_m^2}{p_m} - J_m(\mathbf{g}, \mathbf{p}, \mathbf{z}, t) \right] = extrem. \tag{14.37}$$

The unknowns of this problem are $\mathbf{z}$ and $\mathbf{p}$, and these are *discrete* unknowns. Hence the extremum is attained by merely differentiating the left-hand side of (14.37) with respect to each *independent* unknown and equating the result to zero. This requires that we define the independent unknowns. Since quantities $\mathbf{z}$ are time derivatives of quantities $\mathbf{p}$, obviously both sets of unknowns cannot be independent. Therefore, one or the other set must be chosen for variation. We arbitrarily choose to vary quantities $\mathbf{z}$.

Because of Eq. (14.2), only $N - 1$ of the $\mathbf{z}$ are actually independent. Therefore, the extremum in (14.37) should be effected, formally, by differentiating and setting $\partial K/\partial z_n = 0$ for $n = 1, \ldots, N - 1$. In fact, we find it more convenient to, additionally, set $\partial K/\partial z_N = 0$. That this still gives the correct answer is verified below Eq. (14.54).

Differentiating $\partial/\partial z_n$ Eq. (14.37) for each fixed value $n$ gives

$$2\frac{z_n}{p_n} - \sum_m \frac{\partial J_m}{\partial z_n} = 0, \quad n = 1, \ldots, N. \tag{14.38}$$

The sum arises from the fact that each $J_m$ generally depends upon all $\mathbf{z}$.

*Zero solution*: Using Eqs. (14.34)–(14.36) in Eq. (14.33) gives a requirement

$$\sum_n \left( \frac{z_n^2}{p_n} - \frac{1}{2} J_n \right) = 0. \tag{14.39}$$

This is satisfied by the microlevel EPI condition

$$\frac{z_n^2}{p_n} - \frac{1}{2} J_n = 0, \quad n = 1, \ldots, N. \tag{14.40}$$

*Combined requirement*: According to plan, we seek the common solution $\mathbf{p}$ to Eqs. (14.38) and (14.40). Eliminating their common parameter $p_n$ gives

$$p_n = \frac{2z_n}{\sum_m \partial J_m/\partial z_n} = 2\frac{z_n^2}{J_n}. \tag{14.41}$$

After cancellation this reduces to

$$\sum_m \frac{\partial J_m}{\partial z_n} = \frac{J_n}{z_n}. \tag{14.42}$$

*Particular solution*: The right-hand side of this equation can be regarded as a sifted version of the left. That is, one solution is

$$\frac{\partial J_m}{\partial z_n} = \frac{J_m}{z_m}\delta_{mn} \tag{14.43}$$

in terms of the Kronecker delta $\delta_{mn}$. This solution can be readily verified by back substitution into Eq. (14.42). (Other solutions might exist as well, and are left to future research.)

The requirement (14.43) actually nails down the solution for $J_m$. If $m \neq n$ it gives

$$\frac{\partial J_m}{\partial z_n} = 0, \tag{14.44}$$

implying that each $J_m$ only depends upon $\mathbf{z}$ through $z_m$,

$$J_m(\mathbf{g}, \mathbf{p}, \mathbf{z}, t) = J_m(\mathbf{g}, \mathbf{p}, z_m, t). \tag{14.45}$$

Next, take the case $m = n$. Equation (14.43) now becomes

$$\frac{\partial J_n}{\partial z_n} = \frac{J_n}{z_n}. \tag{14.46}$$

This is easily integrated, giving

$$\ln J_n = \ln z_n + h_n(\mathbf{g}, \mathbf{p}, t) \tag{14.47}$$

for some new functions $h_n$. (As a check, partially differentiating (14.47) with respect to $z_n$ gives back Eq. (14.46).) Solving (14.47) for $J_n$ gives

$$J_n = f_n(\mathbf{g}, \mathbf{p}, t) z_n, \tag{14.48}$$

in terms of new functions $f_n$ that are to be found.

*Combining results*: Using Eq. (14.48) in Eq. (14.40) gives

$$z_n \equiv \dot{p}_n = \tfrac{1}{2} f_n(\mathbf{g}, \mathbf{p}, t) p_n, \tag{14.49}$$

where we used definition (14.34) of the $z_n$. Substituting (14.49) into Eq. (14.2) with $i = 1$ gives

$$\frac{1}{2}\sum_n f_n(\mathbf{g}, \mathbf{p}, t)p_n = 0. \tag{14.50}$$

*Key comparison*: Comparing Eqs. (14.13) and (14.50) shows that a solution for the $f_n$ is

$$f_n(\mathbf{g}, \mathbf{p}, t) = A[g_n(\mathbf{p}, t) + d_n(\mathbf{p}, t)], \quad A = const. \tag{14.51}$$

Using Eq. (14.51) in Eq. (14.49) gives

$$\dot{p}_n = [g_n(\mathbf{p}, t) + d_n(\mathbf{p}, t)]p_n, \quad n = 1, \ldots, N, \tag{14.52}$$

where we took

$$A = 2 \tag{14.53}$$

for simplicity and on empirical grounds. (The resulting growth/transport equations are confirmed experimentally).

Equation (14.52) is identical with the general growth law Eq. (14.10). Hence our goal of deriving Eq. (14.10) via EPI has been met. However, the derivation can be strengthened. Require that both Eqs. (14.13) and (14.50) hold for all possible choices of the $p_n$. Next, take the weighted difference of the two equations, giving a requirement $\sum_n[A(g_n + d_n) - f_n]p_n = 0$, with $A$ some weight. Then this requirement is analogous to that of Eq. (0.12). The solution is, accordingly, the unique one requiring the coefficient $A(g_n + d_n) - f_n$ of each $p_n$ to be zero. This now gives Eq. (14.51) as the unique solution, and then (14.52) as the unique growth law.

We call (14.52) the "discrete" growth equation. A continuous version of the growth equation is useful for the application (4) to the ideal gas. This is formed by taking the continuous, classical limits (14.30b) and (14.30c) of Eq. (14.52). It becomes a single equation,

$$\dot{p} = [g^+(\boldsymbol{r}, \mathbf{v}, t) - g^-(\boldsymbol{r}, \mathbf{v}, t)]p, \quad g^- > 0, \tag{14.54}$$

in continuous $(\boldsymbol{r}, \mathbf{v}, t)$ values. The transition (14.30e) was also taken. We call this result the "continuous" growth equation.

*Verification*: We set out (above Eq. (14.38)) requiring that (a) $\partial K/\partial z_n = 0$, $n = 1, \ldots, N - 1$, but instead carried through a requirement (b) $\partial K/\partial z_n = 0$, $n = 1, \ldots, N$. This led to the solution (14.52). Hence we need to check that requirement (a) is likewise satisfied by the solution (14.52).

Equation (14.37) may be rewritten as

$$K \equiv \sum_{m=1}^{N-1} \frac{z_m^2}{p_m} + \frac{z_N^2}{p_N} - A\sum_{m=1}^{N-1}(g_n + d_n)z_n - A(g_N + d_N)z_N \tag{14.55}$$

after use of Eqs. (14.48) and (14.51). But, by Eq. (14.2) with $i = 1$, $z_N = -\sum_{m=1}^{N-1} z_m$. Using this in Eq. (14.55) gives

$$K \equiv \sum_{m=1}^{N-1} \frac{z_m^2}{p_m} + \frac{\left(\sum_{m=1}^{N-1} z_m\right)^2}{p_N} - A \sum_{m=1}^{N-1} (g_n + d_n) z_n + A(g_N + d_N)\left(\sum_{m=1}^{N-1} z_m\right). \tag{14.56}$$

Note that all the sums are now from $n = 1$ to $n = N - 1$ so that *only* the independent parameters $z_1, \ldots, z_{N-1}$ enter in. Equation (14.56) is then the expression that must be shown to be extremized in these parameters. Differentiating it $\partial/\partial z_n = 0$, $n = 1, \ldots, N - 1$ and then re-using Eqs. (14.52) and (14.2) does give the required zeros. Hence, the growth Eq. (14.52) and related results of the EPI approach succeed in extremizing the information $K$, as required.

It is also necessary to check that the microlevel condition (14.40) is satisfied. This is taken up next.

*Exercise*: Show that use of the definition (14.34) of $z_n$, and Eqs. (14.48) and (14.49), in Eq. (14.40) gives an identity, verifying that condition (14.40) holds for the given solution.

*Applications*: We next specialize EPI discrete and continuous growth solutions (14.52) and (14.54) to the phenomena (1)–(5) described at the outset.

## 14.11 Resulting equations of growth

(1) Substituting the ecological growth changes (14.19) into the discrete growth Eq. (14.52) gives

$$\dot{p}_n = p_n \left( \sum_{k=1}^{N} g_{nk} p_k + d_n \right),$$

verifying Eqs. (14.3).

(2) Substituting the genetic growth changes (14.24) into a (now) doubly subscripted growth Eq. (14.52) gives

$$\dot{p}_{nk} = p_{nk}(w_{nk} - \langle w \rangle), \quad \langle w \rangle \equiv \sum_{nk} w_{nk}\, p_{nk},$$

verifying Eqs. (14.4).

(3) Substituting the atomic population changes (14.25), (14.26), (14.29), and (14.30a) into the discrete growth Eq. (14.52) gives

$$\dot{p}_1 = R(p_2 - p_1) + R'p_2,$$

$$\dot{p}_2 = -R(p_2 - p_1) - R'p_2,$$

verifying Eqs. (14.5).

(4) The general EPI output Eq. (14.54) is the formal statement of the Boltzmann equation (Reif, 1965). The particular changes (14.31) were likewise based upon prior EPI derivations: As discussed below Eq. (14.31), the changes derive from conservation principles of classical mechanics, and the latter was derived by EPI in Chapter 6. Substituting the changes (14.31) into (14.54) gives a solution

$$\dot{p} = \iint d\Omega' \, d^3\mathbf{v}_1 \, V\sigma(p'p_1' - pp_1). \tag{14.57}$$

The left-hand side can be further evaluated, using $p = p(\boldsymbol{r}, \mathbf{v}, t)$. By the chain law of differentiation,

$$\begin{aligned} \dot{p} \equiv \frac{dp}{dt} &= \frac{\partial p}{\partial t} + \frac{dx}{dt}\frac{\partial p}{\partial x} + \cdots + \frac{dv_x}{dt}\frac{\partial p}{\partial v_x} + \cdots \\ &\equiv \frac{\partial p}{\partial t} + v_x \frac{\partial p}{\partial x} + \cdots + \frac{dv_x}{dt}\frac{\partial p}{\partial v_x} + \cdots \\ &\equiv \frac{\partial p}{\partial t} + \mathbf{v} \cdot \frac{\partial p}{\partial \mathbf{r}} + \mathbf{v} \cdot \frac{\partial p}{\partial \mathbf{v}} \end{aligned} \tag{14.58}$$

in handy vector notation. Here vector coordinate $\boldsymbol{r} = (x, y, z)$ and vector velocity $\mathbf{v} = (v_x, v_y, v_z)$. Using this result in (14.57) gives the Boltzmann transport equation

$$\frac{\partial p}{\partial t} + \mathbf{v} \cdot \frac{\partial p}{\partial \mathbf{r}} + \dot{\mathbf{v}} \cdot \frac{\partial p}{\partial \mathbf{v}} = \iint d\Omega' \, d^3\mathbf{v}_1 \, V\sigma(p'p_1' - pp_1),$$

verifying Eq. (14.8).

(5) Substituting the change coefficients for this application as given in Table 14.1 into the solution (14.52) gives

$$\dot{p}_n = p_n\left(\frac{b_n A}{A_n + A} - d_n\right),$$

verifying Eqs. (14.9).

## 14.12 Other transport equations

The Boltzmann transport equation in the general form Eq. (14.7) applies to a wide scope of flow problems, as special cases of its change coefficients $g^{\pm}$. These include (Harris, 1971) the

(a) Poiseuille flow equation of planar flow,
(b) Rankine-Hugoniot equations of shock wave structure,
(c) Navier–Stokes equations of hydrodynamical flow,
(d) Fokker-Planck equation, and the
(e) flow of automobile traffic; others are the
(f) master equation of macroeconomics (Aoki, 1996; Miller, 1979), and the
(g) source-free neutron transport equation of nuclear fission reactors (Loyalka, 1990).

In that the Boltzmann transport equation generates these equations and was derived by EPI, these equations follow from EPI as well.

In addition, the presence of mathematical *chaos* (Scott, 1999) is often analyzed for processes that obey non-linear growth equations of the form (14.10). In these applications the time coordinate can be real or purely imaginary. In the latter case the growth Eq. (14.10) takes the interesting form of a Schrödinger wave equation in the PDF (rather than in the amplitude).

## 14.13 Fisher's theorem of partial change

Darwin's theory of evolution is often paraphrased as "survival of the fittest." One way to quantify this concept is to show that the average fitness increases with time under certain conditions. This is shown next to follow from the Eq. (14.4) of genetic change.

Assume here that the fitness coefficients $w_{nk}$ are *constant* in time. Differentiating $d/dt$ the second Eq. (14.4) and using the first gives directly

$$\frac{d\langle w\rangle}{dt} = \sum_{nk} w_{nk}\dot{p}_{nk} = \sum_{nk} w_{nk}p_{nk}(w_{nk} - \langle w\rangle). \tag{14.59}$$

Next, by the second Eq. (14.4) and normalization property (14.1) for probabilities $p_{nk}$, identically

$$\langle w\rangle \sum_{nk} p_{nk}(w_{nk} - \langle w\rangle) = 0. \tag{14.60}$$

Subtracting this zero from the right-hand side of Eq. (14.59) gives

$$\frac{d\langle w\rangle}{dt} = \sum_{nk} p_{nk}(w_{nk} - \langle w\rangle)^2 \geqslant 0. \tag{14.61}$$

The right-hand side is explicitly the variance in fitness over all genotypes. Since a variance can never be negative, the mean fitness *increases* or remains constant with time. The fittest tend to dominate.

As we noted, this effect holds rigorously for constant fitness coefficients. Then it is also approximately true over time periods for which the fitness changes by small amounts. However, once the fitness changes appreciably with time, the theorem no longer holds. The mean fitness can decrease at times. Other aspects of this effect are discussed in Frieden *et al.* (2001), Ewens (1989), Demetrius (1992), and Frank and Slatkin (1992).

We next treat the *general* growth law (14.10) in the same way. With *constant* change coefficients $g_n$, $d_n$, we operate

$$\frac{d}{dt}\langle g+d\rangle \equiv \frac{d}{dt}\sum_n (g_n+d_n)p_n = \sum_n (g_n+d_n)\dot{p}_n \qquad (14.62)$$

$$= \sum_n (g_n+d_n)^2 p_n \equiv \langle (g+d)^2\rangle \geqslant 0$$

(cf. Eq. (14.61)). The first equation is by definition of the mean. The second is by the constancy of the $g_n$, $d_n$. The third is by the use of Eq. (14.10). Thus, the rate of change of average change coefficient equals the mean-square change coefficient. Since the latter is by definition positive, this means that the mean change monotonically increases with time. Such a monotonic increase describes an unstable system, where there is a runaway growth of certain populations. An ecosystem – application (1) – that obeyed such growth would rapidly break down into a few very dominant species or, even, a monoculture. Thus, mass extinctions would occur. However, this situation is somewhat artificial because in most applications the change coefficients $g_n$, $d_n$ *depend* upon the time (e.g., see Eqs. (14.29) and (14.30a), where the time dependence is through $p_1$ and $p_2$). The time dependence can provide a feedback mechanism that stabilizes the system. This is particularly important in application (g) of Sec. 14.12.

## 14.14 An uncertainty principle of biological growth

The physicist Max Delbruck, widely regarded as the founder of molecular biology, long sought an *uncertainty principle of biology* (Delbruck, 1949). This principle would connect biological growth with physics, such that (i) a parameter of the growth would (ii) obey complementarity with a physical parameter. The uncertainty principle (14.67) or (14.68) below is one such relation. Here the biological parameters are the population changes $(g_n + d_n)$, and the physical parameter is the time $t$ at which the system is observed. It is derived as follows.

Combining Eqs. (14.33), (14.35), (14.36), (14.48), (14.51), and (14.52) gives

$$I = \frac{1}{2}J = \frac{1}{2}\sum_n J_n = \frac{1}{2}\sum_n 2(g_n+d_n)\dot{p}_n = \sum_n (g_n+d_n)^2 p_n \equiv \langle (g+d)^2\rangle. \qquad (14.63)$$

The Fisher information is the mean-square value of all relative population change components per unit time.

Recall that $I = I(t)$, information about the time. The EPI estimation problem (Sec. 14.3) is to form an estimate $\hat{t}_n$ of the time $t$ by observation of a population type $n$. Let this have a mean-squared error $e^2(t)$. In the presence of generally non-zero bias $B$, the Cramer–Rao inequality states that

$$e^2(t)I(t) \geqslant \left(1 + \frac{dB}{dt}\right)^2, \quad e^2 \equiv \langle(\hat{t} - t)^2\rangle. \tag{14.64}$$

Here, $B \equiv B(t)$ is the "bias function" for the estimate, defined as

$$B(t) \equiv \langle(\hat{t}_n - t)\rangle \equiv \sum_n p_n(t)(\hat{t}_n - t). \tag{14.65}$$

By inspection it depends upon the choice of the estimation function $\hat{t}_n$. Equation (14.64) may be derived by the same steps as in Sec. 1.2.3, as follows.

*Exercise*: Starting from a relation

$$\int d\mathbf{y}\, [\hat{\theta}(\mathbf{y}) - \theta]p(\mathbf{y}|\theta) = B(t) \tag{14.66}$$

in place of Eq. (1.3), show that the steps of Sec. 1.2.3 give the result (14.64).

Using Eq. (14.63) in (14.64) gives

$$e^2(t)\langle(g + d)^2\rangle \geqslant \left(1 + \frac{dB}{dt}\right)^2. \tag{14.67}$$

This inequality relation applies to all populations obeying the *generic* growth law (14.10). Hence it describes a wide scope of growth phenomena, including phenomena (1)–(5) described above and those listed in Sec. 14.12. It is one form of the biological uncertainty principle we sought.

The *inequality form* of (14.67) is similar mathematically to that of the Heisenberg *uncertainty relation* Eq. (4.47). The latter expressed *reciprocity* between errors in measured position and momentum values: As one increases the other must decrease.

Analogously, with $dB/dt$ a constant not equal to $-1$, Eq. (14.67) expresses reciprocity between the error $e^2(t)$ in the estimated value of the time and the mean-square relative change $\langle(g + d)^2\rangle$ in population. The smaller the error $e^2$ is in estimating the time, the larger the mean-square change $\langle(g + d)^2\rangle$ tends to be at that time. This implies that the time may be estimated more accurately when the rate of evolutionary change is high than it tends to be during eras of low evolutionary change. This effect is intuitive as well.

The generic uncertainty relation (14.67) was confirmed by numerical simulations of *genetic* growth in particular (see Frieden *et al.*, 2001).

### 14.15 On mass extinctions

Why the dinosaurs became extinct following the apparent asteroid collision of about $6 \times 10^7$ years ago is a much debated question. Many other animals of the day did not. In trying to answer this question, we will devise a framework (14.71) for predicting mass extinctions in general.

To make for a simple analysis, regard all the dinosaurs as comprising a distinct ecological system, called "dinosauria," with resources ignored. The growths in populations of the individual species of dinosauria obey Eq. (14.4). This is a Lotka–Volterra law (14.3) where the general growth coefficients $g_{nk}$ are the fitness coefficients $w_{nk}$. A pair $nk$ denotes the mating of a phenotype $n$ with a phenotype $k$. Also, the depletion coefficients $d_{nk}$ are now $-\langle w \rangle$, where $\langle w \rangle$ is defined in Eqs. (14.4). If a pairing $nk$ is incompatible because $n$ and $k$ describe different *species* of dinosaur, then simply $w_{nk} = 0$.

Aside from obeying a Lotka–Volterra law (14.3), such a system obeys the uncertainty principle (14.67), which, for unbiased estimates $B(t) \equiv 0$, obeys

$$e^2(t)\langle (w - \langle w \rangle)^2 \rangle \geqslant 1. \tag{14.68}$$

Here

$$\langle (w - \langle w \rangle)^2 \rangle \equiv \sum_{nk} (w_{nk} - \langle w \rangle)^2 p_{nk}, \tag{14.69}$$

the mean-square fitness over all dinosauria. This is also called the *fitness variability* since it measures the spread in fitnesses from the mean fitness value.

Equation (14.68) is another form of the *biological uncertainty principle* we sought. Its mathematical counterpart is the Heisenberg principle (4.54) of Chapter 4. As with the latter, it exhibits reciprocity between two system variables. These are here the time and the fitness. In (4.54) it was the time and the energy.

In fact Eq. (14.68) expresses reciprocity between *knowledge* of the evolutionary time of the system (first factor), and *the system's* variability in fitness at that time (second factor). The former is epistemological, i.e., a state of knowledge of the observer, whereas the latter is ontological, i.e., a property of the system. Thus, the principle predicts how well one can learn (about the time) in the presence of a system property (the fitness variability). Equation (14.68) also satisfies Delbruck's requirements (i) and (ii) at the outset of Sec. 14.14: The parameter (i) of the growth is the fitness and the physical parameter (ii) is the time.

However, a fundamental difference between the two uncertainty principles (4.47) and (14.68) is that both left-hand quantities in (4.54) are ontological in nature (spreads in the time and energy obeyed by *the system*) whereas the two in (14.68) are mixed epistemological–ontological as discussed previously. As we want to make *physical* predictions about the biological system, we need to re-express (14.68) as a product of ontological factors. This will be Eq. (14.71) below.

The mean-square error $e^2(t)$ is explicitly a measure of knowledge. However, it is knowledge *about the system*. This permits the knowledge to have both epistemological and ontological components. The tie-in is the simple concept of *relative* error $r$,

$$\frac{\sqrt{e^2(t)}}{t} \approx r = const. \tag{14.70}$$

The relative error is usually at least roughly known, e.g., from the experimental uncertainty in knowledge of the depth of an observed specimen within a geological stratum. Notice that $t$ is a property *of the system*, and hence ontological in character. Cross-multiplying (14.70) allows $e^2(t)$ to be expressed in terms of the ontological factor $t$. Doing so, and substituting it into (14.68), gives

$$t^2\langle(w - \langle w\rangle)^2\rangle \gtrsim 1/r^2. \tag{14.71}$$

This expression provides a basis for computing whether a system, such as dynosauria, is susceptible to extinction at a general system age $t$.

A system is prone to extinction when its fitness variability $\langle(w - \langle w\rangle)^2\rangle$ is so small that no subpopulation exists that can survive a drastic environmental challenge, such as would be caused by an incoming asteroid. (This would be the case, e.g., if dinosauria at the time of the asteroid impact had no members with fur, or which lived underground, so that they could survive the "asteroid winter" that followed the collision. Interestingly, those with feathers – the birds – apparently did survive.) Equation (14.71) shows that the fitness variability can be very small only when the time $t$ is sufficiently large. That is, *for mass extinction of a system to be likely, it is sufficient that the system exist for a long enough time*. The sheer passage of time is one route to extinction of the system. However, even then the extinction is not guaranteed unless the requisite challenge arises.

Of course, the dinosaurs in particular did last for a very long time, on the order of $1.6 \times 10^8$ years. However, to judge whether this exceeds the extinction time value $t$ predicted by Eq. (14.71) requires a model for the fitness variability at that time. This problem is currently being worked on by this author and colleagues R. A. Gatenby and T. L. Vincent.

Is there other evidence for the uncertainty principle (14.68) and its corollary (14.71)? Vincent (2004) has generated long term Lotka–Volterra population solutions at equilibrium. See Fig. 14.1.

Each curve in Fig. 14.1 shows the frequency or probability of a range of phenotype traits over a population (the particular traits are called "strategies" in Fig. 14.1). Each trait value corresponds to a value of fitness. Figure 14.1 shows a *progressive narrowing of the spread or variability in the phenotype traits*, and therefore in the fitness variability $\langle (w - \langle w \rangle)^2 \rangle$, as the time $t$ increases. The narrowing of the curves is also apparent in the progressively smaller numerical values shown for the standard deviation (called "Std"). This progressive narrowing of the fitness variability is consistent with that predicted by (14.71).

Notice that (14.71) is not a simple equality, and hence does not fix the exact dependence of the variability upon the time. However, it indicates that, for fixed error $r$, the phenotype variability $\langle (w - \langle w \rangle)^2 \rangle$ should at least *tend* to decrease with increasing $t$. As we discussed, this is confirmed for the equili-

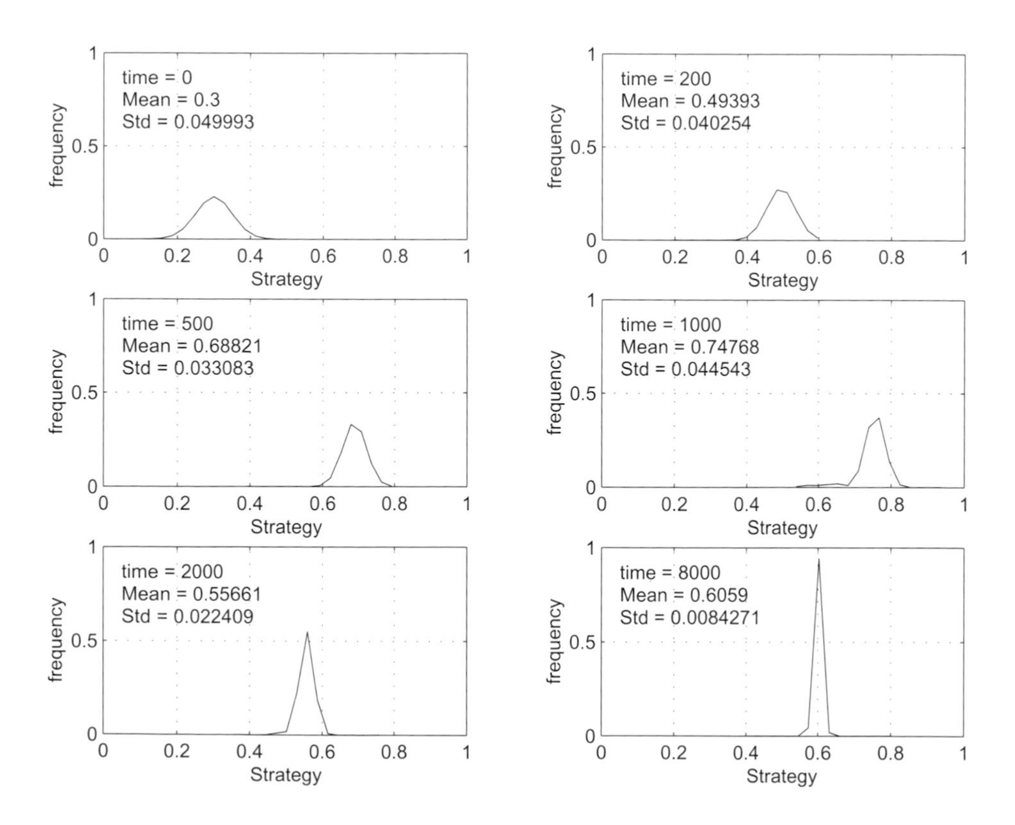


Fig. 14.1. As time passes, the frequency curve (PDF) on survival strategies narrows (Vincent, 2004). The population becomes overspecialized.

brium curves in Fig. 14.1. It also makes sense on the grounds that both the derivation of (14.71) and the curves trace from use of the Lotka–Volterra Eqs. (14.3) and (14.4).

## 14.16 Entangled realities: why we don't observe them for macroscopic systems

In some phenomena – such as quantum mechanics – the measurement actually *causes* the information flow, via perturbation of the object of the measurement (Sec. 10.10). It is also assumed that no outside influences are present (Sec. 1.2.1), and that the object particle's state is perturbed purely by interaction with a probe particle of the measuring system. The probe particle ultimately provides the data of information level $I$ in the data space. Such a system can be in a *coherent*, or pure, state. As a result, it can exhibit entanglement (Chapter 12) between the observed data space provided by one particle (the probe) and the properties of another (the object). Equivalently, the level of information $J$ for the probe particle is the $I$ for the object particle. (This is the *theoretical functional* $I$, since the object particle is not directly observed; the probe is.) The equality of $I$ and $J$ now expresses the total equivalence of the two spaces on the basis of information content. This is commonly called "entangled realities."

However, Albert Einstein notably rejected the possibility of such entanglement. After all, entanglement is not part of our everyday experience. We like to think that we act as individuals, with free will. How is it that entanglement is not commonly encountered?

Most everyday systems we encounter are, in fact, not coherent, i.e., are *decoherent* (Sec. 10.10). These are macroscopic, usually complex, systems, such as the biological systems of types (1) and (2) above. While such a system is being observed, it is typically being perturbed both by *outside influences*, such as an incoming flow of radiation, and by *internal interactions* of the various populations. The systems under study in this chapter have, in fact, been assumed to be influenced by outside influences (Sec. 14.3). These internal and outside influences are "decohering," that is, cause the system to lose coherence.

Such a system is, then, subjected to two sources of perturbation: (i) due to the measurement, through the probe particle, and (ii) due to the decohering influences. However, in these systems usually the latter overwhelm the former in magnitude. It results that the information flow still exists (evidently, since data are still obtained!), with the EPI measurement process still activated. But, these effects are now independent of the *measurement* perturbation. Being so decoupled, such complex systems no longer obey $I = J$. In fact, most obey $I = 0.5J$, i.e., suffer precisely a 50% loss of bound information (as in Chaps.

5, 6 and 14). This loss of Fisher information in the data is the EPI explanation for why quantum entanglement does not take place in macroscopic systems in particular.

### 14.17 A proposed mechanism for genesis

We have noted that both *organic* strands of RNA molecules and *inorganic* molecules obey *a common* growth law (14.52). How could life emulate physical systems in this way? As Bernstein *et al.* (1983) conclude, "life on earth can be viewed as a special case of systems that evolve macroscopically ordered structures." Since RNA is a most primitive life form, these considerations suggest that life might have *originated* as a growth process that was in some way *forced* to replicate a neighboring *non*-living growth process. A current candidate to fill the latter role is a polymer composed of threose nucleic acids (Orgel, 2000), also called "TNA." The TNA polymer would then have acted as *a template* for the original RNA growth. How might the RNA–TNA force have arisen?

A possible mechanism is molecular bonding between corresponding molecules of RNA and TNA. The bonding could, e.g., be electromagnetic, through an exchange of ions and/or virtual photons. It is currently thought that metal ions, such as of magnesium or zinc, could work in this way. Moreover, it is known that pairs of complementary TNA polymers form stable Watson–Crick double helices. It has been hypothesized that these might therefore form stable double helices with *complementary* RNA chains. On this basis, threose would be the template for organic growth that we sought. Those RNA molecules that did not bond in this way with a complementary threose chain would not grow, so that RNA strands that *did* bond would tend to increasingly dominate as time passed.

The latter effect is quantified by the very Lotka–Volterra growth Eq. (14.52). Here a growth coefficient $g_n$ is that of the RNA while a coefficient $g_{nk}$ is proportional to the strength of the bond of the $n$th RNA strand with the $k$th TNA strand. Without such bonding the coefficients $g_{nk}$ would be zero, causing $g_n = 0$ and since, by the second law of thermodynamics, there must be a natural attrition rate $d_n < 0$ present, by Eq. (14.52) each such RNA strand type $n$ would obey $\dot{p}_n \leqslant 0$. It would not grow. On the other hand, with *some weak* (at first) bonding, i.e., $g_{nk}$ finite, the resulting RNA coefficients $g_n$ would weakly emulate those of the growing template, thereby becoming *positive* and permitting some growth of the RNA. Then, if the tendency to increase the bonding were an inherited trait, succeeding generations of RNA would increasingly bond to the template of TNA and, hence, emulate its growth process. Such a "gradualist scenario" has been deemed possible (Orgel, 2000).

It is not known whether TNA or something even more basic was the original template. Whatever it turns out to be, the resulting emulation effect can be interpreted as a learning process. The RNA "learned" that it could grow, out of a *flow of information* to it from the nearby growing template. Presumably the information flow was provided by metal ions – e.g., the magnesium and zinc ions mentioned above – that created the bonding. These carried information about the growth process of the nearby TNA template to the RNA strands. The *double*-stranded nature of the sequences allows, in fact, for error correction during the replication procedure (Eigen, 1992). This also increases the capacity for information transmission. In this way RNA, a most primitive form of life, would initially have grown by imitation, *the most primitive form of learning*.

However, we do not today observe *each* strand of RNA to be accompanied by a nearby strand of TNA. The RNA had to, at some point, "wean itself" from lockstep dependence upon TNA as the growth mechanism. How could this have come about? This question must be left to future work.

## 14.18 Discussion and summary

This chapter shows that the EPI approach can be used to derive both physical and biological growth phenomena. Also, as a departure from previous chapters, the assumption (Sec. 1.2.1) of a closed system is now lifted. Random and/or non-random exterior influences (force fields, ecological resources, etc.) are assumed to be present.

Whether the subject population is of neutrons in a reactor, or plants, animals, and resources (including economic resources) in an environment, these are found to grow in accordance with EPI. Or, if the observation is instead *inward*, e.g., of a cell, a bacterium, or a fungal cell within the observer's body, EPI is found to likewise govern their growth. Thus, whether the observer looks outward at physical or biological phenomena, or inward at the growth of his own cells, he finds that they obey EPI. In effect, an observer exists in a kind of "aether" of Fisher information whose flow governs all of his internal and external growth processes.

A summary of the chapter follows.

In Sec. 14.1.3, we briefly survey some classical growth laws of biology, atomic physics, genetics, and molecular growth, including RNA growth. In Sec. 14.2 we show that these are special cases of a *general* Lotka–Volterra growth Eq. (14.10). The equation is derived in Sec. 14.10, using EPI in conjunction with a general invariance principle (14.13).

Taking the continuous limit (14.7) of the EPI output law (14.52) in space–

velocity phase space, and using change coefficients of type (14.31), gives the Boltzmann transport Eq. (14.8).

Perhaps the most important aspect of the derivation is that its output, the general growth law (14.52), is a *first-order – and not higher –* differential equation. Indeed, the most basic aspect of the Boltzmann transport equation is, likewise, that it is a first-order differential equation; see Reif (1965). Furthermore, the output growth law (14.52) is an *exact* result. It is not, for example, based upon a Taylor expansion to first-order of the hydrodynamical equations as originally used (Lotka, 1956) to derive the Lotka–Volterra equations.

The first-order nature of the output law follows, *mathematically*, from the fact that it is *discrete* in its unknowns $z_n$. This discreteness allows these to be found by merely taking the *first* derivative $dK/dz_n$ of Eq. (14.37), rather than by forming the second-derivative term $d/dt\,(\partial K/\partial z_n)$ that would be required in the Euler–Lagrange solution (0.13) to the corresponding continuous problem. For example, the operation $dK/dz_n$ gives Eq. (14.38), which is explicitly linear in the first-order quantity $z_n \equiv \dot{p}_n$. By comparison, operating $d/dt\,(\partial K/\partial z_n)$ would instead give a second-order term $dz_n/dt \equiv d^2 p_n/dt^2$. Such a term would be incorrect. The problems (1)–(3) and (5) are intrinsically discrete, so this result directly applies to them.

Although we saw that the Boltzmann equation problem likewise is *intrinsically* discrete (Sec. 14.8), it is usually derived in a continuous space. Then the discrete functions $p_n \equiv p((\boldsymbol{r}, \mathbf{v})_n, t)$ would be replaced by functions $p$ that are continuous in the space–velocity variables $(\boldsymbol{r}, \mathbf{v})$. See, e.g., Reif (1965). This transition to a continuum is of course an approximation (see discussion below Eq. (14.8)) that violates the Heisenberg uncertainty principle. It is interesting that, by the preceding paragraph, had we used this continuous space, the output of EPI would have been a *second*-order equation, which is an *incorrect* result. Thus, staying within the rigorously correct discrete phase space of variables $(\boldsymbol{r}, \mathbf{v})_n$ has the important payoff of giving the corrrect first-order growth equation, rather than an incorrect second-order one.

Choices of the change coefficients other than those given in Table 14.1 give rise to *other* transport equations as well. Among these are the equations of hydrodynamics, of shock waves, and the master equation of macroeconomics. Hence these equivalently derive from EPI as well. Likewise, the equations characterizing chaos are of our generic growth form (14.10). See Sec. 14.12.

The EPI derivation gives rise to a new general *uncertainty principle* Eq. (14.64) that applies to classical (Lotka–Volterra type) growth. Its expression in genetics is the uncertainty principle (14.68). A general framework (14.71) for predicting the likelihood of mass extinctions is developed in Sec. 14.15.

The derivation in Sec. 14.10 is based upon the use of a specific information efficiency value of $\kappa \equiv I/J = 1/2$. In these cases one-half of the pre-existing Fisher information in the phenomenon is lost *en route* to data space.

The question of why entangled realities are not observed macroscopically is discussed in Sec. 14.16. The basic reason is that, on the quantum level, $I = J$, so that the information in the observed datum *equals* its value at its source (entanglement), whereas, for macroscopic systems, $I < J$; information is lost (non-entanglement) in transit from the source to the observing device. This information loss traces from a loss of coherence, i.e., *decoherence* (Sec. 10.10), in such systems.

The decoherence follows from the fact that, while such a system is being observed and perturbed by a probe particle (Chaps. 3 and 4), it is also being perturbed by *outside influences*, such as an incoming flow of radiation, and/or *internal interactions* of the various populations. Although the EPI measurement process is still activated by the perturbations, and Fisher information flow still occurs (evidently, since data are obtained), the data fluctuations are now partially decoupled from the source fluctuations. The two sets of fluctuations are no longer fully coherent with one another, and information is lost. Hence (see preceding paragraph) there is no entanglement.

We found that both *organic* strands of RNA molecules and *physical* (non-living) molecules obey *a common* growth law (14.52). Is this merely a coincidence? No, the commonality of the two growth processes is taken to suggest that genesis occurred as a template matching process. Here, atoms of RNA, through *bonding* to corresponding atoms of a neighboring physical substance called TNA,."learned" to grow, i.e. to reproduce, in lockstep imitation of the growth of the TNA. This idea is discussed in Sec. 14.17.

# 15

# Cancer growth

By B. Roy Frieden and Robert A. Gatenby

## 15.1 Introduction

One must understand the enemy before he can be destroyed. Thus, the subject of how cancers grow is of strong current interest. This is growth both in culture, i.e., under ideal conditions (*in vitro*), and in the body, under less-than-ideal conditions (*in situ*). Consequently, a very large number of journals include the subject of cancer growth within their scope. Examples are *Biometrics*, *Cancer Research*, *Cancer*, the *British Journal of Cancer*, the *European Journal of Cancer*, *Oncology*, the *Journal of the National Cancer Institute*, the *Journal of Clinical Oncology*, and the *Journal of Theoretical Biology*. Other such journals are cited in the references below.

A cancer is, by definition, composed of cells that have lost the ability to function normally, but can still reproduce; in fact they do so in uncontrolled fashion. In this chapter we will be primarily concerned with finding the analytical form of the law $p(t)$ governing *in situ* cancer growth. The latter is often termed *carcinogenesis*. (*Note*: For brevity, by "cancer" we will mean *in situ* cancer unless specified otherwise.)

Here $t$ is a general time $t$ after the onset time of the growth, i.e., after the instant at which an invasive cancer cell is initiated; and $p$ represents the relative mass of cancerous tissue in an affected organ (say, the breast) at the time $t$. Since cancer mass density approximates that of healthy tissue, by the law of large numbers $p$ also represents the *probability* that a randomly sampled cell of the organ will be a cancer cell. In this way, the cancer growth problem is recast as a problem of estimating a probability law. This allows it to be attacked by EPI.

## 15.2 Cancer as a random growth process

The probability law $p(t)$ we seek is actually $p(t|C)$, the probability that the event $C$ (cancer) is observed at a time $t$ after onset of the disease. Thus, $t$ is a *random* variable here, as compared with the use of $t$ as a deterministic variable in Chapter 14 (see Eq. (14.0)). By comparison, here the event $C$ is deterministic, since by hypothesis $C$ is observed.

The work of Chapter 14 suggests that, as a growth process, $p(t)$ might obey a Lotka–Volterra equation

$$\dot{p}(t) = p(t)g(p, t), \qquad \dot{p}(t) \equiv dp/dt \tag{15.1}$$

(cf. Eq. (14.10)). Here the change coefficient $g$ is the *net* coefficient $(g + d)$ of Eq. (14.10). If (15.1) is valid, the dependence of the change coefficient $g$ upon $p$ and $t$ has to be found. Given the unknown nature of this quantity, coupled with the fact that (15.1) is only a working hypothesis, it is simpler to ignore the results of Chapter 14 and start all over, *applying EPI to the particular scenario of carcinogenesis*. The resulting growth law will indeed be of the form (15.1), with the functional form $g(p, t)$ identified.

The general use of EPI requires the effect under measurement to be a random one. In fact carcinogenesis has been proposed to be a random process (Gatenby and Frieden, 2002). This is also the main reference for the chapter. The alternative to the hypothesis of randomness is the view that carcinogenesis follows some *predictable*, deterministic sequence of alterations of the gene structure. On the latter basis, if the sequence could be known, the condition could be avoided or cured. However, evidence suggests otherwise. A fixed sequence does not exist, except in certain rare cancer conditions (congenital neoplastic syndromes).

By the random model, clinical cancers emerge out of multiple, fundamentally random genetic pathways. The model seems compelling on the following experimental grounds:

(1) Most mutations occur randomly, i.e., unpredictably. Cancers are always observed to contain many mutations, some on the order of $10^4$ mutations. Results of clinical studies show that the mutation rate increases during tumor growth (Stoler *et al.*, 1999).
(2) The high mutation rate found in cancer is favorable for its survival because it produces a greater cellular diversity and, hence, increased probability of a cancer that will successfully reproduce and proliferate. This is demonstrated as follows.
    (*a*) Bacterial studies have found a 5000-*fold increase* in the mutation rate when culture conditions became more restrictive and, hence, are made to evolve over time (Mao *et al.*, 1997).
    (*b*) Early cancers rely on diffusion and the pre-existing blood supply for support.

However, as the tumor grows, it needs a greater infusion of blood (angiogenesis). The random model predicts that (a) the angiogenic phenotype of cancer will be found far more frequently among "older" tumors that have a high number of mutations, and (b) these will develop far more quickly than in younger tumors with a smaller number of mutations. These predictions have been confirmed (Wodarz and Krakauer, 2001).

(*c*) Increasingly aggressive cellular populations emerge over time during carcinogenesis (Barrett *et al*., 1999).

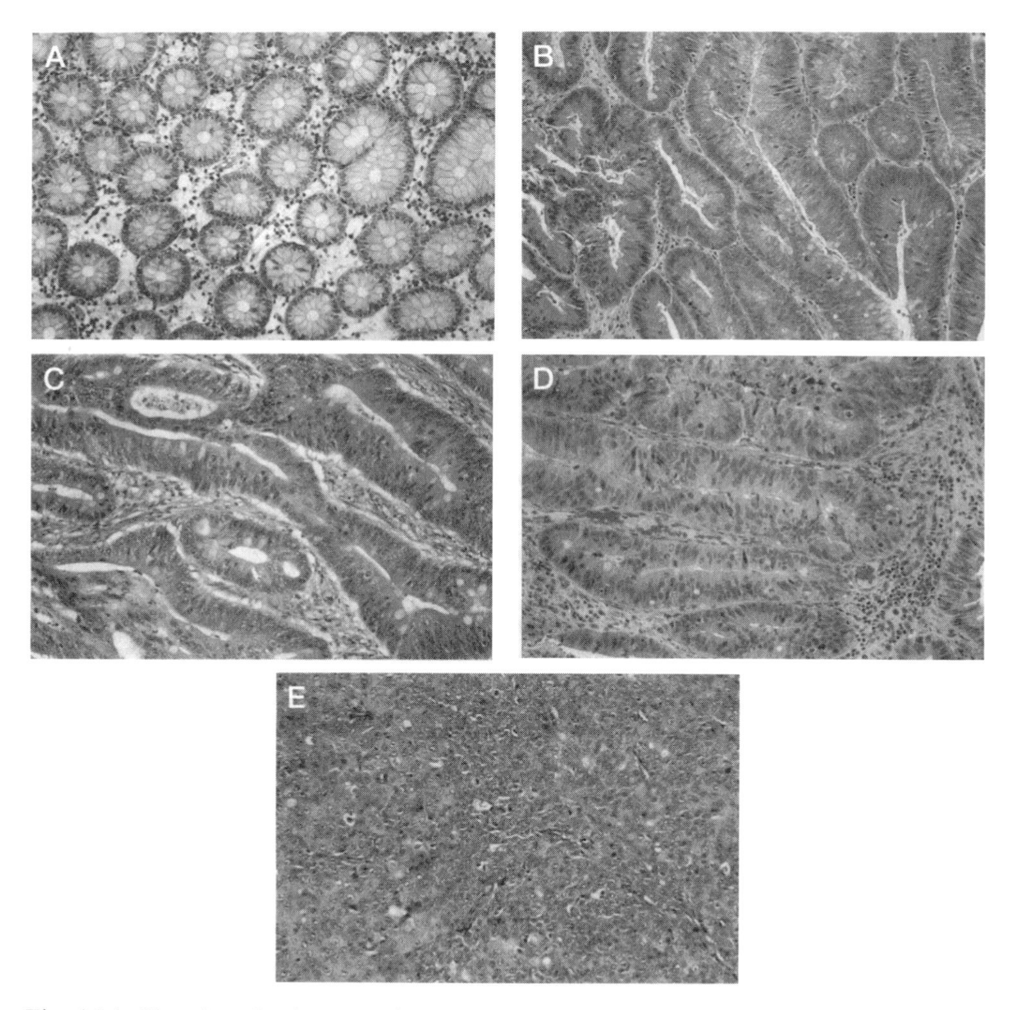


Fig. 15.1. Showing the increase in randomness of cell structure associated with the evolution of colon cancer: (A) normal colon membrane with regular, highly predictable cellular structure and well-organized functional tissue; (B) and (C) colon polyps with low-grade (B) and high-grade (C) abnormality of structure (dysplasia); (D) minimally invasive cancer of the colon, demonstrating further cellular disorder; and (E) invasive colon cancer with highly disordered, heterogeneous cell structure and complete loss of normal colon tissue structure and function.

Finally, if a picture is worth a thousand words, Fig. 15.1 shows the progression from (A) normal, healthy colon tissue through the various progressive stages (B)–(E) of invasive cancer, showing progressively increased randomness in the cell structure phenotype. This reflects increasing randomness on the level of the genes.

## 15.3 Biologically free and bound information

Cancer is an information-dependent "disease of the genes" (Bishop, 1991). It is a complex disease of the storage, processing, and propagation of *cellular information* maintained in the genome (entire set of genes). The information in an organism has been described as being either *bound* or *free* (Kendal, 1990). Historically, this *biological* use of the word "bound" developed independently of the *physically* bound information $J$ that is used in this book. In order to distinguish between these in the following, they are called "biologically bound" or "physically bound" informations, respectively. A comparison is next made between these biological informations and the informations $I$, $J$ used in this book.

*Biologically bound* information is bound information of Brillouin's type (Chapter 3). Thus, it has the mathematical form of a *Shannon* entropy $H$ (Kendal, 1990). By comparison, our *physically bound* information $J$ is, by its definition, a *Fisher* information.

Also, the biologically bound information is encoded in the genome, and serves as a reservoir to be passed from one generation to the next. By comparison, as we find in Chapter 12, the physically bound information $J$ is bound to, or entangled with, the intrinsic physics that gives rise to the observation. Thus, from the point of view of information, the genome operates as the effective *source* of biological function, as might be expected.

*Biologically free* information is contained in organic polymers such as proteins, lipids, fatty acids, and polysaccharides that regulate cellular structure and function. Free information is dependent in a complex way upon the translation of specific subsets of the biologically bound information and their mutual interactions. Free information is also present in the *extracellular space* of multicellular organisms. For these, cellular function is influenced by a flow of both intracellular and extracellular free information.

In our analysis, the total extracellular, or free, information that is produced by an average cancer cell is evaluated as a *Fisher* information of the "physically bound" variety, i.e., a $J$. The manner by which an EPI measurement acquires information about cancer growth is taken up next.

## 15.4 Cancer growth and an EPI measurement

Any measurement system obeys the information flow effect

$$J \to I, \tag{15.2}$$

where $J$ is the level of information in a physical source, $I$ is the received information about the source, and the information is carried from source to data space by some intermediary *probe particles*. A cancer EPI measurement system is illustrated in Fig. 15.2. Existing cancer cells $C_1, C_2, \ldots$ rapidly metabolize glucose, and the principal resulting waste product is positive ions of lactic acid, *protons* providing the positive charge. Randomly, one or more of the cancer cells "illuminate" a nearby healthy cell with the lactic acid ions, along with other organic and inorganic molecules. This "illumination" is called the "cancer event," for reasons that will become apparent. The sudden local increase in the level of lactic acid signals the occurrence of the cancer event to an exterior observer. Associated with the event is the current age $\tau$ of the patient. The age $\tau$ is taken as *a datum* by the observer (say, a clinician). *Why* the age is the datum is discussed next.

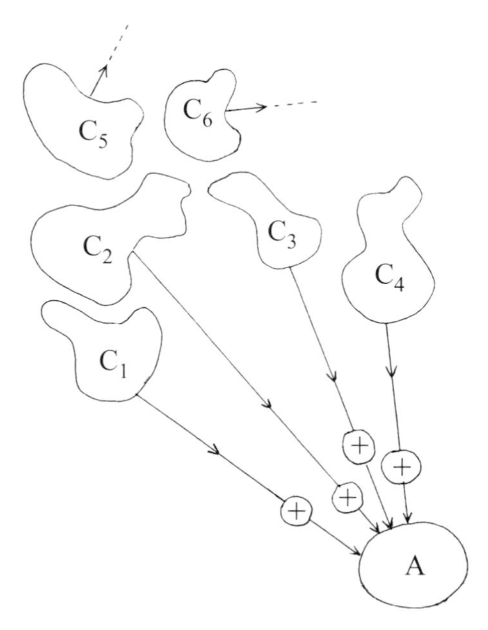


Fig. 15.2. The EPI measurement scenario for cancer growth. A healthy cell A receives messages about the evolutionary age of a neighboring mass of cancer cells $C_1, C_2, \ldots$. Ions indicated by the + signs carry the messages. Many cancer cells randomly contribute to the net information $I$ received by the healthy cell.

### 15.4.1 Knowledge of onset time versus cancer growth

*Time* is sometimes considered "the final frontier" in biology. Biological systems organize time in robust and effective ways. The resulting biological clocks cover a wide range of *time scales*, from the heartbeat to circadian rhythms. In each of these systems, molecular and cell mechanisms underlie the rhythms, and serve to stabilize them. Uncovering these mechanisms is essential to understanding biological systems. Although the complexity of the mechanisms is at first daunting, simple computational models are being sought for describing these systems. A cancer mass is one such biological system. All such systems transmit and process information, and this forms the basis for our temporal system model as follows.

Of central importance to both the growth of the cancer mass and its treatment is the onset age $\theta$, i.e., the time at which the carcinogenesis process *began*. The intercellular transmission of knowledge about the age $\theta$ of a general cell is generally useful for purposes of *regulating* its growth rate. Hence a cancer that transmitted *minimal information* about its age would avoid regulation of its growth. In fact, there is a biological mechanism for such minimal information – telomere length – as discussed below. Furthermore, EPI implies the minimum (Sec. 15.7). Also, the theoretical growth with time that is implied by the minimum is verified by clinical data (Sec. 15.11). One of the chief results of this chapter is that cancer attains this minimum. We discuss it qualitatively next.

The condition of minimum temporal information is apparently an evolutionary device that aids in survival of the cancer. Not only does the minimal information lessen the degree to which its growth is controlled, but knowledge of the evolutionary age of a cancer equates to knowledge of the onset time of the cancer. Minimal information about onset time leads to large uncertainty in the proper choice of medication for suppressing the condition. This further promotes the growth of the disease (See also Sec. 15.10).

The age $\theta$ of a general cell has a biological counterpart in the evolved *telomere length* of the cell. Telomeres are strands of DNA that are at the ends of the chromosomes of healthy cells. Telomere lengths give an indication of cell age, since they linearly become shorter with each generation. By comparison, the telomeres of a cancer do not shorten with time, and therefore ought not give much, if any, information about its age (Fagerberg *et al.*, 1985). This is consistent with the above hypothesis of minimal information.

### *15.4.2 Cancer growth message*

The probe particles are several kinds of biological messengers. Of these, *protons* are dominant (the + signs in Fig. 15.2). It is known that protons occur in the acid that is secreted by the cancer cells. The acid is lactic acid, and cancers secrete excessive amounts of this into the extracellular spaces of the organ. An experimental demonstration of the proton as the biological messenger has been published recently (Volker *et al.*, 2001). An increase in glucose uptake by the cancer accompanies the secretion of the acid (Clarke *et al.*, 1990).

*The lactic acid bathes the healthy cell in a toxic wash that eventually kills it.* This leads to compromise of the functioning tissue and, all too often, death of the host. This mechanism of tumor invasion has been modelled (Gatenby and Gawlinski, 1996). Hence, the proton messengers constantly convey one message to the nearby cell. That message is, in effect, "die!"

As indicated above, a human observer enters the information channel by observing at a time $\tau$ a sudden local increase in the concentration of lactic acid in what was previously considered a healthy patient. Depending upon the instrumentation in use, there is bound to be some *added uncertainty* in determining $\theta$. However, for simplicity we will ignore this. Then, the observer is presumed subject to the same level of uncertainty about the onset time $\theta$ as is the healthy cell. The observer effectively receives the same level of temporal information $I$ as well, although, happily, not the *message*.

More on the last point. This is another case (Sec. 14.16) where the observation does not perturb the source phenomenon as much as do other random, decohering effects. As a result, the observer's reality is decoupled from, i.e., *not entangled* with, that of the source effect. Since he is not so entangled, he does not receive precisely the same message – a strong dose of lactic acid – as does the healthy cell. Hence, he does not "catch" the cancer. An entanglement of realities would definitely not be wanted here. (By comparison, contagious diseases do incur entanglements, although not on the *quantum* level.)

### *15.4.3 Level of data information*

The information $I$ is the total amount of free information that is received by the functional (healthy) cell from the cluster of neighboring cancer cells. As mentioned above, the same level of information $I$ is presumed to reach the external observer as well. Let the average amount of free information that is provided by a single cancer cell be the amount $J$. Then, as usual,

$$I = \kappa J, \qquad \text{but} \qquad \kappa > 1 \tag{15.3}$$

is now possible, since here the efficiency $\kappa$ is the total due to *many* neighboring cancer cells (see Fig. 15.2). A lower bound to the number will be found later.

In this way, cancer growth fits into the EPI measurement model. We next apply EPI to predict cancer growth dynamics. Following this we test the result against clinical data.

## 15.5 EPI approach, general considerations

From the preceding, $\theta$ is the age of the patient at the onset of carcinogenesis, and $\tau$ his age when the cancer event is observed by the clinician. The information $I$ is about the unknown age $\theta$ of onset of the cancer. The time span between the onset and the observation is denoted as $t$. This is unknown and, hence, regarded as random.

The given EPI datum $\tau$ thereby obeys

$$\theta + t \equiv \tau \tag{15.4}$$

(cf. Eq. (1.0)). Here, both left-hand quantities are unknown. That is, the observer (say, a clinician) knows neither the onset time $\theta$ nor the elapsed time $t$ of the tumor growth. This is the usual clinical dilemma. Of course, by the definition (15.4) of $\tau$ and the discussion in Sec. 1.2.1, its size $\tau$ carries *information about* the size of $\theta$. *This allows* $\theta$ *to be estimated from knowledge of the likelihood law* $p(\tau|\theta)$. For example, an estimate of $\theta$ obeying $p(\tau|\theta) = max.$ could be used (Sec. 1.6.2).

Hence our initial aim is to find the likelihood law $p(\tau|\theta)$. However, to a good approximation, a simplification in its form follows. Whether a person is found to have cancer on a Tuesday or the following Thursday should not matter to the cancer's intrinsic growth characteristics. Thus, the likelihood law $p(\tau|\theta)$ should not change shape with a shift in the absolute onset time value $\theta$. This implies that $p(\tau|\theta) = p(t)$, the probability on the elapsed time $t$ (Eq. (1.11)). Under these conditions the Fisher information takes the familiar shift-invariant form Eq. (1.2).

This shift invariance breaks down, however, once the onset age $\theta$ is sufficiently large. For example, very old people – say, beyond age 90 years – have a much lower chance of dying of cancer than do younger people. Evidently, the growth characteristics of cancer do change in the vicinity of that time. Hence, in effect this calculation is for values of $\theta$ that precede about age 90.

Thus, our aim is to find the (approximately) shift-invariant probability law $p(t)$, the probability on the elapsed time $t$ since the onset of the cancer. The time $t$ is regarded as *random* and unknown, and is the Fisher coordinate of the problem. Note that, by comparison, there is no randomness permitted on the level of whether a cancer event or a non-cancer event is observed. *By hypothesis*, the observed event is a cancer event $C$. That is, $p(t) \equiv p(t|C)$.

## 15.6 EPI self-consistent solution

The aim is to find the probability law $p(t)$. As usual, we have to address the issue of how to get the functional $J$. Since we anticipated that $\kappa > 1$, this rules out a unitary space that *preserves* the size of $I$ (and *gives* $J = I$). Hence we have to resort to seeking a *self-consistent* EPI solution (Sec. 3.4.6). This entails seeking the one amplitude function $q(t)$ that obeys both EPI conditions (15.3) and

$$I - J = extrem., \qquad \text{where } I = I[p], \qquad J = J[p], \qquad \text{and } p \equiv q^2. \tag{15.5}$$

The notation $I[p]$, $J[p]$ is meant to indicate that these informations are functionals of the unknown probability law $p$.

It is convenient to work with information densities $i$ and $j$, i.e., the integrands of information functionals

$$I = 4\int dt\, i(t) \qquad \text{and} \qquad J = 4\int dt\, j[q, t]. \tag{15.6}$$

Here, by the shift-invariance property, the information $I$ obeys Eq. (1.24), so that

$$i(t) = \dot{q}^2. \tag{15.7}$$

The extremum condition (15.5) obeys the Euler–Lagrange Eq. (0.13), which, by Eqs. (15.6) and (15.7), becomes

$$\frac{d}{dt}\left(\frac{\partial(i - j)}{\partial\dot{q}}\right) = \frac{\partial(i - j)}{\partial q}. \tag{15.8}$$

The zero-condition (15.3) becomes, by Eqs. (15.6),

$$i - \kappa j = 0, \qquad j \equiv j[q, t], \qquad \text{where} \qquad \kappa > 1 \tag{15.9}$$

is expected. We now seek the common solution $q(t)$ to both conditions (15.8) and (15.9).

### *15.6.1 A power-law solution*

Various possible forms for the *in situ* cancer growth law, including a power-law form, have previously been proposed (Hart *et al.*, 1998). We will show that the power-law form is, in fact, an EPI solution.

Using Eq. (15.7) in Eq. (15.8) and (15.9) gives, respectively, an extremum condition

$$\frac{d}{dt}\left(\frac{\partial(\dot{q}^2 - j[q, t])}{\partial \dot{q}}\right) = \frac{\partial(\dot{q}^2 - j[q, t])}{\partial q} \tag{15.10}$$

and a zero-condition

$$\dot{q}^2 - \kappa j[q, t] = 0. \tag{15.11}$$

Explicitly carrying through the differentiations in (15.10) gives

$$2\frac{d^2 q}{dt^2} = -\frac{\partial j}{\partial q}. \tag{15.12}$$

Equations (15.11) and (15.12) must be simultaneously solved for $j$ and $q$. Notice that each equation involves both $j$ and $q$. The aim is to get *one equation* in *one* of the unknown functions $q$, $j$.

Differentiating $d/dt$ Eq. (15.11) gives

$$2\dot{q}\frac{d^2 q}{dt^2} = \kappa\left[\frac{\partial j}{\partial t} + \frac{\partial j}{\partial q}\dot{q}\right]. \tag{15.13}$$

Equations (15.12) and (15.13) must be solved simultaneously.

Using (15.12) on the left-hand side of (15.13) gives a requirement

$$-\dot{q}\frac{\partial j}{\partial q} = \kappa\left[\frac{\partial j}{\partial t} + \frac{\partial j}{\partial q}\dot{q}\right]. \tag{15.14}$$

Taking the square-root of (15.11) gives

$$\dot{q} = \sqrt{\kappa}\sqrt{j[q, t]} \tag{15.15}$$

where the + sign was chosen because we are seeking a solution that grows with the time. Substituting (15.15) into (15.14) gives directly

$$-\sqrt{\kappa}\sqrt{j}\frac{\partial j}{\partial q} = \kappa\left[\frac{\partial j}{\partial t} + \frac{\partial j}{\partial q}\sqrt{\kappa}\sqrt{j}\right]. \tag{15.16}$$

After a rearrangement of terms and division by $\sqrt{\kappa}$, this becomes

$$\sqrt{j}\frac{\partial j}{\partial q}(1+\kappa) + \sqrt{\kappa}\frac{\partial j}{\partial t} = 0. \tag{15.17}$$

This accomplishes our immediate aim of obtaining one equation in the single unknown function $j$.

Let us seek a separable solution

$$j[q, t] \equiv j_q(q) j_t(t), \tag{15.18}$$

where marginal functions $j_q(q)$ and $j_t(t)$ are to be found. This is a special choice of solution to the problem. Alternative solutions might also exist. Substituting (15.18) into (15.17) gives a requirement

$$\sqrt{j_q}\sqrt{j_t}j_q'j_t(1+\kappa)+\sqrt{\kappa}j_qj_t'=0, \tag{15.19}$$

where the derivatives indicated by primes are with respect to the indicated subscripts (or arguments) $q$ or $t$. Multiplying the equation by $j_q^{-1}j_t^{-3/2}$ gives

$$j_q^{-1/2}j_q'(1+\kappa)+\sqrt{\kappa}j_t^{-3/2}j_t'=0. \tag{15.20}$$

The first term on the left-hand side is only a function of $q$ and the second is only a function of $t$. The only way they can add up to zero (the right-hand side) for all $q$ and $t$ is for the first to equal a constant and the second to equal the negative of that constant. Thus,

$$j_q^{-1/2}j_q'(1+\kappa)=A, \qquad \sqrt{\kappa}j_t^{-3/2}j_t'=-A, \qquad A=const. \tag{15.21}$$

The first equation can be placed into the integrable form

$$j_q^{-1/2}\,dj_q=A(1+\kappa)^{-1}\,dq. \tag{15.22}$$

Integrating and then squaring both sides gives

$$j_q(q)=\frac{1}{4}\left(\frac{Aq}{1+\kappa}+B\right)^2, \qquad A,\,B=const. \tag{15.23}$$

The second Eq. (15.23) can be placed in the integrable form

$$j_t^{-3/2}\,dj_t=-A\kappa^{-1/2}\,dt. \tag{15.24}$$

Integrating this, then solving algebraically for $j_t$, gives

$$j_t(t)=4\left(\frac{At}{\sqrt{\kappa}}+C\right)^{-2}, \qquad C=const. \tag{15.25}$$

Using solutions (15.23) and (15.25) in Eq. (15.18) gives

$$j[q,\,t]=\left(\frac{\dfrac{Aq}{1+\kappa}+B}{\dfrac{At}{\sqrt{\kappa}}+C}\right)^2. \tag{15.26}$$

Having found $j$, we can now use it to find $q$. Using (15.26) in (15.15) gives directly

$$\dot{q}=\frac{\dfrac{Aq}{1+\kappa}+B}{\dfrac{At}{\kappa}+D}, \qquad D=C/\sqrt{\kappa}. \tag{15.27}$$

This is equivalent to the differential form

$$\frac{dq}{\dfrac{Aq}{1+\kappa}+B}=\frac{dt}{\dfrac{At}{\kappa}+D}. \tag{15.28}$$

This may be readily integrated, giving logarithms of functions on each side. After taking antilogarithms the solution is

$$q(t) = \left(\frac{1+\kappa}{A}\right)\left(\frac{At}{\kappa} + D\right)^{\kappa/(1+\kappa)} - (1+\kappa)\frac{B}{A}. \qquad (15.29)$$

This is a general power-law solution in the time $t$. Two of the constants $A$, $B$, $D$ may be fixed from boundary-value conditions, as follows.

### 15.6.2 Boundary-value conditions

At $t = 0$ the cell is presumed not to be cancerous. Cancer has its onset at $t = \epsilon$, $\epsilon > 0$, a small increment beyond 0. This has two boundary-value effects.

(a) At $t = 0$, the information $j$ about cancer in the given cell is zero, $j(0) = 0$.
(b) At $t = 0$ the cancer has zero mass, $q(0) = 0$.

Using these conditions in (15.26) gives the requirement $B = 0$. Using the latter and condition (b) in (15.29) gives either $D = 0$ or $\kappa = -1$. But the latter is ruled out *a priori* by EPI, since $\kappa \geqslant 0$ represents an efficiency measure of information transmission. Hence, $D = 0$. With this choice of constants, (15.29) becomes

$$q(t) = \left(\frac{1+\kappa}{A}\right)\left(\frac{At}{\kappa}\right)^{\kappa/(1+\kappa)}. \qquad (15.30)$$

Hence the amplitude function is a power law in the time. Its square, the required probability law $p(t)$, is likewise a power law.

It is to be noted that the result (15.30) is for a continuous formulation, i.e., where the time $t$ is regarded as continuous. However, of course actual growth occurs at *discrete* generations spaced by $\Delta t$. The discrete effects become more important at $t$ values so small that $t \leqslant \Delta t$, so Eq. (15.30) becomes increasingly inaccurate as $t \to 0$. Equivalently, the constant time offset $D$ in Eq. (15.29) is some unknown, small constant. This means that using (15.30) to reach conclusions about the growth at small $t$ values gives only approximate results.

The power $\kappa/(1+\kappa)$ of the power-law solution (15.30) is determined in Sec. 15.7.

### 15.6.3 Check on solution

It is important to verify that the solution satisfies both EPI requirements (15.11) and (15.12), the zero requirement and extremum requirement, respectively. With $B = C = D = 0$ from the preceding section, the solution (15.26) for $j$ is

$$j = \frac{\kappa}{(1+\kappa)^2} \frac{q^2}{t^2}. \tag{15.30a}$$

Note that this gives rise to an undesirable pole in $j$ at a value $t = 0$ (my thanks to L. Seigel for pointing this out). It is indicative of a breakdown of the continuous-time approach near the origin (see end of preceding section). Hence, Eq. (15.30a) should not be evaluated at values $t \lessapprox \Delta t$.

Subject to this caveat, Eqs. (15.30) and (15.30a) may be shown to obey the EPI requirements (15.11) and (15.12). We recommend the reader to carry through on the verification, which holds at any time $t$ except $t \lessapprox \Delta t$ as previously discussed.

## 15.7 Determining the power by minimizing *I*

Without loss of generality, Eq. (15.30) may be placed in the form

$$q(t) = Et^{\alpha}, \qquad \alpha = \kappa/(1+\kappa), \qquad E = const. \tag{15.31}$$

The power of the law is now $\alpha$. Our overall aim is to find this power.

Since the form (15.31) grows unlimitedly with $t$, for the corresponding $p(t)$ to be normalized, the cancer growth must be presumed to exist over a finite time interval $t = (0, T)$. The size of $T$ is fixed as the survey time over which cancers are observed. The normalization requirement then gives a requirement

$$\int dt\, p(t) = E^2 \int_0^T dt\, t^{2\alpha} \equiv 1, \qquad \text{or} \qquad E^2 = (2\alpha + 1)T^{-(2\alpha+1)} \tag{15.32a}$$

after the integration.

To proceed further, we need a property of $I$ that is suitable for this distinctly biological case. Remarkably, it is implied by *general* EPI theory. We saw in Sec. 3.4.14 that a real, unitless constant of unknown value may be determined by use of the *game corollary*, which, for the real coordinate $t$, is

$$I = min. \tag{15.32b}$$

EPI also requires a "free-field" scenario as well for this condition (Sec. 3.4.14). A *non-free* field would in general be some source or force that is imposed from outside that modifies the evolution of $q(t)$. An example for this problem would be some regimen of chemotherapy. But the given problem is that of a *freely* growing *in situ* tumor. This is by definition free of chemotherapy or any outside influence and, hence, is a free-field case.

Equation (15.32b) has an interesting *biological* consequence. Recall that the Fisher information $I$ of a system increases with both its levels of order (Sec.

1.7) and of complexity (Sec. 2.4.1). Then (15.32b) implies that an *in situ* cancer cell is a minimally ordered and minimally complex system. Biologically, evolution is usually toward ever more complex systems. Therefore a cancer cell represents a regression to a *minimum level of evolutionary complexity*, so minimal that the cell loses its ability to function and, instead, only reproduces.

The minimum will be found through choice of the free parameters of the process. The observation interval $T$ is already fixed, and $E$ is fixed in terms of it by the second Eq. (15.32a). Hence, there is only one free parameter left, the power $\alpha$.

The Fisher information conveyed by the cancer over the entire survey period $(0, T)$ is, by Eqs. (15.6) and (15.7),

$$I = 4\int_0^T dt\, \dot{q}^2. \tag{15.33}$$

This is to be minimized.

Using Eqs. (15.31) and (15.32a) in (15.33) gives, after an integration,

$$I = \frac{4}{T^2}\left(\frac{2\alpha + 1}{2\alpha - 1}\right)\alpha^2. \tag{15.34}$$

This may be extremized through choice of $\alpha$ by ordinary differentiation. Setting $\partial I/\partial\alpha = 0$ results in the algebraic equation

$$4\alpha^2 - 2\alpha - 1 = 0, \qquad \text{or} \qquad \alpha = \tfrac{1}{4}(1 \pm \sqrt{5}) \tag{15.35}$$

as roots. However the negative square-root choice when placed in Eq. (15.34) gives a negative value for $I$, which is inconsistent since $I$ must be positive for this *real*-coordinate $t$ problem. Thus, the answer is a power-law solution

$$p(t) = E^2 t^{2\alpha} \equiv E^2 t^{\gamma}, \qquad \text{where} \qquad \gamma = \tfrac{1}{2}(1 + \sqrt{5}) = 1.618\,034\ldots. \tag{15.36}$$

Differentiating $d/dt$ Eq. (15.36) gives $\dot{p} = \gamma E^2 t^{\gamma - 1} = p \cdot (\gamma/t)$. This verifies the proposed Lotka–Volterra form (15.1) of the growth equation, with change coefficient $g = \gamma/t$. Hence the two EPI routes to solution – via Chapter 14 or via this chapter – give consistent results.

## 15.8 Information efficiency κ

By the second Eq. (15.31), $\kappa = \alpha/(1 - \alpha)$. Then, by the positive root of Eq. (15.35),

$$\kappa = 4.27. \tag{15.37}$$

This is the total efficiency due to the unknown number of cancer cells that are communicating with the functioning cell. Since the efficiency for any single cell can be at most unity, this implies that there are at least *four* neighboring cancer cells that are interacting with the functioning cell. The fractional remainder 0.27 can arise out of an averaging effect: Randomly, sometimes there are four cells interacting, sometimes five, etc. Thus, each normal cell is receiving the information output of at least four cancer cells. Since tumor cells revert to glycolytic metabolism, and lactic acid is the by-product of glycolysis, the *information messengers* conveyed by the cancer cells will in part take the form of *lactic acid*. That more than four cancer cells are involved means that a relatively large amount of lactic acid bathes the healthy cell, eventually killing it. This mechanism of tumor invasion has previously been modelled (Gatenby and Gawlinski, 1996).

## 15.9 Fundamental role played by Fibonacci constant

We note that the power $\gamma$ of the growth process is precisely the Fibonacci "golden mean" or "constant." This number also occurs as the relative increase in an *ideally breeding* population $p_F(t)$ over generation time $\Delta t$

$$\gamma \equiv \frac{p_F(t + \Delta t)}{p_F(t)}. \tag{15.38}$$

(The original subject of Fibonacci's investigations was a colony of rabbits.) By comparison, Eq. (15.36) shows that the relative increase in cancer mass from one generation to the next obeys

$$\frac{p(t + \Delta t)}{p(t)} = \left(\frac{t + \Delta t}{t}\right)^{\gamma} \approx 1 + \gamma \frac{\Delta t}{t}, \tag{15.39}$$

in first-order Taylor series, where $\Delta t \ll t$. Thus, although both populations increase with time, by (15.38) the ideally breeding Fibonacci population maintains a *constant* relative increase $\gamma$ with time, whereas by (15.39) the cancer increase gradually saturates toward 1 (no increase) with time. The prediction is therefore that cancer grows at a *less-than-maximum rate*. This agrees with empirical data, which show that (a) *in vitro* cancer cells maintained in three-dimensional culture conditions grow exponentially (Ducommun *et al.*, 2001), i.e., follow ideal Fibonacci growth, whereas (b) *in situ* cancer grows according to power-law growth, i.e., much slower than Fibonacci growth.

It is also interesting to compare the two forms of growth on the basis of information. Because of the ideal growth effect in Eq. (15.38), a Fibonacci population grows in time according to a pure exponential law

$$p_F(t) = B \exp(t/\tau), \qquad 0 \leqslant t \leqslant T, \qquad \tau = (\gamma + 1)\,\Delta t, \qquad B = const. \tag{15.40}$$

This is to be compared with the slower, power-law growth Eq. (15.36) for *in situ* cancer. Use of the probability law Eq. (15.40) in Eqs. (15.6) and (15.7) gives rise to an ideal Fibonacci information level

$$I_F = \frac{1}{(\Delta t)^2(\gamma + 1)^2}, \qquad \gamma = 1.618\,034\ldots. \tag{15.41}$$

By contrast, the cancer population obeyed the information Eq. (15.34). Comparing the solution (15.34)–(15.35) with the solution (15.41) shows that, since $\Delta t \ll T$, necessarily $I \ll I_F$. That is, *in situ* cancer gives much less Fisher information about the time than does ideally growing tissue.

The similarity of this prediction to the extensive data on telomere shortening in malignant populations is quite evident (Tang *et al.*, 1998). The telomeres of malignant populations do not shorten with time, and therefore *ought not* to give much, if any, information about their age (DePinho, 2000). As we saw, this suspicion is verified by the EPI approach.

## 15.10 Predicted uncertainty in the onset time of cancer

To treat the disease effectively, the clinician would like to know the value of its onset time $\theta$. Instead, by Eq. (15.4), he knows the time $\tau$ at which a cancer event is first observed. The question is, how well can $\theta$ be estimated based upon knowledge of $\tau$?

The Cramer–Rao inequality (1.28) states that any unbiased estimate of $\theta$ suffers a root-mean-square error $e$ that can be no smaller than

$$e_{\min} = 1/\sqrt{I}. \tag{15.42}$$

In our case, $I$ can be computed using Eq. (15.34) with the parameter $\alpha$ given by the positive root of Eq. (15.35),

$$I = \frac{11 + 5\sqrt{5}}{2T^2}. \tag{15.43}$$

Using this in Eq. (15.42) gives a minimum root-mean-square error of size

$$e_{\min} \approx 0.3T. \tag{15.44}$$

As a relative error, this is quite large: 30% of the total time interval $T$ over which cancer events are detected. Thus, cancer has an innate ability to mask its time of onset. This contributes to its prevalence and, hence, might be considered an evolutionary tactic.

Will the approximate nature of the predicted growth at time values $t \leqslant \Delta t$ discussed below Eq. (15.30) compromise the prediction (15.44)? Usually the value of $e_{\min}$ generated by Eq. (15.44) will be on the order of months. This much exceeds the intergenerational time $\Delta t$ for cancer growth, so the accuracy of (15.44) is not significantly affected.

## 15.11 Experimental verification

The theoretically predicted value of

$$\gamma = \tfrac{1}{2}(1 + \sqrt{5}) = 1.618\,034\ldots \tag{15.45}$$

given by Eq. (15.36) can be compared with recently published data on growth determined by mammographic measurements of tumor size in human breast cancers. There are six such studies (Tabar *et al.*, 1992; Fagerberg *et al.*, 1985; Thomas *et al.*, 1984; De Koning *et al.*, 1995; Peer *et al.*, 1994; Burhenne *et al.*, 1992). The tumors in these studies are well suited for testing our "free-field" (Sec. 15.7) predictions, since they were *untreated* cancers growing within the breast and not subject to any apparent clinical or tissue constraints.

The values of $\gamma$ reported in these studies are, respectively,

$$1.72,\ 1.69,\ 1.47,\ 1.75,\ 2.17,\ \text{ and } 1.61 \tag{15.46}$$

(cf. Eq. (15.45)). These have a sample mean of 1.73 and standard deviation of 0.23. Thus, the value (15.45) of the power predicted by application of information theory to cancer growth is easily within one standard deviation of the clinical data. The *relative occurrences* $p_\nu$ of the tumor sizes $\nu$ in all the data from the six studies (save for data indicating indefinite sizes) are plotted as individual points on the log–log plot in Fig. 15.3. This is after translation of the data to the axes of the plot (see Gatenby and Frieden, 2002, for details). The sizes $\nu$ are volumes in units of milliliters. For comparison, the theoretical power-law curve $p(t)$ given by Eq. (15.36) is also shown (solid black line). The agreement between the theoretical curve and the experimental data is visually good. This mirrors the agreement on the power value shown below Eq. (15.46).

## 15.12 Discussion and summary

We have shown that, both according to EPI *theory* and according to clinical *data*, *in situ* cancer growth obeys a power law. This was found to follow as an EPI self-consistency solution Eq. (15.36). The particular value of the power was found to be Fibonacci's constant $\gamma = 1.618\,034\ldots$. This followed from the hypothesis that a cancer cell secretes to its neighbors an absolute *minimum level of information* about its age (Sec. 15.7). The minimal level is consistent with two effects: that (1) a cancer cell cannot function, and that (2) the cell grows uncontrollably. These are discussed next. Following this are specula-

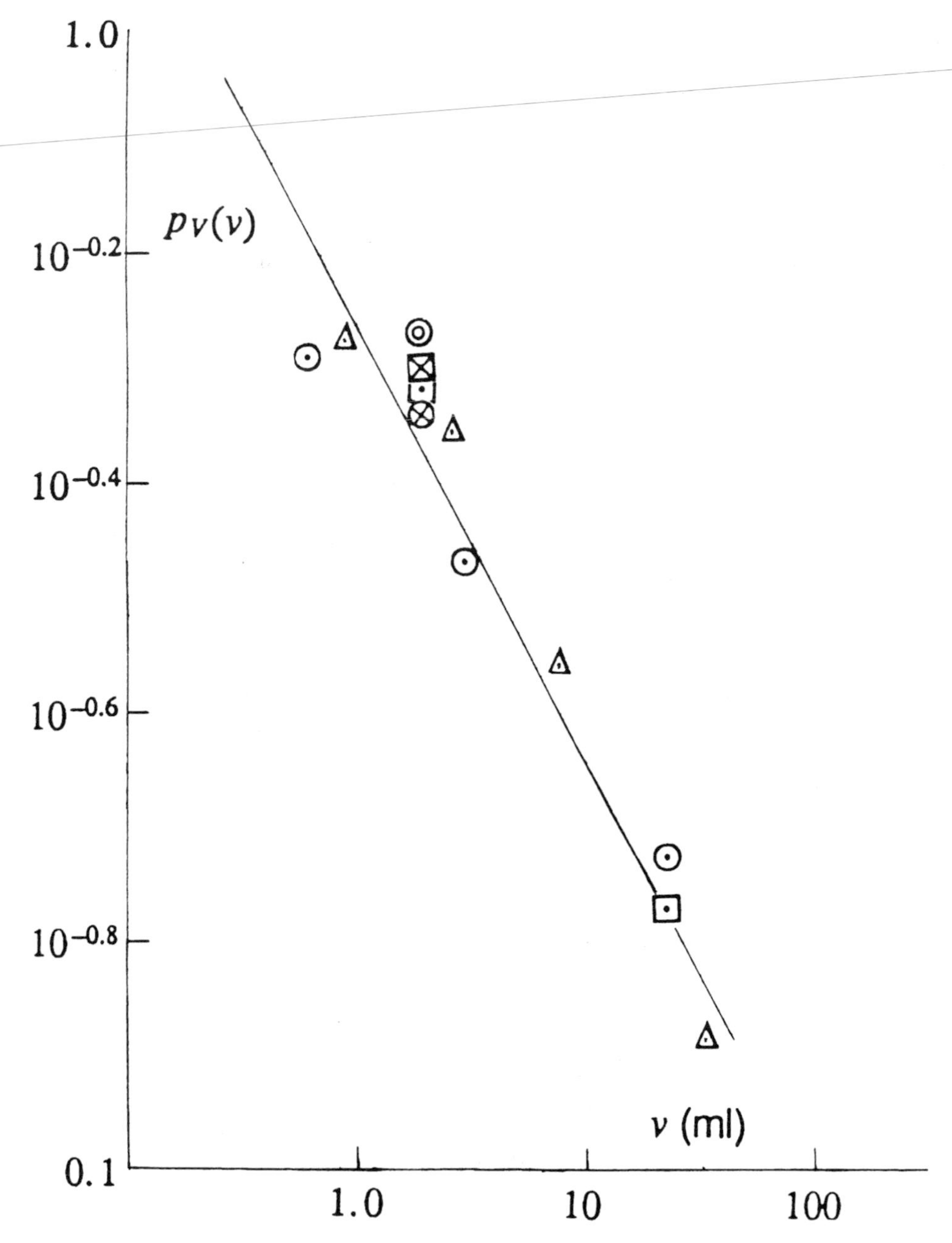


Fig. 15.3. Tumor growth dynamics based on the information-degradation model of carcinogenesis as predicted by EPI (solid straight line) compared with experimental data from six studies: points △ from Tabar *et al.* (1992); ⊡ from Fagerberg *et al.* (1985); ⊙ from Thomas *et al.* (1984); ⊠ from De Koning *et al.* (1995); ◎ from Peer *et al.* (1994); and ⊗ from Burhenne *et al.* (1992). Figure from Gatenby and Frieden (2002).

tions, (3) and (4), on possible routes to cancer treatment that are suggested by the analysis.

(1) Here we argue from the viewpoint of system *complexity*. Fisher information measures complexity (Sec. 2.4.1). Therefore, we use the two terms interchangeably in the following.

All human cells are produced by the mitosis or splitting of special cells called "stem cells." A normal stem cell produces at each mitosis another stem cell and a functioning cell. It is convenient, in the following two paragraphs, to often regard these two daughter cells as a single "cell."

A minimum level of information, as in (1) preceding, implies a cell with minimum complexity. Now, a healthy cell both reproduces and functions (that is, one daughter reproduces and the other functions). On the other hand, *each* such cell activity requires a certain minimum level, or threshold, of complexity. Hence there is one for reproduction and one for function. Each such activity ceases once cell complexity falls below its threshold. For cells in a vital organ, there is obviously zero chance of survival of the individual if reproductive activity is lost, but still some chance if instead function is lost, since this could conceivably be regained in time. Given time, people often recover from illnesses. Therefore evolution must have favored reproductive activity, giving it the *lower* threshold of the two. Moreover, it should be significantly lower, to take advantage of the fact that people recover from illnesses.

Next, consider a stem cell that has become, through mutations, a cancer cell. Its corresponding daughters will be cancerous as well. The daughters comprise the net "cell" mentioned above. The cell was found to obey a growth law consistent with the *minimum* possible level (15.43) of information. Therefore, from the preceding paragraph, that level must correspond to the threshold for reproduction and be significantly lower than the threshold for function. Hence, the cancer cell reproduces but does not function. This is not a proof, of course, but effects suggested by the model.

The medical term for this kind of a cell is an "undifferentiated" one, meaning a cell that is unable to carry out its usual function, to the extent that it is no longer differentiable in function from other normal cells.

(2) The *uncontrolled* nature of the growth of a cancer might result, as well, from another threshold effect. Consider an *initially healthy* cell. Its growth is partially controlled by outside cells (Sec. 15.3). Such control is presumably dependent upon two-way communication between the healthy cell and the outside cells, and hence, in part upon the level of information $I$ that the healthy cell secretes about its state of growth. This implies that if $I$ falls below a critical level, the communication is disrupted, and *the growth is no longer controlled*. Cancer is usually the result of thousands of random mutations in the genes (Sec. 15.2). These mutations *degrade the information* level $I$ such that it falls below the critical level, defining an *information catastrophe*. Thus the initially healthy cell suffers uncontrolled

growth, becoming a cancer.

Equation (15.43) provides further grounds for, and quantifies, this effect. It shows that the information $I$ received by the functioning cell *decreases monotonically* with the total time of observation $T$. This equates to lower and lower levels of information being secreted by the neighboring cancer cells. Thus, as larger and larger observation time intervals are used, the received information becomes ever lower and, by the preceding paragraph, relatively more and more cases of uncontrolled cancer growth must occur over the observed population. Such an acceleration of uncontrolled cancer growth with observation period is confirmed by clinical data.

Some speculative possibilities also emerge from this analysis.

(3) The degraded nature of the information $I$ about the time $\theta$ might act as a *causative* agent of carcinogenesis. While intuitively this seems unlikely, we do note that there have been studies in which telomere dysfunction has been described as resulting in genetic instability and epithelial cancer formation (Artandi *et al.*, 2000; Tang *et al.*, 1998). Conversely, increasing the cellular information $I$ about the onset time by, for example, treatment with telomerase might reduce the possibility of malignancy in at-risk tissues. In fact this seemingly paradoxical therapy has been suggested elsewhere (DePinho, 2000).

(4) Cancer has been modelled as a communication system whose information carriers, lactic acid ions, are vital to its functioning . The system could be suppressed, then, if the chain of communication were broken by suppressing the carriers such as through inhibition of glycolysis. On the other hand, Gatenby and Gawlinski (1996) have shown, ironically, that a sufficiently *high* level of acid inhibits tumor growth. Thus, an alternative strategy is to do the opposite: Purposely flood the resident cancer with acid, decreasing the information content with overwhelming noise.
There are indications that the method might work. First laboratory experiments on nude rats showed a 99.6% *reduction* in mean tumor size when enough hydrochloric acid is injected into the afflicted organ to maintain a pH level of 6.8 for 10 min (Kelley *et al.*, 2002). Of equal importance is that the total exposure to acid was weak enough to *not* cause permanent damage in normally functioning tissue. In any future application of the procedure to humans, of course, some optimum combination of dose and exposure time for maximizing apoptosis in cancer while minimizing damage to normal tissue would be sought.

A summary of the chapter is as follows.

Biologically bound and free information forms are compared with their physical counterparts (Sec. 15.3). The "free" form is utilized, since it has the form of a Fisher information. The EPI measurement is the age (time $\tau$) of a patient at the time of detection of a cancer (Sec. 15.4) by a clinician. The clinician is the exterior observer of this measurement scenario. The ideal parameter value $\theta$ is the onset time of the disease. At the instant of detection by the outside observer, a mass of cancer cells (at least 4.27; see Sec. 15.8) sends biological messengers to a nearby healthy cell. *Protons* of lactic acid are the dominant messengers. The lactic acid bathes the healthy cell in a toxic wash that eventually kills it. Thus, the message conveyed to the cell-observer is "die." Fortunately, the exterior observer does not also receive the message: his existence is not entangled with that of the phenomenon (Sec. 15.4.2).

An EPI self-consistent solution for the probability law (or mass law) $p(t)$ is found in Sec. 15.6. The solution is found via separation of variables Eq. (15.18) to be a power law (15.36). The "free-field" conditions of the problem imply an EPI solution satisfying $I = min$. Enforcing this allows the power to be found, in Eq. (15.45). It is the Fibonacci golden mean $\gamma$, with value 1.618 034 .... The theoretical cancer growth law on this basis is checked against clinical data (Sec. 15.11), and found to fit it quite well.

Some further insight into cancer growth is gained by comparing the growth rates (15.38) and (15.39) for, respectively, ideally growing cancer cells in a test-tube (*in vitro*) environment with realistically growing (*in situ*) cancer cells. Both are positive, but the former is at a *constant* rate $\gamma \approx$ 1.618 034 whereas the latter is at a steadily *decreasing* rate $1 + \gamma\, \Delta t/t$. The latter, in particular, asymptotically increases at just the replacement rate after $t$ has become large enough. What this indicates is that the human body offers a natural degree of resistance to the growth of cancer. Some of this is in the form of local depletion effects. Thus, a cancer must invest energy and information in growing peripheral arterioles to it so as to bring in the extra supplies of blood it needs to sustain its abnormally high growth rate. This, of course, takes "time" and acts to slow down the growth from its *in vitro* (ideal) rate.

The usual EPI uncertainty principle is also determined, with the resulting prediction Eq. (15.44) that the onset time for cancer cannot be known more precisely than with a root-mean-square relative error of 30%. This is a surprisingly large figure. It has at least one ramification to clinical treatment. The prediction is that, if cancer *size* can be determined with *better than* 30% accuracy, a course of treatment that depends upon knowledge of the size might be superior to one that depends upon knowledge of the age.

The good agreement of the main prediction of the approach – the power law Eq. (15.36) – with experimental data suggests that the premise for its derivation is correct. This is that cancer is fundamentally a *non-deterministic* illness, resulting from multiple (thousands of) genetic mutations of initially healthy cells. The mutations force a form of reverse evolution upon the cells. They regress to a state of lowest-possible complexity for living things – cells that can only grow – and, resultingly, secrete information $I$ at a minimum rate.

Finally, a purely *physical* prediction grows out of the growth analysis. The analysis shows that the growth of biomass, either under ideal conditions or under *in situ* conditions, is parameterized by the Fibonacci constant $\gamma$. Under the hypothesis that $\gamma$ might also parameterize *physical* mass growth in the Higgs mass phenomenon, we can form simple empirical *formulae* for the Weinberg and Cabibbo angles of this phenomenon. See Chapter 11 for details. These formulae agree to within 0.3° of current empirical values of the angles.

ROBERT A. GATENBY, M.D., is a Professor of Radiology at the Arizona Health Sciences Center, University of Arizona, Tucson 85 726. Email address: gatenby@radiology.arizona.edu.

# 16

# Summing up

As we have seen, the principle of extreme physical information (EPI) permits physics to be viewed within a unified framework of measurement. Each phenomenon of physics is elicited by a particular real measurement. The following is a synopsis of each chapter, with an emphasis upon the new physical effects, approaches and concepts that arise.

### *Chapter 1*

The Fisher information $I$ in a single measurement of a shift-invariant phenomenon obeys Eq. (1.24),

$$I = 4\int dx \left(\frac{dq(x)}{dx}\right)^2. \tag{16.1}$$

Here $q(x)$ is the *real* probability amplitude of a PDF $p(x) \equiv q^2(x)$, where $x$ is a fluctuation in the measurement. A Cauchy principal value of (16.1) is taken as needed.

In any use of the information (16.1) in the EPI principle, the coordinate $x$ can be chosen at will to be either real or imaginary. In the latter case, $I \leqslant 0$. The use of multiple real and imaginary coordinates in the EPI principle has the benefit of deriving d'Alembertian wave equations.

Under certain conditions, information $I$ obeys an "$I$-theorem" Eq. (1.30),

$$\frac{dI(t)}{dt} \leqslant 0. \tag{16.2}$$

This means that $I$ is a monotonic measure of system disorder. The $I$-theorem probably holds under more general conditions than does the corresponding "$H$-theorem" for the Boltzmann entropy $H$.

In the same manner as that by which a positive increment in thermodynamic *time* is defined by an increase in Boltzmann entropy, a positive increment in a

"Fisher time" is defined by a decrease in the information $I$. (Both times tend to increase as disorder increases.) The two times might not always agree, however, as was found by numerical experiments.

Let $\theta$ identify the phenomenon that is being measured. A Fisher temperature $T_\theta$ corresponding to $\theta$ may be defined as

$$\frac{1}{T_\theta} \equiv -k_\theta \frac{\partial I}{\partial \theta}, \qquad k_\theta = const. \tag{16.3}$$

(Eq. (1.39)). When $\theta$ is taken to be the system energy $E$, the resulting Fisher temperature has analogous properties to those of the (ordinary) Boltzmann temperature. Among these properties is a perfect gas law (Eq. (1.44))

$$\overline{p}V = k_E T_E I \tag{16.4}$$

where $\overline{p}$ is the pressure. It appears that all of thermodynamics can be formulated by the use of information $I$ in place of the usual entropy concept.

## *Chapter 2*

The $I$-theorem derived in Chapter 1 for a single-parameter measurement may be extended to a multiparameter, multi-component measurement scenario. A scalar measure of the information for this general scenario is developed. This information, called $I$, now obeys (Eq. (2.19))

$$I = 4\int d\mathbf{x} \sum_{n=1}^{N} \nabla q_n \cdot \nabla q_n, \tag{16.5}$$

where $q_n \equiv q_n(\mathbf{x})$ is the $n$th component probability amplitude for the four-fluctuation $\mathbf{x} = (x_0, \ldots, x_3)$. $I$ is called the "intrinsic" information, since it is a functional of the probability amplitudes that are intrinsic to the measured phenomenon. $I$ is also called the "channel capacity."

The positivity of each term in the sum (16.5) indicates that $I$ grows with the number $N$ of degrees of freedom $q_n$ of the system, i.e., with its *complexity*.

The information (16.5) obeys a property of *concavity*: The net $I$ for a mixed system is less than or equal to the average of $I$ over its individual subsystems.

This multi-component $I$ still obeys the $I$-theorem and, when used in the EPI principle, gives rise to the major laws of science.

In summary, $I$ is at the same time (i) a thermodynamic measure of disorder; (ii) a concave measure of system complexity; and (iii) a universal measure of information whose variation gives rise to most (perhaps all) of science.

### *Chapter 3*

Where does the information in acquired data come from? It must originate in the physical phenomenon (or system) that is under measurement. Any measurement of physical parameters initiates a *transformation of Fisher information* $J \to I$ connecting the phenomenon with the intrinsic data (Sec. 3.3.2). This information transition occurs in the object, or input, space to a measuring instrument (Sec. 3.8.1). The phenomenological, or "bound," information is denoted as $J$. $J$ also often expresses an *entanglement* of data space with source space.

The acquired information in the intrinsic data is what is called $I$. Information $J$ is ultimately identified by an invariance principle that characterizes the measured phenomenon (Sec. 3.4.5).

Suppose that, due to a measurement, the system is perturbed, causing a perturbation $\delta J$ in the bound information (Secs. 3.3.2, 3.8.1, 3.11). What will be the effect on the data information $I$? In analogy to a thermodynamic model Eq. (3.5) of Brillouin and Szilard, it must be that $\delta I = \delta J$; there is no loss of the perturbed Fisher information in its relay from the phenomenon to the intrinsic data. This is a new conservation law. It is also axiom 1 of an axiomatic approach for deriving physical laws.

Since $\delta I = \delta J$, necessarily $\delta(I - J) = 0$. Hence, if we define $I - J \equiv K$ as a net "physical" information, a variational principle

$$K \equiv I - J = extrem. \tag{16.6}$$

(Eq. (3.16)) results. This is called the EPI variational principle.

The EPI zero principle

$$I - \kappa J = 0, \quad 0 \leqslant \kappa \leqslant 1 \tag{16.7}$$

(Eq. (3.18)) follows, as well, on the grounds that generally $K \neq 0$ so that $I \neq J$ or $I = \kappa J$. That $\kappa \leqslant 1$ follows from the $I$-theorem Eq. (1.30) as applied to the information transition $J \to I$.

Equations (16.6) and (16.7) together define the EPI principle. These equations also follow identically, i.e., independently of the axiomatic approach taken, and of the $I$-theorem, if there is a *unitary transformation* connecting the measurement space with a physically meaningful conjugate space (Secs. 3.8.5 and 3.8.7).

The solution $\mathbf{q}(\mathbf{x})$ to EPI defines the physics of the measured particle or field in the object, or input, space to the measuring instrument (Sec. 3.8.5). (The physics of the *output space* is taken up in Chapter 10.)

It follows that the Lagrangian density for any phenomenon or system is not simply an *ad hoc* construction for producing a differential equation. The

Lagrangian has a definite prior significance. It represents the physical information density $k(\mathbf{x}) \equiv \sum_n k_n(\mathbf{x})$, with components $k_n(\mathbf{x}) \equiv i_n(\mathbf{x}) - j_n(\mathbf{x})$ in terms of data information and bound information components. The integral of $k(\mathbf{x})$ represents the total *physical information* $K$ for the system.

For real coordinates $\mathbf{x}$, the solution $\mathbf{q}(\mathbf{x})$ to such a variational problem represents the payoff in a mathematical game of information "hoarding" between the observer and the measured phenomenon (Secs. 3.4.11–3.4.13). The values of $N$ and of any unknown constants that emerge in the solution $\mathbf{q}(\mathbf{x})$ can be fixed, by use of the *game corollary* (3.21).

The particular requirement (3.22) of invariance of the information to reference frame leads to the *Lorentz transformation* equations (Sec. 3.5.5), and to the *requirement of covariance* for the amplitude functions $\mathbf{q}(\mathbf{x})$. As a corollary, $\mathbf{x}$ must be a four-vector, whose physical nature depends upon that of the measured parameters $\boldsymbol{\theta}$. Since the squared amplitude functions are PDFs, all PDFs that are derived by EPI must obey *four-dimensional* normalization, including integration over time if $\mathbf{x}$ is a four-position. A measurement of time is regarded *a priori* to be as imprecise and random as that of space (Sec. 3.5.8).

The dynamics and kinematics of the system are direct manifestations of the information densities $i_n(\mathbf{x})$, $j_n(\mathbf{x})$ (Sec. 3.7).

The constancy of the physical constants is an expression of the constancy of the corresponding information quantities $J$ for the scenarios that generate them (Sec. 3.4.14). The constants do not change with time, contrary to Dirac's proposal (1937; 1938).

The Von Neumann measurement–wave-function separation effect is derived, by the use of a simple coherent optics measuring setup (Sec. 3.8.3). Photons are the carriers of the information from object to image space.

## *Chapter 4*

An attempt to measure the classical four-position of a known particle gives, via the EPI principle, both the vector Klein–Gordon equation (if the particle is a boson), and the vector Dirac equation (if the particle is a fermion). The Klein–Gordon result follows from Eq. (3.16) of EPI, while the Dirac result follows from Eq. (3.18) of EPI. The rank of the vector defines the particle's spin value.

The Schrödinger wave equation follows either as the non-relativistic limit of the Klein–Gordon result (Appendix G) or, independently, in a one-dimensional EPI derivation (Appendix D).

The Weyl Eq. (4.45a) for the neutrino is derived as a special case of the

Dirac Eq. (4.41). The game corollary is used to predict that the neutrino mass $m$ is finite.

The Yukawa potential (4.30c) follows as a particular solution of the Klein–Gordon Eq. (4.26).

The Rarita–Schwinger wave Eq. (4.46l) for fermions of general spin is seen to follow from the Dirac Eq. (4.41).

The intrinsic information $I$ in the four-position of a free-field particle is proportional to the square of its relativistic energy $mc^2$ (Eq. (4.18)).

The equivalence Eq. (4.17) of energy, momentum, and mass is a consequence (not an assumption) of the theory, as is the Compton "resolution length" Eq. (4.21).

The constancy of the particle rest mass $m$ and Planck's constant $\hbar$ is implied by the constancy of information $J$.

The well-known problem of a variable normalization integral in ordinary Klein–Gordon theory is solved by using the four-dimensional approach afforded by EPI.

Derivation of the Dirac equation is not by a variational principle, but by satisfying the EPI zero-condition Eq. (3.18).

The Heisenberg uncertainty principle Eq. (4.53) is usually regarded as a quantum effect *par excellence*. But, in fact, we find that it is simply an expression of the Cramer–Rao inequality Eq. (1.1) of *classical* measurement theory, as applied to a space-time determination. Furthermore, such use of the Cramer–Rao inequality gives a generalization of the uncertainty principle – it defines the spread in *any function* of the space-time measurement.

The four dimensional amplitude functions $\psi_n(x, y, z, t)$ of EPI are shown to give rise to positive probability laws $p_n(x, y, z, t)$ that also are normalizable (Sec. 4.4 and Appendix I).

## *Chapter 5*

The vector wave equation of electromagnetic theory arises, via EPI, as the reaction to an attempt to determine a four-position within an electromagnetic field. The four-position is intrinsically uncertain and is characterized by random errors $\mathbf{x}$.

The PDF on these errors varies as the square of the four-potential (Eq. (5.52)). In general, this PDF is unique and normalizable (Secs. 5.1.25–5.1.28). In the special case of zero sources the PDF becomes simply that for the position of a photon within a coarse-grained field (with "grain" size on the order of the wavelength).

That classical electromagnetic theory should arise out of a statistical approach is, perhaps, surprising. The tie-in is the correspondence Eq. (5.48) between the probability amplitudes for the measurements (statistical quantities) and the electromagnetic four-potential (physical quantities). This correspondence is based on the recognition that the given sources are electromagnetic.

With the vector wave equation so derived, Maxwell's equations follow as a consequence. In this manner, classical electromagnetic theory derives from the EPI principle. Ultimately, it derives from the admission that any measurement of the electromagnetic field must suffer from random errors in position.

## *Chapter 6*

The weak-field equation of classical gravity arises as a reaction to an attempt at four-position measurement within a gravitational field.

The entire EPI derivation follows along analogous lines to that in Chapter 5. Thus, EPI unifies the two classical theories of electromagnetism and gravitation.

The PDF on four-position fluctuation that is implied by the theory is the square of the weak-field metric. In the special case of zero energy-stress tensor this becomes the PDF on position of a graviton. The marginal PDF on the time is consistent with the hypothesis that gravitons can be created and absorbed (Sec. 6.3.25).

The Planck length is derived (Sec. 6.3.22), and the cosmological constant $\Lambda$ is found to be either zero or proportional to $G$, the gravitational constant. (Sec. 6.3.23).

With the weak-field equation so derived, it is a simple matter to "riff up" to the general Einstein field equation.

## *Chapter 7*

The Boltzmann energy distribution law and the Maxwell–Boltzmann velocity law are derived by considering a four-dimensional space of energy-momentum fluctuations $(iE, c\boldsymbol{\mu})$. For the purposes of these derivations, this four-dimensional space replaces the usual six-dimensional phase space $(\boldsymbol{\mu}, \mathbf{x})$ of momentum-coordinate values that is conventionally used.

The computed Boltzmann and Maxwell–Boltzmann PDFs are the equilibrium laws for this problem. The EPI approach finds, as well, a set of PDF laws on velocity $\mathbf{x} = (x, y, z)$ for which $I$ is stationary but not at its absolute minimum value. These obey Eq. (7.59), a weighted superposition of Hermite–Gauss functions

$$p(\mathbf{x}) = p_0 e^{-|\mathbf{x}|^2/(2a_0^2)} \left\{ 1 + \sum_{n=1}^{N} 2^{-n} \times \left[ \sum_{\substack{ijk \\ i+j+k=n}} b_{nijk} H_i(x/a_0\sqrt{2}) H_j(y/a_0\sqrt{2}) H_k(z/a_0\sqrt{2}) \right]^2 \right\} \tag{16.8}$$

with parameters $a_0$, $p_0$, $b_{nijk} = const$. This solution is confirmed by previous work of Rumer and Ryvkin (1980). By the game corollary (Sec. 7.4.8), the lowest-order Hermite–Gauss function – a simple Gaussian – gives the equilibrium solution. Higher-order Hermite–Gauss functions in (16.8) define *non-equilibrium* solutions $p(\mathbf{x})$.

The analysis, when extended to allow higher momentum values, leads to anharmonic oscillator solutions and the prediction of *solitons* (Sec. 7.4.13).

The second law of thermodynamics states that the time rate of change $dH/dt$ of entropy must exceed or equal 0. The zero is, then, a lower bound. Is there an upper bound? Equation (7.87) is

$$\left(\frac{dH(t)}{dt}\right)^2 \leqslant I(t) \int d\boldsymbol{r} \, \frac{\boldsymbol{P} \cdot \boldsymbol{P}}{p}, \tag{16.9}$$

where $\boldsymbol{P}$ is the flow vector for the phenomenon. This shows that an upper bound to $dH/dt$ exists, and, is proportional to the square root of the Fisher information for the system. In other words, "entropy shall increase, but not by too much."

Like inequalities of this type follow in other fields as well. In application to electromagnetics, Eq. (7.99) gives the upper bound to the entropy rate for charge density $\rho(\boldsymbol{r}, t)$ in terms of the Fisher information of the charge density distribution. Also, Eq. (7.101) gives a like expression for the entropy of the scalar potential $\phi(\boldsymbol{r}, t)$.

There is a similar entropy bound for gravitational flow phenomena, as discussed.

An upper bound to the entropy rate $dH/dt$ for electron flow is given by (Eq. (7.114))

$$\left(\frac{1}{c}\frac{\partial H}{\partial t}\right)^2 \leqslant I \quad \text{or} \quad c \geqslant \frac{(\partial H/\partial t)}{\sqrt{I}}. \tag{16.10}$$

The speed of light $c$ is an upper bound to how rapidly we can learn relative to how much we already know.

### Chapter 8

$1/f$-type power spectral noise is shown to follow as an equilibrium state of a measurement procedure. The measurements are the complex amplitudes of one or more pure tones (in the language of acoustics).

The $1/f$ answer holds over an infinite range of values ($\kappa$, $N$) of the EPI theory. This accounts for the fact that $1/f$ noise characterizes a wide variety of phenomena.

### Chapter 9

EPI predicts that the logarithms of the fundamental physical constants $x$ obey a uniform probability law. This is equivalent to a $1/x$ law on the constants *per se*. The predicted effect is independent of the choice of units, i.e., whether m.k.s., c.g.s., f.p.s., etc. It is also independent of inversion or combination operations on any number of the constants. The currently known constants *confirm* the $1/x$ law quite well (chi-square $\alpha = 0.95$).

Theoretically, the median value of the physical constants should be 1 (Sec. 9.3.9). Again, this effect is independent of the choice of units. The currently known constants confirm this value fairly well.

An implication of the $1/x$ law is that, *a priori*, small constants are more probable than larger ones. This implies that, given a number $N$ of experimental determinations of any constant, the maximum probable estimate $x_{\text{MP}}$ of the constant is *not* merely the arithmetic average $\overline{x}$ of the data; it must be corrected *downward* from the average via (Eq. (9.38))

$$x_{\text{MP}} = \tfrac{1}{2}\left(\overline{x} + \sqrt{\overline{x}^2 - 4\sigma^2/N}\right). \tag{16.11}$$

Quantity $\sigma^2$ is the variance of error in one determination.

### Chapter 10

Previous chapters showed how to derive the physical laws that occur at the *input space* to a measuring instrument. Here, we want to modify the approach in order to acquire, as well, the physical law that occurs at the instrument's output, measurement space. A one-dimensional analysis of this kind was given in Sec. 3.8, but was severely limited in scope. Accordingly, a new, covariant approach to quantum measurement is developed. Consistently with the preceding EPI principle (Eqs. (3.16) and (3.18)), this approach treats time on an equal basis with the three space variables, i.e., as a quantity that is only imperfectly (statistically) known.

How are observed data values accomodated by the theory? Given our premise that the observer is part of the phenomenon under measurement, it follows that *observation* of a data value affects the *physical state* of the overall observer–phenomenon system (Sec. 10.10). This suggests that we can accomodate real data values in the overall EPI approach by simply adding Lagrange constraint terms to the EPI variational principle Eq. (3.16), two for each data value (Eq. (10.13)). The effect of each datum is to re-initialize the EPI output wave equation.

On this basis, a Klein–Gordon equation with data constraints is found (Eq. (10.14)) as well as a Schrödinger Eq. (10.17). Solving the latter at the particular time of data collection gives rise to the Von Neumann separation effect, Eq. (10.22).

The analysis defines, as well, the sequence of physical events that define EPI as a physical process (Sec. 10.10). The sequence defines an "epistemactive" process. That is, a question is asked (what is the value of $\boldsymbol{\theta}$?), and the response is not only a measured value of $\boldsymbol{\theta}$, but also the mechanism for determining the measured value. This mechanism is the probability law from which the measured value is randomly sampled.

A system that loses coherence is describable by a density function $\rho(x, x')$ (Eq. (10.41)). The information capacity $I$ may be directly related to $\rho(x, x')$ by Eq. (10.43) in a case $P_i = 1/n = const.$ of maximum ignorance of system state.

## *Chapter 11*

Here we describe some problems that are currently being attacked using EPI. One of these is quantum gravity. We unite quantum mechanical and gravitational effects, through the activity of measuring the metric tensor over all of four-space. The derivation is completely analogous to that of the Klein–Gordon equation of Chapter 4.

Thus, we postulate a momentum space that is the Fourier conjugate (Eq. (11.18)) to measurement space. The existence of such a unitary space implies that EPI rigorously holds for this measurement scenario (see Sec. 11.2.12). The result is the Wheeler–DeWitt equation of quantum gravity (Eq. (11.32)), for a pure radiation universe.

The point of departure of this derivation from that of the Klein–Gordon equation in Chapter 4 is replacement of the ordinary derivatives and integrals of the Klein–Gordon derivation by *functional* derivatives and integrals.

Interestingly, the latter are naturally in the form of Feynman path integrals. Past derivations of the Wheeler–DeWitt equation have had to *postulate* the

Feynman approach. A generalization of the approach is suggested which will result in a *vector* wave equation of gravity.

The turbulence problem is being attacked using a quasi-EPI approach. Probability laws on fluid density and velocity have been found for a situation of low velocity flow of a nearly incompressible medium. The acquired laws agree well with published data arising out of detailed simulations of such turbulence.

The EPI approach has not yet been applied to particle physics in the large. Instead, it has been applied to selected topics on the subject.

The *Weinberg angle* $\theta_W$ is a key parameter of the Higgs mass phenomenon. The standard model of elementary particles does not provide an analytical formula for it. We find its empirical value to obey the relation (11.54a),

$$\tan^2 \theta_W = \tfrac{1}{2}(\gamma - 1) = \tfrac{1}{4}(\sqrt{5} - 1) = 0.309\,02 \tag{16.12}$$

to an accuracy of 0.3°. Parameter $\gamma$ is the Fibonacci "golden mean," and Eq. (15.45) was used to form the second equality. The formula is suggested by some analogous properties that mass growth and biomass growth are known to share.

The Weinberg angle is one characteristic constant of electroweak interactions. Another is the *Cabibbo* angle $\theta_C$. As with the Weinberg angle, the standard model of elementary particles does not provide an analytical formula for the Cabibbo angle. We find its empirical value to obey the relation

$$\tan^2(4\theta_C) = \gamma = 1.618\,034, \tag{16.13}$$

again to an accuracy of 0.3° (as with the formula (16.12) for the Weinberg angle).

Eliminating $\gamma$ between the two Eqs. (16.12) and (16.13) relates the two angles as

$$\tan^2(4\theta_C) = 1 + 2\tan^2 \theta_W. \tag{16.14}$$

The standard model does not predict any analytical relation between the two angles.

In Sec. 11.4.2 it is shown that the EPI zero-condition predicts the existence of *bound* quark–anti quark pairs called "mesons." A known property of "asymptotic freedom" of the quarks is utilized, i.e., the independent behavior of the quarks at small separations. The EPI measurement is of the space-time of a composite particle composed of a quark and an anti-quark – a meson.

In Sec. 11.5 it is shown that the output probability amplitude functions $\psi$ of EPI are either *ordinary* or *operator* functions, depending upon their ranks. The operator functions can subsequently be used to form a second-quantized theory.

### Chapter 12

The EPR–Bohm effect is historically the paradigm that demonstrates the phenomenon of *entangled realities*. In the experiment, a "mother" particle that is a diatomic molecule decays into a pair of identical spin-$1/2$ particles called (1) and (2) that head off in opposite directions. Let particle (1) have spin value $S_a$ and (2) have spin $S_b$. Qualitatively, the phenomenon of entangled realities is that an observer of particle (1)'s spin $S_a$ is able with high probability to predict the spin $S_b$ of particle (2) without observing it. Quantitatively, the effect is described by probabilities $P(S_a S_b|x)$ of the four possible spin pair values $S_a S_b = (++), (--), (+-), (-+)$ that obey Eq. (12.2),

$$P(+\,+|x) = P(-\,-|x) = \tfrac{1}{2}\sin^2(x/2), \qquad (16.15)$$

$$P(+\,-|x) = P(-\,+|x) = \tfrac{1}{2}\cos^2(x/2).$$

Here $+$ means "up" and $-$ means "down," and $x$ is the general angle between the Stern–Gerlach planes of detection of the two spins. It is shown that Eqs. (16.15) can be derived using EPI. The general approach is to first find the probability law $P(S_a|S_b, x)$ using EPI, then use the latter probability in Eq. (12.20) to obtain the required law $P(S_a S_b|x)$.

The EPI problem is that of estimating the angle $x$. The datum is the one particle spin value $S_a$, the other spin $S_b$ remaining unobserved (conditional). The likelihood law of the scenario is then $P(S_a|S_b, x)$, and as usual this is the law that the EPI approach attempts to compute.

This probability law is found as an EPI self-consistent solution (Sec. 12.3.5). All unitless constants are found by use of the game corollary (Sec. 12.4) and an orthogonality condition (Sec. 12.5). Then the required law $P(S_a S_b|x)$ is found as described above.

The Fisher information capacity $I_1$ at solution is found (Eq. (12.31b)). From this the Cramer–Rao inequality is used to compute the minimum possible uncertainty in the estimated angle value $x$. This is the value 0.28 rad (Eq. (12.32)).

Entanglement enters in the assumption that the bound information for the observed particle *equals* the Fisher information in the *un*observed particle (Eq. (12.9)). The degree of entanglement is measured by the size of the bound information $J$ relative to that of $I$. Equality of the two means a total equivalence of the spaces of the two particles on the level of information: entangled realities. Although the entanglement is generally statistical and not deterministic, it becomes fully deterministic, with complete predictability of spin $S_b$, when the angle $x = 0$ or $\pi$ (Sec. 12.2).

Information $J$ is also found to obey many of the requirements of the enigmatic "active information" postulated by D. Bohm.

The polarization experiment that is the optical analog to the particle-based EPR–Bohm experiment is analyzed in the same way, and with similar results.

## *Chapter 13*

We simultaneously attack the problems of finding PDFs describing economic valuations and classical statistical mechanics. Prior knowledge in the form of *macroscopic data constraints* is assumed. The activity of purchasing a security falls within the measurement framework (Chapter 1) of EPI. The purchase price $y$ may be viewed as the sampling or "observation" of a valuation datum $y$ in the presence of an unknown ideal valuation $\theta$. The difference $y - \theta \equiv x$ is the error in valuation. The unknown PDF $p(x)$ is perturbed by the purchase and all others that are made during the purchase time interval $(t, t + dt)$. This perturbs the corresponding informations $I$ and $J$, causing the EPI variational principle to execute, and enabling its PDF $p(x)$ to be computed.

Ignoring microscopic knowledge in mechanics, or correspondingly, ignoring "fundamentals" in economic valuations, is equivalent to having zero source information $J$ (Eq. (13.4)). It results that the EPI output $p(x)$ for the problem is an approximation. The EPI problem then becomes effectively that of minimum Fisher information (MFI) with added Lagrange constraints, Eq. (13.5). A benefit is that this problem is relatively easy to solve. Its solution is the well-explored differential Eq. (13.9),

$$q''(x) = \frac{q(x)}{4} \sum_{m=1}^{M} \lambda_m f_m(x), \qquad p(x) \equiv q^2(x), \tag{16.16}$$

which has the form of a SWE. Here $q(x)$ is the unknown probability amplitude on the error $x$ in valuation, the $\lambda_m$ are undetermined Lagrange multipliers, and the $f_m(x)$ are known constraint kernel functions.

Using the same data, corresponding PDF solutions from the maximum entropy principle Eq. (13.6) are also formed. These allow comparisons to be made. The maximum entropy solution has the well-known exponential solution form Eqs. (13.13a, b). The EPI solution to the differential Eq. (16.16) is formed using the Numerov method (Appendix K).

The two alternative approaches to finding PDFs are used to find the PDFs of (1) the term structure of interest rates; (2) a perpetual annuity; and (3) option prices. The EPI solutions are visibly smoother than corresponding maximum entropy solutions, and quantitatively smoother by most measures. Also, in example (3) the EPI solution using but three data constraints well approximates the skewed nature of the log-normal valuation law of the Black–Scholes model. Thus, three (or more) data are "strong" enough to permit good reconstruction of the underlying PDF despite ignoring important microscale information.

The EPI approach also provides a natural measure of the volatility in a security: By the Cramer–Rao inequality, this is the minimized mean-square error of valuation $1/I$. Here $I$ is formed as $I[p(x)]$ out of the EPI solution $p(x)$.

## *Chapter 14*

Both physical transport and biological growth effects describe populations $p_n = p_n(t)$ of neutrons, molecules, lasing atoms, DNA chains, polymer chains, ecosystems, etc. These change in time $t$ according to generic Lotka–Volterra Eqs. (14.10),

$$\dot{p}_n = p_n(g_n + d_n), \quad n = 1, \ldots, N. \tag{16.17}$$

The $g_n \equiv g_n(\mathbf{p}, t)$ and $d_n \equiv d_n(\mathbf{p}, t)$ are called "growth" and "depletion" coefficients, respectively.

This chapter uses EPI to derive these equations. Thus, whether the observer looks outward at physical or biological phenomena, or inward at the growth of his own cells, he finds that they obey EPI. A sentient being exists in a kind of aether of Fisher information flow that governs all internal and external growth processes he can observe.

In Sec. 14.1.3, we briefly survey some classical growth laws of biology, atomic physics, genetics, and molecular growth, including RNA growth. In Sec. 14.2 we show that these are special cases of a *general* Lotka–Volterra growth Eq. (14.10). The equation is derived in Sec. 14.10, using EPI in conjunction with a general invariance principle (14.13).

Taking the continuous limit (14.7) of the EPI output law (14.52) in space-velocity phase space, and using change coefficients of type (14.31), gives the Boltzmann transport Eq. (14.8).

Perhaps the most important aspect of the derivation is that its output, the general growth law (14.52), is a first-order – *and not higher* – differential equation. Furthermore, the output growth law (14.52) is an *exact* result. It is not, for example, based upon a Taylor expansion to first order of the hydrodynamical equations, as was originally used (Lotka, 1956) to derive the Lotka–Volterra equations.

The first-order nature of the output law follows, *mathematically*, from the fact that it is *discrete* in its unknowns $z_n$. This discreteness allows these to be found by merely taking the *first* derivative $dK/dz_n$ of Eq. (14.37), rather than by forming the second-derivative term $d/dt\,(\partial K/\partial z_n)$ that would be required in the Euler–Lagrange solution (0.13) to the corresponding continuous problem.

Choices of the change coefficients other than those given in Table 14.1 give rise to *other* transport equations as well. Among these are the equations of hydrodynamics, of shock waves, and the master equation of macroeconomics. Hence, these equivalently derive from EPI as well. Likewise, the equations characterizing chaos are of our generic growth form (14.10). See Sec. 14.12.

The EPI derivation gives rise to a new general *uncertainty principle* Eq. (14.67) that applies to classical (Lotka–Volterra type) growth. Its expression in *genetics* is the uncertainty principle (14.71),

$$t^2 \langle (w - \langle w \rangle)^2 \rangle \geqslant 1/r^2. \tag{16.18}$$

The indicated averages are over all mating types $nk$. Equation (16.18) states that, for a fixed relative error $r$ in knowledge of the age $t$ of a species type $nk$, the longer $t$ the species has existed the smaller can be the mean-square variability $\langle (w - \langle w \rangle)^2 \rangle$ of genotype traits in its offspring. This effect has been independently verified in long-term Lotka–Volterra population solutions (Vincent, 2004), which show a *progressive narrowing of traits* with time $t$ in equilibrium populations. See Fig. 14.1. An important result of such an overspecialization of traits is that the species becomes ever more probable of becoming extinct at the next environmental challenge. This effect may have contributed to the famous dinosaur extinction of about $6 \times 10^7$ years ago (Sec. 14.15).

The derivation in Sec. 14.10 is based upon use of the classical information efficiency value $\kappa \equiv I/J = 1/2$. In these cases one-half of the pre-existing Fisher information in the phenomenon is lost *en route* to data space.

The question of why entangled realities are not observed *macroscopically* is discussed in Sec. 14.16. The basic reason is that, whereas on the *quantum* level $I = J$, so that the information in the observed datum *equals* its value at its source (entanglement), on the macroscopic level $I < J$, so that information is lost (non-entanglement) in transit from the source to the observing device. The information loss ultimately traces from a loss of coherence, i.e., *decoherence* (Sec. 10.10), in macroscopic systems.

The decoherence of a macroscopic system follows from the fact that, while an object in such a system is being observed and perturbed by a probe particle (Chaps. 3 and 4), it is also being perturbed by *outside influences*, such as an incoming flow of radiation, and/or *internal interactions* of the various populations. Although the EPI measurement process is still activated by the perturbations, and Fisher information flow still occurs (evidently, since data are obtained), the data perturbations from the object source are overwhelmed by those due to the outside influences and internal interactions. The result is that the data fluctuations are no longer fully coherent with those of the object, and information is lost.

We found that both *organic* strands of RNA molecules and *physical* (non-living) molecules obey *a common* growth law (14.52). Is this merely a coincidence? No, the commonality of the two growth processes is taken to suggest that genesis occurred as a template matching process. Here, atoms of RNA, through bonding to corresponding atoms of a neighboring physical substance called TNA, "learned" to grow, i.e., to reproduce, in one-to-one imitation of the growth of the TNA. This idea is discussed in Sec. 14.17.

## *Chapter 15*

Both according to EPI *theory* and according to clinical *data*, *in situ* cancer growth obeys a power law

$$p(t) = \left(\frac{1+\gamma}{T}\right)\left(\frac{t}{T}\right)^{\gamma}, \quad \gamma = \frac{1}{2}(1+\sqrt{5}) = 1.618\,034\ldots \tag{16.19}$$

is the Fibonacci constant (or "golden mean"). Parameter $T$ is the total time over which cancers are observed. The cancer mass at time $t$ is proportional to the PDF (16.19).

The EPI measurement is the age (time $\tau$) of a patient at the time of detection of a cancer (Sec. 15.4) by a clinician. The clinician is the exterior observer of this measurement scenario. The ideal parameter value $\theta$ is the onset time of the disease. At the instant of detection by the outside observer, a mass of cancer cells (at least 4.27; see Sec. 15.8) sends biological messengers to a nearby healthy cell. *Protons* of lactic acid are the dominant messengers. The lactic acid bathes the healthy cell in a toxic wash that eventually kills it. Thus, the message conveyed to the cell-observer is "die." Fortunately, the observer-clinician does not also receive the message: his existence is not entangled with that of the phenomenon (Sec. 15.4).

The EPI solution (16.19) for the probability law (or mass law) $p(t)$ is found in Sec. 15.6 as a self-consistent solution. The particular power $\gamma$ in (16.19) follows from the hypothesis that a cancer cell secretes to its neighbors an absolute *minimum level of information* about its age (Sec. 15.7). The minimal level has two major ramifications: (1) that a cancer cell cannot function, and (2) that the cell grows uncontrollably. These points are discussed next.

(1) Fisher information measures both the degree of order and the complexity of a system (Sec. 2.4.1). Therefore, a minimal level of information, as in the previous

paragraph, implies a minimally complex system. Now, a healthy stem cell produces by mitosis a cell daughter-pair that both reproduces and functions. This dual property requires a higher level of cell complexity, and hence information, than if only one of the two properties were obeyed. But, the daughter pairs in a vital organ of a living individual must, at least, reproduce, if the individual is to continue living. Therefore, reproduction must require a lower complexity level than function. It follows that the *minimum* level of complexity attained above is obeyed by a cell daughter-pair that has lost its ability to *function*. This is consistent with the observation that cancer cells cannot function; they only reproduce.

The medical term for this kind of a daughter cell is an "undifferentiated" one, meaning a cell that is unable to carry out its intended function.

(2) The *uncontrolled* nature of the growth of a cancer might result, as well, from the minimum level of temporal (growth) information that it secretes. Consider an *initially healthy* cell. Its growth is partially controlled by outside cells (Sec. 15.3). Such control is presumably dependent upon two-way communication between the healthy cell and the outside cells, and hence, upon the level of information $I$ that the healthy cell secretes about its state of growth. This implies that, if $I$ falls below a critical level, the communication is disrupted, and *the growth is no longer controlled*. Cancer is usually the result of thousands of random mutations of the genes (Sec. 15.2). These mutations *degrade the information* level $I$ such that it falls below the critical level, defining an *information catastrophe*. Thus the initially healthy cell suffers uncontrolled growth, becoming a cancer.
Equation (15.43) provides further grounds for, and quantifies, this effect. It shows that the information $I$ received by the functioning cell *decreases monotonically* with the total time of observation $T$,

$$I = \frac{11 + 5\sqrt{5}}{2T^2}. \tag{16.20}$$

This equates to lower and lower levels of information being secreted by the neighboring cancer cells as $T$ increases. Then, by the preceding paragraph, relatively more and more cases of uncontrolled cancer growth must occur in the nearby cell masses. Such an acceleration of uncontrolled cancer growth with observation period is confirmed by clinical data.

The power-law dependence of the growth in (16.19) is notable in being considerably slower than the exponential growth that occurs under *in vitro*, or ideal, growth conditions. This indicates that the human body offers a natural degree of resistance to the growth of cancer. Some of this is in the form of local depletion effects. Thus, a cancer must invest energy and information in growing peripheral arterioles to it so as to bring in the extra supplies of blood it needs to sustain its abnormally high growth rate. This, of course, takes "time" and acts to slow down the growth from its *in vitro* rate.

The usual EPI uncertainty principle is also determined, with the surprising prediction Eq. (15.44) that the onset time for cancer cannot be known more precisely than with a root-mean-square relative error of 30%. This is a large figure. It has at least one ramification to clinical treatment. The prediction is that, if cancer *size* can be determined with *better than* 30% accuracy, a course of treatment that depends upon knowledge of the size might be superior to one that depends upon knowledge of the age.

The good agreement of the main prediction of the approach – the power law Eq. (16.19) – with experimental data in Fig. 15.3 implies that the premise for its derivation is correct. This is that cancer is fundamentally a *non-deterministic* illness, resulting from multiple (thousands of) genetic mutations of initially healthy cells. The mutations force a form of reverse evolution upon the cells. They regress to a state of lowest-possible complexity for living things – cells that can only grow – and, resultingly, secrete information $I$ at a minimum rate.

A purely *physical* prediction grows out of the biological analysis. The latter shows that the growth of biomass, either under ideal conditions or *in situ* conditions, is parameterized by the Fibonacci constant $\gamma$. Under the hypothesis that $\gamma$ might also parametrize *physical* mass growth in the Higgs mass phenomenon, we can form simple empirical *formulae* for the Weinberg and Cabibbo angles of this phenomenon. See Sec. 11.4. These formulae agree to within 0.3° of current empirical values of the two angles.

Finally, some speculative possibilities for controlling cancer growth also emerge from this analysis.

(1) The degraded nature of the information $I$ about the time $\theta$ might act as a *causative* agent of carcinogenesis. While intuitively this seems unlikely, we do note that there have been studies in which telomere dysfunction has been described as resulting in genetic instability and epithelial cancer formation (Artandi *et al.*, 2000; Tang *et al.*, 1998). Conversely, increasing the cellular information $I$ about the onset time by, for example, treatment with telomerase might reduce the possibility of malignancy in at-risk tissues. In fact this seemingly paradoxical therapy has been suggested elsewhere (DePinho, 2000).

(2) Cancer has been modelled as a communication system whose information carriers, lactic acid ions, are vital to its functioning. The system could be suppressed, then, if the chain of communication were broken by suppressing the carriers such as through inhibition of glycolysis. On the other hand, Gatenby and Gawlinski (1996) have shown that a sufficiently high acid level inhibits tumor growth. The tumors commit suicide (apoptosis). Thus, an alternative strategy is to do the opposite:

*overwhelm the resident cancer with externally applied lactic acid*, decreasing the information content with high noise. First laboratory experiments on rats indicate that this will work if a pH level of 6.8 is maintained for 10 min (Kelley *et al.*, 2002). A possible side effect in humans is significant damage to functioning cells owing to the acid environment, although this might not occur over the requisite short exposure time of 10 min. Further reducing exposure time, isolating vital organs, or finding some optimum combination of dose and exposure time might mitigate this problem.

## *The Last Word*

After all is said and done, this can only be a repetition of the remarks of Professor J. A. Wheeler (1990) that were quoted at the beginning of the book. And so we come full circle:

> All things physical are information-theoretic in origin and this is a participatory universe ... Observer participancy gives rise to information; and information gives rise to physics.

## *Acknowledgements*

This project could not have progressed nearly as far as it has without the aid of many people. I want to thank the following.

*H. Cabezas*, for widening my view to include problems of ecology, in particular the application of Fisher information to the problem of sustainable technology.

*W. J. Cocke*, my co-worker on the connection between EPI and classical gravitation, for many engaging and instructive sessions on covariance, gravitation, and the universe at large.

*B. Ebner*, whose probing questions have been a real stimulus.

*H. A. Ferwerda*, for carefully reading sections of the manuscript and clarifying many issues that arose.

*R. A. Gatenby*, for opening my mind to the possibility of applying EPI to cancer growth phenomena and providing valuable insights on the subject.

*P. W. Hawkes*, for continued support and encouragement of the research. It is much appreciated.

*R. J. Hawkins*, my colleague on problems of economic security valuation, who carried through most of the numerical EPI applications of Chapter 13, and continues to advocate the use of EPI for such problems.

*R. J. Hughes*, my co-worker in research on the $1/f$ law, for many imaginative and stimulating conversations by email on subjects ranging from the fundamental nature of the physical constants to cosmic radiation.

*A. S. Marathay*, for help in electromagnetic theory and other topics.

*B. Nikolov*, my co-worker in establishing links between entropy and Fisher information.

*A. Plastino*, for immeasurable encouragement, and contributions to statistical mechanics and quantum mechanics.

*R. N. Silver*, for encouragement, constructive criticism of the initial theory, and insights on basic quantum theory.

*B. H. Soffer*, my "spiritual" mentor overall, and co-worker in research on the knowledge acquisition game. His encouragement, perceptive remarks, thoughts and observations have also been of immeasurable help to the overall research. He introduced me to the work of J. A. Wheeler and insisted, ever so gently (but correctly), that EPI had a lot in common with Wheeler's views of measurement theory. Many of his ideas have been of fundamental importance, including the notion that unitarity is fundamental to the overall theory, and the idea that EPI is intimately connected with the Szilard–Brillouin model of statistical physics.

*E. M. Wright*, for bringing to my attention many aspects of modern quantum theory, especially its wonderful mysteries and paradoxes, bringing me up to date on the subject of measurement theory, and (hopefully) constraining my enthusiastic forays of theory to stay within physical bounds.

# Appendix A

## Solutions common to entropy and Fisher *I*-extremization

Maximum entropy (ME) and extreme physical information (EPI) are two variational principles for finding an unknown probability amplitude function $q(x)$. Equation (1.23) then gives the PDF. Suppose that both are subject to the same constraints. For what constraints do the two solutions coincide? And what are the coincident solutions?

### *A.1 Coincident solutions*

The two variational problems are as follows. By Eq. (2.19) and (3.16), the EPI approach for a single unknown amplitude $q(x)$ is

$$4\int dx\, q'^2 + \sum_{k=1}^{K} \lambda_k \int dx\, q^\alpha f_k(x) = extrem., \qquad q \equiv q(x). \tag{A1}$$

Information $J$ generally has the form of the above sum for some values of $\alpha$ and $k$. (*Note*: To aid in making comparisons, in this appendix we often refer to the sum $J$ as "constraint terms.")

The constrained ME approach is, by Eq. (13.6),

$$-\int dx\, q^2 \ln(q^2) + \sum_{k=1}^{K} \mu_k \int dx\, q^\alpha f_k(x) = extrem. \tag{A2}$$

We want to know what constraint kernels $f_k(x)$ cause the same solution $q(x)$ to both problems (A1) and (A2). To obtain solutions, we regard (A1) and (A2) as variational problems, and so use the Euler–Lagrange Eq. (0.13)

$$\frac{d}{dx}\left(\frac{\partial \mathscr{L}}{\partial q'}\right) = \frac{\partial \mathscr{L}}{\partial q} \tag{A3}$$

to define the solution. The Lagrangian $\mathscr{L}$ is the total integrand of the integral to be extremized.

First consider the case $\alpha = 2$, where $J$ is then an expectation in Eq. (A1). From the Lagrangian of (A1), solution (A3) to EPI obeys

$$q'' - q \sum_k \lambda_k f_k(x) = 0. \tag{A4}$$

We have renamed the $\lambda_k$ for convenience. (This is permitted since they are, at this point, variables to be fixed by the constraint Eqs. (1.48b).)

Likewise, from the Lagrangian of (A2) the Euler–Lagrange solution to the ME approach obeys

$$q = \exp\left( -\frac{1}{2} + \sum_k \mu_k f_k(x) \right). \tag{A5}$$

The $\mu_k$ were renamed for convenience, as in the preceding.

Since we are seeking a common solution $q(x)$, we want to substitute the solution (A5) into solution (A4). This will produce a requirement on the constraint functions $f_k(x)$. Two differentiations of (A5) give

$$q'' = q\left[ \sum_k \mu_k f_k'' + \left( \sum_k \mu_k f_k' \right)^2 \right]. \tag{A6}$$

Substituting this into (A4) gives the requirement on the $f_k(x)$ that

$$\sum_k \mu_k f_k'' + \left( \sum_k \mu_k f_k' \right)^2 = \sum_k \lambda_k f_k. \tag{A7}$$

There are two main problem areas to consider.

## A.2 Moment constraint information

One possibility is for the constraints to be moments, with

$$f_k(x) = x^k, \quad k = 0, 1, \ldots, K. \tag{A8}$$

(Note that $k = 0$ corresponds to normalization.) What values $K$ allow a common solution?

The general approach is to balance equal powers of $x$ on both sides of Eq. (A7) after substitution (A8). Doing this, we find the following.

For $K = 0$, $\lambda_0 \equiv 0$ with $\mu_0$ any value. This means that there is a normalization constraint for ME, but none for EPI. This violates the premise of the same moments for both approaches and, hence, is a negative result.

For $K = 1$, $\lambda_1 \equiv 0$. Therefore there is a zeroth- and first-moment constraint for ME, but no first moment for EPI. This is again a negative result, since it violates the premise of the same moments in both approaches. It also indicates that the solution to ME with up to a first-moment constraint is the same as that for EPI with just the zeroth-moment constraint. The common solution is an exponential or Boltzmann law; see Sec. 7.3.4.

For $K = 2$, we obtain the requirements

$$\begin{aligned} x^2&: \lambda_2 = 4\mu_2^2, \\ x^1&: \lambda_1 = 4\mu_1\mu_2, \\ x^0&: \lambda_0 = \mu_1^2 + 2\mu_2 \end{aligned} \tag{A9}$$

resulting from the indicated powers of $x$. This is a *positive* result since the relations may be satisfied by generally non-zero values of the $\lambda_i$ and the $\mu_i$. There is now no requirement that a moment of EPI be deleted. Hence, both ME and EPI now allow the three moments as constraints, and with the above relations (A9) satisfied they will give the same answer for $q(x)$.

For cases $K > 2$ the result is that coefficients $\mu_k \equiv 0$, $k > 1 + K/2$. Once again the constraints cannot be the same in the two approaches. What this is saying is that only the above case $K = 2$ gives the same solution for the same moment constraints.

### *A.3 Maxwell–Boltzmann case*

We now observe that the common solution $q(x)$ to the $K = 2$ problem is, by Eq. (A5), a normal law. Therefore its square, the PDF $p(x)$, is normal as well. In statistical mechanics, the PDF on the velocity of particles subject to conditions of zero mean velocity and fixed mean-square velocity (kinetic energy) due to a fixed temperature is a normal law with mean zero, the Gaussian law. These physical "conditions" correspond mathematically to the power-law constraints for the case $K = 2$. Hence, solving for the unknown PDF on velocity by *either* ME or EPI subject to these constraints would produce the same, correct answer.

### *A.4 Constraints imposed by information J*

In EPI, the constraints are physically effected by a functional $J$ (Eq. (3.4)) defining the information that is intrinsic, or "bound," to the phenomenon under measurement (see Secs. 3.1 and 3.3). In all physical applications of EPI, with the exception of the preceding Maxwell–Boltzmann one, $J$ is defined as a *single* term. Hence, we now compare EPI and ME approaches (A1) and (A2) for a case $K = 1$, seeking the form of a common single constraint kernel $f(x)$ that leads to the same solution by the two approaches. This is done as follows.

Equation (A7) in the case $K = 1$ is

$$\mu f'' + \mu^2 f'^2 = \lambda f. \tag{A10}$$

This differential equation has the unique solution (Kamke, 1948)

$$f'^2 = Ce^{-2\mu f} - \frac{\lambda}{2\mu^3}(1 - 2\mu f), \quad C = const. \tag{A11}$$

This cannot be directly solved for $f(x)$, but may be solved for $x(f)$ by taking the square-root of both sides, solving for $dx$, and then integrating (if need be, numerically).

By (A1), this solution corresponds to a "bound" information case

$$J = \int dx\, q^2(x) f(x), \tag{A12}$$

where $f$ is given by the solution to Eq. (A11). We do not know of a scenario that is describable by this specific $J$, although it would be interesting to find it.

### A.5 Constraint case $\alpha = 1$

From Eq. (A4) onwards, a constraint exponent case $\alpha = 2$ was assumed. This corresponds to expectation constraints. However, in certain scenarios (Chaps. 5 and 6) the constraint term $J$ is actually *linear* in $q$, i.e., $\alpha = 1$. We now consider this case. Again address the possibility of obtaining common solutions to the two approaches. The same analysis as the preceding was followed. This shows that (a) regarding the possibility of moment constraint (A8) solutions in common, there are now none; and (b) for a single general constraint kernel $f(x)$ a very specialized integral condition must be satisfied, and this probably does not define a physically realizable $J$ functional.

In summary, there are only two possible scenarios for which EPI and ME give the same solution: where the constraints are moments of $p(x)$ up to the quadratic one, or where $J$ obeys Eqs. (A11) and (A12). The former case corresponds to the Maxwell–Boltzmann velocity dispersion law while the latter corresponds to an, as yet, unknown scenario. This scenario is probably unphysical.

The Maxwell–Boltzmann answer is an equilibrium distribution. Hence, equilibrium statistical mechanics is the common meeting ground of the EPI and ME approaches to estimating PDFs. Next, consider the more general circumstance of non-equilibrium statistics. The solution must obey the Boltzmann transport differential equation, and the solution to this is a general superposition of Hermite–Gauss functions (Rumer and Ryvkin, 1980). In fact, EPI generates these solutions as subsidiary minima in $I$, with the absolute minimum attained by the Maxwell–Boltzmann solution (see Sec. 7.4.8). This is under the constraints of normalization and mean kinetic energy (Sec. 7.4.11).

However, under the same constraint inputs ME gives only the equilibrium, Maxwell–Boltzmann answer (Jaynes, 1957a, 1957b); it fails to produce any of the higher-order Hermite–Gauss solutions. Basically, this is because multiple solutions follow from differential equations, which ME cannot produce; see Sec. 1.3.1. Hence, EPI and ME coincide only at the most elemental level of statistical mechanics, that of equilibrium statistics. Beyond this level ME does not apply.

# Appendix B

## Cramer–Rao inequalities for vector data

Here we consider the case of vector data

$$\mathbf{y} = \boldsymbol{\theta} + \mathbf{x}, \tag{B1}$$

where each vector has $N$ components. Each estimator $\hat{\theta}_i(\mathbf{y})$ is allowed to be a general function of all the data $\mathbf{y}$. Also, it is given to be unbiased,

$$\int d\mathbf{y}\, \hat{\theta}_i(\mathbf{y}) p \equiv \theta_i, \quad i = 1, \ldots, N,$$
$$p \equiv p\,(\mathbf{y}|\boldsymbol{\theta}). \tag{B2}$$

We want to find a relation that defines the mean-square error

$$e_i^2 \equiv \int d\mathbf{y}\, [\hat{\theta}_i(\mathbf{y}) - \theta_i]^2 p, \quad i = 1, \ldots, N \tag{B3}$$

in each estimate in terms of an information quantity of some kind. The derivation is entirely different from that of the scalar result (1.1), but equally simple (Van Trees, 1968).

### *B.1 Fisher information matrix*

First we establish the relation for $e_1^2$. Form an auxiliary vector

$$\mathbf{v} \equiv [(\hat{\theta}_1(\mathbf{y}) - \theta_1)(\partial \ln p/\partial\theta_1) \ldots (\partial \ln p/\partial\theta_N)]^{\mathrm{T}}, \tag{B4}$$

where T denotes the transpose. Next, form the matrix

$$[M] \equiv \langle \mathbf{v}\mathbf{v}^{\mathrm{T}} \rangle, \tag{B5}$$

where $\langle\ \rangle$ denotes an expectation over all data $\mathbf{y}$. By the use of (B4) we get

$$[\mathrm{M}] = \begin{bmatrix} e_1^2 & 1 & 0 & \dots & 0 \\ 1 & F_{11} & F_{12} & \dots & F_{1N} \\ 0 & F_{21} & F_{22} & \dots & F_{2N} \\ \vdots & \vdots & \vdots & \vdots & \vdots \\ 0 & F_{N1} & F_{N2} & \dots & F_{NN} \end{bmatrix}. \tag{B6}$$

Elements $F_{ij}$ obey

$$F_{ij} \equiv \int d\mathbf{y} \, \frac{\partial \ln p}{\partial \theta_i} \frac{\partial \ln p}{\partial \theta_j} \, p. \tag{B7}$$

With $i$ and $j$ ranging independently from 1 to $N$, the $F_{ij}$ form a matrix [F] called the "Fisher information matrix." Notice that [F] is a straightforward generalization of the scalar information quantity (1.9).

In (B6), the elements of [F] derive in straightforward fashion. Also, the (1, 1) element is obviously $e_1^2$ as given. We next derive the elements that are 0 or 1. Consider element

$$M_{12} \equiv \int d\mathbf{y} \, [\hat{\theta}_1(\mathbf{y}) - \theta_1] p \, \partial \ln p / \partial \theta_1. \tag{B8}$$

This is directly

$$M_{12} = M_1 - M_2, \quad M_1 \equiv \int d\mathbf{y} \, \hat{\theta}_1 p \, \partial \ln p / \partial \theta_1, \quad M_2 \equiv \theta_1 \int d\mathbf{y} \, p \, \partial \ln p / \partial \theta_1. \tag{B9}$$

Using the formula for the derivative of a logarithm gives

$$M_1 = \int d\mathbf{y} \, \hat{\theta}_1 p \frac{1}{p} \frac{\partial p}{\partial \theta_1} = \frac{\partial}{\partial \theta_1} \int d\mathbf{y} \, \hat{\theta}_1 p = \frac{\partial \theta_1}{\partial \theta_1} = 1, \tag{B10}$$

by unbiasedness condition (B2). Regarding $M_2$,

$$\int d\mathbf{y} p \, \frac{\partial \ln p}{\partial \theta_1} = \int d\mathbf{y} p \frac{1}{p} \frac{\partial p}{\partial \theta_1} = \frac{\partial}{\partial \theta_1} \int d\mathbf{y} p = \frac{\partial}{\partial \theta_1} 1 = 0. \tag{B11}$$

Hence $M_2 = 0$. Therefore by the first Eq. (B9), $M_{12} = 1$, as it was required to show.

### *B.2 Exercise*

By the same steps, show that $M_{13} = 0$. The proof is easily generalized to $M_{1i} = 0$, $i = 4, \dots, N$. The top row of (B6) is now accounted for; as is the first column, by the symmetry in [M].

### B.3 [F] is diagonal for independent data

Next, consider the special case of independent data $\mathbf{y}$. Then by definition

$$p(\mathbf{y}|\boldsymbol{\theta}) = \prod_i p_i(y_i|\boldsymbol{\theta}) = \prod_i p_i(y_i|\theta_i), \tag{B12}$$

the latter by the fact that, by the form of (B1), for independent data $\mathbf{y}$ a change in $\theta_j$ does not influence in any way $y_i$ for $i \neq j$. Then taking the logarithm and differentiating gives

$$\frac{\partial \ln p}{\partial \theta_i} = \frac{\partial \ln p_i}{\partial \theta_i}. \tag{B13}$$

Then by Eqs. (B12) and (B13) and definition (B7),

$$F_{ij} = \int \prod_k dy_k\, p_k(y_k|\theta_k) \frac{\partial \ln p_i}{\partial \theta_i} \frac{\partial \ln p_j}{\partial \theta_j}. \tag{B14}$$

This evaluates differently according to whether $i \neq j$ or $i = j$. In the former case we get

$$F_{ij} = F_i F_j, \quad F_i \equiv \int dy_i\, p_i \frac{\partial \ln p_i}{\partial \theta_i}, \quad p_i \equiv p_i(y_i|\theta_i), \tag{B15}$$

since all other integrals $dy_k$ integrate out to unity by normalization. But then, by (B15),

$$F_i = \int dy_i \frac{\partial p_i}{\partial \theta_i} = \frac{\partial}{\partial \theta_i} \int dy_i\, p_i = \frac{\partial}{\partial \theta_i} 1 = 0. \tag{B16}$$

Then by the first Eq. (B15), $F_{ij} = 0$, $i \neq j$. This proves the proposition that [F] is diagonal in this case.

The diagonal elements are evaluated by taking $i = j$ in Eq. (B14). This gives

$$F_{ii} = \int dy_i\, p_i \left( \frac{\partial \ln p_i}{\partial \theta_i} \right)^2 \tag{B17}$$

since the remaining integrals $dy_k$ integrate out to unity, by normalization. The information (B17) is precisely of the form (1.9) for one-dimensional data.

### B.4 Cramer–Rao inequalities for general data

We now use the fact that matrix [M], by its definition (B5), is positive-definite. Hence its determinant is greater than or equal to zero. Expanding form (B6) for [M] by cofactors along its top row, we get

$$\det[M] = e_1^2 \det[F] - 1 \cdot \mathrm{cof}[F_{11}] \geqslant 0. \tag{B18}$$

Therefore

$$e_1^2 \geqslant \frac{\text{cof}[F_{11}]}{\det[F]} = [F]_{11}^{-1}, \tag{B19}$$

the (1, 1) element of the inverse matrix to [F]. This is the Cramer–Rao inequality obeyed by error $e_1^2$.

The Cramer–Rao inequality for the general error component $e_i^2$, $i = 1, \ldots, N$ may be derived entirely analogously to the preceding steps (B4)–(B19). The result is

$$e_i^2 \geqslant [\mathrm{F}]_{ii}^{-1}, \tag{B20}$$

the $i$th element along the diagonal of the inverse matrix to [F].

### *B.5 Cramer–Rao inequalities for independent data*

We found in Sec. B.3 that, when the data are independent, the Fisher information matrix [F] is diagonal. Now, the inverse of a diagonal matrix [F] is a diagonal matrix whose elements are the reciprocals of the respective diagonal elements of [F]. Then the general Cramer–Rao inequality (B20) becomes

$$e_i^2 \geqslant 1/F_{ii}, \quad i = 1, \ldots, N, \tag{B21}$$

where $F_{ii}$ is given by Eq. (B17). This verifies Eqs. (2.6) and (2.7).

# Appendix C

## Cramer–Rao inequality for an imaginary parameter

The basis for the EPI approach is the Cramer–Rao inequality (1.1). It is derived in Sec. 1.2.3 for the case of a purely *real* parameter $\theta$ and real data $\mathbf{y}$. However, by definition the rigorous EPI approach is covariant (Sec. 3.5), i.e., based upon the use of four-coordinates. Hence, in rigorous use of EPI (Chaps. 3–7, 10 and 11), one or more parameters are purely *imaginary*,

$$\theta \equiv ia = e^{i\pi/2}a, \qquad a \text{ real}. \tag{C1}$$

Note that $\theta$ now has a non-zero phase, of value $\pi/2$. We address the issue of extending the definitions of the Fisher information and its corresponding mean-squared error to this imaginary case. There will be choice in the matter, and we will choose the one most suitable to use of EPI.

Since the parameter $\theta$ is imaginary, its estimate must be imaginary as well,

$$\hat{\theta} = i\hat{a}, \qquad \hat{a} \text{ real}. \tag{C2}$$

Also, it is convenient (although not essential) to the analysis to regard the "data" $\mathbf{y}$ as imaginary,

$$\mathbf{y} \equiv i\mathbf{w}, \qquad \mathbf{w} \text{ real}. \tag{C3}$$

Here the actual data, i.e., observables, are the real $\mathbf{w}$, not the imaginary $\mathbf{y}$. The estimator function $\hat{\theta}(\mathbf{w})$ depends upon the observables as usual, that is, upon the $\mathbf{w}$.

The arguments of *any* PDF must be real. In particular, the probability law for an imaginary random variable is by definition the probability law for its (real) amplitude. Hence, in our imaginary case, the likelihood law *p(data|parameter)* is identically the probability $p(\mathbf{w}|a)$. This takes the alternative forms

$$p(\mathbf{w}|a) = p(\mathbf{w}|\theta/i) \equiv p \tag{C4}$$

for brevity. Equation (C1) was also used.

## C.1 Analysis

In the corresponding all-real case of Sec. 1.2.3, the concept of Fisher information arose in derivation of an error inequality, specifically the Cramer–Rao inequality Eq. (1.8). The latter directly implies the Heisenberg uncertainty principle (Chapter 4), and also is fundamental to EPI. Therefore we take it as axiomatic that

*Any extension of the concept of Fisher information to the imaginary case should likewise obey a Cramer–Rao error inequality.*

It turns out that there are two different ways of doing this. These are as follows.

*First candidate definition.* The first approach (ultimately *rejected*) is to demand that *analogous steps* to (1.3)–(1.10) of Sec. 1.2.3 yield the new error inequality. Thus, assuming $p$ to be a differentiable function of $\theta$ permits Eq. (1.3) to be differentiated $\partial/\partial\theta$ as before, and the steps giving rise to Eq. (1.7) are repeated. At this point the modulus-square, rather than the ordinary square, is taken. Then use of the complex form of the Schwarz inequality (squares replaced by modulus-squares) gives

$$e_0^2 I_0 \geqslant 1, \tag{C5}$$

where

$$I_0 \equiv \left\langle \left| \frac{\partial \ln p}{\partial \theta} \right|^2 \right\rangle, \qquad e_0^2 \equiv \langle |\hat{\theta}(\mathbf{y}) - \theta|^2 \rangle. \tag{C6}$$

The modulus-square operations in (C6) arose from the Schwarz inequality. The subscripts 0 are added to distinguish this choice of definition from one below.

Equations (C6) show that modulus-square operations *replace* simple squares in the previous definitions (1.9) and (1.10). Thus, unless $\theta$ is real, the new definitions (C6) do not analytically coincide with the original definitions (1.9) and (1.10). One way to accept both sets of definitions is to use (C6) as the general ones, since these become (1.9) and (1.10) in the all-real case.

However, a problem with the definitions (C6) is that they lose vital phase information. In cases where $\theta$ and $\hat{\theta}$ are complex, the phases of the operation $\partial/\partial\theta$ in the expression for $I_0$ and of $\hat{\theta}$ and $\theta$ in the expression for $e_0^2$ are lost in the modulus operations (C6). For example, in our imaginary case (C1) the phase has the value $\pi/2$, and this would be lost. In fact the phase of $\partial/\partial\theta$ is needed in many physical applications. See (3) below. This brings us to the second possible definition.

*Second candidate definition.* The original definitions (1.9) and (1.10) utilize ordinary squaring operations, so they *do not lose* the phase information mentioned above. Hence, on this basis, we are tempted to keep these definitions. However, we have to check that these definitions will satisfy an error inequality, as required at the outset of this appendix. This question is addressed next.

Assume that $\ln p$ is a differentiable function of the imaginary parameter $\theta$. Applying definition (1.9) to this case gives

$$I \equiv \left\langle \left[ \frac{\partial \ln p}{\partial \theta} \right]^2 \right\rangle = \int d\mathbf{w} \left[ \frac{\partial \ln p(\mathbf{w}|\theta/i)}{\partial \theta} \right]^2 p(\mathbf{w}|\theta/i)$$

$$= \int d\mathbf{w} \left[ \frac{\partial \ln p(\mathbf{w}|a)}{i\partial a} \right]^2 p(\mathbf{w}|a) = -\int d\mathbf{w} \left[ \frac{\partial \ln p(\mathbf{w}|a)}{\partial a} \right]^2 p(\mathbf{w}|a). \tag{C7}$$

The second equality is by (C1), (C4) and the definition of the given expectation (to review, the arguments of *any* PDF are always limited to *real* random variables); the third is by (C1) and (C4); and the fourth is by the squaring of the imaginary $i$. Note that the final minus sign originates in the phase $\pi/2$ of $\theta$. That is, the operation $(\partial/\partial\theta)^2$ gives rise to the multiplier $(e^{-i\pi/2})^2 = -1$.

Likewise, the definition (1.10) gives

$$e^2 \equiv \langle [\hat{\theta}(\mathbf{w}) - \theta]^2 \rangle = \int d\mathbf{w}\, [\hat{\theta}(\mathbf{w}) - \theta]^2 p(\mathbf{w}|\theta/i)$$

$$= \int d\mathbf{w}\, [i\hat{a}(\mathbf{w}) - ia]^2 p(\mathbf{w}|a) = -\int d\mathbf{w}\, [\hat{a}(\mathbf{w}) - a]^2 p(\mathbf{w}|a). \tag{C8}$$

The second equality is by definition of the given expectation; the third is by (C1), (C2) and (C4); and the fourth is by the squaring of $i$. As with the minus sign in Eq. (C7), the minus sign in (C8) originates in the phase $\pi/2$ of $\theta$ (see material below Eq. (C7)).

Another way to account for the minus signs in (C7) and (C8) is to note that, by their definitions, both $I$ and $e^2$ are squared distance measures, i.e., $L^2$ norms, and squared imaginary distances are, of course, negative. The central question (see above) is whether the two negative quantities $I$ and $e^2$ can still obey a Cramer–Rao inequality. This is considered next.

Multiplication of Eqs. (C7) and (C8) gives

$$e^2 I = \int d\mathbf{w}\, [\hat{a}(\mathbf{w}) - a]^2 p(\mathbf{w}|a) \int d\mathbf{w}' \left[ \frac{\partial \ln p(\mathbf{w}'|a)}{\partial a} \right]^2 p(\mathbf{w}'|a) \tag{C9}$$

directly. The minus signs have multiplied out. By Eq. (C4) each integral in (C9) is an expectation, so that

$$e^2 I = \langle [\hat{a}(\mathbf{w}) - a]^2 \rangle \left\langle \left[ \frac{\partial \ln p(\mathbf{w}|a)}{\partial a} \right]^2 \right\rangle$$

$$= \langle |\hat{\theta}(\mathbf{y}) - \theta|^2 \rangle \left\langle \left| \frac{\partial \ln p}{\partial \theta} \right|^2 \right\rangle. \tag{C10}$$

The second equality is by Eqs. (C1) and (C2) and the fact that $a$, $\hat{a}$, and $p$ are all real. Then, by (C5) and (C6),

$$e^2 I = e_0^2 I_0 \geqslant 1. \tag{C11}$$

This shows that $I$ and $e^2$ as defined by (C7) and (C8) satisfy a Cramer–Rao inequality

$$e^2 I \geqslant 1, \tag{C12}$$

even in this imaginary case. Therefore *we accept defining* Eqs. (1.9) and (1.10) as generally valid for all physical applications. These definitions have a valuable property, discussed next.

*Role played by negative informations.* As in Eq. (1.28) for a real parameter, let an *efficient* estimator for the imaginary parameter be defined as attaining equality in its error inequality, here the outside inequality (C11). Thus

$$e^2_{\text{min}} = 1/I. \tag{C13}$$

As discussed next, this has important ramifications to derivations where the imaginary parameter $\theta$ is one of *many* Fisher coordinates.

Equation (C13) gives rise to a negative contribution (by (C7)) in the Stam information form Eq. (2.9). Equation (2.9) gives rise to the channel capacity information defined in Eq. (2.10). The negative term is therefore incorporated as well into that information. Where there are many imaginary parameters $\theta$ there are many such negative terms. This channel capacity information is used as the basis for all the derivations of wave equations in this book. Therefore the negative terms contribute to these as well. The totality of these negative terms and other positive ones gives rise to the Lorentz covariant, *d'Alembertian* parts of these wave equations; see Chaps. 4–6. This is a valuable property of the extended definitions (C7) and (C8) of the information and error.

Finally, consider the scalar case $\mathbf{w} = w$ of $N = 1$ datum and shift invariance (1.11), which states that $p(w|a) = p_X(w - a) \equiv p(w - a)$. Use of these in (C7) gives, as in Sec. 1.2.4, a Fisher information

$$I = -\int dx\, \frac{p'^2(x)}{p(x)} = -4\int dx\, q'^2(x), \quad x \equiv w - a, \quad p \equiv q^2. \tag{C14}$$

This "$q$-form" of the information is the counterpart to Eq. (1.24) for a corresponding real parameter. Negative informations of the form (C14) are used in the d'Alembertian derivations mentioned previously, as well as for the *single* Fisher variable case Eq. (7.9) in deriving the Boltzmann energy law.

# Appendix D

## EPI derivations of Schrödinger wave equation, Newtonian mechanics, and classical virial theorem

EPI is by construction a relativistically covariant theory (Sec. 3.5). Thus, *non*-covariant uses of EPI, specifically with Fisher coordinates whose dimension is less than four, give *approximate* answers. However, these are often easier to obtain than by a fully covariant EPI approach. We use EPI in this manner to derive the following effects: the one-dimensional, stationary Schrödinger wave equation, Newton's second law of classical mechanics, and the classical virial theorem. All are recognizably approximations in one sense or another.

### D.1 Schrödinger wave equation

Here, a one-dimensional analysis is given. The derivation runs parallel to the fully covariant EPI derivation in Chap. 4 of the Klein–Gordon equation. We point out corresponding results as they occur.

The position $\theta$ of a particle of mass $m$ is measured as a value $y = \theta + x$ (see Eq. (2.1)), where $x$ is a random excursion whose probability amplitude law $\mathbf{q}(x)$ is sought. Since the time $t$ is ignored, we are in effect seeking a stationary solution to the problem. Notice that the one-dimensional nature of $x$ violates the premise of covariant coordinates as made in Sec. 4.1.2.

Since the approach is no longer covariant, it is being used improperly. However, a benefit of EPI is that it gives approximate answers when used in a projection of four-space. The approximate answer will be the *non-relativistic*, Schrödinger wave equation.

Assume that the particle is moving in a conservative field of scalar potential $V(x)$. Then the total energy $W$ is conserved. This is assumed in the derivation below.

Again using definition (2.24) to define new, complex wave functions $\psi_n(x)$, the information expression (2.19) now becomes Eq. (3.38), which in one dimension is

$$I = 4N \sum_{n=1}^{N/2} \int dx \left| \frac{d\psi_n(x)}{dx} \right|^2. \tag{D1}$$

As in Eq. (4.4), we define a Fourier transform space consisting of functions $\phi_n(\mu)$ of momentum $\mu$ obeying

$$\psi_n(x) = \frac{1}{\sqrt{2\pi\hbar}} \int d\mu\, \phi_n(\mu) \exp(i\mu x/\hbar). \tag{D2}$$

The unitary nature of this transformation and the physical reality of $\mu$ as momentum guarantees the validity of the EPI variational procedure (Sec. 3.8.7).

Since $\psi$ and $\phi$ are Fourier mates, we may use Parseval's theorem to get

$$I = \frac{4N}{\hbar^2} \int d\mu\, \mu^2 \sum_n |\phi_n(\mu)|^2 \equiv J. \tag{D3}$$

This corresponds to Eq. (4.9), and is the invariance principle for the given measurement problem.

Equation (2.25) defines the PDF $p(x)$. Using its normalization and Eq. (D2) shows that the sum in (D3) is a probability density $P(\mu)$ on $\mu$. Then (D3) is actually an expectation

$$I \equiv J = \frac{4N}{\hbar^2} \langle \mu^2 \rangle \tag{D4}$$

(cf. Eq. (4.13)). Now we use the specifically non-relativistic approximation that the kinetic energy $E_{\text{kin}}$ of the particle is $\mu^2/(2m)$. Then

$$J = \frac{8Nm}{\hbar^2} \langle E_{\text{kin}} \rangle = \frac{8Nm}{\hbar^2} \langle [W - V(x)] \rangle. \tag{D5}$$

This may be re-expressed in $x$-coordinate space as

$$J = \frac{8Nm}{\hbar^2} \int dx\, [W - V(x)] \sum_n |\psi_n(x)|^2 \tag{D6}$$

since the latter sum is the PDF $p(x)$ by Eq. (2.25). Equation (D6) is the bound information functional $J[\mathbf{q}] \equiv J[\psi]$ for the problem. By Eqs. (D3) and (3.18), parameter $\kappa = 1$ for this problem.

Use of Eqs. (D1) and (D6) in definition (3.16) of $K$ gives a variational problem

$$K = N \sum_{n=1}^{N/2} \int dx \left[ 4 \left| \frac{d\psi_n(x)}{dx} \right|^2 - \frac{8m}{\hbar^2} [W - V(x)] |\psi_n(x)|^2 \right] = \textit{extrem}. \tag{D7}$$

The Euler–Lagrange equation for the problem is

$$\frac{d}{dx}\left( \frac{\partial \mathscr{L}}{\partial \psi^*_{nx}} \right) = \frac{\partial \mathscr{L}}{\partial \psi^*_n}, \qquad n = 1, \ldots, N/2, \qquad \psi^*_{nx} \equiv \partial \psi^*_n / \partial x. \tag{D8}$$

Note that, during the variation of the Lagrangian (integrand) $\mathscr{L}$ of Eq. (D7), the value of $W$ remains constant, i.e., obeys $\delta W = 0$. In fact, this is the basis for a related variational approach, that of successively approximating $\psi$ subject to required boundary conditions (Morse and Feshbach, 1953, Part II, p. 1114).

Using the integrand of (D7) as the Lagrangian $\mathscr{L}$ in (D8) gives a solution

$$\psi_n''(x) + \frac{2m}{\hbar^2}[W - V(x)]\psi_n(x) = 0, \quad n = 1, \ldots, N/2. \quad \text{(D9)}$$

This is the Schrödinger wave equation without the time, also called the "Schrödinger energy eigenvalue equation."

We see from (D4) that $I = min.$ in $N$ for a choice $N = 2$ (the term $n = 1$ in (D7)). Therefore, by the game corollary, the scenario admits of $N = 2$ degrees of freedom $q_n(x)$ or, by Eq. (2.24), one complex degree of freedom $\psi(x)$. This is the usual result that the SWE defines a single complex wave function.

A noteworthy aspect of this derivation is that it works with a purely real Fisher coordinate $x$. This implies that the information transfer game of Sec. 3.4.12 is played here. The payoff of the game is the Schrödinger wave equation.

## D.2 Newton's second law from EPI extremum principle

We found previously that $N = 2$, so that Eq. (D4) gives

$$I = \frac{8}{\hbar^2}\langle\mu^2\rangle. \quad \text{(D10)}$$

Since Newton's law is in the domain of classical physics, we next consider a classical limit of Eq. (D10).

Consider a classical, non-relativistic particle of mass $m$ that has a definite trajectory $x(t)$ defining position $x$ at time $t$ over a total time interval $(-T, T)$. Its linear momentum (Goldstein, 1950) obeys $\mu = \mu(t) = m\dot{x}(t)$, the dot denoting a time derivative.

The average in (D10) is explicitly over all momentum values. Let the particle's momentum values obey *ergodicity*, so that the average is equivalently over *all time*,

$$I = \frac{8}{\hbar^2}\langle m^2\dot{x}^2(t)\rangle = \frac{8m^2}{\hbar^2}\lim_{T\to\infty}\int_{-T}^{T} dt\, p(t)\dot{x}^2(t). \quad \text{(D11)}$$

Here $p(t)$ is the density function ruling time values. Regard time in the usual classical sense. That is, each time interval $(t, t + dt)$ occurs once and only once, so that these are *uniformly* dense over the total time interval, i.e.,

$$p(t) = (2T)^{-1} \quad \text{for} \quad -T \leqslant t \leqslant T, \quad \text{or 0 for other } t. \quad \text{(D12)}$$

Using this in (D11) gives

$$I = \frac{8m^2}{\hbar^2}\langle\dot{x}^2(t)\rangle = \frac{8m^2}{\hbar^2}\lim_{T\to\infty}\frac{1}{2T}\int_{-T}^{T} dt\,\dot{x}^2(t). \quad \text{(D13)}$$

A well-known classical limit of quantum mechanics is obtained by letting $\hbar \to 0$. Parameter $T$ is already approaching infinity in (D13). To obtain a definite limit $\hbar^2 T$ in (D13), let

$$\hbar \to 0 \quad \text{and} \quad T \to \infty, \quad \text{such that } \hbar^2 T = A^2, \quad \text{(D14)}$$

a finite constant. Then (D13) becomes

$$I = \frac{4m^2}{A^2}\int_{-T}^{T} dt\,\dot{x}^2(t). \tag{D15}$$

(All integration limits are regarded as large but finite unless an explicit limit is indicated.) With $I$ known, we now seek the functional $J$.

The quantum answer to the particle measurement experiment was the *temporally stationary* Eq. (D9). This suggests that the information source for determining particle positions does not explicitly depend upon the time. Rather, the dependence upon time is *implicit*, through $x(t)$. Hence, represent the bound information as

$$J \equiv \frac{8m}{A^2}\int_{-T}^{T} dt\, j(x), \qquad x \equiv x(t) \tag{D16}$$

with $j(x)$ some unknown information density. Then, by (D15) and (D16), the EPI extremum condition (3.16) is

$$K \equiv I - J = \frac{4m^2}{A^2}\int_{-T}^{T} dt\left[\dot{x}^2 - \frac{2j(x)}{m}\right] = extrem. \tag{D17}$$

The extremum is through variation of the trajectory $x(t)$. With the Lagrangian the integrand of (D17), the single-coordinate Euler–Lagrange Eq. (0.27) is here

$$\frac{d}{dt}\left(\frac{\partial \mathscr{L}}{\partial \dot{x}}\right) = \frac{\partial \mathscr{L}}{\partial x}, \qquad \text{with} \quad \mathscr{L} = \dot{x}^2 - \frac{2j(x)}{m}. \tag{D18}$$

Combining these gives directly the equation of motion for the particle,

$$m\ddot{x} = -\frac{\partial j(x)}{\partial x} = -\frac{dj(x)}{dx}, \tag{D19}$$

the latter since $j$ is not an explicit function of $t$. This is *Newton's second law*

$$m\ddot{x} = -\frac{dV(x)}{dx} \tag{D20}$$

for a scalar potential function $V(x)$ defined as

$$j(x) \equiv V(x) + C, \qquad x \equiv x(t). \tag{D21}$$

The constant $C$ is to be determined (see next section).

The identification (D21) once again indicates (as in Sec. 3.7.3) that EPI views potential functions as information densities (up to an additive constant).

## D.3 Virial theorem from EPI zero principle

The derivations in this book are consistent with the hypothesis that, in any measurement scenario, *both* the extremum condition and the zero-condition of EPI – either distinctly or in tandem – always give rise to valid physical laws. We previously showed that use of the EPI *extremum* condition (D17) in a classical limit (D14) gives

Newton's second law (D20). What law, then, does the EPI *zero*-condition give in the same classical limit?

Use of the zero principle requires a definite value for the information efficiency $\kappa$. This was already fixed as value

$$\kappa = 1. \tag{D22}$$

See below Eq. (D6). Although $\kappa = 1/2$ is normally used to define classical limits in EPI problems, the required classical limit has already been enforced by the special limit (D14). Equation (D22) stipulates that the classical limit is of a specifically quantum problem.

We next form the functionals $I$ and $J$. By Eqs. (D15), (D16), and (D21),

$$I = \frac{8m^2 T}{A^2} \frac{1}{2T} \int_{-T}^{T} dt\, \dot{x}^2(t) \tag{D23a}$$

and

$$J = \frac{16mT}{A^2} \frac{1}{2T} \int_{-T}^{T} dt [V(x) + C]. \tag{D23b}$$

Both equations were multiplied and divided through by factor $2T$.

Equation (D23a) is evaluated as follows. Assume the usual conditions (Goldstein, 1950) for validity of the virial theorem: a particle motion that is either periodic or with positions $x$ and speed $\dot{x}$ that remain *finite* at all times within $(-T, T)$. Then, with $T$ very large, necessarily

$$\frac{x(t)\dot{x}(t)|_{-T}^{T}}{T} \to 0. \tag{D24a}$$

Using this during integration by parts of the integral (D23a) gives

$$\frac{1}{2T} \int_{-T}^{T} dt\, \dot{x}^2(t) = \frac{x(t)\dot{x}(t)|_{-T}^{T}}{2T} - \frac{1}{2T} \int_{-T}^{T} dt\, x(t)\ddot{x}(t) = \frac{1}{2T} \int_{-T}^{T} dt\, x(t) \frac{1}{m} \frac{dV(x)}{dx} \tag{D24b}$$

after the use of Eq. (D20). Using (D24b) in (D23a) gives

$$I = \frac{8m^2 T}{A^2} \frac{1}{2T} \int_{-T}^{T} dt\, x(t) \frac{1}{m} \frac{dV(x)}{dx}. \tag{D24c}$$

Equation (D23b) is directly an expectation,

$$J = \frac{16mT}{A^2} \langle V(x) + C \rangle, \quad \text{where} \quad x \equiv x(t), \tag{D25}$$

by the definition (D13) of the time average.

Next we need to evaluate the constant $C$. By Eq. (D21), $C$ is a constant with the unit of energy. The only such constants of the scenario are $\langle V(x) \rangle$, $\langle E_{\text{kin}}(x) \rangle$, and the conserved total energy $W$. Of course the three are related by conservation of mean energy, $\langle V(x)\rangle + \langle E_{kin}(x)\rangle = W$, so that only two are *a priori independent* quantities. Choosing a particular pair, $C$ is then some superposition

$$C = a\langle V(x)\rangle + b\langle E_{\text{kin}}(x)\rangle, \tag{D26}$$

the constants $a$, $b$ to be found.

Substituting (D26) into (D25) yields

$$J = \frac{16mT}{A^2}\langle V(x) + a\langle V(x)\rangle + b\langle E_{\text{kin}}(x)\rangle\rangle. \tag{D27}$$

The constants $a$, $b$ are fixed by the requirement that taking the classical limit (D14) should not change the level $J$ of the source information. Thus, classical information (D27) must equal its corresponding quantum mechanical value (D5). Equating the two $J$ values,

$$\frac{16mT}{A^2}\langle V(x) + a\langle V(x)\rangle + b\langle E_{\text{kin}}(x)\rangle\rangle \equiv \frac{16m}{\hbar^2}\langle E_{\text{kin}}(x)\rangle \tag{D28}$$

for this $N = 2$ degrees-of-freedom problem. Since, as we discussed, $\langle V(x)\rangle$ and $\langle E_{\text{kin}}(x)\rangle$ are *independent* quantities, in general the two sides cannot be equal unless the left-hand terms in $\langle V(x)\rangle$ drop out. Since identically $\langle V(x) + a\langle V(x)\rangle\rangle = (a+1)\langle V(x)\rangle$, this requires

$$a = -1. \tag{D29}$$

Using this in (D28) gives the required cancellation, and a requirement on $b$ of

$$\frac{16mTb}{A^2} = \frac{16m}{\hbar^2}. \tag{D30}$$

Then using definition (D14) of $A^2$ in (D30) gives

$$b = 1. \tag{D31}$$

In summary, by Eqs. (D26), (D29), and (D31),

$$C = \langle E_{\text{kin}}(x)\rangle - \langle V(x)\rangle. \tag{D32}$$

(As a check, use of (D32) in (D25) gives (D5) again. This also defines an alternative route to computing $C$.)

Putting everything together, the EPI zero-condition is now

$$I - \kappa J \equiv 0$$

$$= \frac{8mT}{A^2}\left\{\frac{1}{2T}\int_{-T}^{T} dt\, x(t)\frac{dV(x)}{dx} - 2\frac{\kappa}{2T}\int_{-T}^{T} dt\,[V(x) + \langle E_{\text{kin}}(x)\rangle - \langle V(x)\rangle]\right\}. \tag{D33}$$

We used Eqs. (D24c), (D23b), and (D32). With use of Eq. (D22) and recognizing time averages as in (D13), (D33) becomes

$$I - \kappa J \equiv 0 = \frac{8mT}{A^2}\left[\left\langle x(t)\frac{dV(x)}{dx}\right\rangle - 2\langle V(x) + \langle E_{\text{kin}}(x)\rangle - \langle V(x)\rangle\rangle\right]. \tag{D34}$$

The potential $V(x)$ directly drops out during the averaging, giving

$$\left\langle x(t)\frac{dV(x)}{dx}\right\rangle = 2\langle E_{\text{kin}}(x)\rangle. \tag{D35}$$

This is the virial theorem (Goldstein, 1950). Hence, whereas the classical limit of the EPI extremum condition gives Newton's second law, the classical limit of the zero condition gives the virial theorem. We saw this kind of dual use of EPI in Chap. 4 as well, where the extremum condition gave the Klein–Gordon equation and the zero-condition gave the Dirac equation.

Equation (D35) is actually the virial theorem for a single particle. All three problems of this appendix have been restricted to the case of a single particle. Generalizing the preceding virial derivation to a system with multiple particles seems to present no insurmountable obstacles, since the summation index $n$ in the information form (D1) can refer to different particles as well as different degrees of freedom. This is left to future work.

Although we derived the virial theorem using the zero principle of EPI, it may of course be alternatively derived by direct use of Newton's law (D20) and the assumptions preceding (D24a) (Goldstein, 1950).

# Appendix E

## Factorization of the Klein–Gordon information

The aims here are (a) to establish result (4.39), the factorization property of the helper vectors $\mathbf{v}_1$, $\mathbf{v}_2$; and (b) to show that $S_4 + S_5 = 0$ for $\mathbf{v}_1 = 0$ or $\mathbf{v}_2 = 0$. These require that we evaluate all terms in the dot product indicated in Eq. (4.40). Commutation rules for the matrices $[\alpha]$, $[\beta]$ will result.

Accordingly, using definitions (4.37), we form

$$\mathbf{v}_1 \cdot \mathbf{v}_2 = \sum_{n=1}^{N/2} \left( i \sum_{m=1}^{3} \sum_{j=1}^{N/2} \alpha_{mnj} \psi_{mj} - \sum_{j=1}^{N/2} \beta_{nj} \eta \psi_j + i\lambda \psi_{4n} \right)$$

$$\times \left( i \sum_{l=1}^{3} \sum_{k=1}^{N/2} \alpha_{lkn} \psi_{lk}^* + \sum_{k=1}^{N/2} \beta_{kn} \eta \psi_k^* - i\lambda \psi_{4n}^* \right). \tag{E1}$$

The notation is as follows. Quantity $\alpha_{lmn}$ denotes the $(m, n)$ element of matrix $[\alpha_l]$, $l = 1, 2, 3$ corresponds to components $x$, $y$, $z$, respectively, and

$$\partial\psi_j/\partial x \equiv \psi_{1j}, \quad \partial\psi_j/\partial y \equiv \psi_{2j}, \quad \partial\psi_j/\partial z \equiv \psi_{3j}, \quad \partial\psi_j/\partial t \equiv \psi_{4j}. \tag{E2}$$

We also assume that the matrices are Hermitian,

$$\alpha_{lnk}^* \equiv \alpha_{lkn}, \quad \beta_{nk}^* \equiv \beta_{kn}. \tag{E3}$$

We proceed to evaluate all individual terms in Eq. (E1).

### *Product of first terms in (E1)*

This is

$$S_1 \equiv - \sum_{n=1}^{N/2} \sum_{l,m=1}^{3} \sum_{j,k=1}^{N/2} \alpha_{lkn} \alpha_{mnj} \psi_{mj} \psi_{lk}^*$$

$$= - \sum_{\substack{j,k,l,m,n \\ m<l}} \alpha_{lkn} \alpha_{mnj} \psi_{mj} \psi_{lk}^* - \sum_{\substack{j,k,l,m,n \\ m\leq l}} \alpha_{mkn} \alpha_{lnj} \psi_{lj} \psi_{mk}^* \tag{E4}$$

after renaming $l \to m$, $m \to l$ in the second sum. Rearranging orders of summation and combining the two sums in (E4) gives

$$S_1 = -\sum_{\substack{j,k,l,m \\ m<l}} \left( \psi_{mj}\psi^*_{lk} \sum_n \alpha_{lkn}\alpha_{mnj} + \psi_{lj}\psi^*_{mk} \sum_n \alpha_{mkn}\alpha_{lnj} \right)$$

$$-\sum_{j,k,m} \psi_{mj}\psi^*_{mk} \sum_n \alpha_{mkn}\alpha_{mnj}, \tag{E5}$$

where the last sum is for $m = l$. Now suppose that

$$\sum_n \alpha_{lkn}\alpha_{mnj} = -\sum_n \alpha_{mkn}\alpha_{lnj}, \quad \text{all } j,\ k,\ l,\ m, \quad l \neq m. \tag{E6}$$

In matrix notation this means that

$$[\alpha_l][\alpha_m] = -[\alpha_m][\alpha_l], \quad l \neq m. \tag{E7}$$

The matrices $[\alpha_x]$, $[\alpha_y]$, $[\alpha_z]$ anticommute. The effect on Eq. (E5) is directly

$$S_1 = -\sum_{\substack{j,k,l,m \\ m\leqslant l}} \sum_n \alpha_{mkn}\alpha_{lnj}(\psi_{lj}\psi^*_{mk} - \psi_{mj}\psi^*_{lk}) - \sum_{j,k,m} \psi_{mj}\psi^*_{mk} \sum_n \alpha_{mkn}\alpha_{mnj}. \tag{E8}$$

When $S_1$ is integrated $d\mathbf{r}\,dt$, the first double sum contributes zero. This is shown in Appendix F, Eqs. (F1)–(F5). Hence, the first double sum effectively contributes zero to $S_1$ of (E4).

Regarding the second double sum in (E8), suppose that matrices $[\alpha_m]$ obey

$$\sum_n \alpha_{mkn}\alpha_{mnj} = \delta_{jk}.$$

This is equivalent to the matrix statement

$$[\alpha_m]^2 = [1], \quad m = 1, 2, 3, \tag{E9}$$

where [1] denotes the diagonal unit matrix. Then directly the second double sum becomes

$$-\sum_{j,k,m} \psi_{mj}\psi^*_{mk}\delta_{jk} = -\sum_{m,j} \psi_{mj}\psi^*_{mj} = -\sum_{j=1}^{N/2} \nabla\psi_j \cdot \nabla\psi^*_j. \tag{E10}$$

The latter is by use of the notation (E2). Equation (E10) gives the first right-hand term in the required Eq. (4.39), i.e.,

$$\iint d\mathbf{r}\,dt\, S_1 = -\iint d\mathbf{r}\,dt \sum_j \nabla\psi^*_j \cdot \nabla\psi_j. \tag{E11}$$

### *Product of second terms in (E1)*

This is

$$S_2 \equiv -\eta^2 \sum_{j,k,n} \beta_{kn}\beta_{nj}\psi_j\psi^*_k. \tag{E12}$$

Suppose that matrix $[\beta]$ obeys

$$\sum_n \beta_{kn}\beta_{nj} = \delta_{jk}.$$

This is equivalent to a matrix requirement

$$[\beta]^2 = [1]. \tag{E13}$$

Then $S_2$ becomes directly

$$S_2 = -\eta^2 \sum_{j,k} \psi_j \psi_k^* \delta_{jk} = -\eta^2 \sum_{j=1}^{N/2} \psi_j \psi_j^*. \tag{E14}$$

This has the form of the third right-hand term in form (4.39), as required.

### *Cross-products of first and second terms*

These are

$$S_3 \equiv i\eta \sum_{j,k,l} \psi_k^* \psi_{lj} \sum_n \beta_{kn} \alpha_{lnj} - i\eta \sum_{j,k,l} \psi_j \psi_{lk}^* \sum_n \alpha_{lkn} \beta_{nj}. \tag{E15}$$

We renamed dummy summation index $m$ to $l$ in the first sum. Suppose that

$$\sum_n \alpha_{lkn} \beta_{nj} = -\sum_n \beta_{kn} \alpha_{lnj}. \tag{E16}$$

In matrix form this states that

$$[\alpha_l][\beta] = -[\beta][\alpha_l]. \tag{E17}$$

That is, matrix $[\beta]$ *anticommutes with matrices* $[\alpha_x]$, $[\alpha_y]$, $[\alpha_z]$.

Using identity (E16) in Eq. (E15) gives

$$S_3 = -i\eta \sum_{j,k,l,n} \alpha_{lkn} \beta_{nj} (\psi_k^* \psi_{lj} + \psi_j \psi_{lk}^*). \tag{E18}$$

It is shown in Appendix F that the integral $d\mathbf{r}\, dt$ of this sum vanishes,

$$\iint d\mathbf{r}\, dt\, S_3 = 0. \tag{E19}$$

(In particular, see Eqs. (F6)–(F9).)

### *Cross-products of second and third terms*

The product of these terms is a sum

$$S_4 \equiv i\eta\lambda \sum_{k,n} (\psi_k^* \psi_{4n} \beta_{kn} + \psi_k \psi_{4k}^* \beta_{nk}), \tag{E20}$$

after renaming the dummy index $j$ to $k$. Replacing $\beta_{kn}$ by $\beta_{nk}^*$ (Hermiticity property (E3)) gives

$$S_4 = i\eta\lambda \sum_{k,n} (\psi_k^* \psi_{4n} \beta_{nk}^* + \psi_k \psi_{4n}^* \beta_{nk}). \tag{E21}$$

This is a purely imaginary term. It will cancel out another imaginary term at the solution for $\mathbf{v}_1$, $\mathbf{v}_2$. That term is found next.

### *Cross-products of first and third terms*

This is a sum

$$S_5 \equiv -\lambda\left(\sum_{k,l,n} \alpha^*_{lnk}\psi^*_{lk}\psi_{4n} - \sum_{k,l,n} \alpha_{lnk}\psi_{lk}\psi^*_{4n}\right) \tag{E22}$$

after renaming dummy indices $j$, $m$ to $k$, $l$, respectively, and replacing $\alpha_{lkn}$ by $\alpha^*_{lnk}$, Hermiticity property (E3). The first sum in (E22) is the complex conjugate of the second. Therefore, the difference is purely imaginary. Hence, $S_5$ is purely imaginary.

### *Product of last terms*

These two terms multiply each other to give a contribution

$$S_6 \equiv \lambda^2 \sum_n \psi_{4n}\psi^*_{4n},$$

or

$$S_6 = \lambda^2 \sum_n \frac{\partial \psi_n}{\partial t}\frac{\partial \psi^*_n}{\partial t} \tag{E23}$$

by notation (E2).

### *Total sum of products*

Using sum results (E11), (E14), (E19), (E21), (E22), and (E23) in Eq. (E1), we have

$$\begin{aligned}\iint d\mathbf{r}\,dt\,\mathbf{v}_1\cdot\mathbf{v}_2 &\equiv \iint d\mathbf{r}\,dt \sum_{j=1}^{6} S_j \\ &= \iint d\mathbf{r}\,dt \sum_{n=1}^{N/2}\left[-(\nabla\psi)^*\cdot\nabla\psi_n + \lambda^2\left(\frac{\partial\psi_n}{\partial t}\right)^*\frac{\partial\psi_n}{\partial t} - \eta^2|\psi_n|^2\right] \\ &\quad + i\iint d\mathbf{r}\,dt\,(S_4+S_5).\end{aligned} \tag{E24}$$

This accomplishes the required goal (a) of the appendix.

### *Evaluation at* $\mathbf{v}_1 = \mathbf{0}$

We can rearrange the terms of Eqs. (E21) and (E22) to give

$$\begin{aligned}S_4 + S_5 &= \lambda\sum_n \psi_{4n}\left(i\eta\sum_k \psi^*_k\beta^*_{nk} - \sum_{k,l}\psi^*_{lk}\alpha^*_{lnk}\right) \\ &\quad + \lambda\sum_n \psi^*_{4n}\left(i\eta\sum_k \psi_k\beta_{nk} + \sum_{k,l}\psi_{lk}\alpha_{lnk}\right).\end{aligned} \tag{E25}$$

By Eqs. (4.36) and (4.37), we can express the fact that the $n$th component $v_{1n} = 0$ as

$$v_{1n} \equiv i \sum_k \left[ \alpha_{1nk} \frac{\partial \psi_k}{\partial x} + \alpha_{2nk} \frac{\partial \psi_k}{\partial y} + \alpha_{3nk} \frac{\partial \psi_k}{\partial z} \right] - \eta \sum_k \beta_{nk} \psi_k + i\lambda \frac{\partial \psi_n}{\partial t} \equiv 0. \tag{E26}$$

Using the derivative notation (E2) gives

$$v_{1n} = i \sum_{l,k=1}^{3} \alpha_{lnk} \psi_{lk} - \eta \sum_k \beta_{nk} \psi_k + i\lambda \psi_{4n} = 0. \tag{E27}$$

Multiplying this by $i$ gives

$$i\eta \sum_k \beta_{nk} \psi_k + \sum_{l,k=1}^{3} \alpha_{lnk} \psi_{lk} = -\lambda \psi_{4n}. \tag{E28}$$

The left-hand side is the last factor in Eq. (E25). Moreover, its complex conjugate is the first factor. The result is that (E25) becomes

$$S_4 + S_5 = \lambda \sum_n \psi_{4n} \lambda \psi_{4n}^* + \lambda \sum_n \psi_{4n}^* (-\lambda \psi_{4n}) = 0 \tag{E29}$$

identically.

***Evaluation at* $\mathbf{v}_2 = 0$**

Because of relation (4.38) between $\mathbf{v}_1$ and $\mathbf{v}_2$, result (E29) must hold if the alternative solution $\mathbf{v}_2 = 0$ is used. We leave this as an exercise for the reader.

Hence, goal (b) of this appendix has been accomplished.

## Generalization to arbitrary fermion spin

The factorization in this appendix has been for an arbitrary value of $N$. The particular value $N = 8$ corresponds to a spin value of $1/2$, by Sec. 4.2.9, while a general value of $N$ corresponds to a fermion spin value $s$ obeying Eq. (4.46h). Thus the factorization procedure holds for *all values* of the spin $s$. With $N$ general, the corresponding matrices $[\alpha_m]$, $[\beta]$ are no longer the particular $4 \times 4$ Dirac matrices, since these arise only in the case $N = 8$. Rather, they are of a more general category called *Bhabha-type* matrices (Roman, 1960). See also Sec. 4.2.12.

# Appendix F

## Evaluation of certain integrals

The aims are to show that (a) the integral of the first sum in Eq. (E8) is zero; and (b) the integral of $S_3 = 0$.

To show (a), arbitrarily consider the case $l = 3$, $m = 1$. The integrals in question are then (for arbitrary $j$, $k$)

$$T = \iint d\boldsymbol{r}\, dt\, (\psi_{3j}\psi^*_{1k} - \psi_{1j}\psi^*_{3k}) \equiv T_{zx} - T_{xz}. \tag{F1}$$

It is convenient to work in frequency space. Represent each $\psi_j(\boldsymbol{r}, t)$ by its Fourier transform, Eq. (4.4). Then, by Parseval's theorem,

$$T_{zx} \equiv \iint d\boldsymbol{r}\, dt\, \frac{\partial\psi_j}{\partial z}\frac{\partial\psi^*_k}{\partial x} = \iint d\boldsymbol{\mu}\, dE\, \phi_j(\boldsymbol{\mu}, E)\phi^*_k(\boldsymbol{\mu}, E)\mu_3\mu_1. \tag{F2}$$

Likewise, we get

$$T_{xz} \equiv \iint d\boldsymbol{r}\, dt\, \frac{\partial\psi_j}{\partial x}\frac{\partial\psi^*_k}{\partial z} = \iint d\boldsymbol{\mu}\, dE\, \phi_j(\boldsymbol{\mu}, E)\phi^*_k(\boldsymbol{\mu}, E)\mu_1\mu_3. \tag{F3}$$

Comparing Eqs. (F2) and (F3) shows that

$$T_{xz} = T_{zx}. \tag{F4}$$

Then, by (F1),

$$T = 0. \tag{F5}$$

But the integral of the first double sum in Eq. (E8) is a weighted sum of integrals of the form (F1). Hence, its value is zero. This was goal (a) of the appendix.

By Eq. (E18), goal (b) will be accomplished if we can show that

$$\iint d\boldsymbol{r}\, dt\, (\psi_j\psi^*_{lk} + \psi^*_k\psi_{lj}) = 0, \quad \text{all } j,\ k,\ l. \tag{F6}$$

First consider the case $l = 3$, with $j$, $k$ arbitrary. Then the integral is

$$\iint d\mathbf{r}\,dt\,(\psi_j\psi_{3k}^* + \psi_k^*\psi_{3j}). \tag{F7}$$

Again working in frequency space, we have, by Parseval's theorem,

$$\iint d\mathbf{r}\,dt\,\psi_j\psi_{3k}^* \equiv \iint d\mathbf{r}\,dt\,\psi_j\,\frac{\partial\psi_k^*}{\partial z} = \frac{i}{\hbar}\iint d\boldsymbol{\mu}\,dE\,\phi_j(\boldsymbol{\mu})\phi_k^*(\boldsymbol{\mu})\mu_3. \tag{F8}$$

Likewise,

$$\iint d\mathbf{r}\,dt\,\psi_k^*\psi_{3j} \equiv \iint d\mathbf{r}\,dt\,\psi_k^*\,\frac{\partial\psi_j}{\partial z} = -\frac{i}{\hbar}\iint d\boldsymbol{\mu}\,dE\,\phi_k^*(\boldsymbol{\mu})\phi_j(\boldsymbol{\mu})\mu_3, \tag{F9}$$

the negative of (F8). Therefore, the sum of (F8) and (F9) is zero, for this case $l$. But obviously the same derivation can be carried through for any chosen $l$. Therefore, task (b) is accomplished.

# Appendix G

## Schrödinger wave equation as a non-relativistic limit

It is straightforward to derive the equation as the non-relativistic limit of the Klein–Gordon equation (Schiff, 1955). Multiply out the factors and dot products in the Klein–Gordon Eq. (4.28), bringing all time-derivative terms to the left and all space-derivative terms to the right:

$$\left(\hbar^2 \frac{\partial^2}{\partial t^2} + 2i\hbar e\phi \frac{\partial}{\partial t} - e^2\phi^2 + i\hbar e \frac{\partial \phi}{\partial t}\right)\psi$$

$$= [c^2\hbar^2 \nabla^2 - 2iec\hbar(\boldsymbol{A} \cdot \nabla) - iec\hbar(\nabla \cdot \boldsymbol{A}) - e^2|\boldsymbol{A}|^2 - m^2c^4]\psi. \quad \text{(G1)}$$

We used the previously found scalar nature of the wave function, $\psi_n = \psi$, and the easily established identity

$$\nabla \cdot (\boldsymbol{A}\psi) = \psi \nabla \cdot \boldsymbol{A} + \boldsymbol{A} \cdot \nabla\psi. \quad \text{(G2)}$$

Use, now, the non-relativistic representation for $\psi$ of

$$\psi(\boldsymbol{r}, t) = \psi'(\boldsymbol{r}, t)e^{-imc^2 t/\hbar}. \quad \text{(G3)}$$

Effectively, the rest energy $mc^2$ has been separated out of $\psi$ to form a new wave function $\psi'$. Note that this step violates relativistic covariance by separating out a time $t$ component $\exp(-imc^2 t/\hbar)$ from the rest of the wave function. This is part of the non-relativistic limit taken.

The overall aim is to substitute the form Eq. (G3) into Eq. (G1). The left-hand side of the latter requires time derivatives of $\psi$. Differentiating Eq. (G3) once, and then twice, gives

$$\frac{\partial \psi}{\partial t} = \left(\frac{\partial \psi'}{\partial t} - \frac{imc^2}{\hbar}\psi'\right)e^{-imc^2 t/\hbar} \quad \text{(G4)}$$

and

$$\frac{\partial^2 \psi}{\partial t^2} = \left(\frac{\partial^2 \psi'}{\partial t^2} - 2\frac{imc^2}{\hbar}\frac{\partial \psi'}{\partial t} - \frac{m^2c^4}{\hbar^2}\psi'\right)e^{-imc^2 t/\hbar}. \quad \text{(G5)}$$

The first right-hand terms in these equations can be neglected. Then, ignoring as well the last two terms on the left-hand side (LHS) of Eq. (G1), substitute the resulting Eqs. (G4) and (G5) into the LHS of Eq. (G1). The result is

$$\text{LHS} = -\left(2imc^2\hbar\,\frac{\partial\psi'}{\partial t} + m^2c^4\psi'\right)e^{-imc^2t/\hbar} - 2i\hbar e\phi(-imc^2/\hbar)\psi' e^{-imc^2t/\hbar}. \quad \text{(G6)}$$

We now substitute this for the LHS of Eq. (G1). A term $-m^2c^4\psi'\exp(-imc^2t/\hbar)$ common to both the LHS and the RHS cancels out. Multiplying both sides of the result by $-1/(2mc^2)$ and renaming $\psi' = \psi$ gives a result

$$i\hbar\,\frac{\partial\psi}{\partial t} - e\phi\psi = -\frac{\hbar^2}{2m}\nabla^2\psi + \frac{ie\hbar}{mc}\boldsymbol{A}\cdot\nabla\psi + \frac{ie\hbar}{2mc}(\nabla\cdot\boldsymbol{A})\psi + \frac{e^2|A|^2}{2mc^2}\psi. \quad \text{(G7)}$$

This is the Schrödinger wave equation (SWE) in the presence of general electromagnetic fields $\boldsymbol{A}$, $\phi$. In the usual case of a purely scalar field $\phi$, it becomes the standard expression

$$i\hbar\,\frac{\partial\psi}{\partial t} = -\frac{\hbar^2}{2m}\nabla^2\psi + e\phi\psi. \quad \text{(G8)}$$

This is the SWE with the time.

# Appendix H

## Non-uniqueness of potential A for finite boundaries

The proof is due to Ferwerda (1998). Any one component $g = g(\boldsymbol{r}, t)$ of Eq. (5.51) is of the form

$$\Box g = h, \tag{H1}$$

where $h = h(\boldsymbol{r}, t)$ is an arbitrary source function. The question we address is whether $g$ is unique, given that it obeys either Dirichlet or Neumann conditions on (now) a closed *four*-dimensional surface $M_0$ that encloses space $(\boldsymbol{r}, t)$. *Note*: The Dirichlet condition is that $g = 0$ on $M_0$; the Neumann condition is that $\boldsymbol{\nabla} g = 0$ on $M_0$.

Assume, contrariwise, that $g$ is not unique, so that Eq. (H1) has two different solutions $g_1$ and $g_2$. The difference $g_{12} \equiv g_1 - g_2$ then satisfies, by Eq. (H1),

$$\Box g_{12} = 0. \tag{H2}$$

We would like to show that the only solution to this problem is that $g_{12} = 0$ within the bounding four-dimensional surface $M_0$. However, this will not be generally forthcoming.

Start from Green's first identity (Jackson, 1975) in four-space $(\boldsymbol{r}, ct)$,

$$c\int_V d\boldsymbol{r}\, dt\, (\boldsymbol{\nabla}\phi \cdot \boldsymbol{\nabla}\psi + \phi\,\Box\psi) = \int_{M_0} d\boldsymbol{\sigma} \cdot (\boldsymbol{\nabla}\psi)\,\phi,$$

$$\text{where} \quad \boldsymbol{\nabla} \equiv \left(\nabla, \frac{1}{ic}\frac{\partial}{\partial t}\right), \qquad d\boldsymbol{\sigma} \equiv (d\boldsymbol{\sigma}_3, icdt). \tag{H3}$$

Quantity $d\boldsymbol{\sigma}$ is a four-component "surface" element with space part $d\boldsymbol{\sigma}_3$, and V is the four-volume enclosed by surface $M_0$. Functions $\phi$ and $\psi$ are arbitrary but differentiable.

Now take $\phi = \psi = g_{12}$ in Eq. (H3). Then, using Eq. (H2) as well, we obtain

$$c\int_V d\boldsymbol{r}\, dt\, \boldsymbol{\nabla} g_{12} \cdot \boldsymbol{\nabla} g_{12} = \int_{M_0} d\boldsymbol{\sigma} \cdot (\boldsymbol{\nabla} g_{12})\, g_{12}. \tag{H4}$$

But then, by the Dirichlet or Neumann condition, the right-hand side is zero. Since volume V in the left-hand side integral is arbitrary, its integrand must be zero,

$$\nabla g_{12} \cdot \nabla g_{12} = 0, \quad \text{i.e.,} \quad (\nabla g_{12})^2 - \left(\frac{1}{c}\frac{\partial g_{12}}{\partial t}\right)^2 = 0. \tag{H5}$$

This equation is satisfied by any constant or any function $g_{12}(x \pm ct)$, $g_{12}(y \pm ct)$, $g_{12}(z \pm ct)$, or $g_{12}(r \pm ct)$, as substitution will show. These represent travelling waves of arbitrary shape. Hence, Eq. (H5) is not satisfied only by $g_{12} = 0$, as we had hoped.

In conclusion, Eq. (H1), and hence, Eq. (5.51), does not have a unique solution in the case of a finite space $(\boldsymbol{r}, t)$ and Dirichlet or Neumann conditions on the four-dimensional boundary. As discussed in Sec. 5.1.26, separate *Cauchy* conditions on the time accomplish a unique solution.

# Appendix I

## Four-dimensional normalization

A brief description of this appendix is as follows. The EPI output probability amplitudes $\psi_n(\boldsymbol{r}, t)$ of quantum mechanics describe the joint random behavior of *four* variables $\boldsymbol{r}$, $t$ (Secs. 3.1.2, 4.1.2 and 4.4). By comparison, the amplitudes $\psi_{3Dn}(\boldsymbol{r}|t)$ of conventional theory describe the joint random behavior of but the *three* variables $\boldsymbol{r}$ in the presence of a *fixed* value of the fourth, $t$. The aim here is to show that, if the $\psi_n(\boldsymbol{r}, t)$ are properly defined, they give rise to normalizable probability laws despite the required extra integration over the time $t$. For this purpose we first develop a hypothetical definition Eq. (I4a) that expresses the $\psi_n(\boldsymbol{r}, t)$ in terms of the conventional $\psi_{3Dn}(\boldsymbol{r}|t)$. Then we show that the definition (I4a) gives *finite* four-dimensional normalization integrals for solutions $\psi(\boldsymbol{r}, t)$ to, in turn, *non-relativistic-limit* "wave packet" solutions of the Klein–Gordon equation (Sec. I.3), wave packet solutions of the *relativistic* Dirac equation (Sec. I.4), and solutions of the *relativistic* Klein–Gordon equation (Sec. I.5).

First consider the EPI Klein–Gordon Eq. (4.28) for a particle of mass $m$. EPI interprets its four-dimensional output function $\psi(\boldsymbol{r}, t)$ as describing the *joint* behavior of the *four* random variables $\boldsymbol{r}$, $t$. The PDF on space-time is taken *by EPI* to be simply $|\psi(\boldsymbol{r}, t)|^2$. (*Note*: This is *not the conventional choice* of a Klein–Gordon *PDF*, which is Eq. (4.60) of Chap. 4. In fact the latter is well known to be unsuitable as a PDF because it can go negative, as discussed below Eq. (4.62), although this negativity is not the issue in this appendix.)

This EPI four-dimensional PDF $|\psi(\boldsymbol{r}, t)|^2$ should have a *finite* four-dimensional normalization integral over all physically allowed space and time values,

$$c \int\int d\boldsymbol{r}\, dt\, |\psi(\boldsymbol{r}, t)|^2 \equiv N_0. \tag{I1}$$

The aim is to show that $N_0$ does indeed have a finite value.

The scalar nature of $\psi$ is indicated by suppression of the subscript $n$ in the $\psi_n$ of Eq. (4.26). The finite nature of $N_0$ will be established by showing that it increases close to linearly with the maximum possible value $\mu_0$ of the momentum of the Klein–Gordon particle (Eq. (I19)). Thus, since in any experiment a finite value of $\mu_0$ is known *a priori* to the observer, $N_0$ must be finite, allowing normalization. Finally, it will be shown that if the total interval $T$ of time fluctuations is known as well (Eq. (I2b)), then the value of $N_0$ is not only finite, but also computable.

The matter of units will be important to the calculation. Quantity $N_0$ is, of course, to be a pure, i.e., unitless, number. Then (I1) shows that $\psi(\boldsymbol{r}, t)$ must have units of *length*$^{-2}$.

## I.1 Three-dimensional versus four-dimensional amplitude functions

The *standard*, Born interpretation of classical quantum mechanics (Schiff, 1955, p. 22) regards the three coordinates $\boldsymbol{r}$ as random, in the presence of an arbitrary but fixed time $t$. The conventional statistical notation for a PDF on a random variable $\boldsymbol{r}$ of this type is $p(\boldsymbol{r}|t)$, that is, "$p$ of $\boldsymbol{r}$ if $t$." Also, in non-relativistic problems $p(\boldsymbol{r}|t)$ is taken to be the modulus-square of an amplitude function. A suitable notation for this *amplitude* function is therefore $\psi_{3\mathrm{D}}(\boldsymbol{r}|t)$, with a corresponding PDF

$$p_{3\mathrm{D}}(\boldsymbol{r}|t) \equiv |\psi_{3\mathrm{D}}(\boldsymbol{r}|t)|^2, \quad \text{and} \quad \int d\boldsymbol{r}\, |\psi_{3\mathrm{D}}(\boldsymbol{r}|t)|^2 \equiv 1 \tag{I2a}$$

as the normalization requirement. The subscript 3D identifies the three-dimensionality of the random variable $\boldsymbol{r}$. By the normalization requirement (I2a), the units of $\psi_{3\mathrm{D}}(\boldsymbol{r}|t)$ are *length*$^{-3/2}$. *Note*: Even though the time $t$ in $\psi_{3\mathrm{D}}(\boldsymbol{r}|t)$ is statistically regarded as fixed, it is mathematically an ordinary variable, so that $\psi_{3\mathrm{D}}(\boldsymbol{r}|t)$ may be differentiated $d/dt$, integrated $dt$, etc. Analogously, the probabilistically "fixed" parameter $\theta$ of the PDF $p(\mathbf{y}|\theta)$ in Sec. 1.2.3 was differentiated $\partial/\partial\theta$.

*By comparison with* this standard interpretation, the EPI view is to regard $t$, as well as $\boldsymbol{r}$, as a random variable (Secs. 3.1.2 and 3.5.8).

An important problem of *notation* now arises. The standard notation for the three-dimensional wave function is $\psi(\mathbf{x})$, where $\mathbf{x} \equiv (\boldsymbol{r}, ct)$. Notice the comma between the $\boldsymbol{r}$ and $ct$. In statistical notation this would mean the *and* operation, implying that the event $\mathbf{x} \equiv (\boldsymbol{r}, ct)$ describes the *joint random* event $(\boldsymbol{r}, ct)$. But, as we saw (first Eq. (I2a), the value of $t$ is considered to be fixed in the conventional approach, not random, so its wave function should more clearly be indicated as $\psi_{3\mathrm{D}}(\boldsymbol{r}|t)$. We used this notation in Eq. (I2a) and in the following.

## I.2 Time localization

The coordinate $t$ is by definition an *uncertainty* in the measurement time of the particle (Secs. 3.1.2 and 3.5.8). Any experiment is carried through over some fixed, finite time interval. Call this $T$. Assuming, for simplicity, a symmetric interval of time fluctuations, $t$ then obeys

$$-T/2 \leqslant t \leqslant +T/2. \tag{I2b}$$

Normalization will hold even if the value of $T$ is allowed to approach infinity, as is shown below.

## I.3 Defining relation between three-dimensional and four-dimensional amplitude functions

As discussed above, there are *four jointly random* variables $(\boldsymbol{r}, t)$ by the EPI view. Hence the amplitude function lists them (without the "if" notation of (I2a)) as

$$\psi(\boldsymbol{r},\ t). \tag{I3a}$$

Let $\psi_{3D}(\boldsymbol{r}|t)$ be a conventional solution to the Klein–Gordon Eq. (4.28). Because of the homogeneity of the Klein–Gordon equation, any constant *times* this solution is *also* a solution. Then, since our four-dimensional amplitude function $\psi(\boldsymbol{r},\ t)$ *likewise* satisfies the Klein–Gordon Eq. (4.28), the two solutions $\psi_{3D}(\boldsymbol{r}|t)$ and $\psi(\boldsymbol{r},\ t)$ must be functions of $\boldsymbol{r}$ and $t$ that are proportional,

$$\psi(\boldsymbol{r},\ t) = K\psi_{3D}(\boldsymbol{r}|t). \tag{I3b}$$

The constant $K$ is to be found. We will seek the value that allows finite four-dimensional normalization (I1).

A requirement on the choice of proportionality constant $K$ is that it impart the required unit of $length^{-2}$ to $\psi(\boldsymbol{r},\ t)$, after $K$ has multiplied $\psi_{3D}(\boldsymbol{r}|t)$ with its unit of $length^{-3/2}$. Clearly the constant must have the unit of $length^{-1/2}$. An example is

$$K = 1/\sqrt{cT}. \tag{I3c}$$

The use of this is shown next to satisfy four-dimensional normalization in a particular case.

In *non-relativistic* cases the Klein–Gordon equation goes over into the Schrödinger wave equation (Appendix G). We consider "wave packet" solutions (Schiff, 1955) $\psi_{3D}(\boldsymbol{r}|t)$ of the SWE. These obey normalization over all space, i.e.,

$$\int d\mathbf{r}\, |\psi_{3D}(\boldsymbol{r}|t)|^2 = 1 \tag{I3d}$$

at each $t$. (This includes most elementary applications, such as the free particle in a box, the harmonic oscillator, the hydrogen atom, etc.) Suppose that we use (I3c), defining the general (now) $n$th component as

$$\begin{aligned}\psi_n(\boldsymbol{r},\ t) &\equiv \frac{1}{\sqrt{cT}}\psi_{3Dn}(\boldsymbol{r}|t), \quad \text{for} \quad |t| < T/2, \\ &= 0 \quad \text{for} \quad |t| > T/2\end{aligned} \tag{I4a}$$

(cf. Eq. (I3b)). (The scalar case currently under consideration is the component case $n = 1$ or, equivalently, $n$ suppressed.) Then the *four*-dimensional integral (I1) over all space and time using this $\psi(\boldsymbol{r},\ t)$ is

$$N_0 = c\frac{1}{cT}\int_{-T/2}^{T/2} dt \int d\mathbf{r}\, |\psi_{3D}(\boldsymbol{r}|t)|^2 = c\frac{1}{cT}\int_{-T/2}^{T/2} dt \cdot 1 = c\frac{1}{cT}T = 1, \tag{I4b}$$

as required. Equation (I3d) was used. Notice that the four-dimensional normalization holds even in the limit $T \to \infty$, since $T$ cancels out in the calculation. It is presumed that in any experiment, the value of $T$ is known as some *finite* (possibly large) number. Then $\psi(\boldsymbol{r},\ t)$ as defined in Eq. (I4a) is finite as well, i.e, not of vanishing amplitude.

Hence *the definition* (I4a) *of* $\psi_n(\boldsymbol{r},\ t)$ *achieves four-dimensional normalization in these non-relativistic cases*. Therefore, we provisionally take (I4a) to define $\psi_n(\boldsymbol{r},\ t)$ *in general*. It remains to see whether this gives a normalizable PDF in *relativistic* cases.

## I.4 Relativistic Dirac equation

Let the vector amplitude $\boldsymbol{\psi}(\boldsymbol{r}, t)$ be any four-dimensional EPI solution to the Dirac Eq. (4.42). Let Eq. (I4a) define the relation between *each four-dimensional component* $\psi_n(\boldsymbol{r}, t)$ of $\boldsymbol{\psi}(\boldsymbol{r}, t)$ and a corresponding three-dimensional component $\psi_{3\mathrm{D}n}(\boldsymbol{r}|t)$ of the Dirac $\boldsymbol{\psi}_{3\mathrm{D}}(\boldsymbol{r}|t)$. Suppose that the three-dimensional Dirac solutions $\boldsymbol{\psi}_{3\mathrm{D}}(\boldsymbol{r}|t)$ are "wave packets." Then they obey normalization (Schiff, 1955),

$$\int d\boldsymbol{r}\, \boldsymbol{\psi}_{3\mathrm{D}}(\boldsymbol{r}|t)^{\dagger} \boldsymbol{\psi}_{3\mathrm{D}}(\boldsymbol{r}|t) \equiv \int d\boldsymbol{r} \sum_n |\psi_{3\mathrm{D}n}(\boldsymbol{r}|t)|^2 = 1, \tag{I4c}$$

where $^{\dagger}$ denotes the Hermitian adjoint. Then the four-dimensional normalization integral for $\boldsymbol{\psi}(\boldsymbol{r}, t)$ can be evaluated as

$$\begin{aligned} N_0 &\equiv c \iint dr\, dt\, \boldsymbol{\psi}(\boldsymbol{r}, t)^{\dagger} \boldsymbol{\psi}(\boldsymbol{r}, t) = c \frac{1}{cT} \int_{-T/2}^{T/2} dt \int d\mathbf{r} \sum_n |\psi_{3\mathrm{D}n}(\boldsymbol{r}|t)|^2 \\ &= c \frac{1}{cT} \int_{-T/2}^{T/2} dt \cdot 1 = c \frac{1}{cT} T = 1. \end{aligned} \tag{I4d}$$

Equation (I4a) was used in the second equality. Hence normalization also holds for these four-dimensional EPI solutions $\boldsymbol{\psi}(\boldsymbol{r}, t)$ to the vector Dirac equation. Also, as in the non-relativistic cases preceding, the time interval $T$ can be finite or infinite.

The remaining case to be considered is that of four-dimensional solutions to the *relativistic* Klein–Gordon equation. This is a slightly more challenging problem.

## I.5 Relativistic Klein–Gordon equation

The chief background reference for this section is Roman (1961). He of course works from the paradigm of a conventional three-dimensional wave function $\psi_{3\mathrm{D}}(\boldsymbol{r}|t)$, with $t$ regarded as *fixed*, compared with our four-dimensional function $\psi(\boldsymbol{r}, t)$ with $t$ *random*. Consequently, Roman's function $\psi(\mathbf{x})$ is our $\psi_{3\mathrm{D}}(\boldsymbol{r}|t)$. See Sec. I.1. Also, our indicated function $\psi(\boldsymbol{r}, t)$ will always mean the four-dimensional amplitude law Eq. (I4a).

Roman (1961, p. 193) regards the Fourier Eq. (4.4) of Chap. 4 as holding identically for what *he* calls $\psi(\mathbf{x})$ – our $\psi_{3\mathrm{D}}(\boldsymbol{r}|t)$ by the preceding. The EPI approach likewise regards Eq. (4.4) as holding for $\psi(\boldsymbol{r}, t)$ (Sec. 4.1.7).

*We next use Roman's solution* $\psi_{3\mathrm{D}}(\boldsymbol{r}|t)$, *convert it into* $\psi(\boldsymbol{r}, t)$ *by the use of Eq.* (I4a), *and then evaluate the integral* (I1) *on this basis.*

Roman (1961, p. 185) shows that the only non-trivial *scalar* solution to the free-field Klein–Gordon equation is Eq. (4.4), with the choice

$$\phi_n(\boldsymbol{\mu}, E) \equiv \phi(\boldsymbol{\mu}, E) = B\epsilon(\boldsymbol{\mu}, E)\delta(E^2 - c^2\mu^2 - m^2c^4), \quad B \equiv const. \tag{I5}$$

The subscript $n$ is suppressed once again, since we are working with a scalar solution. The size of $B$ is *finite*, its absolute size immaterial for our purposes. The function $\epsilon(\boldsymbol{\mu}, E)$ is a type of Heaviside function

$$\epsilon(\boldsymbol{\mu}, E) \equiv \begin{cases} +1 & \text{for } E > 0, \\ -1 & \text{for } E < 0 \end{cases}. \tag{I6}$$

Taking Roman's (conventional) view that Eq. (4.4) represents $\psi_{3D}(\boldsymbol{r}|t)$, and using Eqs. (I4a) and (I5), allows us to represent

$$\psi(\boldsymbol{r}, t) \equiv \frac{1}{\sqrt{cT}} \psi_{3D}(\boldsymbol{r}|t)$$

$$= \frac{1}{\sqrt{cT}} \frac{B}{(2\pi\hbar)^2} \iint d\boldsymbol{\mu}\, dE\, \epsilon(\boldsymbol{\mu}, E) \delta(E^2 - c^2\mu^2 - m^2c^4) e^{i(\boldsymbol{\mu}\cdot\boldsymbol{r} - Et)/\hbar}. \tag{I7}$$

The integral is sometimes called "Schwinger's Δ-function." Roman notes that all other covariant solutions $\psi_n(\boldsymbol{r}, t)$ can be obtained from (I7) by differentiating it and/or multiplying it by appropriate constants. We return to this point later.

To proceed further the integral Eq. (I7) must be evaluated. Might the $\delta(E^2 - c^2\mu^2 - m^2c^4)$ factor in the integrand lead to a singularity and hence a non-convergent integral? If so, the normalization integral (I1) would not be convergent. However, this will prove not to be the case.

The delta function obeys, by Eq. (0.72),

$$\delta(E^2 - c^2\mu^2 - m^2c^4) = \frac{\delta(E - R)}{2R} + \frac{\delta(E + R)}{2R}, \tag{I8a}$$

$$R \equiv R(\mu) \equiv +\sqrt{c^2\mu^2 + m^2c^4}. \tag{I8b}$$

Using (I6) and (I8a, b) in (I7) and duly keeping track of $\epsilon(\boldsymbol{\mu}, E)$ as defined by (I6), gives

$$\psi(\boldsymbol{r}, t) = \frac{1}{\sqrt{cT}} \frac{B}{(2\pi\hbar)^2} \int d\boldsymbol{\mu} \left[ \int_0^\infty dE\, \frac{\delta(E - R)}{2R} e^{i(\boldsymbol{\mu}\cdot\boldsymbol{r} - Et)} - \int_{-\infty}^0 dE\, \frac{\delta(E + R)}{2R} e^{i(\boldsymbol{\mu}\cdot\boldsymbol{r} - Et)} \right], \tag{I9}$$

where only one of the two right-hand-side delta functions in (I8a) contributes to each integral $dE$, according to whether the integral is over positive or negative values of $E$. Next, the sifting property of each delta function in (I9) gives

$$\psi(\boldsymbol{r}, t) = \frac{1}{\sqrt{cT}} \frac{B}{(2\pi\hbar)^2} \int d\boldsymbol{\mu}\, \frac{e^{i(\boldsymbol{\mu}\cdot\boldsymbol{r} - Rt)} - e^{i(\boldsymbol{\mu}\cdot\boldsymbol{r} + Rt)}}{2R} = -2i \frac{1}{\sqrt{cT}} \frac{B}{(2\pi\hbar)^2} \int d\boldsymbol{\mu}\, \frac{e^{i\boldsymbol{\mu}\cdot\boldsymbol{r}} \sin(Rt)}{2R}. \tag{I10}$$

The required normalization integral (I1) is then, by (I10) and the finite range for $t$,

$$N_0 = \frac{4B^2}{(2\pi\hbar)^4} \frac{1}{T} \int_{-T/2}^{T/2} dt \int d\boldsymbol{\mu}\, \frac{\sin(Rt)}{R} \int d\boldsymbol{\mu}'\, \frac{\sin(R't)}{R'} \int d\boldsymbol{r}\, e^{i\boldsymbol{r}\cdot(\boldsymbol{\mu} - \boldsymbol{\mu}')}, \tag{I11a}$$

$$R' \equiv R(\mu') = +\sqrt{c^2\mu'^2 + m^2c^4} \tag{I11b}$$

(the latter by (I8b)). We interchanged orders of integration in order to bring all integration $d\boldsymbol{r}$ to the far right in (I11a). The latter integral is the familiar Fourier representation $(2\pi)^3\delta(\boldsymbol{\mu} - \boldsymbol{\mu}')$ of the Dirac delta function. Using its sifting property in (I11a) gives

$$N_0 = \frac{2B^2}{\pi\hbar^4}\frac{1}{T}\int_{-T/2}^{T/2} dt \int d\boldsymbol{\mu}\, \frac{\sin^2(Rt)}{R^2}, \qquad R \equiv +\sqrt{c^2\mu^2 + m^2c^4}. \tag{I12}$$

In any real experiment the particle momentum must be bounded above by some *finite* value $\mu_0$. In most scenarios the *momentum value mc* given by Eq. (4.54a) is the effective bound, but any finite multiple of this momentum will obviously suffice for keeping $\mu_0$ finite. Clearly no real experiment will experience an infinite value of $\mu_0$. Finally, regardless of the size of $\mu_0$, in practice no detected momentum value can exceed some finite value that is fixed by the dynamical range of the detection equipment.

Using this finite bound in Eq. (I12), and also $d\boldsymbol{\mu} = 4\pi\mu^2\, d\mu$ by the azimuthal symmetry in $\mu$ of its integrand, gives

$$N_0 = \frac{8B^2}{\hbar^4}\frac{1}{T}\int_{-T/2}^{T/2} dt \int_0^{\mu_0} d\mu\, \mu^2 \frac{\sin^2(Rt)}{R^2}. \tag{I13}$$

Switching orders of integration gives

$$N_0 = \frac{8B^2}{\hbar^4}\frac{1}{T}\int_0^{\mu_0} d\mu\, \frac{\mu^2}{R^2}\int_{-T/2}^{T/2} dt\, \sin^2(Rt). \tag{I14}$$

At this point we find an upper bound to this double integral. This will establish that normalization is at least possible. Later we evaluate the integral, relating $N_0$ to $\mu_0$ and $T$.

Since $\sin^2(Rt) \leqslant 1$, (I14) gives

$$N_0 \leqslant \frac{8B^2}{\hbar^4}\int_0^{\mu_0} d\mu\, \frac{\mu^2}{R^2}, \tag{I15}$$

the dependence upon the time interval $T$ having dropped out once again. Since by (I8b) $R^2 \geqslant c^2\mu^2$, the integrand of (I15) is bounded above by $1/c^2$, and integrating this gives

$$N_0 \leqslant \frac{8B^2\mu_0}{c^2\hbar^4}. \tag{I16}$$

All right-hand quantities are finite since, as we discussed, *any momentum interval size $\mu_0$ is finite*.

*Therefore the normalization integral* (I1) *is finite*, as we set out to prove. In addition, the upper bound to the integral is independent of the size of the finite detection time interval $T$.

We next proceed to evaluate the integral (I14) to see exactly how $N_0$ depends upon $\mu_0$. Using the identity

$$\sin^2(Rt) = \tfrac{1}{2}[1 - \cos(2Rt)] \tag{I17}$$

allows the time integral to be done, giving

$$N_0 = \frac{4B^2}{\hbar^4}\left(\int_0^{\mu_0} d\mu \, \frac{\mu^2}{R^2} - \frac{1}{T}\int_0^{\mu_0} d\mu \, \frac{\mu^2}{R^3}\sin(RT)\right) \tag{I18}$$

with the function $R \equiv R(\mu)$ given in Eqs. (I12). The first integral is analytically known (see below), and the second integral may, of course, be evaluated numerically for any values of $\mu_0$ and $T$. This integral is *finite* for any such finite values since, by (I8b), $R_{\min} = mc^2 > 0$. In this way, $N_0$ is known from known values of $\mu_0$ and $T$, and is finite.

However, analytical answers are preferred. Interestingly, one may be obtained by evaluating the second integral in (I18) in the lim $T \to \infty$. Its magnitude is seen to be bounded above by $\int d\mu \, \mu^2/R^3$, and this is independent of $T$. Then, since the integral is multiplied by $1/T$, in the lim $T \to \infty$ the integral effectively contributes zero to $N_0$. The result is that only the first integral remains, with the analytical answer

$$N_0 = \frac{4mB^2}{c\hbar^4}\left[\frac{\mu_0}{mc} - \tan^{-1}\left(\frac{\mu_0}{mc}\right)\right]. \tag{I19}$$

This simplifies further if $\mu_0 = mc$, the Compton momentum (see above),

$$N_0 = \frac{4mB^2}{c\hbar^4}\left(1 - \frac{\pi}{4}\right). \tag{I20}$$

It should be remarked that Eqs. (I16), (I18), (I19) and (I20) express a real, *physical* normalization, by utilizing the *a priori* knowledge of an upper bound $\mu_0$ to all particle momenta.

*Note on prior knowledge*: A probability law is generally determined from *all* knowledge about its random variables. This includes both knowledge based upon observable data and knowledge in the form of prior information. For example, an exponential probability law $\exp(-|E|)$ must be *chosen* to be either one-tailed or two-tailed on the basis of prior knowledge of the physical nature of $E$; thus, if it is an energy, the law is one-tailed. To ignore such prior knowledge is to violate a basic premise of probability theory, not to mention the "observer participancy" premise of this book.

We presumed for simplicity a scalar solution $\psi(\mathbf{r}, t)$ obeying (I7). Roman notes that all other covariant solutions $\psi_n(\mathbf{r}, t)$, $n > 1$, can be obtained from (I7) by differentiating it and/or multiplying it by appropriate (finite) constants. Certainly multiplication by a finite constant will not change the normalizability of $N_0$. Differentiation of (I7) will merely bring down factors of $\mu$ or $E$ as multipliers in the integrand, and these do not, of course, introduce singularities into the integrand over the range of integration. Hence, going through the same steps as before again leads to finite and computable $N_0$. In the interest of brevity we leave this as an exercise for the motivated reader.

Normalization may alternatively be shown to follow, by the same steps, for a *general* function $\epsilon(\boldsymbol{\mu}, E)$ in Eq. (I5) that is square-integrable, i.e., whose integrals

$$\int_{\boldsymbol{\mu}_0} d\boldsymbol{\mu}\, |\epsilon(\boldsymbol{\mu}, \pm R)|^2 \tag{I21}$$

are finite over a *finite* momentum space $\boldsymbol{\mu}_0$ of magnitude $|\boldsymbol{\mu}_0| = \mu_0$ (as in the above).

# Appendix J

## Transfer matrix method

To generalize the amplitude-matching that we used in Sec. 13.7.3 we use a transfer matrix method that is employed often for formally similar problems in quantum electronics. This presentation was adapted, in part, from unpublished optoelectronics notes of Professor David A. B. Miller. Given a set of discount factors $\{D(T_m)\}$, we seek a solution to

$$\frac{d^2q(t)}{dt^2} = \frac{q(t)}{4}\left[\lambda_0 + \sum_{m=1}^{M} \lambda_m \Theta(t - T_m, 0)\right]. \tag{J1}$$

Between tenors, say $T_{m-1}$ and $T_m$, it is straightforward to show that the probability amplitude $q(t)$ is given by

$$q(t) = A_m e^{i\omega_m(t-T_m)} + B_m e^{-i\omega_m(t-T_m)}, \tag{J2}$$

where $A_m$ and $B_m$ are coefficients to be determined by the matching conditions, $i = \sqrt{-1}$, and

$$\omega_m = \frac{1}{2}\sqrt{\sum_{j=0}^{m} \lambda_j}, \tag{J3}$$

with $\{\lambda_j\}$ being the Lagrange undetermined multipliers. We now consider the match across the $m$th tenor. Since $t - T_m = 0$, continuity of $q(t)$ gives

$$q(T_M) = A_L + B_L = A_{m+1} + B_{m+1}, \tag{J4}$$

where the subscript L denotes the value of the coefficients of $q(t)$ immediately to the left of $T_m$. Continuity of the first derivative of $q(t)$ gives

$$A_L - B_L = \Delta_m(A_{m+1} + B_{m+1}), \tag{J5}$$

where $\Delta_m = \omega_{m+1}/\omega_m$. These two equations combine to give

$$\begin{bmatrix} A_\mathrm{L} \\ B_\mathrm{L} \end{bmatrix} = \frac{1}{2}\begin{bmatrix} 1+\Delta_m & 1-\Delta_m \\ 1-\Delta_m & 1+\Delta_m \end{bmatrix}\begin{bmatrix} A_{m+1} \\ B_{m+1} \end{bmatrix}. \tag{J6}$$

Since $A_m = A_\mathrm{L}\exp(-i\omega_m\tau_m)$ and $B_m = A_L\exp(i\omega_m\tau_m)$, where $\tau_m = T_{m+1} - T_m$, it follows that the coefficients for $q(t)$ in layer $m$ are related to those in layer $m+1$ by

$$\begin{bmatrix} A_m \\ B_m \end{bmatrix} = \frac{1}{2}\begin{bmatrix} e^{-i\omega_m\tau_m} & 0 \\ 0 & e^{i\omega_m\tau_m} \end{bmatrix}\begin{bmatrix} 1+\Delta_m & 1-\Delta_m \\ 1-\Delta_m & 1+\Delta_m \end{bmatrix}\begin{bmatrix} A_{m+1} \\ B_{m+1} \end{bmatrix}. \tag{J7}$$

Given this relationship, we can generate the function $q(t)$ sequentially, beginning with the region beyond the longest tenor that we shall take to be $T_{m+1}$. In this region we require that the solution be a decaying exponential $\exp(-\omega_{m+1}t)$, which implies that $B_{m+1} = 0$ and that $\omega_{m+1}$ be a positive real number. Given a $\{\lambda_m\}$ and setting $A_{m+1} = 1$, it is now a straightforward matter to calculate the fixed income functions described in Sec. 13.7.1. This transfer matrix procedure maps $\{\lambda_m\}$ into a discount function and, in a straightforward manner, into coupon bond prices.

# Appendix K

## Numerov method

Our choice of the Numerov method to solve Eq. (13.9) follows from the popularity of this method for solving this equation in quantum mechanics (Koonin, 1986). In this appendix we present a brief overview of the method and its application to the option problem discussed in Sec. 13.7.3.

The Numerov method provides a numerical solution to

$$\frac{d^2 q(x)}{dx^2} + G(x)q(x) = 0, \tag{K1}$$

that begins by expanding $q(x)$ in a Taylor series

$$q(x \pm \Delta x) = \sum_{n=0}^{\infty} \frac{(\pm\Delta x)^n}{n!} \frac{d^n q(x)}{dx^n}. \tag{K2}$$

Combining these two equations and retaining terms to fourth order yields

$$\frac{q(x+\Delta x) + q(x-\Delta x) - 2q(x)}{(\Delta x)^2} = \frac{d^2 q(x)}{dx^2} + \frac{(\Delta x)^2}{12}\frac{d^4 q(x)}{dx^4} + \mathcal{O}((\Delta x)^4). \tag{K3}$$

The second-derivative term in Eq. (K3) can be cleared by substituting Eq. (K1); and the fourth-derivative term can be cleared by differentiating Eq. (K1) two more times,

$$\begin{aligned}\frac{d^4 q(x)}{dx^4} &= -\frac{d^2(G(x)q(x))}{dx^2} \\ &\cong -\frac{G(x+\Delta x)q(x+\Delta x) + G(x-\Delta x)q(x-\Delta x) - 2G(x)q(x)}{(\Delta x)^2},\end{aligned} \tag{K4}$$

and substituting this result into Eq. (K3). On performing these substitutions and rearranging terms, we arrive at the Numerov algorithm:

$$q_{n+1} = \frac{2\left(1 - \frac{5(\Delta x)^2}{12} G_n\right)}{\left(1 + \frac{(\Delta x)^2}{12} G_{n+1}\right)} q_n - \frac{\left(1 + \frac{(\Delta x)^2}{12} G_{n-1}\right)}{\left(1 + \frac{(\Delta x)^2}{12} G_{n+1}\right)} q_{n-1} + \mathscr{O}((\Delta x)^6). \qquad \text{(K5)}$$

To apply this to the option problem, it is useful to note that the calculation can be separated into two stages: first, the calculation of the $\{\lambda_0, q(x)\}$ pair that solves the eigenvalue problem for a fixed set of $\{\lambda_m\}_{m>0}$, and second, the search for the set $\{\lambda_m\}_{m>0}$ for which $q(x)$ of the corresponding $\{\lambda_0, q(x)\}$ pair reproduces the observed option prices.

The $\{\lambda_0, q(x)\}$ pair for a given $\{\lambda_m\}_{m>0}$ is obtained by integrating $q(x)$ from both the lower and upper limits of the $x$ range and varying $\lambda_0$ until the logarithmic derivatives of each solution match at an arbitrary mid point. In this example that point was $x = 1$ and $\lambda_0$ was found with a one-dimensional minimization algorithm using the difference between the logarithmic derivatives as a penalty function and the minimum of the function $\sum_{m=1}^{M} \lambda_m \max(x - k_m, 0)/4$ as the initial guess for $\lambda_0$.

Once the logarithmic derivatives have been matched, the option prices can be calculated and compared with the observed option prices. The results of this comparison form the penalty function for a multidimensional variation of the $\{\lambda_m\}_{m>0}$ to obtain agreement between the observed and calculated option prices.

# References


Abramowitz, A. and Stegun, I. (1965) *Handbook of Mathematical Functions*. Washington, D.C.: National Bureau of Standards.

Adams, K. J. and Van Deventer, D. R., (1994) *J. Fixed Income* 52.

Aitchison, I. J. and Hey, A. J. (1982) *Gauge Theories in Particle Physics*. Bristol: Adam Hilger.

Akhiezer, A. I. and Berestetskii, V. B. (1965) *Quantum Electrodynamics*. New York: Wiley.

Allen, C. W. (1973) *Astrophysical Quantities*. 3rd edn. London: Athlone Press.

Allen, L. and Eberly, J. H. (1987) *Optical Resonance and Two-Level Atoms*. New York: Dover.

Amari, S. (1985) *Differential-Geometrical Methods in Statistics, Lecture Notes in Statistics, vol. 28*. New York: Springer-Verlag.

Aoki, M. (1996) *New Approaches to Macroeconomic Modeling*. Cambridge: Cambridge University Press.

Aoki, M. (1998) *New Approaches to Macroeconomic Modeling: Evolutionary Stochastic Dynamics, Multiple Equilibria, and Externalities as Field Effects*. New York: Cambridge University Press.

Aoki, M. (2001) *Modeling Aggregate Behavior & Fluctuations in Economics: Stochastic Views of Interacting Agents*. New York: Cambridge University Press.

Artandi, S., Chang, S., Lee, S., Alson, S., Gottlieb, G., Chin, L., and DePinho, R. (2000) *Nature* **406**, 641–645.

Arthurs, E. and Goodman, M. (1988) *Phys. Rev. Lett.* **60**, 2447.

Avellaneda, M. (1998) *Int. J. Theo. Appl. Finance* **1**, 447.

Bachelier, L. (1900) *Theory of Speculation*, Ph.D. thesis. Paris: Faculty of Sciences of the Academy of Paris.

Badii, R. and Politi, A. (1997) *Complexity: Hierarchical Structure and Scaling in Physics*. Cambridge: Cambridge University Press.

Barnes, J. A. and Allan, D. W. (1966) *Proc. IEEE* **54**, 176.

Barrett, M. T., Sanchez, C. A., Prevo, L. J., Wong, D. J., Galipeau, P. C., Paulson, T. G., Rabinovitch, P. S., and Reid, B. J. (1999) *Nat. Genet.* **22**, 106–109.

Barrow, J. D. (1991) *Theories of Everything*. Oxford: Clarendon Press, p. 101.

Batchelor, G. K. (1956) *The Theory of Homogeneous Turbulence*. Cambridge: Cambridge University Press.

Beckner, W. (1975) *Ann. Math.* **102**, 159.

Bell, D. A. (1980) *J. Phys.* C**13**, 4425.

Bernstein, H., Byerly, H. C., Hopf, F. A., Michod, R. A. and Vemulapalli, G. K. (1983) *Quarterly Rev. Biol.* **58**, 185.

Bhaduri, R. K. (1992) *Models of the Nucleon*. New York: Addison-Wesley.

Białynicki-Birula, I. (1975) *Quantum Electrodynamics*. Oxford: Pergamon Press.

Białynicki-Birula, I. (1996) *Progress in Optics, vol. XXXVI*, ed. E. Wolf. Amsterdam: Elsevier.

Binder, P. M. (2000) *Phys. Rev.* E**61**, R3303.

Bishop, J. M. (1991) *Cell* **64**, 235–248.

Blachman, N. M. (1965) *IEEE Trans. Inf. Theory* **IT-11**, 267.

Black, F. and Scholes, M. (1973) *J. Political Economy* **3**, 637.

Bohm, D. (1951) *Quantum Theory*. Englewood Cliffs, New Jersey: Prentice-Hall, pp. 591–598; 614–619.

Bohm, D. (1952a) *Phys. Rev.* **85**, 180; first 3 pages of article.

Bohm, D. (1952b) *Phys. Rev.* **85**, 166.

Bohm, D., and Hiley, B. J. (1993) *The Undivided Universe*. London: Routledge.

Born, M. (1926) *Z. Phys.* **37**, 863.

Born, M. (1927) *Nature* **119**, 354.

Born, M. and Wolf, E. (1959) *Principles of Optics*. New York: MacMillan.

Bouchard, J.-P. and Potters, M. (2000) *Theory of Financial Risks: from Statistical Physics to Risk Management*. Cambridge: Cambridge University Press.

Bracewell, R. (1965) *The Fourier Transform and its Applications*. New York: McGraw-Hill.

Bradshaw, P., ed. (1978) *Turbulence, Topics in Applied Physics, vol. 12*. New York: Springer-Verlag.

Braunstein, S. L. and Caves, C. M. (1994) *Phys. Rev. Lett.* **72**, 3439.

Brill, M. H. (1999) private communication.

Brillouin, L. (1956) *Science and Information Theory*. New York: Academic Press.

Brody, D. and Meister, B. (1995) *Phys. Lett.* A**204**, 93.

Brody, D. C. and Hughston, L. P. (2001) *Proc. R. Soc. Lond.* A **457**, 1343.

Brody, D. C. and Hughston, L. P. (2002) *Quantitative Finance* **2**, 70.

Buchen, P. W. and Kelley, M. (1996) *J. Financial Quantitative Anal.* **31**, 143.

Buck, B., and Macaulay, V. A., eds (1990) *Maximum Entropy in Action: A Collection of Expository Essays*. Oxford: Oxford University Press.

Burhenne, L. J., Hislop, T. G. and Burhenne, H. J. (1992) *Am. J. Roent.* **158**, 45–49.

Caianiello, E. R. (1992) *Riv. Nuovo Cim.* **15**, 7.

Campbell, M. J. and Jones, B. W. (1972) *Science* **177**, 889.

Campbell, J. Y., Lo, A. W. and MacKinlay, A. C. (1997) *The Econometrics of Financial Markets*, Princeton, New Jersey: Princeton University Press.

Carter, B. (1974) *Confrontation of Cosmological Theories with Observational Data*, ed. M. S. Longair, IAU Symposium, p. 291.

Caves, C. M. (1986) *Phys. Rev.* D**33**, 1643.

Caves, C. M. and Milburn, G. J. (1987) *Phys. Rev.* A**36**, 5543.

Clarke, R., Dickson, R. B., and Brunner, N. (1990) *Ann. Oncol.* **I**, 401–407.

Cocke, W. J. (1996) *Phys. Fluids* **8**, 1609.

Cocke, W. J. and Frieden, B. R. (1997) *Found. Physics* **243**, 1397.

Cocke, W. J. and Spector, M. D. (1998) *Phys. Fluids* **10**, 2055.

Cohen-Tannoudji, C., Diu, B., and Laloe, F. (1977) *Quantum Mechanics* (English transl.). New York: Wiley.

Compton, A. H. (1923) *Phys. Rev.* **21**, 715; **22**, 409.

Cook, R. J. (1982) *Phys. Rev.* A**26**, 2754.

Cootner, P. H., ed. (1964) *The Random Character of Stock Market Prices*. Cambridge, Massachusetts: MIT Press.
Cover, T. A. and Thomas, J. M. (1991) *Elements of Information Theory*. New York: Wiley.
Crow, J. F. and Kimura, M. (1970) *Introduction to Population Genetics*. New York: Harper & Row.
Cvitanovic, P., Percival, I., and Wirzba, A., eds. (1992) *Quantum Chaos – Quantum Measurement*. Dordrecht: Kluwer Academic.
DeGroot, S. R. and Suttorp, L. G. (1972) *Foundations of Electrodynamics*. Amsterdam: North-Holland.
De Koning, H. J., Fracheboud, J., Boer, R., Verbeek, A. L., Collette, H. J., Hendriks, J. H., van Ineveld, B. M., de Bruyn, A. E., and van der Maas, P. J. (1995) *Int. J. Cancer* **60**, 777–780.
Delbruck, M. (1949) "A physicist looks at biology," *Trans. Connecticut Acad. Arts Sci.* **38**, 173–190.
Demetrius, L. (1992) *Theor. Popul. Biol.* **41**, 208–236.
DePinho, R. A. (2000) *Nature* **408**, 248–254.
Deutsch, I. H. and Garrison, J. C. (1991) *Phys. Rev.* A**43**, 2498.
DeWitt, B. S. (1967) *Phys. Rev.* **160**, 1113.
DeWitt, B. S. (1997) "The quantum and gravity: the Wheeler–DeWitt equation," *Proc. of 8th Marcel Grossman Conf.*, Jerusalem, eds. R. Ruffini and T. Piran. Singapore: World Scientific.
Dicke, R. H. (1961) *Nat. Lett.* **192**, 440.
Dirac, P. A. M. (1928) *Proc. R. Soc. Lond.* **A117**, 610; also (1947) *Principles of Quantum Mechanics*, 3rd edn. Oxford, New York.
Dirac, P. A. M. (1937) *Nature* **139**, 23; also (1938) *Proc. R. Soc. Lond.* A**165**, 199.
Ducommun, P., Bolzonella, I., Rhiel, M., Pugeaud, P., von Stockar, U., and Marison, I. W. (2001) *Biotechnol. Bioeng.* **72,** 515–522.
Duering, E., Otero, D., Plastino, A., and Proto, A. N. (1985) *Phys. Rev.* A**32**, 2455.
Edelman, D. (2003) "The minimum local cross-entropy criterion for inferring risk-neutral price distributions from traded options prices," University College Dublin Graduate School of Business, Centre for Financial Markets, Working Paper 2003–47.
Ehlers, J., Hepp, K., and Weidenmuller, H. A., eds. (1972) *Statistical Models and Turbulence, Lecture Notes in Physics, vol. 12*. New York: Springer-Verlag.
Eigen, M. (1992) *Steps toward Life*. Oxford: Oxford University Press, Chap. 8.
Einstein, A. (1905) *Ann. Phys.* **17**, 891; also (1956) *The Theory of Relativity*, 6th edn. London: Methuen.
Einstein, A., Podolsky, B. and Rosen, N. (1935) *Phys. Rev.* **47**, 777.
Eisberg, R. M. (1961) *Fundamentals of Modern Physics*. New York: Wiley.
Eisele, J. A. (1969) *Modern Quantum Mechanics*. Wiley: New York.
Ewens, W. J. (1989) *Theor. Popul. Biol.* **36**, 167–180.
Fagerberg, G., Baldetorp, L., Grontoft, O., Lundstrom, B., Manson, J. C., and Nordenskjold, B. (1985) *Acta Radiol. Oncol.* **24**, 465–473.
Fath, B. D., Cabezas, H., and Pawlowski, C. W. (2003) *J. Theor. Biol.* **222**, 517.
Ferwerda, H. A. (1998) personal communication.
Feynman, R. P. (1948) *Rev. Mod. Phys.* **20**, 367.
Feynman, R. P. and Hibbs, A. R. (1965) *Quantum Mechanics and Path Integrals*. New York: McGraw-Hill, p. 26.
Feynman, R. P. and Weinberg, S. (1993) *Elementary Particles and the Laws of Physics*. Cambridge: Cambridge University Press.

Fisher, M., Nychken, D., and Zervos, D. (1995) *FEDS* **95-1**. Washington, D.C.: Federal Reserve Board.
Fisher, R. A. (1922) *Phil. Trans. R. Soc. Lond.* **222**, 309.
Fisher, R. A. (1943) *Ann. Eugenics* **12**, 1.
Fisher, R. A. (1959) *Statistical Methods and Scientific Inference*, 2nd edn. London: Oliver and Boyd.
Fisher Box, J. (1978) *R. A. Fisher, the Life of a Scientist*. New York: Wiley; the author is his daughter.
Fomby, T., and Hill, R. C., eds. (1997) *Applying Maximum Entropy to Econometric Problems, Advances in Econometrics, vol. 12*. Greenwich, Connecticut: JAI Press Inc.
Frank, S. A. and Slatkin, M. (1992) *Trends Ecol. Evol.* **7**, 92–95.
Frieden, B. R. (1986) *Found. Phys.* **16**, 883.
Frieden, B. R. (1988) *J. Mod. Opt.* **35**, 1297.
Frieden, B. R. (1989) *Am. J. Phys.* **57**, 1004.
Frieden, B. R. (1990) *Phys. Rev.* A**41**, 4265.
Frieden, B. R. (1995) in *Advances in Imaging and Electron Physics*, ed. P. W. Hawkes. Orlando, Florida: Academic Press.
Frieden, B. R. (2001) *Probability, Statistical Optics and Data Testing*, 3rd edn. Berlin: Springer-Verlag.
Frieden, B. R. (2002) *Phys. Rev.* A**66**, 022107.
Frieden, B. R. and Cocke, W. J. (1996) *Phys. Rev.* E**54**, 257.
Frieden, B. R. and Hughes, R. J. (1994) *Phys. Rev.* E**49**, 2644.
Frieden, B. R. and Plastino, A. (2000) *Phys. Lett.* A**272**, 326.
Frieden, B. R. and Plastino, A. (2001a) *Phys. Lett.* A**278**, 299.
Frieden, B. R. and Plastino, A. (2001b) *Phys. Lett.* A**287**, 325.
Frieden, B. R. and Soffer, B. H. (1995) *Phys. Rev.* E**52**, 2247.
Frieden, B. R. and Soffer, B. H. (2002) *Phys. Lett.* A**304**, 1–7.
Frieden, B. R., Plastino, A., Plastino, A. R., and Soffer, B. H. (1999) *Phys. Rev.* E**60**, 48.
Frieden, B. R., Plastino, A., and Soffer, B. H. (2001) *J. Theor. Biol.* **208**, 49.
Frieden, B. R., Plastino, A., Plastino, A. R., and Soffer, B. H. (2002a) *Phys. Rev.* E**66**, 046128.
Frieden, B. R., Plastino, A., Plastino, A. R., and Soffer, B. H. (2002b) *Phys. Lett.* A**304**, 73.
Frishling, V. and Yamamura, J. (1996) *J. Fixed Income* 97.
Gardiner, C. W. (1985) *Handbook of Stochastic Methods*, 2nd edn. Berlin: Springer-Verlag.
Gardiner, C. W. (1991) *Quantum Noise*. Berlin: Springer-Verlag.
Gatenby, R. A. and Frieden, B. R. (2002) *Cancer Res.* **62**, 3675–3684.
Gatenby, R. A. and Gawlinski, E. T. (1996) *Cancer Res.* **56**, 5745.
Glansdorff, P. and Prigogine, I. (1971) *Thermodynamic Theory of Structure, Stability and Fluctuation*. London: Wiley.
Golan, A., Miller, D., and Judge, G. G. (1996) *Maximum Entropy Econometrics: Robust Estimation with Limited Data, Financial Economics and Quantitative Analysis*. New York: John Wiley & Sons.
Goldstein, H. (1950) *Classical Mechanics*. Cambridge, Massachusetts: Addison-Wesley.
Good, I. J. (1976) in *Foundations of Probability Theory, Statistical Inference and*

*Statistical Theories of Science, Vol. II*, eds. W. L. Harper and C. A. Hooker. Boston: Reidel.
Gottfried, K. (1966) *Quantum Mechanics, vol. I: Fundamentals*. New York: Benjamin, pp. 170–171.
Granger, C. W. J. (1966) *Econometrics* **34**, 150.
Groom, D. E. *et al*. (2000) *Euro. Phys. J.* C**15**, 73.
Hall, M. J. (2000) *Phys. Rev.* A**62**, 012107.
Halliwell, J. J., Perez-Mercader, J., and Zurek, W. H., eds. (1994) *Physical Origins of Time Asymmetry*. Cambridge: Cambridge University Press.
Handel, P. H. (1971) *Phys. Rev.* A**3**, 2066.
Harris, S. (1971) *An Introduction to the Theory of the Boltzmann Equation*. New York: Holt, Rinehart and Winston.
Hart, D., Shochat, E., and Agur, Z. (1998) *Brit. J. Cancer* **78**, 382–387.
Hartle, J. B. and Hawking, S. W. (1983) *Phys. Rev.* D**28**, 2960.
Hawking, S. W. (1979) in *General Relativity: An Einstein Centenary Survey*, ed. W. Israel. Cambridge: Cambridge University Press.
Hawking, S. W. (1988) *A Brief History of Time*. Toronto: Bantam Books, p. 129.
Hawkins, R. J. (1997) in *Applying Maximum Entropy to Econometric Problems*, eds. T. Fomby and R. C. Hill, *Advances in Econometrics, vol. 12*, Greenwich, Connecticut: JAI Press Inc., p. 277.
Hawkins, R. J., Rubinstein, M., and Daniell, G. (1996) in *Maximum Entropy and Bayesian Methods*, eds K. M. Hanson and R. N. Silver, *Fundamental Theories of Physics, vol. 79*. Santa Fe, New Mexico: Kluwer Academic Publishers, p. 1.
Hawkins, R. J. and Frieden, B. R. (2004) *Phys. Lett.* A**322**, 126
Heinz, U. and Jacob, M. (2000) Evidence for a new state of matter: an assessment of the results from the CERN lead beam programme, CERN press release, February 10.
Heitler, W. (1984) *The Quantum Theory of Radiation*. New York: Dover; reprint of 3rd edn. (1954). Oxford: Oxford University Press
Hirschman, I. I. (1957) *Am. J. Math.* **79**, 152.
Hooge, F. N. (1976) *Physica* **8B**, 14.
Holden, A. V. (1976) *Models of the Stochastic Activity of Neurons*. Berlin: Springer-Verlag.
Huber, P. J. (1981) *Robust Statistics*. New York: Wiley, pp. 77–86.
Hughes, R. J. (1994) suggested by discussions with.
Jackson, J. D. (1975) *Classical Electrodynamics*, 2nd edn. New York: Wiley, pp. 536–541, 549.
Jackwerth, J. (1999) *J. Derivatives* **7**, 66.
Jauch, J. M. and Rohrlich, F. (1955) *Theory of Photons and Electrons*. Cambridge, Massachusetts: Addison-Wesley, p. 23
Jaynes, E. T. (1957a) *Phys. Rev.* **106**, 620.
Jaynes, E. T. (1957b) *Phys. Rev.* **108**, 171.
Jaynes, E. T. (1968) *IEEE Trans. System Sci. Cybernetics* **SSC-4** 227.
Jaynes, E. T. (1985) in *Maximum Entropy and Bayesian Methods in Inverse Problems*, eds. C. R. Smith and W. T. Grandy. Dordrecht: Reidel.
Jeffreys, H. H. (1961) *Theory of Probability*, 3rd edn. London: Oxford University Press.
Jensen, H. J. (1991) *Phys. Scr.* **43**, 593.
Johnson, J. B. (1925) *Phys Rev.* **26**, 71.
Kahre, J. (2000) *The Mathematical Theory of Information*. Boston: Kluwer.

Kajita, T. (2000) "Latest Results from Super-Kamiokande," *AIP Conf. Proc.* **539**, 31–40.
Kamke, E. (1948) *Differentialgleichungen Lösungsmethoden und Lösungen*. New York: Chelsea, p. 551.
Kelley, S. T., Menon, C., Buerk, D. G., Bauer, T. W., and Fraker, D. L. (2002) *Surgery* **132**, 252.
Kelvin, Lord (1889) *Popular Lectures and Addresses, Vol. I*, Epigraph to Chap. 37. London: MacMillan. We quote: "When you cannot measure [what you are speaking about], your knowledge is of a meagre and unsatisfactory kind: . . . you have scarcely . . . advanced to the stage of science."
Kendal, W. (1990) *Math. Biosci.* **100**, 143–159.
Keshner, M. S. (1982) *Proc. IEEE* **70**, 212.
Koonin, S. E. (1986) *Computational Physics*. New York: Addison-Wesley.
Korn, G. A. and Korn, T. M. (1968) *Mathematical Handbook for Scientists and Engineers*, 2nd edn. New York: McGraw-Hill.
Kuchar, K. (1973) in *Relativity, Astrophysics and Cosmology*, ed. W. Israel. Dordrecht: Reidel, p. 268.
Kullback, S. (1959) *Information Theory and Statistics*. New York: Wiley.
Kuwahara, H. and Marsh, T. (1994) *Jap. J. Financial. Econ.* **1**, 33.
Lagrange, J. L. (1788) *Mécanique analytique*; republished in (1889) *Œuvres de Lagrange*. Paris: Gauthier-Villars.
Landau, L. D. and Lifshitz, E. M. (1951) *Classical Theory of Fields*. Cambridge, Massachusetts: Addison-Wesley, p. 258.
Landau, L. D. and Lifshitz, E. M. (1977) *Quantum Mechanics (Non-relativistic Theory)*, 3rd edn, *Course of Theoretical Physics, vol. 3*. New York: Pergamon Press.
Landauer, R. (1961) *IBM J. Res. Development* **5**, 183.
Lawrie, I. D. (1990) *A Unified Grand Tour of Theoretical Physics*. Bristol: Adam Hilger.
Lindblad, G. (1983) *Non-equilibrium Entropy and Irreversibility*. Boston: Reidel.
Lindsay, R. B. (1951) *Concepts and Methods of Theoretical Physics*. New York: Van Nostrand.
Louisell, W. H. (1970) in *Quantum Optics*, eds. S. M. Kay and A. Maitland. New York: Academic Press.
Lotka, A. J. (1956) *Elements of Mathematical Biology*. New York: Dover.
Lowen, S. B. and Teich, M. C. (1989) *Electron. Lett.* **25**, 1072.
Loyalka, S. K. (1990) "Transport theory," in *Encyclopedia of Physics*, 2nd edn, eds. R. Lerner and G. L. Trigg. Weinheim: VCH Publishers.
Luo, S. (2002) *Found. Phys.* **32**, 1757.
Luo, S. (2003) *Phys. Rev. Lett.* **91**, 180403.
Maasoumi, E. (1993) *Econometric Rev.* **12**, 137.
Mandelbrot, B. B. (1983) *The Fractal Geometry of Nature*. New York: Freeman.
Mandelbrot, B. B. and Wallis, J. R. (1969) *Water Resour. Res.* **5**, 321.
Mantegna, R. N. and Stanley, H. E. (2000) *An Introduction to Econophysics: Correlations and Complexity in Finance*, Cambridge: Cambridge University Press.
Mao, E. F., Lane, L., Lee, J., and Miller, J. U. (1997) *J. Bacteriol.* **179**, 417–422.
Marathay, A. S. (1982) *Elements of Optical Coherence Theory*. New York: Wiley.
Martens, H. and de Muynck, W. M. (1991) *Phys. Lett.* A**157**, 441.
Martin, B. R. and Shaw, G. (1992) *Particle Physics*, New York: Wiley.

Matheron, G. (1973) *Adv. Appl. Probab.* **5**, 439.
Matveev, V. B. and Salle, M. A. (1991) *Darboux Transformations and Solitons*. Berlin: Springer-Verlag.
McCullough, J. H. (1975) *J. Finance* **30**, 811.
Mensky, M. B. (1979) *Phys. Rev.* D**20**, 384.
Mensky, M. B. (1993) *Continuous Quantum Measurements and Path Integrals*. Bristol: Institute of Physics Publishing.
Merton, R. C. (1974) *Bell J. Economics Management Sci.* **7**, 141.
Merzbacher, E. (1970) *Quantum Mechanics*, 2nd edn. New York: Wiley.
Meystre, P. (1984) in *Quantum Electrodynamics and Quantum Optics*, ed. A. O. Barut. New York: Plenum.
Miller, R. E. (1979) *Dynamic Optimization and Economic Applications*, New York: McGraw-Hill, p. 178.
Milonni, P. W. and Eberly, J. H. (1988) *Lasers*. New York: Wiley, pp. 200–215.
Misner, C. W., Thorne, K. S., and Wheeler, J. A. (1973) *Gravitation*. New York: Freeman.
Morgenstern, O. and von Neumann, J. (1947) *Theory of Games and Economic Behavior*. Princeton: Princeton University Press.
Morse, P. M. and Feshbach, H. (1953) *Methods of Theoretical Physics*. New York: McGraw-Hill.
Motchenbacher, C. D. and Connelly, J. A. (1993) *Low-Noise Electronic System Design*. New York: Wiley.
Musha, T. and Higuchi, H. (1976) *Jap. J. Appl. Phys.* **15**, 1271.
Muta, T. (1998) *Foundations of Quantum Chromodynamics*. Singapore: World Scientific.
Nikolov, B. (1992) personal communication.
Nikolov, B. and Frieden, B. R. (1994) *Phys. Rev.* E**49**, 4815.
O'Neill, E. L. (1963) *Introduction to Statistical Optics*. London: Addison-Wesley.
Orgel, L. (2000) *Science* **290**, 1306.
Panofsky, W. K. H. and Phillips, M. (1955) *Classical Electricity and Magnetism*. Reading, Massachusetts: Addison-Wesley.
Papoulis, A. (1965) *Probability, Random Variables and Stochastic Processes*. New York: McGraw-Hill.
Pauli, W. (1927) *Z. Phys.* **43**, 601.
Peer, P. G., Holland, R., Hendriks, J. H., Mravunac, M., and Verbeek, A. L. (1994) *J. Nat. Cancer Inst.* **86**, 436–441.
Perlmutter, S. *et al.* (1998) *Nature* **391**, 51.
Pike, E. R. and Sarkar, S., eds. (1987) *Quantum Measurement and Chaos*. New York: Plenum.
Plastino, A. R. and Plastino, A. (1995) *Phys. Rev.* E**52**, 4580.
Plastino, A. R. and Plastino, A. (1996) *Phys. Rev.* E**54**, 4423.
Pumir, A. (1994) *Phys. Fluids* **6**, 2071.
Pumir, A. (1996) *Phys. Fluids* **8**, 1609.
Reginatto, M. (1998) *Phys. Rev.* A**58**, 1775.
Reif, F. (1965) *Fundamentals of Statistical and Thermal Physics*. New York: McGraw-Hill.
Reza, F. M. (1961) *An Introduction to Information Theory*. New York: McGraw-Hill.
Risken, H. (1996) *The Fokker–Planck Equation: Methods of Solution and Applications*, 2nd edn, *Springer Series in Synergetics, vol. 18*. New York: Springer-Verlag.

Robertson, B. (1973) *Am. J. Phys.* **41**, 678.
Roman, P. (1960 and 1961) *Theory of Elementary Particles*. Amsterdam: North-Holland.
Rosen, G. (1996) *Hadronic J.* **19**, 95.
Rumer, I. B. and Ryvkin, M. Sh. (1980) *Thermodynamics, Statistical Physics and Kinetics*. Moscow: Mir Publishers, p. 526
Ryder, L. H. (1987) *Quantum Field Theory.* Cambridge: Cambridge University Press.
Sagdeev, R. Z., Usikov, D. A., and Zaslavskii, G. M. (1988) *Nonlinear Physics: From the Pendulum to Turbulence and Chaos*. New York: Harwood Academic.
Savage, L. J. (1972) *Foundations of Statistics*. Englewood Cliffs, New Jersey: Dover, p. 236.
Schiff, L. I. (1955) *Quantum Mechanics*, 2nd edn. New York: McGraw-Hill.
Schrödinger, E. (1926) *Ann. Phys.* **79**, 361.
Scott, A. (1999) *Nonlinear Science*. Oxford: Oxford University Press.
Sengupta, J. K. (1993) *Econometrics of Information and Efficiency.* Dordrecht: Kluwer Academic Press.
Shannon, C. E. (1948) *Bell Syst. Tech. J.* **27**, 379; 623.
Shea, G. S. (1985) *J. Finance* **40** 319.
Shirai, H. (1998) *Found. Phys.* **28**, 1633.
Silver, R. N. (1992) in *E. T. Jaynes: Physics and Probability*, eds. W. T. Grandy and P. W. Milonni, Cambridge: Cambridge University Press.
Sipe, J. E. (1995) *Phys. Rev.* A**52**, 1875.
Soffer, B. H. (1993) personal communication.
Soffer, B. H. (1994) personal communication.
Solo, V. (1992) *SIAM J. Appl. Math.* **52**, 270.
Stam, A. J. (1959a) *Some Mathematical Properties of Quantities of Information*, Ph.D. dissertation. Technical University of Delft.
Stam, A. J. (1959b) *Information and Control* **2**, 101.
Stein, E. M. and Weiss, G. (1971) *Fourier Analysis on Euclidean Spaces*. Princeton, New Jersey: Princeton University Press.
Stoler, D. L., Chen, N., Basik, M., Kahlenberg, M. S., Rodriguez, M. A., Petrelli, N. J., and Anderson, G. R. (1999) *Proc. Nat. Acad. Sci. USA* **96**, 15121–15126.
Stutzer, M. (1994) in *Differential Equations, Dynamical Systems, and Control Science – A Festschrift in Honor of Lawrence Markus*, eds. R. Elworth, W. Everitt, and E. Lee, *Lecture Notes in Pure and Applied Mathematics, vol. 152*. New York: Marcel Dekker, p. 321.
Stutzer, M. (1996) *J. Finance* **51** 1633.
Szilard, L. (1929) *Z. Phys.* **53**, 840.
Tabar, L., Fagerberg, G., Duffy, S. W., Day, N. E., Gad, A., and Grontoft, O. (1992) *Radiol. Clin. North Am.* **30**, 187–210.
Takayasu, H. (1987) *J. Phys. Soc. Japan* **56**, 1257.
Tang, R., Cheng, A. J., Wang, J., and Wang, T. C. (1998) *Cancer Res.* **58**, 4052–4054.
Thomas, B. A., Price, J. L., Boulter, P. S., and Gibbs N. M. (1984) *Recent Results Cancer Res.* **90**, 195–199.
Thomas, S. (2002) *Phys. Rev. Lett.* **89**, 081301.
Van Trees, H. L. (1968) *Detection, Estimation, and Modulation Theory, Part I.* New York: Wiley.
Vasicek, O. A. and Fong, G. H. (1982) *J. Finance* **37**, 339.
Vilenkin, A. (1982) *Phys. Lett.* **117B**, 25.
Vilenkin, A. (1983) *Phys. Rev.* D**27**, 2848.

Villani, C. (1998) *J. Math. Pure Appl.* **77**, 821.
Vincent, T. L. and Brown, J. S. (in press) *Evolutionary Game Theory, Natural Selection and Darwinian Dynamics*. Cambridge: Cambridge University Press.
Vincent, T. L. and Grantham, W. J. (1997) *Nonlinear and Optimal Control Systems*. New York: Wiley.
Voit, J. (2001) *The Statistical Mechanics of Financial Markets*. New York: Springer-Verlag.
Volker, J., Klump, H., and Breslauer, K. (2001) *Proc. Natl. Acad. Sci. USA* **98**, 7694–7699.
Von Mises, R. (1936) *Wahrscheinlichkeit, Statistik und Wahrheit*. Vienna: Springer-Verlag OHG.
Von Neumann, J. (1955) *Mathematical Foundations of Quantum Mechanics* (English transl.). Princeton, New Jersey: Princeton University Press.
Voss, R. F. and Clarke, J. (1978) *J. Acoust. Soc. Am.* **63**, 258.
Vstovsky, G. V. (1995) *Phys. Rev.* E**51**, 975.
Wheeler, J. A. (1962) *Geometrodynamics*. New York: Academic Press.
Wheeler, J. A. (1990) in *Proceedings of the 3rd International Symposium on Foundations of Quantum Mechanics, Tokyo, 1989*, eds. S. Kobayashi, H. Ezawa, Y. Murayama, and S. Nomura. Tokyo: Physical Society of Japan, p. 354.
Wheeler, J. A. (1994) in *Physical Origins of Time Asymmetry*, eds. J. J. Halliwell, J. Perez-Mercader, and W. H. Zurek. Cambridge: Cambridge University Press, pp. 1–29.
Wodarz, D. and Krakauer, D. C. (2001) *Oncol. Rep.* **8**, 1195–1201.
Woodroofe, M. and Hill, B. (1975) *J. Appl. Prob.* **12**, 425.
Wootters, W. K. (1981) *Phys. Rev.* D**23**, 357.
Wyss, W. (1986) *J. Math. Phys.* **37**, 2782.
Yndurain, F. J. (1999) *Quantum Chromodynamics*, 3rd edn. New York: Springer-Verlag.
Zeh, H.-D. (1992) *Physical Basis of the Direction of Time*. Berlin: Springer-Verlag.

# Index